Springer Series in

OPTICAL SCIENCES 122

Founded by H.K.V. Lotsch

Springer Series in
OPTICAL SCIENCES

The Springer Series in Optical Sciences, under the leadership of Editor-in-Chief *William T. Rhodes*, Georgia Institute of Technology, USA, provides an expanding selection of research monographs in all major areas of optics: lasers and quantum optics, ultrafast phenomena, optical spectroscopy techniques, optoelectronics, quantum information, information optics, applied laser technology, industrial applications, and other topics of contemporary interest.

With this broad coverage of topics, the series is of use to all research scientists and engineers who need up-to-date reference books.

The editors encourage prospective authors to correspond with them in advance of submitting a manuscript. Submission of manuscripts should be made to the Editor-in-Chief or one of the Editors. See also www.springer.com/series/624

Editor-in-Chief

William T. Rhodes
Georgia Institute of Technology
School of Electrical and Computer Engineering
Atlanta, GA 30332-0250, USA
E-mail: bill.rhodes@ece.gatech.edu

Editorial Board

Ali Adibi
School of Electrical and Computer Engineering
Van Leer Electrical Engineering Building
Georgia Institute of Technology
777 Atlantic Drive NW
Atlanta, GA 30332-0250, USA
E-mail: adibi@ece.gatech.edu

Toshimitsu Asakura
Hokkai-Gakuen University
Faculty of Engineering
1-1, Minami-26, Nishi 11, Chuo-ku
Sapporo, Hokkaido 064-0926, Japan
E-mail: asakura@eli.hokkai-s-u.ac.jp

Theodor W. Hänsch
Max-Planck-Institut für Quantenoptik
Hans-Kopfermann-Strasse 1
85748 Garching, Germany
E-mail: t.w.haensch@physik.uni-muenchen.de

Takeshi Kamiya
Ministry of Education, Culture, Sports
Science and Technology
National Institution for Academic Degrees
3-29-1 Otsuka, Bunkyo-ku
Tokyo 112-0012, Japan
E-mail: kamiyatk@niad.ac.jp

Ferenc Krausz
Ludwig-Maximilians-Universität München
Lehrstuhl für Experimentelle Physik
Am Coulombwall 1
85748 Garching, Germany
and
Max-Planck-Institut für Quantenoptik
Hans-Kopfermann-Strass 1
85748 Garching, Germany
E-mail: ferenc.krausz@mpq.mpg.de

Bo Monemar
Department of Physics
and Measurement Technology
Materials Science Division
Linköping University
58183 Linköping, Sweden
E-mail: bom@ifm.liu.se

Herbert Venghaus
Heinrich-Hertz-Institut
für Nachrichtentechnik Berlin GmbH
Einsteinufer 37
10587 Berlin, Germany
E-mail: venghaus@hhi.de

Horst Weber
Technische Universität Berlin
Optisches Institut
Strasse des 17. Juni 135
10623 Berlin, Germany
E-mail: weber@physik.tu-berlin.de

Harald Weinfurter
Ludwig-Maximilians-Universität München
Sektion Physik
Schellingstrasse 4/III
80799 München, Germany
E-mail: harald.weinfurter@physik.uni-muenchen.de

Yuriy K. Sirenko   Staffan Ström
Nataliya P. Yashina

Modeling and Analysis of Transient Processes in Open Resonant Structures

New Methods and Techniques

Springer

Yuriy K. Sirenko
Institute of Radiophysics and Electronics
National Academy of Sciences of Ukraine
12 Academician Proskura Street
Kharkov 61085
Ukraine
yks@ire.kharkov.ua

Nataliya P. Yashina
Institute of Radiophysics and Electronics
National Academy of Sciences of Ukraine
12 Academician Proskura Street
Kharkov 61085
Ukraine
nataliya@lin.com.ua

Staffan Ström
Division of Electromagnetic Theory
Alfven Laboratory
Royal Institute of Technology
SE-100 44 Stockholm
Sweden
staffan.strom@alfvenlab.kth.se
staffan.strom@ee.kth.se

Library of Congress Control Number: 2006922415

ISSN 0342-4111
ISBN-10: 0-387-30878-4 e-ISBN-10: 0-387-32577-8
ISBN-13: 978-0387-30878-4 e-ISBN-13: 978-0387-32577-4

Printed on acid-free paper.

9 8 7 6 5 4 3 2 1

springer.com

Preface

The focus of electromagnetic theory is initial boundary value and boundary value problems for the Maxwell equations. Those are the initial models, from which, by applying mathematical methods, we should extract physical results. The modern computer-aided research process can be divided into several stages: qualitative mathematical analysis of the initial problem, the development of algorithms and implementation of the problem in software, problem-oriented computational experiments, and physical interpretation of the results. The success of the study depends in many aspects on whether sufficiently high standards of investigation can be maintained at all these stages and whether there is an "intellectual core" in these investigations that enables us to gain new scientific knowledge [1]. As an example of a successful implementation of such an approach that has settled a long-standing conflict between theory and experiment, we can cite the development of the theory of resonant wave scattering in the frequency domain. These results have been reported (see [2–16] and the bibliographies contained in those references), and they have served as a basis for the development of a number of essentially new functional units and devices in millimeter and submillimeter radio engineering, vacuum electronics, and solid-state electronics, optics, and spectroscopy.

The modern theory of transient electromagnetic fields is still lacking achievements that may be compared with those existing in the frequency domain, neither by the profoundness of the study, nor by the intensity of the study of electromagnetic phenomena and, as a result, by their applications. However, the process of accumulation of potentialities for a breakthrough is a process still going on. This book is devoted to just these problems; for the most part, it is focused on the development and implementation of robust and efficient mathematical models for transient electromagnetic theory.

Chapter 1 has the character of a survey. Here, we have collected information that may be useful for the problems we consider. We have also, up to our abilities, tried to analyze the state of the art in time-domain electromagnetics, as it existed at the time the book was written. We focused our analysis on methods based on spatio-frequency and spatio-temporal representations.

The content of Chapters 2 to 4 has been inspired by the statement from S.V. Georgakopoulos et al. [17]:

Therefore, it becomes clear that the boundary conditions are an integral part of a PDE (partial-differential-equation) problem, and should always accompany the FDTD (finite-difference time-domain) formulation of it. This inflicts particular concerns when the problem under examination is so-called "open" space or unbounded problem, e.g., radiating, scattering, etc., meaning that the domain of interest is unbounded in one or more spatial-coordinate directions. For such problems, there are no exact boundary conditions known.

However, since some time back, exact absorbing conditions providing efficient limitation of the computational domain of finite-difference methods do exist. They have already taken their proper place in electromagnetic simulation in fundamental and applied electromagnetics. In 1986, scientists from Moscow State University, A.R. Maikov, A.G. Sveshnikov, and S.A. Yakunin, published their paper [18]. In this paper, the exact nonlocal conditions for virtual boundaries in regular semi-infinite hollow waveguides (with constant cross-section), serving as channels for signals propagating from certain resonant junctions, were formulated for the first time. Later on (see, e.g., references [19–21]), the approach suggested in [18], which is based on the utilization of radiation conditions for spatio-temporal amplitudes of outgoing modes, has been modified for various electromagnetic problems: antenna design, analysis and synthesis problems for quasi-optical open resonators with dispersive elements, electromagnetic monitoring of the human environment, and others. For several particular cases, the problems of nonlocality and the problem of corner points, points of intersection of coordinate boundaries, have been resolved in a rigorous way. The efficiency and correctness of the approaches, based on the application of the exact absorbing conditions, are verified and proved by specific tests and numerical experiments.

The major attention of the book is focused on the general theory and details of the technique that are the basis for construction of exact conditions. These conditions are incorporated into the algorithms of the solutions of initial boundary value problems, and they enable an efficient and accurate modeling of a rather wide class of problems of transient electromagnetic scattering. The exact absorbing conditions form a natural complement to the list of conditions that have already become classical ones, those of B. Engquist, A. Majda, G. Mur, J.-P. Berenger, and others [22–26].

The analytic results of Chapters 2 to 5 are oriented at the application of the finite-difference methods in the final computational algorithms. The FDTD method [27] appeared in 1966 (see the canonical paper by K.S. Yee [28]) and actually has created a boom in computational electromagnetics. It may serve as a perfect example of an excellent elaborated implementation of well-known principles for the discretization of the curl-type Maxwell equations. This scheme allowed a huge body of computational experiments to be performed, as required by electromagnetic engineering.

Preparatory work, based sometimes on sophisticated analytic derivations, can transform the algorithm based on the FDTD method into not only a universal and powerful but also an accurate and efficient tool, providing researchers with new knowledge and profound insight into transient phenomena (see, e.g., [27,29]). We hope that the results presented in Chapters 6 and 7 may be considered as support for this conclusion.

Chapter 5 deals with new algorithmic schemes for simple (canonical) and complicated (chain of the junctions of simple problems) initial boundary value problems of the electromagnetic theory of waveguides. The approaches suggested in this chapter are based on the description of the scattering properties of inhomogeneities of regular waveguides in terms of the transform operators of the signals' evolutionary basis [21,30] that is qualitatively equal for all guiding structures. The corresponding approaches in the frequency domain are widely known. The modification of these prototypes and the solution to the technical and methodological problems arising due to the specific character of the time domain are the major issues highlighted in this chapter.

In Chapters 6 and 7, we put forward our ideas about the most efficient approaches to the study of physical peculiarities of resonant scattering both in the time and frequency domains. We also discuss methods for solving complicated applied problems of model synthesis of resonant quasi-optics. We describe these ideas for real models that have as a principal element a periodic diffraction grating.

Electromagnetic analysis of open periodic resonators and waveguides (gratings) is always associated with the solution of important applied problems; the efficiency and further development of some promising topics substantially depend on how deep the characteristics of such scattering and wave-directing objects are studied. Special attention should be given to theoretical studies that significantly simplify or even make experimental analysis unnecessary and that can be used for model synthesis and hardware optimization. This timely problem is partially solved in terms of the wave diffraction theory for the frequency domain, which enabled one to efficiently analyze many cases that are interesting from a practical point of view, and in the process to gather a lot of experience concerning abnormal and resonance scattering modes [2,3,6,11,14].

However, many problems in the electromagnetic theory of gratings still remain unsolved, and the traditional approaches and methods are not capable of providing the solution to them. One such problem is concerned with studying the physical nature and analytic description of various resonance and anomalous spatio-frequency and spatio-temporal field transformations. Its efficient solution is provided by a mathematical analysis of the peculiar features of the analytic continuation of the solutions to the elliptic boundary value problems into the domain of complex, usually physically unrealizable, parameter values. Here, unlike the traditional problems of the frequency domain, the closest attention should be paid not to the regular points in the intervals of parameter variation, where the corresponding operators are boundedly invertible, but rather mostly to the complementary sets, that is, spectra, and the analysis of such phenomena and their behavior in complex space.

Maybe the very first attempt to use complex frequencies in electromagnetics should be dated back to 1884, when J.J. Thomson analyzed free field oscillations in the exterior of a perfectly conducting sphere. The oscillations that satisfied the condition of "outgoing radiation" increased exponentially in space, which was a reason for criticism: H. Lamb alleged the problem to be ill posed. The effect of the "exponential catastrophe" is still keeping many researchers busy solving nonself-adjoint spectral boundary value problems, although the question is completely settled by turning to spatio-temporal representations in the analysis: every divergent oscillation is associated with an exponential time-dependent factor that covers the coordinate-dependence in any space point.

Studies of dispersion relations and spectrum analysis are now a dominating trend in many areas of physics. It is concentrated on analytic properties of various functions describing physical phenomena, as they are extended into the domain of complex, nonphysical argument values. This means that, although the physical meaning is inherent only in such notions as frequency and energy, nevertheless, the complex values of these parameters, which are never actually realized, are used in the analysis of the behavior of the system.

Dispersion relations result from the self-adjoint or nonself-adjoint theory of open (exterior) elliptic boundary value problems. In the first case, the solution is sought in the space $\mathbf{L}_2$ (the $\mathbf{L}_2$-theory), and in the second case, the solution is subject to the conditions of "outgoing radiation" (along the real axis). C.L. Dolph has emphasized that there are no strict rules for using one theory or another, and at the same time, he notes that the analysis of nonself-adjoint problems contributes substantially to the understanding of the case and provides a solid ground for many existing scattering theories. The $\mathbf{L}_2$-theory seems to be more convenient and elegant in use from the mathematical point of view, but the more generalized approach based on the analysis of nonself-adjoint problems proves to be a more powerful and capable tool for yielding useful, physically relevant representations.

Diffraction gratings are among the most popular objects in classical electromagnetic theory. What is the reason for that? There are several, but as the principal one, we may consider the dispersive property of the grating: the diffraction grating is the most universal dispersive (frequency-selective) element within the total frequency range that is exploited nowadays. As a result, the design and modeling of gratings of various configurations are always in demand. And, as a personal reason, we can add that we think that the electromagnetic theory of diffraction gratings is beautiful, describing a large variety of physical phenomena and their promising applications. The history of the study of diffraction gratings is a fascinating story and deserves a special book to be written about it.

There are many prominent scientists and scientific schools that made contributions that cannot be overestimated nowadays. We should like to emphasize here that the results presented in this book, especially in Chapters 6 and 7, are based on the results of and inspired by the Kharkov (Ukraine) scientific school that made considerable contributions to the mathematical background and profound study of electromagnetic phenomena of diffraction theory. This book should, in particular, be seen as an effort to make that work better known outside the former Soviet Union.

Yuriy K. Sirenko
Staffan Ström
Nataliya P. Yashina

Contents

1 Numerical Analysis of Transient Processes: Fundamental Results of the Theory, Methods, and Problems

1.1. Introduction

In this chapter, we describe the state of the art in the development of general approaches and algorithms that serve as bases for solving fundamental and applied problems in the theory of transient electromagnetic fields.

Section 1.2 of this chapter is devoted to the description of the initial boundary value problems under consideration (main equations, the geometry of the analyzed domains, general and fundamental solutions, correctness classes, etc.) and the characteristic features of the mathematical methods employed (elements of the theory of generalized functions).

Section 1.3 considers the achievements and problems associated with the attempts to apply the approaches and methods based on the spatio–frequency field representations and Laplace and Fourier integral transforms to the analysis of problems in transient wave theory.

The subject of Section 1.4 is methods and algorithms for numerical analysis of transient processes that are based directly on the spatio–temporal field representations. The following methods are outlined in following order: the finite-difference method, the integral equations method, the method of partial separation of variables, and the directional decomposition method. Furthermore, some problems that are associated with using so-called absorbing boundary conditions (ABCs) for limiting the computation area of the finite-difference methods are discussed in that section.

The references to the sources used in this chapter, which is essentially a survey chapter, as well as the enumeration of the achievements and problems are certainly open to discussion. The reasons for choosing the indicated sources are quite conventional: preference was given to those works that we found to be thought-provoking and a source of inspiration, that have taught us something, and that, to our minds, can be useful to others.

1.2. Main Equations and Fundamental Results of the Theory

1.2.1. Maxwell Equations

Consider the complete system of Maxwell equations in differential form, which is supplemented with the equation of conservation of charge

$$\eta_0 \mathrm{rot}\vec{H} = \varepsilon \frac{\partial \vec{E}}{\partial t} + \sigma \vec{E} + \vec{J}, \quad \mathrm{rot}\vec{E} = -\eta_0 \frac{\partial \vec{H}}{\partial t}, \tag{1.1}$$

$$\mathrm{div}\vec{H} = 0, \quad \mathrm{div}(\varepsilon \vec{E}) = \tilde{\rho}, \tag{1.2}$$

$$\mathrm{div}\vec{J} + \mathrm{div}(\sigma \vec{E}) + \frac{\partial \tilde{\rho}}{\partial t} = 0. \tag{1.3}$$

Here, $\vec{E} \equiv \vec{E}(g,t)$ and $\vec{H} \equiv \vec{H}(g,t)$ are the vectors of the electrical and magnetic field strengths; $\eta_0 = (\mu_0/\varepsilon_0)^{1/2}$ is the free space impedance; ε_0 and μ_0 are the electric and magnetic vacuum constants; $\vec{J} = \eta_0 \vec{j}$, $\vec{j} \equiv \vec{j}(g,t)$ is the extraneous current density; $\sigma = \eta_0 \sigma_0$; $\varepsilon \equiv \varepsilon(g) \geq 1$ and $\sigma_0 \equiv \sigma_0(g) \geq 0$ are the relative dielectric permittivity and the specific conductivity of a locally inhomogeneous (isotropic, nonmagnetic, and nondispersive) medium in which the waves propagate; $\tilde{\rho} = \rho_0/\varepsilon_0$, $\rho_0(g,t)$ is the volume density of the induced and external electric charges; the time t has the dimension of length (it is the product of the natural time and the velocity of the propagation of light in vacuum); $g = \{x, y, z\}$ or $g = \{\rho, \phi, z\}$, $g = \{r, \vartheta, \phi\}$ is a point in the space $\mathbf{R}^3$; x, y, and z are the Cartesian coordinates; ρ, ϕ, z, and r, ϑ, ϕ are spatial cylindrical and spherical coordinates.

If $\tilde{\rho}$ in (1.2), (1.3) is presented as the sum of two terms ρ_1 and ρ_2, denoting correspondingly the induced and external electric charge, then the continuity equation can be rewritten for each term separately: the induced charge ρ_1 corresponds to the conductivity current $\sigma \vec{E}$, and external charge ρ_2 corresponds to external current $\vec{J}$. In the absence of external charges and currents the induced electric charges and currents in homogeneous conducting media disappear rather quickly (relaxation).

The first equation in (1.2) follows from the second one in (1.1), only if $\mathrm{div}\vec{H} = 0|_{t=0}$. The second equation in (1.2) follows from the first equation in (1.1) and equation (1.3), only if $\mathrm{div}(\varepsilon \vec{E}) = \tilde{\rho}|_{t=0}$. Equation (1.3) follows from the first equation in (1.1) and the second one in (1.2). That means that the divergence equations in (1.1) to (1.3) are, in essence, conditions imposed on the initial data of the problem, but initial data for $\vec{E}$ and $\tilde{\rho}$ have to be consistent [31,32]. Equation (1.3) makes the sources consistent (external electric charges and electric currents), generating electromagnetic fields for all observation times t. Generally there are no formal reasons to assume that all the above-mentioned conditions are satisfied. That is why the system of equations describing electromagnetic processes is conventionally written in its complete form (1.1) to (1.3).

Following such a formulation of the problem, all six components of the vectors $\vec{E}$ and $\vec{H}$ are defined by curl equations (1.1), which in Cartesian coordinates have the form:

$$\begin{cases} \dfrac{\partial H_z}{\partial y} - \dfrac{\partial H_y}{\partial z} = \varepsilon \eta_0^{-1} \dfrac{\partial E_x}{\partial t} + \sigma_0 E_x + j_x \\ \dfrac{\partial H_x}{\partial z} - \dfrac{\partial H_z}{\partial x} = \varepsilon \eta_0^{-1} \dfrac{\partial E_y}{\partial t} + \sigma_0 E_y + j_y \\ \dfrac{\partial H_y}{\partial x} - \dfrac{\partial H_x}{\partial y} = \varepsilon \eta_0^{-1} \dfrac{\partial E_z}{\partial t} + \sigma_0 E_z + j_z \end{cases}, \quad \begin{cases} \dfrac{\partial E_z}{\partial y} - \dfrac{\partial E_y}{\partial z} = -\eta_0 \dfrac{\partial H_x}{\partial t} \\ \dfrac{\partial E_x}{\partial z} - \dfrac{\partial E_z}{\partial x} = -\eta_0 \dfrac{\partial H_y}{\partial t} \\ \dfrac{\partial E_y}{\partial x} - \dfrac{\partial E_x}{\partial y} = -\eta_0 \dfrac{\partial H_z}{\partial t} \end{cases}.$$

In cylindrical and spherical coordinates we have, respectively,

$$\begin{cases} \dfrac{1}{\rho} \dfrac{\partial H_z}{\partial \phi} - \dfrac{\partial H_\phi}{\partial z} = \varepsilon \eta_0^{-1} \dfrac{\partial E_\rho}{\partial t} + \sigma_0 E_\rho + j_\rho \\ \dfrac{\partial H_\rho}{\partial z} - \dfrac{\partial H_z}{\partial \rho} = \varepsilon \eta_0^{-1} \dfrac{\partial E_\phi}{\partial t} + \sigma_0 E_\phi + j_\phi \\ \dfrac{1}{\rho} \left[\dfrac{\partial (\rho H_\phi)}{\partial \rho} - \dfrac{\partial H_\rho}{\partial \phi} \right] = \varepsilon \eta_0^{-1} \dfrac{\partial E_z}{\partial t} + \sigma_0 E_z + j_z \end{cases},$$

$$\begin{cases} \dfrac{1}{\rho} \dfrac{\partial E_z}{\partial \phi} - \dfrac{\partial E_\phi}{\partial z} = -\eta_0 \dfrac{\partial H_\rho}{\partial t} \\ \dfrac{\partial E_\rho}{\partial z} - \dfrac{\partial E_z}{\partial \rho} = -\eta_0 \dfrac{\partial H_\phi}{\partial t} \\ \dfrac{1}{\rho} \left[\dfrac{\partial (\rho E_\phi)}{\partial \rho} - \dfrac{\partial E_\rho}{\partial \phi} \right] = -\eta_0 \dfrac{\partial H_z}{\partial t} \end{cases},$$

and

$$\begin{cases} \dfrac{1}{r \sin \vartheta} \left[\dfrac{\partial (H_\phi \sin \vartheta)}{\partial \vartheta} - \dfrac{\partial H_\vartheta}{\partial \phi} \right] = \varepsilon \eta_0^{-1} \dfrac{\partial E_r}{\partial t} + \sigma_0 E_r + j_r \\ \dfrac{1}{r} \left[\dfrac{1}{\sin \vartheta} \dfrac{\partial H_r}{\partial \phi} - \dfrac{\partial (r H_\phi)}{\partial r} \right] = \varepsilon \eta_0^{-1} \dfrac{\partial E_\vartheta}{\partial t} + \sigma_0 E_\vartheta + j_\vartheta \\ \dfrac{1}{r} \left[\dfrac{\partial (r H_\vartheta)}{\partial r} - \dfrac{\partial H_r}{\partial \vartheta} \right] = \varepsilon \eta_0^{-1} \dfrac{\partial E_\phi}{\partial t} + \sigma_0 E_\phi + j_\phi \end{cases},$$

$$\begin{cases} \dfrac{1}{r \sin \vartheta} \left[\dfrac{\partial (E_\phi \sin \vartheta)}{\partial \vartheta} - \dfrac{\partial E_\vartheta}{\partial \phi} \right] = -\eta_0 \dfrac{\partial H_r}{\partial t} \\ \dfrac{1}{r} \left[\dfrac{1}{\sin \vartheta} \dfrac{\partial E_r}{\partial \phi} - \dfrac{\partial (r E_\phi)}{\partial r} \right] = -\eta_0 \dfrac{\partial H_\vartheta}{\partial t} \\ \dfrac{1}{r} \left[\dfrac{\partial (r E_\vartheta)}{\partial r} - \dfrac{\partial E_r}{\partial \vartheta} \right] = -\eta_0 \dfrac{\partial H_\phi}{\partial t} \end{cases}.$$

1.2.2. Wave and Telegraph Equations

The geometry of most model problems (see, for instance, Fig. 1.1) are taken to be rather simple. These are the problems that we use to test the details of the mathematical techniques and methods of efficient calculation, to reveal and analyze the key regularities and characteristics of the spatio–temporal field transformations. This fact allows us in some cases and under certain conditions to reduce the three-dimensional vector problems for equations (1.1) to two-dimensional scalar problems. Thus, for instance, for the structures shown in Figures 1.1A–D, with their sections of the plane $y0z$, and when the sources of the excitation field are homogeneous along the x-axis, we get $\partial/\partial x \equiv 0$, and the general problem (1.1) splits into two mutually complementary problems: for E-($E_y = E_z = H_x = j_y = j_z \equiv 0$) and H-polarized ($H_y = H_z = E_x = j_x \equiv 0$) fields. In the case of E-polarization (this is mainly the case that is considered in detail in Chapter 1 and in the rest of this book)

$$\left[-\varepsilon\frac{\partial^2}{\partial t^2} - \sigma\frac{\partial}{\partial t} + \frac{\partial^2}{\partial y^2} + \frac{\partial^2}{\partial z^2}\right] E_x = F, \tag{1.4}$$

$$\frac{\partial H_y}{\partial t} = -\eta_0^{-1}\frac{\partial E_x}{\partial z}, \qquad \frac{\partial H_z}{\partial t} = \eta_0^{-1}\frac{\partial E_x}{\partial y}. \tag{1.5}$$

Here, $\varepsilon \equiv \varepsilon(g)$, $\sigma = \eta_0\sigma_0 \equiv \sigma(g)$, $E_x \equiv E_x(g, t)$, $F = \eta_0\partial j_x/\partial t \equiv F(g, t)$ and $g = \{y, z\}$ is a point in the space $\mathbf{R}^2$. When $\varepsilon =$ const and $\sigma_0 =$ const, equations of the same kind can be derived also for the components of the H-polarized field:

$$\left[-\varepsilon\frac{\partial^2}{\partial t^2} - \sigma\frac{\partial}{\partial t} + \frac{\partial^2}{\partial y^2} + \frac{\partial^2}{\partial z^2}\right] H_x = F \equiv \frac{\partial j_y}{\partial z} - \frac{\partial j_z}{\partial y}, \tag{1.6}$$

$$\varepsilon\frac{\partial E_y}{\partial t} + \sigma E_y + \eta_0 j_y = \eta_0\frac{\partial H_x}{\partial z}, \qquad \varepsilon\frac{\partial E_z}{\partial t} + \sigma E_z + \eta_0 j_z = -\eta_0\frac{\partial H_x}{\partial y}. \tag{1.7}$$

In cylindrical coordinates ρ, ϕ, x, the E-polarized fields ($E_\rho = E_\phi = H_x = j_\rho = j_\phi \equiv 0$; see Fig. 1.1C) are described by the following scalar equations.

$$\left[-\varepsilon\frac{\partial^2}{\partial t^2} - \sigma\frac{\partial}{\partial t} + \frac{1}{\rho}\frac{\partial}{\partial \rho}\rho\frac{\partial}{\partial \rho} + \frac{1}{\rho^2}\frac{\partial^2}{\partial \varphi^2}\right] E_x = F \equiv \eta_0\frac{\partial j_x}{\partial t}, \tag{1.8}$$

$$\frac{\partial H_\rho}{\partial t} - \eta_0^{-1}\frac{1}{\rho}\frac{\partial E_x}{\partial \phi}, \qquad \frac{\partial H_\phi}{\partial t} = \eta_0^{-1}\frac{\partial E_x}{\partial \rho}. \tag{1.9}$$

In the H-polarization case ($H_\rho = H_\phi = E_x = j_x = 0$), when $\varepsilon =$ const and $\sigma_0 =$ const, we have

$$\left[-\varepsilon\frac{\partial^2}{\partial t^2} - \sigma\frac{\partial}{\partial t} + \frac{1}{\rho}\frac{\partial}{\partial \rho}\rho\frac{\partial}{\partial \rho} + \frac{1}{\rho^2}\frac{\partial^2}{\partial \phi^2}\right] H_x = F \equiv \frac{1}{\rho}\left[\frac{\partial j_\rho}{\partial \phi} - \frac{\partial(\rho j_\phi)}{\partial \rho}\right], \tag{1.10}$$

$$\varepsilon\frac{\partial E_\rho}{\partial t} + \sigma E_\rho + \eta_0 j_\rho = \frac{\eta_0}{\rho}\frac{\partial H_x}{\partial \phi}, \qquad \varepsilon\frac{\partial E_\phi}{\partial t} + \sigma E_\phi + \eta_0 j_\phi = -\eta_0\frac{\partial H_x}{\partial \rho}. \tag{1.11}$$

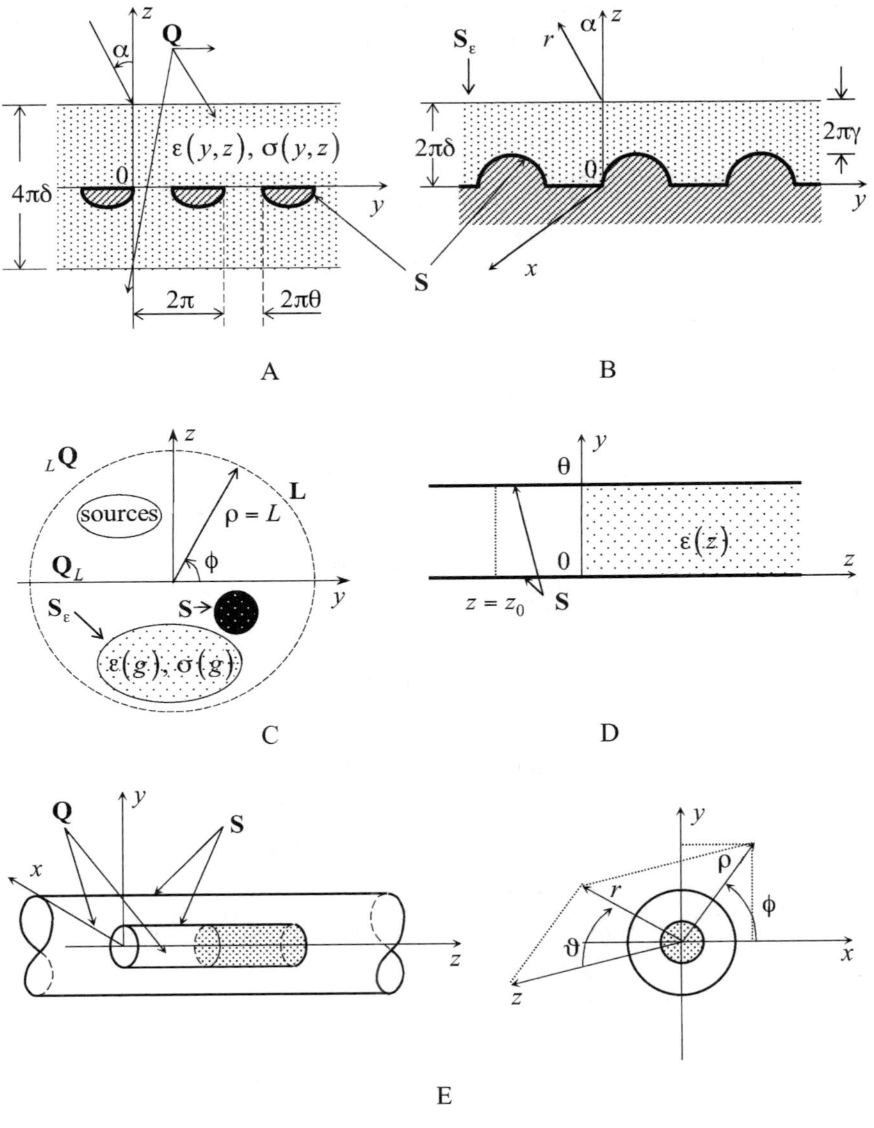

FIGURE 1.1. The geometry of model problems in right-handed coordinate systems: (A) Semi-transparent and (B) reflecting gratings, $\varepsilon(y + 2\pi, z) = \varepsilon(y, z)$, $\sigma(y + 2\pi, z) = \sigma(y, z)$. (C) Compact object in free space. Scattering inhomogeneities in (D) plane-parallel and (E) circular waveguides. Structures A to D are homogeneous along the x-axis.

For an axially symmetric structure (an example of such a structure is shown in Fig. 1.1E), with $\partial/\partial\phi \equiv 0$, we can analyze the TE_0-($E_\rho = E_z = H_\phi = j_\rho = j_z \equiv 0$) and TM_0-waves ($H_\rho = H_z = E_\phi = j_\phi \equiv 0$) separately. For the TE_0-waves from (1.1), we have

$$\left[-\varepsilon\frac{\partial^2}{\partial t^2} - \sigma\frac{\partial}{\partial t} + \frac{\partial^2}{\partial z^2} + \frac{\partial}{\partial\rho}\left(\frac{1}{\rho}\frac{\partial}{\partial\rho}\right)\rho\right]E_\phi = F \equiv \eta_0\frac{\partial j_\phi}{\partial t}, \tag{1.12}$$

$$\frac{\partial H_\rho}{\partial t} = \eta_0^{-1}\frac{\partial E_\phi}{\partial z}, \qquad \frac{\partial H_z}{\partial t} = -\eta_0^{-1}\frac{1}{\rho}\frac{(\partial E_\phi)}{\partial\rho}, \tag{1.13}$$

and for the TM_0-waves, when $\varepsilon = \text{const}$ and $\sigma_0 = \text{const}$,

$$\left[-\varepsilon\frac{\partial^2}{\partial t^2} - \sigma\frac{\partial}{\partial t} + \frac{\partial^2}{\partial z^2} + \frac{\partial}{\partial\rho}\left(\frac{1}{\rho}\frac{\partial}{\partial\rho}\rho\right)\right]H_\phi = F \equiv \frac{\partial j_z}{\partial\rho} - \frac{\partial j_\rho}{\partial z}, \tag{1.14}$$

$$\varepsilon\frac{\partial E_\rho}{\partial t} + \sigma E_\rho + \eta_0 j_\rho = -\eta_0\frac{\partial H_\phi}{\partial z}, \qquad \varepsilon\frac{\partial E_z}{\partial t} + \sigma E_z + \eta_0 j_z = \frac{\eta_0}{\rho}\frac{\partial(\rho H_\phi)}{\partial\rho}. \tag{1.15}$$

In spherical coordinates the TE_0-waves ($\partial/\partial\phi \equiv 0$, $E_r = E_\vartheta = H_\phi = j_r = j_\vartheta \equiv 0$) are described by the equations

$$\left[-\varepsilon\frac{\partial^2}{\partial t^2} - \sigma\frac{\partial}{\partial t} + \frac{1}{r}\frac{\partial^2}{\partial r^2}r + \frac{1}{r^2}\frac{\partial}{\partial\vartheta}\left(\frac{1}{\sin\vartheta}\frac{\partial}{\partial\vartheta}(\sin\vartheta)\right)\right]E_\phi = F \equiv \eta_0\frac{\partial j_\phi}{\partial t}, \tag{1.16}$$

$$\frac{\partial H_r}{\partial t} = -\frac{1}{\eta_0 r\sin\vartheta}\frac{\partial(E_\phi\sin\vartheta)}{\partial\vartheta}, \qquad \frac{\partial H_\vartheta}{\partial t} = \frac{1}{\eta_0 r}\frac{\partial(rE_\phi)}{\partial r}, \tag{1.17}$$

and the TM_0-waves for $\varepsilon = \text{const}$ and $\sigma_0 = \text{const}$, respectively, by

$$\left[-\varepsilon\frac{\partial^2}{\partial t^2} - \sigma\frac{\partial}{\partial t} + \frac{1}{r}\frac{\partial^2}{\partial r^2}r + \frac{1}{r^2}\frac{\partial}{\partial\vartheta}\left(\frac{1}{\sin\vartheta}\frac{\partial}{\partial\vartheta}\sin\vartheta\right)\right]H_\phi$$
$$= F \equiv \frac{1}{r}\left[\frac{\partial j_r}{\partial\vartheta} - \frac{\partial(rj_\vartheta)}{\partial r}\right], \tag{1.18}$$

$$\varepsilon\frac{\partial E_r}{\partial t} + \sigma E_r + \eta_0 j_r = \frac{\eta_0}{r\sin\vartheta}\frac{\partial(H_\phi\sin\vartheta)}{\partial\vartheta},$$
$$\varepsilon\frac{\partial E_\vartheta}{\partial t} + \sigma E_\vartheta + \eta_0 j_\vartheta = -\frac{\eta_0}{r}\frac{\partial(rH_\phi)}{\partial r}. \tag{1.19}$$

1.2.3. The Borgnis Functions

In the general case the vector functions $\vec{E}$ and $\vec{H}$ can be found either directly from (1.1), or from the equations, following from system (1.1) after certain transformations (introducing various potentials, etc.). In this book we usually consider the following vector problems, that are equivalent to (1.1).

$$\left\{\begin{array}{l}\left[\Delta - \text{grad div} - \varepsilon(g)\dfrac{\partial^2}{\partial t^2} - \sigma(g)\dfrac{\partial}{\partial t}\right]\vec{E}(g,t) = \dfrac{\partial}{\partial t}\vec{J}(g,t) \equiv \vec{F}(g,t) \\ \dfrac{\partial}{\partial t}\vec{H}(g,t) = -\dfrac{1}{\eta_0}\text{rot}\vec{E}(g,t); \quad g \in \mathbf{R}^3\end{array}\right. ; \tag{1.20}$$

Δ is the Laplace operator, having in Cartesian coordinates the form

$$\Delta \equiv \frac{\partial^2}{\partial x^2} + \frac{\partial^2}{\partial y^2} + \frac{\partial^2}{\partial z^2}.$$

In connection with the construction of exact absorbing conditions (see Chapters 2 to 4) we deal with the part of $\mathbf{R}^3$-space where the charges $\tilde{\rho}$ vanish for $t = 0$, whereas $\varepsilon = 1$, $\sigma = 0$, and $\vec{J} = 0$. Here div $\vec{E} = 0$, and problems (1.20) are simplified to

$$\left\{\begin{array}{l}\left[\Delta - \dfrac{\partial^2}{\partial t^2}\right]\vec{E}(g,t) = 0 \\ \dfrac{\partial}{\partial t}\vec{H} = -\dfrac{1}{\eta_0}\text{rot}\vec{E}\end{array}\right. . \tag{1.21}$$

Using the scalar Borgnis functions [32,33] $U^E(g,t)$ and $U^H(g,t)$ such that $[\Delta - \partial^2/\partial t^2][\partial U^{E,H}(g,t)/\partial t] = 0$, we can represent the general solution to problems (1.21) in the form

$$\left\{\begin{array}{l}E_x = \dfrac{\partial^2 U^E}{\partial x \partial z} - \dfrac{\partial^2 U^H}{\partial y \partial t}, \quad E_y = \dfrac{\partial^2 U^E}{\partial y \partial z} + \dfrac{\partial^2 U^H}{\partial x \partial t}, \quad E_z = \dfrac{\partial^2 U^E}{\partial z^2} - \dfrac{\partial^2 U^E}{\partial t^2}; \\ \eta_0 H_x = \dfrac{\partial^2 U^E}{\partial y \partial t} + \dfrac{\partial^2 U^H}{\partial x \partial z}, \quad \eta_0 H_y = -\dfrac{\partial^2 U^E}{\partial x \partial t} + \dfrac{\partial^2 U^H}{\partial y \partial z}, \\ \eta_0 H_z = \dfrac{\partial^2 U^H}{\partial z^2} - \dfrac{\partial^2 U^H}{\partial t^2}\end{array}\right. . \tag{1.22}$$

The field $\{\vec{E}, \vec{H}\}^E$ determined by (1.22) with $U^H \equiv 0$ is said to be the field of TM-waves with respect to the z-axis. The longitudinal component H_z of the magnetic vector $\vec{H}$ of this field is equal to zero. By assuming $U^E \equiv 0$ in (1.22), we obtain the field $\{\vec{E}, \vec{H}\}^H$ of TE-waves (here $E_z = 0$).

The Borgnis functions can be equally efficiently used in two other basic coordinate systems: cylindrical (the fields of TM- and TE-waves with respect to the z-axis) and spherical (the fields of TM- and TE-waves with respect to the r-axis) [32]. The relevant representations are valid in more general cases too (divergent fields). The presence of external currents is also accepted ($J_x = J_y \equiv 0$ in Cartesian rectangular coordinates; similar restrictions are imposed on the currents in cylindrical and spherical coordinates). However, in this situation, a homogeneous wave equation with respect to U^E is replaced by a nonhomogeneous one.

1.2.4. Domains of Analysis, Boundary, and Initial Conditions

In 3-D (2-D) vector (scalar) problems the domain of analysis $\mathbf{Q}$ is part of the space $\mathbf{R}^3$ (plane $\mathbf{R}^2$), bounded by surfaces (contours) $\mathbf{S}$ that are the boundaries of the domains int $\mathbf{S}$, filled with a "perfect conductor": $\mathbf{Q} = \mathbf{R}^3 \backslash \overline{\mathrm{int}\,\mathbf{S}}$ ($\mathbf{Q} = \mathbf{R}^2 \backslash \overline{\mathrm{int}\,\mathbf{S}}$).

An adequate mathematical model for the actual physical situation must, in addition to the main equations, contain the initial and boundary conditions (at the boundaries $\mathbf{S}$). On the perfect conductor surface the tangential component of the electric field density vector is equal to zero in every observation time t:

$$E_{tg}(g,t)\,|_{g\in\mathbf{S}} = 0; \qquad t \geq 0. \tag{1.23}$$

Besides, on the surfaces (in 3-D problems) or contours (in 2-D problems) $\mathbf{S}_\varepsilon$ of discontinuities of the material properties of the medium (i.e., of the functions $\varepsilon(g)$ and $\sigma(g)$, $g \in \mathbf{R}^3$ or $g \in \mathbf{R}^2$), as well as all over the domain $\mathbf{Q}$, the tangential components E_{tg} and H_{tg} of the vectors of strength of the electrical ($\vec{E}$) and magnetic ($\vec{H}$) fields should be continuous. The normal components $\varepsilon\varepsilon_0 E_{nr}(g,t)$ and $\mu_0 H_{nr}(g,t)$ of the vectors of electric and magnetic flux density are also continuous here. From (1.23) and (1.1) follows

$$H_{nr}(g,t)|_{g\in\mathbf{S}} = 0, \qquad \left.\frac{\partial H_{tg}(g,t)}{\partial \vec{n}}\right|_{g\in\mathbf{S}} = 0; \quad t \geq 0;$$

$\vec{n}$ is the outward normal with respect to the domain $\mathbf{Q}$. The function $H_{tg}(g,t)|_{g\in\mathbf{S}}$ defines so-called surface currents, generated on $\mathbf{S}$ by external electromagnetic field $\{\vec{E}, \vec{H}\}$.

The initial conditions, which we assume to be bounded up to the time $t = 0$, define the reference state of the system that develops at times $t > 0$, in compliance with the differential equations and the boundary conditions. The reference states $\vec{E}(g,0)$ and $\vec{H}(g,0)$ in the system (1.1) are the same as $\vec{E}(g,0)$ and $[\partial \vec{E}(g,t)/\partial t]|_{t=0}$ ($\vec{H}(g,0)$ and $[\partial \vec{H}(g,t)/\partial t]|_{t=0}$) in the differential forms of the second order (in the terms of t), to which (1.1) is transformed if the vector $\vec{H}$ (vector $\vec{E}$) is eliminated. Thus, (1.20) should have initial conditions of the kind

$$\vec{E}(g,0) = \vec{\varphi}(g), \qquad \left.\frac{\partial}{\partial t}\vec{E}(g,t)\right|_{t=0} = \vec{\Psi}(g); \quad g \in \bar{\mathbf{Q}}. \tag{1.24}$$

The functions $\vec{\varphi}(g)$ and $\vec{\Psi}(g)$ (the source functions) usually have limited support in the closure of $\mathbf{Q}$. By using them, one can model almost any of the ways of exciting scattering objects, in particular, the excitation by a pulsed signal $\vec{U}^i(g,t)$, and then $\vec{\varphi}(g) = \vec{U}^i(g,0)$ and $\vec{\Psi}(g) = [\partial \vec{U}^i(g,t)/\partial t]|_{t=0}$. The pulsed signal $\vec{U}^i(g,t)$ itself should satisfy the corresponding wave equation and the causality principle. One should also make sure that until the time $t = 0$ the pulsed signal does not make contact with the scattering object.

The latter is obviously impossible if infinite structures (see Fig. 1.1A and B) are excited by plane pulsed waves that propagate in a direction different from the normal one: $\alpha \neq 0$, $|\alpha| < \pi/2$. Such waves are able to sweep up a part of the scatterer's surface by any moment of time. As a result the mathematically correct modeling of the process becomes impossible: the input data required for the initial boundary problem formulation, are defined, as a matter of fact, by the solution of this problem.

In the frequency domain (see [3,9,11]), there is nothing to prevent placing a grating into a plane wave field arriving at the structure on an arbitrary angle α. The nondegenerate boundary conditions on the planes $y = 0$ and $y = 2\pi$ that are conditioned by the 2π-periodicity of structure and the 2π-quasi-periodicity of the excitation wave field allow the reduction of the domain of analysis to $\mathbf{Q}_{new} = \{g \in \mathbf{Q} : 0 < y < 2\pi\}$.

And, in the time domain problems, the domain of analysis can be reduced to $\mathbf{Q}_{new}$. The object of analysis is in this case not quite physical (a complex plane wave). However, by simple mathematical transformations of the result, all the values can be presented in the customary, physically correct form. There are many reasons why the modeling of a physically realizable situation in initial boundary value problems in the electromagnetic theory of gratings should start with the analysis of the Fourier images $f(y, z, t, \Phi)$ of the functions $f_{true}(y, z, t)$ describing the true fields and sources:

$$f_{true}(y, z, t) = \int_{-\infty}^{\infty} \tilde{f}(z, t, \Phi) e^{i\Phi y} d\Phi = \int_{-0.5}^{0.5} \left[\sum_{n=-\infty}^{\infty} \tilde{f}(z, t, n + \Phi) e^{i(n+\Phi)y} \right] d\Phi$$
$$= \int_{-0.5}^{0.5} f(y, z, t, \Phi)\, d\Phi. \tag{1.25}$$

It follows from (1.25) that

$$f\left\{\frac{\partial f}{\partial y}\right\}(y + 2\pi, z, t, \Phi) = e^{i2\pi\Phi} f\left\{\frac{\partial f}{\partial y}\right\}(y, z, t, \Phi), \tag{1.26}$$

and thus, in terms of the functions f, the statement of the initial boundary value problems is in good agreement with the classical statement of the problems in the frequency domain. Therefore we can make full use of the mathematical and physical results obtained there. If $y = 0$, (1.26) transforms into boundary conditions that reduce the domain of analysis to $\mathbf{Q}_{new} \subset \mathbf{R} = \{g \in \mathbf{R}^2 : 0 < y < 2\pi\}$. In grating theory, the band $\mathbf{R}$ with the conditions (1.26) at its boundaries is usually called the Floquet channel. There is no principal difference between this channel and the conventional treatment of closed waveguides. Hence, most of the results obtained by analyzing waveguide inhomogeneities are transferred to the grating problems and vice versa, practically without any changes. The analytic relation between these two types of problems is described in detail in [6].

Going back to Figures 1.1A and B, we should make one more remark. Generally the initial boundary value problems for the periodic structures are stated and considered in dimensionless space–time coordinates, where the spacing length of the grating is 2π. Here we agree with the traditional choice for the frequency domain that simplifies all mathematical presentations considerably and allows the analysis to concentrate on the main parameter, that is the dimensionless frequency parameter κ that characterizes the relation between the true spacing length l of the grating and the excitation wavelength λ. All the essential results of the electromagnetic grating theory are formulated in terms of κ, δ, and other related dimensionless parameters. These parameters are accepted among researchers solving applied problems. Hence, for the sake of convenience, and because we wish to conform to the results of both the frequency and the time domain, we also use them in our study.

In all other problems, based on the main equations (1.1) to (1.3) we retain the dimensions of the system of physical units SI, except for the time t that has the length dimensions ($[m]$, meters). Hereafter t stands for the product of the real time and the velocity of light in vacuum; that is, an excitation in free space travels one unit of distance during one unit of time.

In numerical experiments, it is rather inconvenient to give the dimensions in meters. In this case, for the sake of simplicity, one can introduce an arbitrary length dimension 1 [un] ([un], unit), that can be equal, for example, to r [m], where r is a random number, and then all operations will be performed in the system of this unit. For this purpose, we substitute all dimensions in meters for the $1/r$ dimensions in units. Thus, for example, in such a system, the wavelength λ [un] corresponds to the frequency parameter $k = 2\pi/\lambda$[rad/un], and if we write the real circular frequency as ω[rad/s] ($[s]$ are seconds), and the real time as τ[s], then $\omega\tau = kt = k\tau v$, where the velocity v of the disturbance has the dimension of [un/s].

1.2.5. *Formulation of Initial Boundary Value Problems: Generalized Functions and Generalized Solutions*

Thus, for the geometries presented in Figure 1.1C (all objects and sources are homogeneous along the x-axis; that is, $\partial/\partial x \equiv 0$), the problems describing the transient states of the E-polarized field can be reduced to the solution $U(g,t) = E_x$ of the two-dimensional telegraph equation

$$P_{\varepsilon,\sigma}[U] \equiv \left[-\varepsilon(g)\frac{\partial^2}{\partial t^2} - \sigma(g)\frac{\partial}{\partial t} + \frac{\partial^2}{\partial y^2} + \frac{\partial^2}{\partial z^2}\right] \times U(g,t) = F(g,t); \quad g \in \mathbf{Q},\ t > 0 \tag{1.27}$$

that satisfies the Dirichlet boundary conditions (see (1.23), the boundary $\mathbf{S}$ is assumed to be sufficiently smooth)

$$U(g,t)|_{g\in\mathbf{S}} = 0; \quad t \geq 0 \tag{1.28}$$

and the initial data (see (1.24))

$$U(g,0) = \varphi(g), \quad \frac{\partial}{\partial t}U(g,t)|_{t=0} = \Psi(g); \quad g \in \bar{\mathbf{Q}}. \tag{1.29}$$

It is known (see, e.g., [34]) that for any $F(g,t) \in \mathbf{C}^2(t \geq 0)$, $\varphi(g) \in \mathbf{C}^3(\mathbf{R}^2)$, and $\Psi(g) \in \mathbf{C}^2(\mathbf{R}^2)$, the classical solution $U(g,t) \in \mathbf{C}^2(t > 0) \cap \mathbf{C}^1(t \geq 0)$ of the Cauchy problem for the wave equation in $\mathbf{R}^2$ (equations (1.27) to (1.29) with $\mathbf{Q} = \mathbf{R}^2$ and $\varepsilon(g) \equiv 1, \sigma(g) \equiv 0$) does exist, is unique, and is described by the Poisson formula

$$U(g,t) = \frac{1}{2\pi}\left\{ -\int_0^t \int_{\mathbf{S}(g,t-\tau)} \frac{F(p,\tau)\,dp\,d\tau}{\sqrt{(t-\tau)^2 - |g-p|^2}} + \int_{\mathbf{S}(g,t)} \frac{\Psi(p)\,dp}{\sqrt{t^2 - |g-p|^2}} + \frac{\partial}{\partial t}\int_{\mathbf{S}(g,t)} \frac{\varphi(p)\,dp}{\sqrt{t^2 - |g-p|^2}} \right\}.$$

This formula gives us the explicit solution to the problem about the excitation of a homogeneous space $\mathbf{R}^2$, that is, the simplest problem in the electromagnetic theory of transient waves. Similar solutions also can be obtained for problems of excitation of regular plane-parallel waveguides [34] and regular ($\mathbf{Q}_{new} = \mathbf{R}$) Floquet channels [9].

In the case of the classical statement of the problems (1.27) to (1.29) (all the equations are satisfied in each point of the relevant domain), the solution $U(g,t)$ is originally assumed to be smooth enough (it should have as many continuous derivatives as are present in the equation), and that implies strict limitations on the smoothness for all the entries. The generalized statements and generalized solutions are more suitable for physical phenomena that are described by differential equations and make the analysis of the problem much simpler.

A generalized function is a generalization of the classical concept of a function. Roughly speaking, the generalized function is determined by its average values near each point, and this enables us to get a mathematically correct description of many of the idealized notions, for example, intensity of a momentary point source, and the like.

A generalized function denotes any linear continuous functional (f, γ) on the space of functions $\mathbf{D} = \mathbf{D}(\mathbf{R}^n)$, that is, on the space of all finite infinitely differentiable functions γ in $\mathbf{R}^n$. The linear set $\tilde{\mathbf{D}} = \tilde{\mathbf{D}}(\mathbf{R}^n)$ of all generalized functions, with weak convergence of the sequence of functionals, becomes a complete space.

The generalized function f vanishes in the domain $\mathbf{G} \subset \mathbf{R}^n$ if $(f, \gamma) = 0$ for all $\gamma \in \mathbf{D}(\mathbf{G})$. Accordingly, we study equality of generalized functions in the domain $\mathbf{G}$ and generalized solutions of the differential equations, with boundary value and initial boundary value conditions. Thus, for instance, for the generalized solutions U of (1.27), the values of the functional $(P_{\varepsilon,\sigma}[U] - F, \gamma)$ should vanish for all $\gamma \in \mathbf{D}(\mathbf{Q} \times (0;\infty))$. Several basic operations with the generalized functions are determined by simple equalities:

1. By the equality $(f(Ap+b), \gamma(p)) = |\det A|^{-1}(f(g), \gamma[A^{-1}(g-b)])$, determining a linear substitution ($g \in \mathbf{R}^n$, $p \in \mathbf{R}^n$; $g = Ap + b$, $\det A \neq 0$, a nonsingular linear transformation of the space $\mathbf{R}^n$ with respect to itself);
2. By the equality $(\beta f, \gamma) = (f, \beta\gamma)$, determining the product βf of the generalized function $f \in \tilde{\mathbf{D}}$ with an infinitely differentiable function β;

3. By the equality $(\partial^\alpha f, \gamma) = (-1)^{|\alpha|}(f, \partial^\alpha \gamma)$, determining the generalized derivative $\partial^\alpha f$ of the generalized function $f \in \tilde{\mathbf{D}}(\mathbf{R}^n)$ ($\alpha = \{\alpha_i\}, i = 1, \ldots, n$) is multi-index; $\partial^\alpha \gamma$ stands for the derivative of the function γ of the order $|\alpha| = \alpha_1 + \ldots + \alpha_n$);
4. By the equality $(f_1(g) \times f_2(p), \gamma) = (f_1(g), (f_2(p), \gamma(g, p)))$, determining the direct product $f_1 \times f_2$ of the generalized functions $f_1(g) \in \tilde{\mathbf{D}}(\mathbf{R}^n)$ and $f_2(p) \in \tilde{\mathbf{D}}(\mathbf{R}^m)$, $\gamma(g, p) \in \mathbf{D}(\mathbf{R}^{n+m})$; and
5. By the equality $((f_1 * f_2), \gamma) = (f_1(g) \times f_2(p), \gamma(g + p))$, determining the convolution $(f_1 * f_2)$ of two generalized functions $f_1(g)$ and $f_2(g)$ from $\tilde{\mathbf{D}}(\mathbf{R}^n)$, $\gamma \in \mathbf{D}(\mathbf{R}^n)$; and so on.

A consistent, detailed, and comprehensive description of the properties of generalized functions and operations with them is given in [34]. Hereafter we refer frequently to this source.

Among all the generalized functions, our interest mostly concerns the simplest of them, that is, the regular generalized functions. Regular generalized functions are related to functions $f(g)$, locally integrated in $\mathbf{R}^n$ (in $\mathbf{G} \subset \mathbf{R}^n$), and are determined by the formula

$$(f, \gamma) = \int_{\mathbf{R}^n(\mathbf{G})} f(g)\,\gamma(g) dg; \quad \gamma \in \mathbf{D}(\mathbf{R}^n) \quad (\gamma \in \mathbf{D}(\mathbf{G})). \tag{1.30}$$

There is a one-to-one correspondence between locally integrated functions and regular generalized functions. Therefore, the latter can be treated as conventional point functions, which is more suitable for the purposes of functional analysis and the theory of boundary value problems [35]. In the class $\tilde{\mathbf{D}}_r(\mathbf{G})$ of regular generalized functions, not all of the elements are infinitely differentiable. Due to their differential properties, they can be considered as elements of different functional spaces, in particular, the spaces $\mathbf{W}_m^l(\mathbf{G})$, consisting of functions $f(g) \in \mathbf{L}_m(\mathbf{G})$; $g \in \mathbf{G}$, that have generalized derivatives up to the order l from $\mathbf{L}_m(\mathbf{G})$, and others (see the appendix for a list of the symbols and abbreviations).

Among all the singular generalized functions (viz. nonregular ones), we use only the Dirac δ-function ($\delta(g)$) and its generalized derivatives. However, we make the following reservation. In cases when the standard mathematical operations (numerical implementation of computational schemes, finite-difference approximation of the initial boundary value problems, etc.) cannot be performed correctly because of the presence of such functions, then the regularization is achieved by substituting an appropriate locally-integrated δ-approximating function for the δ-function (the “cap” function $\omega_\varepsilon(g)$, etc.; see [34]). In the general case, an arbitrary generalized function $f(g) \in \tilde{\mathbf{D}}$ is regularized according to the scheme $f_\varepsilon(g) = (f * \omega_\varepsilon) = (f(p), \omega_\varepsilon(g - p))$: the infinitely differentiable function $f_\varepsilon(g)$ obtained by a convolution converges to $f(g)$ when $\varepsilon \to 0$ (the convergence is in $\tilde{\mathbf{D}}$).

In some chapters of this book, we often refer to the notion of a fundamental solution (principal function, Green's function) of the differentiation operator $B[U]$: the generalized function $G(g) \in \tilde{\mathbf{D}}(\mathbf{R}^n)$ is a fundamental solution of the operator $B[U]$, if $B[G] = \delta(g)$. Using the generalized function G, one can construct the solution of the equation $B[U] = f$ with an arbitrary right-hand part $f : U = (G * f)$. This solution is unique in the above-mentioned class of generalized functions from $\tilde{\mathbf{D}}(\mathbf{R}^n)$ that are obtained by convolution with G. This scheme can also be applied to a partial inversion of the differentiation operator of a problem, followed by an equivalent presentation of the latter as an integro differential equation. The most detailed analytic description of the fundamental solutions for concrete classic differentiation operators can be found in [32,34,36,37]. Here we list the ones that are useful for solving problems in nonstationary electromagnetics:

1. $$\left[-\frac{\partial^2}{\partial t^2} + \frac{\partial^2}{\partial y^2} + \frac{\partial^2}{\partial z^2}\right] G(g,t) = \delta(g,t),$$
$$G(g,t) = -\frac{\chi(t-|g|)}{2\pi\sqrt{t^2-|g|^2}}, \quad g = \{y,z\};$$

2. $$\left[-\frac{\partial^2}{\partial t^2} + \frac{\partial^2}{\partial x^2} + \frac{\partial^2}{\partial y^2} + \frac{\partial^2}{\partial z^2}\right] G(g,t) = \delta(g,t),$$
$$G(g,t) = -\frac{\chi(t)}{2\pi}\delta(t^2-|g|^2), \quad g = \{x,y,z\};$$

3. $$\left[-\frac{\partial^2}{\partial t^2} + \frac{\partial^2}{\partial z^2} - m^2\right] G(z,t) = \delta(z,t),$$
$$G(z,t) = -\frac{\chi(t-|z|)}{2} J_0\left(m\sqrt{t^2-z^2}\right);$$

4. $$\left[-\frac{\partial^2}{\partial t^2} + a^2\left(\frac{\partial^2}{\partial y^2} + \frac{\partial^2}{\partial z^2}\right) - 2m\frac{\partial}{\partial t}\right] G(g,t) = \delta(g,t),$$
$$G(g,t) = -\frac{e^{-mt}\chi(at-|g|)}{2\pi a^2\sqrt{t^2-|g|^2/a^2}} \operatorname{ch}\left(m\sqrt{t^2-|g|^2/a^2}\right), \quad g = \{y,z\};$$

5. $$\left[-\frac{\partial^2}{\partial t^2} + a^2\left(\frac{\partial^2}{\partial x^2} + \frac{\partial^2}{\partial y^2} + \frac{\partial^2}{\partial z^2}\right) - 2m\frac{\partial}{\partial t}\right] G(g,t) = \delta(g,t),$$
$$G(g,t) = -\frac{\chi(at)e^{-mt}\delta(a^2t^2-|g|^2)}{2\pi a} + \frac{me^{-mt}\chi(at-|g|)I_1\left(m\sqrt{t^2-|g|^2/a^2}\right)}{4\pi a^3\sqrt{t^2-|g|^2/a^2}}, \quad g = \{x,y,z\};$$

6. $$\left[\frac{d^2}{dt^2} + a^2\right] G(t) = \delta(t), \quad G(t) = \chi(t)\frac{\sin(at)}{a};$$

7. $$\left[\frac{d}{dt} + a\right] G(t) = \delta(t), \quad G(t) = \chi(t)e^{-at}.$$

Here, χ is the Heaviside step function, J_m is the Bessel function, and $I_1(b) = -iJ_1(ib)$. In conclusion, we would like to draw attention to the publication [38] that is considered to be an important step toward the general solution to this problem: there, using quite simple assumptions about the smoothness of the functions ε and σ_0, the author has analyzed the structure of the singular and regular parts of the fundamental solution to the differentiation operator that satisfies the set of the Maxwell curl equations (1.1).

1.2.6. Fundamental Results of the Theory

Let us assume that the source functions $\varphi(g)$, $\Psi(g)$, and $F(g,t)$ (for all $t > 0$) of the problem (1.27) to (1.29) are finite in $\overline{\mathbf{Q}}$, and that the functions $\partial\varepsilon(g)/\partial y$, $\partial\varepsilon(g)/\partial z$, and $\sigma(g)$, $g \in \mathbf{Q}$ are bounded. The following statement is true [35].

Statement 1.1. *Let* $F(g,t) \in \mathbf{L}_{2,1}(\mathbf{Q}^T)$, $\varphi(g) \in \overset{\circ}{\mathbf{W}}{}_2^1(\mathbf{Q})$, $\Psi(g) \in \mathbf{L}_2(\mathbf{Q})$, $\mathbf{Q}^T = \mathbf{Q} \times (0;T)$, $(0;T) = \{t : 0 < t < T < \infty\}$. *Then problem (1.27) to (1.29) for all* $t \in [0;T]$ *has a generalized solution from the energy class, and the uniqueness theorem is true in this class.*

By a generalized solution from the energy class we understand a function $U(g,t)$, belonging to $\overset{\circ}{\mathbf{W}}{}_2^1(\mathbf{Q})$, for any $t \in [0;T]$ and depending continuously on t in the norm $\mathbf{W}_2^1(\mathbf{Q})$. Furthermore the derivative $\partial U/\partial t$ should exist as an element of the space $\mathbf{L}_2(\mathbf{Q})$ for any $t \in [0;T]$ and vary continuously with t in the norm $\mathbf{L}_2(\mathbf{Q})$. The initial conditions (1.29) should be continuous in the spaces $\overset{\circ}{\mathbf{W}}{}_2^1(\mathbf{Q})$ and $\mathbf{L}_2(\mathbf{Q})$, respectively, and (1.27) should be satisfied in terms of the identity

$$\int_{\mathbf{Q}^T} \left\{ \varepsilon \left(\frac{\partial}{\partial t}U\right)\left(\frac{\partial}{\partial t}\gamma\right) - \sigma\left(\frac{\partial}{\partial t}U\right)\gamma - \left(\frac{\partial}{\partial y}U\right)\left(\frac{\partial}{\partial y}\gamma\right) - \left(\frac{\partial}{\partial z}U\right)\left(\frac{\partial}{\partial z}\gamma\right)\right\} dgdt$$

$$= + \int_{\mathbf{Q}} \varepsilon\Psi\gamma(g,0)\,dg = \int_{\mathbf{Q}^T} F\gamma dgdt.$$

Here, $\gamma = \gamma(g,t)$ is an arbitrary element from $\mathbf{W}_{2,0}^1(\mathbf{Q}^T)$ such that $\gamma(g,T) = 0$. This identity is derived in a formal way from the following identity

$$\left(P_{\varepsilon,\sigma}[U] - F, \gamma\right) = \int_{\mathbf{Q}^T} \left(P_{\varepsilon,\sigma}[U] - F\right)\gamma dg\,dt = 0$$

by means of single partial integration of the terms, containing second-order derivatives of the function $U(g,t)$. In [35] it was proven that such a definition makes sense and is actually a generalized notion of the classic solution.

Using practically the same assumptions, the unique solvability of the problem (1.27) to (1.29) and problems (1.27), (1.29) with impedance-type boundary conditions has been demonstrated in [35]. The class of generalized solutions, which has been called the energy class, is somewhat narrower than $\mathbf{W}_2^1(\mathbf{Q}^T)$. It is worth a more detailed study because of two factors [35]. First, this class is the only one where the following specific feature of hyperbolic equations can be determined, namely, the proof that the solution $U(g, t)$ has the same differential features that are assumed to be satisfied at the initial moment of time (continuable initial conditions). Second, the energy class of generalized solutions is, in a certain sense, the main one among those directly associated with the energy conservation law: for a $U(g, t)$ from this class the following energy relation is satisfied,

$$\int_{\mathbf{Q}} \left(\varepsilon \left(\frac{\partial U}{\partial t} \right)^2 + |\mathrm{grad} U|^2 \right) dg \Bigg|_0^T + 2 \int_{\mathbf{Q}^T} \left(\sigma \left(\frac{\partial U}{\partial t} \right)^2 + \left(F \frac{\partial U}{\partial t} \right) \right) dg\, dt = 0. \tag{1.31}$$

From relations of this kind and restrictions on the factors in initial boundary value problems of different types, "energy" estimates for U^2, $(\partial U/\partial t)^2$, and $|\mathrm{grad} U|^2$ are derived. They are of considerable importance for proving uniqueness theorems.

In [35], the author has also studied the increased smoothness of the solutions and determined when such solutions have derivatives included in the equation; the respective smoothness of input data and proper order of harmonization for the initial conditions, boundary conditions, and the equation are required. In the case of Statement 1.1, the harmonization condition of zero order is satisfied, which implies that the function $\varphi(g)$ should be equal to zero at the boundary $\mathbf{S}$ (for $\varphi(g) \in \overset{\circ}{\mathbf{W}}{}_2^1(\mathbf{Q})$). The next order of harmonization assumes the vanishing of the function $\Psi(g)$ at $\mathbf{S}$. In the second order of harmonization we should speak about the consistency on the set of points $g \in \mathbf{S}$ of the boundary condition, the initial conditions, and the equation (1.27). All the remaining orders imply the corresponding increased requirements on the input data of the problem: vanishing of the higher derivatives of U at $\mathbf{S}^T = \mathbf{S} \times (0; T)$, and so on.

In connection with the formulation of the problem (1.27) to (1.29), we should mention the well-known characteristic feature of classic solutions to hyperbolic equations, that is, the finiteness of the velocity of propagation of a disturbance. The initial disturbance that is concentrated in a certain finite subdomain $\mathbf{Q}_a$ of the domain $\mathbf{Q}$, has at time t not propagated farther than the distance $(\mu)^{1/2}\, t$ from $\mathbf{Q}_a$. The value μ is determined by the largest value of the function $\varepsilon^{-1}(g)$ at the points g of the domain of analysis.

The beginning of the practical work of solving concrete problems in the theory of transient oscillations can be associated with the development in the 1930s of the method of functional-invariant solutions (V.I. Smirnov and S.L. Sobolev). It was expected to help find more efficient approaches to the analysis of canonical problems of dynamic elasticity theory. Without giving any estimates of the effectiveness of this method, we can now state that, along with the Fourier method and the contour integral method, it belongs to the classic ones. We note that, being implemented, it allowed us, after the results achieved by H. Bateman [39], to obtain new classes of solutions to the wave equation with two and three spatial variables. The references [40,41] give examples that are relevant to the cases considered in this book. As is known, any sufficiently smooth function $U_0(\tau)$, $\tau = t + \alpha y \pm (1 - \alpha^2)^{1/2} z + \eta(\alpha)$, where $\eta(\alpha)$ is an arbitrary function, satisfies for any complex α the homogeneous wave equation

$$P_{1,0}\left[U_0\right]=0 \tag{1.32}$$

and corresponds to complex plane waves propagating toward decreasing (plus) or increasing (minus) values of z. From this follows the statement [40] that the solutions to the wave equation (1.32) are analytic functions $U_0(\alpha)$ of the complex variable α, related to the physical variables y, z, and t through the equation $t+\alpha y\pm(1-\alpha^2)^{1/2}z+\eta(\alpha)=0$, where $\eta(\alpha)$ is an arbitrary analytic function.

Let now $U_0^\beta(Y,Z,T)$ be a β-grade homogeneous function with respect to Y, Z, T that satisfies the wave equation in coordinates Y, Z, T; that is,

$$\left[Y\frac{\partial}{\partial Y}+Z\frac{\partial}{\partial Z}+T\frac{\partial}{\partial T}-\beta\right]U_0^\beta=0,\qquad \left[\frac{\partial^2}{\partial Y^2}+\frac{\partial^2}{\partial Z^2}-\frac{\partial^2}{\partial T^2}\right]U_0^\beta=0.$$

Then [41], if $Y=y$, $Z=z$, $T=t+\alpha(t^2-r^2)$, $\alpha=\text{const}$, $r=(y^2+z^2)^{1/2}$, the function $U_0=U_0^\beta(y,\ z,\ t+\alpha(t^2-r^2))[(1+\alpha t)^2-\alpha^2r^2]^{-\beta-1/2}$ is the solution of the wave equation (1.32).

There are two reasons why such results are considered to be the fundamental theses of the theory. First, the analytic description of the class of solutions enables one to formulate diverse representation theorems. These theorems, as seen from the theory of elliptic equations [42], serve as a basis for a variety of computation methods and enrich the theory with qualitative fundamental results. Second, the explicit analytic form of the solution to the wave equation in various regular domains is the key to the correct limitation of the computational space in the exterior problems at any values of t by using various kinds of partial radiation conditions and absorbing boundary conditions. In view of the latter, the problem of obtaining the corresponding results for regular directing structures (plane-parallel, cylindrical, horn, sector, etc.) becomes more relevant.

The state of the art of the problem was determined, to our mind, in the references [18,43–46] that are concerned with the implementation of the method of partial separation of variables [43] in equations of the second order derived from the set of Maxwell's equations. Usually in this method one or two spatial variables are separated that are transverse to the direction of propagation of the disturbance. In order to determine the functions depending on t and the other spatial variables, we should solve a hyperbolic equation of the second order. In rectangular coordinates it is the well-studied Klein–Gordon equation. The monograph [44] is devoted to the development of one of the possible approaches to the question about separation of variables on the basis of studying the Lie algebra of the symmetry of the equation and on the representation theory for this algebra. All the solutions in the separated variables were found for the Klein–Gordon equation: a total of ten different variants with a discrete and continuous spectrum, with these variants covering all feasible situations. In [32,45], this type of equations are solved by the Riemann method, which allows us in certain practically interesting cases to give an explicit description of the field in the spatio–temporal presentation. In [18] it is shown how the technique of the Fourier cosine transform of the Klein–Gordon equation with respect to the spatial variable yields an expression that essentially determines the strict form of the radiation condition for outgoing waves in any virtual transverse plane of a regular directing structure. We apply a similar technique repeatedly in this book to construct conditions that efficiently limit the analysis area in fundamentally open problems but which, however, do not distort the physics of the modeled processes of propagation and scattering of electromagnetic waves.

1.3. Methods of Solving Initial Boundary Problems on the Basis of Spatio-Frequency Representations

1.3.1. The Laplace and Fourier Integral Transformation Methods

Bearing in mind the energy relation (1.31) and the energy estimates that follow from it, Statement 1.1 can be easily reformulated in terms of the space

$$\mathbf{W}_2^1(\mathbf{Q}^\infty, \beta) \equiv \left\{ \{U(g,t)\} : U(g,t)\exp(-\beta t) \in \mathbf{W}_2^1(\mathbf{Q}^\infty) \right\}.$$

It allows us [35,47], basically, to apply correctly the direct and inverse Laplace transform (image $\leftrightarrow$ original)

$$\tilde{f}(s) = L[f](s) \equiv \int_0^\infty f(t)e^{-st}dt \leftrightarrow f(t) = L^{-1}\left[\tilde{f}\right](t) \equiv \frac{1}{2\pi i}\int_{\alpha-i\infty}^{\alpha+i\infty} \tilde{f}(s)e^{st}\,ds \tag{1.33}$$

and to connect the solutions of the initial boundary value problem (1.27) to (1.29) and the following elliptic problem,

$$\tilde{P}_{\varepsilon,\sigma}\left[\tilde{U}\right] \equiv \left[\frac{\partial^2}{\partial y^2} + \frac{\partial^2}{\partial z^2} + \tilde{\varepsilon}k^2\right]\tilde{U}(g,k) = \tilde{f}(g,k); \qquad g \in \mathbf{Q}, \tag{1.34}$$

$$\tilde{U}(g,k)\big|_{g\in\mathbf{S}} = 0, \quad \tilde{\varepsilon} = \varepsilon + i\sigma/k, \quad \tilde{f} = \tilde{F} + ik\tilde{\varepsilon}\varphi - \varepsilon\Psi, \quad s = -ik. \tag{1.35}$$

As is known [8–10,15,35,47], for $\operatorname{Im} k > 0$ for any $\tilde{f}$ from $\mathbf{L}_2(\mathbf{Q})$, problem (1.34), (1.35) is uniquely solvable in $\mathbf{W}_2^1(\mathbf{Q})$, and its solution $\tilde{U}(g,k)$ is an analytic function of the parameter k. This is also true for elliptic problems for any geometry (see Fig. 1.1). However, for the case of periodic structures, $\mathbf{Q}$ should be substituted for $\mathbf{Q}_{new}$ and the problem should be complemented with a quasi-periodicity condition similar to (1.26). If $\operatorname{Re} s > \beta \geq 0\,(\operatorname{Im} k > \beta)$, and the function $\tilde{U}(g,k)$ is absolutely integrable over $\operatorname{Re} k$ on $\mathbf{R}^1$ for a certain $\operatorname{Im} k = \alpha > \beta$, then under the conditions of Statement 1.1, the solution $U(g,t)$ of (1.27) to (1.29) from the energy class and solution $\tilde{U}(g,t)$ of (1.34), (1.35) from $\mathbf{W}_2^1(\mathbf{Q})$ are connected by the relations

$$U(g,t) = \frac{1}{2\pi}\int_{i\alpha-\infty}^{i\alpha+\infty} \tilde{U}(g,k)e^{-ikt}\,dk, \qquad \tilde{U}(g,k) = \int_0^\infty U(g,t)e^{ikt}\,dt. \tag{1.36}$$

The transforms (1.36) can be used to analyze initial boundary value problems of very diverse kinds that are substantially different both regarding the final computational schemes and concerning the data that have to be provided on the analytic features and numerical values of the function $\tilde{U}(g, k)$ of the complex variable k. One of the simplest variants is the numerical implementation of the inverse Laplace transform, using the fast Fourier transform (FFT, the standard tool for numerical Fourier analysis [48]). The required provision includes sufficiently reliable fast computation algorithms for the calculation of $\tilde{U}(g, k)$ at practically any values of k, as well as the results of a preliminary qualitative analysis of the possible effects of various frequency and time samplings, subject to the relative importance of separate sections of the frequency spectrum of the restorable transient wave.

Among the drawbacks of this approach are the great amount of computations, the absence of reliable criteria for optimal parameter selection and estimation of the validity of results obtained, and the impossibility of making a detailed physical analysis of the role of various components in the scattered field. Theoretically, the performance of the scheme can be improved by a successful equivalent deformation of the contour of integration in (1.36) and an optimal smoothing approximation of the functions of the sources of the transient signal (reduction of the frequency domain that effectively influences $\tilde{f}(g, k)$). The trivial procedure that allows us to improve the decrease of the function $\tilde{U}(g, k)$ as a function of k implies the preliminary transformation of (1.27) to (1.29) to a problem with homogeneous initial conditions or even to one where

$$\varphi = \Psi = \left(\partial^m F / \partial t^m\right)\Big|_{t=0} = 0; \qquad m = 0, 1, \ldots, M.$$

The general and most complicated problem for all methods based on the spatio-frequency representations is the extraction of reliable data on the analytic features of the resolvent operator function of the problem (1.34), (1.35) over the whole natural region of variation of the complex frequency parameter k. The first serious attempts to solve this problem are relatively new and they have been undertaken within the so-called spectral theory of open resonators. The results of this theory for resonators with noncompact boundaries extending to infinity (gratings and discontinuities of waveguide channels) are illustrated in Chapter 6 by a comprehensive set of examples. Here we cite only the key statements concerning compact scattering bodies (see Fig. 1.1C; the boundary $\mathbf{S}$ can have edges and be open, dielectric bodies can be adjoined to metal bodies, etc.).

Let us complement the problem (1.34), (1.35) with the radiation condition

$$\tilde{U}(g, k) = \sum_{n=-\infty}^{\infty} a_n H_n^{(1)}(k\rho) e^{in\phi} \tag{1.37}$$

($H_n^{(1)}$ denotes the Hankel function; ρ and ϕ are polar coordinates) that is satisfied for its solutions (if $\operatorname{Im} k > 0$) in the domain ${}_a\mathbf{Q} = \mathbf{Q} \backslash \bar{\mathbf{Q}}_a$, $\mathbf{Q}_a = \{g \in \mathbf{Q} : |g| < a\}$, that is free from scatterers and sources. This condition is equivalent to the Sommerfeld condition for real $k > 0$, and generalizes it correctly by extending the diffraction problem into the region of complex k. The natural boundaries of this extension are determined by the infinite-sheeted Riemann surface $\mathbf{K}$ of the analytic extension of the fundamental solution to the Helmholtz equation, or, what is the same, of the function Ln k.

Statement 1.2 [8,10]. *Let $\tilde{f}(g,k) \in \mathbf{L}_2(\mathbf{Q}_a)$. Then the solution of (1.34), (1.35), (1.37) exists and is unique in $\mathbf{W}_2^1(\mathbf{Q}_b)$ ($\mathbf{Q}_b$ is an arbitrary limited subdomain of the domain $\mathbf{Q}$) all over $\mathbf{K}$, except for the countable set Ω_k of the elements $\bar{k} \in \mathbf{K}$ without finite accumulation points. On the elements of set Ω_k, the solution $\tilde{U}(g,k)$, which is a meromorphic function of the complex parameter k (in the local variables on the surface $\mathbf{K}$), has poles of finite order, and the exterior homogeneous problem (1.34), (1.35), (1.37) has a nontrivial solution. These nontrivial solutions $\tilde{U}_n(g,\bar{k}_n)$ describe free oscillations of the electromagnetic field in the corresponding compact open resonators at the eigenfrequencies $\bar{k}_n$.*

The poles of the resolvent operator function of problem (1.34), (1.35), (1.37) for a metal ($\varepsilon \equiv 1,\ \sigma \equiv 0$) compact resonator, limited by a sufficiently smooth convex contour $\mathbf{S}$ with an everywhere strictly positive and finite curvature radius, stay away from the real axes at least logarithmically in the lower part of the first sheet of the surface $\mathbf{K}$ [47,49]. The region

$$\operatorname{Im} k > -\operatorname{const}(\mathbf{S})\,[1+\ln(1+|\operatorname{Re} k|)]; \qquad \operatorname{Re} k \neq 0 \tag{1.38}$$

is free from the singularities of the Green's function $\tilde{G}(g,p,k)$ ($\tilde{G}(g,p,k)$ is the solution to (1.34), (1.35), (1.37) at $\tilde{f}(g,k) = \delta(g-p)$). Here (see [50]),

$$\begin{aligned}&\left|\tilde{G}(g,p,k) + \frac{i}{4} H_0^{(1)}(k\,|g-p|)\right| \\ &\quad \leq \frac{\operatorname{const}(\mathbf{S})}{(d(p)+d(g)+|g-p|)^{1/2}} e^{-\operatorname{Im} k(|g|+|p|)+4D|\operatorname{Im} k|}\end{aligned} \tag{1.39}$$

for all points $g, p \in \mathbf{Q}$($d(p)$ is the distance from the point p to $\mathbf{S}$, D is the maximal curvature radius of the contour $\mathbf{S}$). By using (1.38) and (1.39) and allowing for the special status of the point $k=0$ for problems in $\mathbf{R}^2$, we obtain the following result [51].

Statement 1.3. *Let $F(g,t)=0$, $\psi(g) \in \mathbf{C}^{1+j}(\bar{\mathbf{Q}})$, $\varphi(g) \in \mathbf{C}^{2+j}(\bar{\mathbf{Q}})$ and assume that they vanish together with their derivatives up to the order $1+j$ and $2+j$ on $\mathbf{S}$ belonging to $\mathbf{C}^{2+j}$. Then, if the contour $\mathbf{S}$ is strictly convex, the derivatives with respect to t of the solution $U(g,t)$ of (1.27) to (1.29) are continuous and for large t the following estimate holds.*

$$\left|\frac{\partial^m U(g,t)}{\partial t^m}\right| \leq \frac{\operatorname{const}(j,\mathbf{S},\varphi,\psi)\ \operatorname{In}(2+|g|)}{t^{1/2}(1-|t-|g||)^{1/2+m}\ \operatorname{In} t\ \operatorname{In}(2+|t-|g||)};$$
$$g \in \mathbf{Q}, \quad m = 0,1,\ldots,j. \tag{1.40}$$

The estimate (1.40) exemplifies the behavior of a scattered field that is characteristic for a two-dimensional space obtained by exciting the resonator with a momentary source. In three-dimensional spaces and for scattering objects that do not generate trapped waves, the energy in any compact subdomain of the domain **Q** decreases exponentially [52]. Note that there are a huge number of publications devoted to the study of the asymptotic behavior of the steady-state ($\bar{U}(g,k)$, $|k| \to \infty$) and transient ($U(g,t)$, $t \to \infty$) fields as solutions to the different types of boundary value problems, in spaces of various dimensions. However, to learn the facts (essential results, the techniques applied), it is enough to concentrate on a few of them. We have chosen the publications [47,49–55].

In the theory of transient oscillations, the approach to the numerical inversion of the Laplace transform based on the general concepts and methods of the Tikhonov regularization [56,57] is not widely recognized yet. Nevertheless, despite the obvious difficulties occurring in transferring theoretically transparent regularization schemes into practical concrete calculations, this process is worthy of a close study, first of all, for the following reason. The fundamental notion of the regularizing operator itself makes the schemes for solving ill-posed problems with incomplete, imperfect, or fluctuating data clear and restores them as formally controllable computation procedures.

The most consequent implementation of such an approach, using the predecessors' experience, can be found in [58]. The problem of the numerical inversion of the Laplace transform is reduced here to a finite moments problem (the function $\tilde{f}(s)$ is determined by its values at a finite equidistant system of real points s). For a good choice of parameters, a satisfactory approximation for the normal solution (solution with the minimum norm) can be obtained in some analytically completely evident situations. In view of this, note that the focus on a qualitatively predicted result allows us to substantially decrease the nonefficient efforts of searching for optimal parameters of the method by means of preliminary testing it on a definitely solvable problem.

The original algorithm of numerical integration in the complex k-plane for passing from the frequency domain into the time domain has been suggested in [59] and generalized in [60]. According to the authors, it provides high computational accuracy with small operational expense. The efforts required for software support are minimal if compared to analogous algorithms with corresponding computational accuracy. These characteristics are confirmed by the solutions to some test problems. The required efficiency can be achieved by approximating the integrand $\exp(st)$ in the inverse Laplace transform by a two-parameter family of functions

$$\exp(st) \approx E(st,a,\lambda) = e^{a}\left[e^{a-st} - \lambda e^{-(a-st)}\right]^{-1}; \quad a > \alpha t, \quad \operatorname{Im}\lambda = 0 \tag{1.41}$$

followed by substituting (1.41) in (1.33) by the Mittag–Leffler and series (in terms of $\exp(st)$) expansions. As a result we obtain two representations for the two-parameter family of functions:

$$f_0(t,a,\lambda) = \frac{1}{2\pi i}\int\limits_{\alpha-i\infty}^{\alpha+i\infty} \tilde{f}(s)E(st,a,\lambda)ds,$$

whose linear combinations for various λ lead to the following computing formulas.

$$f(t) = f_0(t, a, -1) + O(|f(t)|e^{-2(a-Kt)}),$$
$$f(t) = \frac{1}{2}[f_0(t, a, 1) + f_0(t, a, -1)] + O(|f(t)|e^{-4(a-Kt)}),$$
$$f(t) = \frac{8}{3}f_0\left(t, a, \frac{1}{4}\right) - 2f_0\left(t, a, \frac{1}{2}\right) + \frac{1}{3}f_0(t, a, 1) + O(|f(t)|e^{-6(a-Kt)})$$

(and so on with the sequential exponential decrease of the absolute value of the components determining the approximation accuracy). It is assumed that for the function $f(t)$ the bound $|f(t)| \leq M\exp(Kt)$ is satisfied for certain constants M and K and all $t > 0$, and that $\tilde{f}(s)$ has all features required for transformation for $\mathrm{Re}\, s \geq \alpha$.

The mathematical problems occurring with the Fourier transform (image ↔ original)

$$\tilde{f}(k) = F[f](k) \equiv \frac{1}{2\pi}\int_{-\infty}^{\infty} f(t)e^{ikt}dt \leftrightarrow f(t) = F^{-1}[\tilde{f}](t) \equiv \int_{-\infty}^{\infty} \tilde{f}(k)e^{-ikt}dk \tag{1.42}$$

for the solutions to the initial boundary value problems of the type (1.27) to (1.29) are the same as for the Laplace transform. Nominally the transition to the elliptic problems

$$\begin{cases} \tilde{P}_{\varepsilon,\sigma}\left[\tilde{U}\right] \equiv \left[\dfrac{\partial^2}{\partial y^2} + \dfrac{\partial^2}{\partial z^2} + \tilde{\varepsilon}k^2\right]\tilde{U}(g, k) = \tilde{f}(g, k); \quad g \in \mathbf{Q} \\ \tilde{U}(g, k)\big|_{g\in \mathbf{S}} = 0, \quad \tilde{U}^s(g, k) = \sum\limits_{n=-\infty}^{\infty} a_n H_n^{(1)}(k\rho)e^{in\phi}; \quad g \in {}_a\mathbf{Q} \\ \tilde{\varepsilon} = \varepsilon + i\sigma/k, \quad \tilde{f}(g, k) \leftrightarrow F(g, t) \\ \tilde{U}(g, k) = \tilde{U}^i(g, k) + \tilde{U}^s(g, k), \quad \tilde{U}^i(g, k) \leftrightarrow U^i(g, t) \end{cases} \tag{1.43}$$

is realized by Fourier transform of the right-hand part of (1.27) and the excitation wave $U^i(g, t)$. In order to return to the time domain, one should apply the inverse transform (1.42): $U(g, t) = F^{-1}\left[\tilde{U}(g, k)\right](t)$.

1.3.2. Natural Resonances and the Singularity Expansion Method

The general methodological background (i.e., application of the analytic properties of the function $\tilde{U}(g, k)$ by integrating in the complex k-plane) and the objectives (i.e., obtaining the "natural mode" representations for $U(g, t)$) of the singularity expansion method in the electromagnetic theory of transient waves are featured in a great number of publications (see, e.g., [61–66]). The solution techniques for the separate stages have many different variants. The common feature for all these variants is that none of the sequences of steps can be sufficiently strictly motivated.

Several canonical problems (solved in a closed form) and problems that are simplified due to the presence of large (small) parameters constitute exceptions to this statement. Thus, for instance, in [47] the asymptotic expansions of the solutions to the initial boundary value problems have been obtained for convex compact bodies in spaces of different dimensions. These expansions are similar to the Fourier series for mixed problems in a bounded domain but, for obvious reasons, they include not only the oscillating but also the exponentially decreasing components as $t \to \infty$. The frequencies here are the poles of the analytic continuation of the resolvent of the corresponding elliptic problem.

One of two main directions in the implementation of the general concept of this method is based on the use of expansions for $\tilde{U}(g, k)$ that are fit for a direct transformation according to (1.36). Let us exemplify this. Let $\tilde{f}(s)$ be a meromorphic function of the variable $s \in \mathbf{C}$. According to the Mittag–Leffler theorem [67], it can always be presented as

$$\tilde{f}(s) = E(s) + E_0(s - a_0) + \sum_{n=1}^{\infty} [E_n(s - a_n) - h_n(s)]. \tag{1.44}$$

Here, $a_0, a_1 \ldots$ are the poles of $\tilde{f}(s)$ numbered in ascending order of their moduli, $E_j(s - a_j)$ are the main parts of $\tilde{f}(s)$ at these poles, $E(s)$ is an entire function, and $h_n(s)$ are the polynomials that should provide a uniform convergence of the series in any finite circle (on throwing away the components with the poles within this circle). In the case of simple poles, when $\tilde{f}(s)$ is bounded on the system of regular contours $\mathbf{C}_r$ in $\mathbf{C}$ (the radius r of $\mathbf{C}_r$ grows without bounds) and the series $\sum_{n=1}^{\infty} |\mathrm{Res}\,\tilde{f}(a_n)\|a_n|^{-m-1}$ converges, the form of (1.44) becomes much simpler. However, only if the condition $|\tilde{f}(s)| \to 0$ on $\mathbf{C}_r$ at $r \to \infty$ is satisfied, is it possible to obtain an expansion that can be easily converted into the time domain:

$$\tilde{f}(s) = \sum_{n=0}^{\infty} \mathrm{Res}\,\tilde{f}(a_n)(s - a_n)^{-1} \leftrightarrow f(t) = \sum_{n=0}^{\infty} \mathrm{Res}\,\tilde{f}(a_n)\,\mathrm{e}^{a_n t}.$$

An alternative (see [13]) is based on using the representations

$$\begin{cases} b(k) = \sum\limits_{n}\sum\limits_{j=1}^{K_n} c_{n,j}(k) b_{n,j}(k); \quad n = 1, 2, \ldots \\ h(k) = \sum\limits_{n}\sum\limits_{j=1}^{K_n} d_{n,j}(k) b_{n,j}(k) \\ c_{n,K_n} = d_{n,K_n}\lambda_n^{-1}, \quad c_{n,j} = (d_{n,j} - c_{n,j+1})\lambda_n^{-1}; \quad j = K_n - 1, \ldots, 1 \end{cases} \tag{1.45}$$

for the solution $b(k)$ to the problem $D(k)\,b(k) = h(k)$. Here, $\{b_{n,j}(k)\}_{n,j}$ is the orthonormal system of the principal (eigen and adjoint) vectors (modes) of the operator $D(k) : \mathbf{H} \to \mathbf{H}$ ($\mathbf{H}$ is the appropriate Hilbert space):

$$[D(k) - \lambda_n(k)]^j b_{n,j}(k) = 0. \tag{1.46}$$

The operator equation with respect to the vector $b(k)$ is an equivalent form of presenting the boundary value problem (1.34), (1.35), (1.37). Its solution, the function $\tilde{U}(g, k)$, is determined by the vector $b(k)$ according to explicit formulas. It is clear that any of the variety of possible algorithms solving the original elliptic problem generates its own operator $D(k)$, but none of them can, in principle, be self-adjoint. This fact certainly complicates the analysis and it is obvious that the situation when the representation (1.45), (1.46), supposedly the basic property of the system $\{b_{n,j}(k)\}$ in $\mathbf{H}$, could be completely justified, is quite unusual. Still, formally the transition from (1.45) through (1.36) to the "natural-mode" representation for $U(g, t)$ is possible due to the obvious property of the eigenvalues $\lambda_n(k)$: at least one of them should vanish at any of the eigenfrequencies $\bar{k}_n \in \Omega_k$ of the respective open resonator.

The second main direction in the implementation of the concept of the method is associated with the deformation of the integration contour in (1.36) providing for the contribution of the specific features of $\tilde{U}(g, k)$ as a function of k, that come to the transition zones, and the estimation of the residuals formed by the integrals that have not vanished in certain contours in $\mathbf{K}$. The analytic support needed to provide the proper performance is basically the same as for the first alternative considered. The most serious problems for this method as a whole are associated with obtaining reliable estimates of the behavior of $|\tilde{U}(g, k)|$ as $|k| \to \infty$. Other problems are associated with the qualitative analysis of the specific features of $\tilde{U}(g, k)$ (poles and branch points, their localization and possible concentration areas, dependence on parameters), and further with expanding $\tilde{U}(g, k)$ in terms of free oscillations $\tilde{U}_n(g, \bar{k}_n)$ at the eigenfrequency spectrum $\bar{k}_n \in \Omega_k$ and in terms of eigenmodes of the operators $D(k)$ related to the diffraction problem. Furthermore, there are problems associated with the development of reliable and efficient algorithms for numerical solution to the spectral problems. These are not just the problems of this method; these problems are common in mathematical diffraction theory. Until there is a satisfactory solution to these problems, the singularity expansion method cannot be considered to be sufficiently rigorous and well founded.

It can hardly be said that these problems have not received due attention (see, e.g., the already mentioned references [47,49–55]). However, substantial analytic efforts are still required to obtain at least partial results in this area. As a reference point here classical studies [68,69] (theorems about the basis property and completeness of the system of the root vectors of some classes of nonself-adjoint operators, in particular, dissipative operators that correspond most exactly to the physical nature of the problems considered) and [70] (spectral analysis of the finitely meromorphic Fredholm operator functions) can be mentioned. Some interesting ideas have been expressed in [13], devoted to the generalization of the eigenoscillation method in wave diffraction theory with a nonstandard approach to the selection of spectral parameters. The state of the art in algorithmic support for calculating the spectral characteristics (the eigenfrequencies $\bar{k}_n$ from Ω_k and associated free oscillations $\tilde{U}_n(g, \bar{k}_n)$) of open two-dimensional compact and periodic resonators, as well as waveguide resonators, can be characterized as relatively successful. The analytic regularization method applied in [8–10,71] reduces equivalently the homogeneous ($\tilde{f}(g, k) = \tilde{U}^i(g, k) \equiv 0$) problems (1.34), (1.35), (1.37), and (1.43) for $k \in \mathbf{K}$ to the definition of the nontrivial solutions to the operator equation of the second kind $[\boldsymbol{I} + \mathrm{A}(k)]\ a = 0$ with a finite meromorphic operator function $A(k)$. In most cases, the infinite matrix functions $A(k) : l_2 \to l_2$ generate a trace-type operator or the Koch matrix [10,68]. This enables one to construct algorithms for searching for spectrum components on the basis of an approximate solution to the characteristic equation

$$h(k) = \det [\boldsymbol{I} + A(k)] = 0; \quad k \in \mathbf{K}. \tag{1.47}$$

The multiplicity of the root $\bar{k}_n$ of the scalar equation (1.47) determines the multiplicity of the eigenfrequency $\bar{k}_n$ of an open resonator ($\bar{k}_n$ is not a pole of the operator function $A(k)$), that is, the value $N = N_1 + \ldots + N_m + m$, where m is the number of linearly independent free oscillations of the field $\tilde{U}_{n,j}(g, \bar{k}_n)$, and N_j is the number of the adjoint functions for the eigenfunction with the number j. The order of the pole of the resolvent $[\boldsymbol{I} + \mathrm{A}(k)]^{-1}$ (of the Green's function $\tilde{G}(g, p, k)$ of the original problem) at $k = \bar{k}_n$ is determined by the maximum value $N_j + 1$. The algorithms of the analytical regularization method have proven to be efficient in computational experiments. The results of these experiments have served as a basis for determining the main regularities of the eigenfrequencies' dynamics, the physical results of the intertype interactions of free oscillations, and the mechanisms of forming the resonant responses of structures placed in the field of plane waves and waves of lumped sources [8,9,71–73].

1.4. Time-Domain Methods

1.4.1. The Finite-Difference Method

The FDTD method (finite-difference time-domain method) [27] is now undoubtedly a most widespread computational approach. Paper [74] confirms this evident statement.

The finite-difference method reduces various types of problems for differential equations to an algebraic set of equations where the unknowns are the values of the mesh functions in the mesh points covering the area of analysis, and to the study of the limit for vanishing mesh size. The desired results can be obtained, provided that the mesh functions in the limit give the solution to the problem. Such reduction of the problem to an infinite sequence of auxiliary finite problems determining the approximate solutions (mesh functions) is ambiguous for problems of a certain type and nonuniform for problems of other types.

The method was first applied to solve ordinary differential equations and is known in mathematics as the Euler polyline method. It was studied and applied to PDEs (partial differential equations) only in the 20th century. A significant contribution to the theoretical development of the method (the principles for constructing convergent difference schemes, stability studies, the rate of convergence, the coefficient freezing principle, etc.) was made by I.G. Petrovsky, O.A. Ladyzhenskaya, A.A. Samarsky, S.K. Godunov, V.S. Ryaben'ky, V.Ya. Rivkind, F. John, P. Lax, L. Nirenberg, G.G. O'Brien, M.A. Hyman, S. Kaplan, and others.

The simplest variant of the finite-difference form of the problems considered here (see (1.27) to (1.29) for scatterers whose geometry is illustrated by Fig. 1.1C; $\sigma \equiv 0, t \in [0; T]$) consists of determining the mesh function $u = U(y_j, z_k, t_m) = U(j, k, m)$ that satisfies the difference equations

$$\left[-\varepsilon(j,k) D_+^t D_-^t + D_+^y D_-^y + D_+^z D_-^z\right] u = F(j,k,m) \tag{1.48}$$

at the mesh points $(g_{jk} = \{y_j, z_k\} \in \mathbf{Q}(h,T)$ on the time layers $t_m = ml, m = 0, 1, \ldots, M-1 = T/l$. They are complemented by the equations

$$\begin{cases} U(j,k,0) = \varphi(j,k), \quad U(j,k,1) = \varphi(j,k) + l\psi(j,k); \\ g_{jk} \in \mathbf{Q}(h,T) \\ U(j,k,m) = 0; \quad g_{jk} \in S(h,T), \quad m = 0, 1, \ldots, M-1 \end{cases} \tag{1.49}$$

(i.e., the difference form of the initial and boundary conditions). Here,

$$D_+^y[u] = h^{-1}[U(j+1,k,m) - U(j,k,m)]$$

and

$$D_-^y[u] = h^{-1}[U(j,k,m) - U(j-1,k,m)]$$

are the standard operators of the right- and left-hand difference derivatives (the same with obvious changes is true also for $D_\pm^z[u]$, $D_\pm^t[u]$); $y_j = jh$, $z_k = kh$, $j, k = 0, \pm 1, \ldots$. All mesh functions $f(j,k)$ at the points $(g_{jk} \in \mathbf{Q}(h,T)$ are constructed with respect to $f(g)$; $g \in \mathbf{Q}$ as the averages

$$f(j,k) = h^{-2} \int_{\omega_h(j,k)} f(g)dg;$$

$\omega_h(j,k) = \{g\colon jh < y < (j+1)h;\ kh < z < (k+1)h\}$; $\mathbf{Q}(h,T)$ is the union of cells $\omega_h(j,k)$ belonging to $\mathbf{Q}(T)$; $\mathbf{S}(h,T)$ is the boundary of $\mathbf{Q}(h,T)$; and $\mathbf{Q}(T)$ is the cut of the cone of influence of sources F, φ, and ψ in the region $\mathbf{Q}$ at the time $\tau > T$. It is obvious that (1.48) and (1.49) uniquely determine u, and u can be calculated without inversion of any matrix operators (i.e., through an explicit scheme).

The central part of a theoretical analysis of any finite-difference scheme is the solution to the stability problem. The scheme is considered to be stable (in a certain norm) if for the approximate solutions u a bound can be determined that is uniform with respect to the lengths of h and l. From the stability follows the intrinsic convergence of the sequence $\{u\}_{h,l}$, and the limiting function u will be the solution to the original initial boundary value problem provided that this problem is approximated by finite-difference equations. The latter are satisfied for (1.27) to (1.29), and (1.48), (1.49). As for the stability of the considered scheme, it is most convenient, as in most other cases, to analyze it in the energy spaces where the original problem is well posed. In [35] the validity of the following statement has been proven on the basis of differential analogues of the energy inequalities.

Statement 1.4. *For the functions* $F(g,t)$, $\varphi(g)$, $\psi(g)$, *and* $\varepsilon(g) - 1$, *that are finite in the region* $\mathbf{Q}$, *let* $F(g,t) \in \mathbf{L}_{2,1}(\mathbf{Q}^T)$, $\varphi(g) \in \mathbf{W}_2^1(\mathbf{Q})$, $\psi(g) \in \mathbf{L}_2(\mathbf{Q})\xi \leq \varepsilon^{-1}$

$(g) \leq \eta$; $g \in \mathbf{Q}$ and derivatives $\partial\varepsilon(g)/\partial y$ and $\partial\varepsilon(g)/\partial z$ be bounded. Then the $\mathbf{W}_2^1(\mathbf{Q}^T)$ norms of the continuous polylinear complements $\tilde{u}$ of solutions u in (1.48), (1.49) (the interpolations of the mesh functions u that are linear for each variable) are uniformly limited for any h and l that satisfy one of the following conditions,

$$\frac{\eta\sqrt{2}}{\sqrt{\xi}}\frac{l}{h} < 1, \quad 2\sqrt{\eta}\frac{l}{h} < 1. \tag{1.50}$$

The sequence $\{\bar{u}\}_{h,l}$ converges weakly in the space $\mathbf{W}_2^1(\mathbf{Q}^T)$ as $h, l \to 0$ and strongly in $\mathbf{L}_2(\mathbf{Q}^T)$ to the solution $U(g, t)$ of the problem (1.27) to (1.29).

In mathematics, the finite-difference method is not just a tool for obtaining approximate solutions to the initial boundary value and boundary value problems. It is also applied to prove theorems about the existence of generalized solutions, and to analyze the increase in their smoothness with the increasing smoothness of all parameters and the growing order of harmonization [75]. Moreover, it is precisely this method that first made it possible to study the solvability of hyperbolic problems in various function spaces, with reasonable data assumptions. The finite-difference method can be applied to a wide spectrum of initial boundary value problems with all kinds of classic boundary conditions. This spectrum and the well-developed theory of generalized solutions also encompass the problems of the theory of electromagnetic wave scattering.

In the full wave analysis in electromagnetic theory, the six-component Maxwell vector equations are discretized as a rule. Thus, for instance, the classic algorithm by Yee [28] is a central-difference approximation of the Maxwell curl equations. The mutual "location" of the electric and magnetic components of the field strength is determined by the so-called Yee-cell. The structure of the latter is well suited for the purposes of practical computation. In a variety of studies undertaken after [28], the Yee scheme stimulated an active expansion of the method to many engineering applications of electromagnetic theory; the scheme was modified in accordance with the class of geometries and wave propagation media considered. The operational characteristics of the method were improved and its computational efficiency enhanced. We mention here some of the papers describing the characteristic applications of this method.

The method is most frequently used in computations of microstrip lines and various types of functional units based on them [76–78]. The time-domain solution is here, as a rule, an intermediate one and serves then to obtain the frequency-dependent characteristics. Here we refer to [76], which basically presents a powerful variant of the method aimed at the analysis of planar structures, whose geometry allows modeling a variety of interesting practical applications. The application of the method for the analysis of scattering properties of two- and three-dimensional bounded perfectly conducting objects, for modeling of two- and three-dimensional horn antennas, as well as open waveguide resonators with dielectric embeddings, is considered in [79,80], and [81,82], respectively.

The solution of the problem in the time domain does not by itself provide a qualitative physical analysis of the modeled process. Arrays of numbers that are not "packed" into minimum sets of key characteristics, comprehensively describing the phenomenon, are quite a cumbersome result as far as physical interpretation is concerned.

As there are no conventional standards regarding the description of the wave scattering processes in the time domain, many authors just avoid this difficulty preferring the usual formalism of the frequency domain. This is evidently one of the reasons why the electromagnetic theory of transient processes to a large extent lacks its own physical concepts and results, although it describes much more diverse (in comparison with the electromagnetic theory of time-harmonic waves) electromagnetic processes among which we focus particularly on resonance effects and phenomena. The existence of several successes in this scientific area (see, e.g., [77] where it is illustrated how a 90° turn of a strip line at a certain cut configuration practically does not at all disturb the passage of a Gaussian pulse) only confirms the statement: problem-oriented research might provide us with many more successful results.

Modification of the method in accordance with the geometry of the problem and the wave propagation media is considered in papers [78,80,83–86]. In the first three the elementary cells that do not fit to the standard coordinate grid at the curvilinear boundary of scattering inhomogeneities are changed. By switching to the contour FDTD method, the integral Ampere and Faraday laws become the basic ones. The corresponding procedures are intended to reduce the required RAM capacities of the PC without loss of accuracy of the results obtained (the boundary conditions can be satisfied without reducing the cell sizes). The publications [84–86], although keeping the essence of the method unchanged, adapt it to the analysis of dispersive and gyrotropic materials and media.

The standard implementation of the FDTD method requires, as a rule, vast volumes of online storage and cannot always be carried out within acceptable terms. The problem of saving computer resources and enhancing the computation efficiency of the method is well understood. The main potential here consists:

In reducing the number of the associated unknown field components that are counted within an elementary cell [82,87];

In a flexible correction of the FDTD method scheme (e.g., in the resonance cases [88]);

In the adaptation of the cell configuration to the concrete noncoordinate boundaries [80,83];

In the reduction of the total computation space due to the use of imaginary boundaries that cause only a slight distortion of the modeled process (see below); and

In using the advantages of parallel virtual computers.

Paper [87] presents a comparison of the operational characteristics of the Yee algorithm, the equivalent FDTD algorithm for vector wave equations, and the FDTD algorithm for scalar wave equations (zero divergence fields), describes the construction of a hybrid computation scheme (considering parts of the computation area), and draws our attention to an additional problem, common to all of computational electromagnetics, namely: the only explanation for the fact that the computation time was reduced twice and the storage was reduced by a third by the simplest means, without losses in the quality of the results 25 years after the basic algorithm had been published for the first time, is that computer science professionals had not been involved in this problem at all.

An efficient reduction of the computation space in the analysis of initial boundary value problems in unbounded regions is one of the most important elements, not only in the FDTD method but also for other direct computation schemes, that are conceptually close to it. The known variants of the solution to this problem are based on several basically different approaches. The first one [22,23,89–95], and the most widely used for practical computations, implies limiting the region of source localization and part of scattering objects that are of fundamental importance by imaginary boundaries with so-called absorbing boundary conditions (ABCs) on it.

For the solutions U_0 to the homogeneous wave equation (1.32) that are considered in the region $z \geq 0$, the simplest of such ABCs [22,23] have the form

$$\left(\frac{\partial}{\partial z} - \frac{\partial}{\partial t}\right) U_0\Big|_{z=0} = 0$$

(the first-order approximation with respect to the angle of arrival of the wave to the boundary $z = 0$) and

$$\left(\frac{\partial^2}{\partial z \partial t} - \frac{\partial^2}{\partial t^2} + \frac{1}{2}\frac{\partial^2}{\partial y^2}\right) U_0\Big|_{z=0} = 0$$

(the second-order approximation). These conditions are derived from the physically transparent idea about the propagation of a plane complex wave beam in free space toward decreasing z and they make the reflection coefficient with respect to the partial beam component normally incident on the boundary vanish exactly. When the angle of arrival deviates from the normal one, the reflection coefficient for the corresponding partial component increases; but this increase is less pronounced for higher approximation orders. Actual situations usually exhibit many additional factors deviating from the model cases used to obtain the local classic ABCs [22,23,93–95]. Therefore the latter are seldom used in their original form. Rather, they are modified subject to the requirements of the specific cases (the expected computation accuracy, acceptable computation area, characteristics of the discontinuity of material parameters at the imaginary boundaries). Such modifications are usually based on heuristic considerations about the structure of the field interacting with the boundary.

The validity of such hypotheses is indirectly confirmed by numerical experiments that are also used to test the accuracy and efficiency of algorithms and to select the optimal free parameters of the method. Thus, in [89], a new technique (super absorption) has been suggested, that improves all known classic absorbing conditions. The latter should be used twice for different tangential field components that are separated in the elementary cell of the FDTD method, and the result is then recomputed according to a certain scheme (as it turned out, with intercompensation of errors) in the narrow boundary layer involved in the modification. The achieved effect is demonstrated in numerous examples. The authors of [90–92] have implemented their original ideas by using, as a matter of fact, schemes that are basically similar. The idea to multiply the absorption factor that is, for example, provided by the ABC of the first order, is realized in a condition that can be formally given as

$$\left[\prod_{j=1}^{N}\left(\frac{\partial}{\partial z}-\beta_j\frac{\partial}{\partial t}-\alpha_j\right)\right]U_0\Big|_{z=0}=0. \tag{1.51}$$

Here, β_j, α_j are free parameters. It was noted [91] that already at $N = 2$ and a reasonable choice of free parameters, the ABC (1.51) is much more favorable than the classic one with the second-order approximation. In spite of the advantages of this new ABC, we would like to note a substantial gap, in our opinion, in the mathematical foundations of the steps undertaken. Thus, for instance, already for $N = 3$ (and, naturally, for larger N) the study of the stability of the finite-difference approximation of the initial boundary value problem in the limited region with specified conditions (1.51) for parts of the boundary becomes quite problematical. As long as this problem remains unsolved, the corresponding FDTD scheme cannot be accepted as a sufficiently reliable one.

Examples of other, principally different, approaches to limiting the computation region can be found in [24–26,96–99] and [18,19,100–102]. The first direction is associated with developing a theory of a perfectly matched layer (with respect to free space) that absorbs the arriving electromagnetic waves, and leads to approximate conditions at the imaginary boundaries of the area of analysis. In the second case it is proposed to use as the exact ABCs the rigorous radiation conditions for secondary fields referred to some virtual boundary, usually to coordinate boundaries enveloping the scatterers. As they use no heuristic hypotheses about a thin field structure near the boundary and exactly conform to the physical nature of the numerically modeled process, they (at least theoretically) do not distort this process and do not have any awkward free parameter such as the total size of the computational space. In [18,19] the radiation conditions are presented through relations whose analogues are often used in this book, so we do not comment upon them further here. In [100,101] ABCs are constructed on the basis of strict relations of the theorem about the representation of the radiation field in free space through equivalent currents on the surface containing the primary and secondary (scatterers) sources. Unlike the condition used in [18,19], the boundary for the exact nonlocal ABCs suggested in [100,101] can be arbitrarily close to the scatterers' surface.

Summing up all the above, we note once again that the algorithms of the finite-difference method, together with all their additional constructions, are very powerful tools for obtaining concrete information on complicated electromagnetic objects. The possibility of entering additional information on the qualitative aspect and physical features of the modeled processes, especially under potential resonance wave scattering, makes it possible to adequately adapt the finite-difference schemes for computations, enhancing the reliability and validity of the numerical data obtained. The FDTD method is a kind of "big guns" in electromagnetic theory. It is efficient for concrete targets that are preliminarily prospected and placed into the hard coordinate system. To pave the way for wider use is the task for other, more refined, approaches and methods, allowing us to determine uniquely the effect of various factors, to react soundly on the eventual changes in the course of numerical experiments, and to accumulate knowledge about the physics of modeled processes. The FDTD method serves electromagnetic theory, however, to the same extent, electromagnetic theory itself has to serve this method.

1.4.2. The Integral Equation Method

In our short review, we certainly are not able to give full coverage of the role and significance of integral equations in present-day computational electromagnetic theory. As an excellent manual on this issue we can recommend [62,103–106]. They reveal both the general problems of mathematical modeling and some specific problems of implementation of models based on integral equations. The latter are characterized, in particular, by such parameters as computational efficiency and the volume of machine memory. Thus, for instance, in [105] the author suggests how one can substantially reduce the amount of computations by implementing integral equation method models in the time domain for large nonresonant scatterers irradiated by short pulses. The essence of the method is in the realization of the principle of "here and now", which implies a physically reasonable division of the scatterer's surface into active and passive sections, that is, sections whose influence on the computed characteristics can be considered separately.

The integral equations that constitute the basis for developing direct numerical algorithms can be obtained either directly in the time domain by using the concept of the fundamental solution of the operator or in the frequency domain, while applying the inverse Laplace transform to the classic equations of potential theory. Let us discuss more precisely the last variant and exemplify it by a compact perfectly conducting scatterer (see Fig. 1.1C; $\varepsilon - 1 = \sigma \equiv 0$).

Let $k \in \mathbf{K}$. It is known [10,71] that the radiation condition of type (1.37) is necessary and sufficient to provide that the solution to the boundary value problem (1.34), (1.35) with respect to $\tilde{U}(g,k) = \tilde{G}(g,g_0,k)$ for $\tilde{f}(g,k) = \delta(g-g_0)$ (the generalization to arbitrary $\tilde{f}(g,k)$ is trivial), if it exists, can be given as

$$\tilde{U}(g,k) = \int\limits_{\mathbf{S}} \tilde{\mu}(p)\tilde{G}_0(g,p,k)ds_p + \tilde{G}_0(g,g_0,k); \quad g \in \mathbf{Q}, \quad g_0 \in \mathbf{Q}_a. \quad (1.52)$$

Here, $\tilde{\mu}(p)$ is certain function on $\mathbf{S}$ that is proportional to the jump of the normal derivative of $\tilde{U}$ at $\mathbf{S}$, and $\tilde{G}_0(g,p,k) = -(i/4)H_0^{(1)}(k|g-p|)$. Making (1.52) subject to the boundary conditions, we derive for the definition $\tilde{\mu}(p)$ the boundary singular integral equation of the first kind

$$\int\limits_{\mathbf{S}} \tilde{\mu}(p)\tilde{G}_0(g, p, k)ds_p = -\tilde{G}_0(g, g_0, k), \quad g \in \mathbf{S}. \tag{1.53}$$

Similarly, using the potential of the double layer at **S**,

$$\tilde{U}(g, k) = \int\limits_{\mathbf{S}} \tilde{\eta}(p)\frac{\partial}{\partial n_p}\tilde{G}_0(g, p, k)\, ds_p + \tilde{G}_0(g, g_0, k); \quad g \in \mathbf{Q}, \quad g_0 \in \mathbf{Q}_a,$$

we obtain the boundary Fredholm integral equation

$$\tilde{\eta}(g) + 2\int\limits_{\mathbf{S}} \tilde{\eta}(p)\frac{\partial}{\partial n_p}\tilde{G}_0(g, p, k)\, ds_p = -2\tilde{G}_0(g, g_0, k); \quad g \in \mathbf{S}. \tag{1.54}$$

The pair of relations (1.52) and (1.53), by inversion according to (1.36), result in the following problem in time domain,

$$U(g, t) = \int\limits_{\mathbf{S}}\int\limits_{0} \mu(p, \tau)G_0(g, p, t-\tau)\, d\tau dp + G_0(g, g_0, t); \quad g \in \mathbf{Q}, \quad g_0 \in \mathbf{Q}_a; \tag{1.55}$$

$$\int\limits_{\mathbf{S}}\int\limits_{0} \mu(p, \tau)G_0(g, p, t-\tau)\, d\tau dp = -G_0(g, g_0, t); \quad g \in \mathbf{S}. \tag{1.56}$$

Here, $G_0(g, p, t) = -(2\pi)^{-1}\chi(t - |g-p|)[t^2 - |g-p|^2]^{-1/2}$ is the fundamental solution of the wave operator, that is, $P_{1,0}[G_0(g, p, t)] = \delta(g-p, t)$; $\mu(p, \tau) \leftrightarrow \tilde{\mu}(p, k)$ is an intermediate unknown surface potential used to calculate, according to (1.55), the field $U(g, t)$, generated by an open compact resonator. Applying similar transformations to (1.54), we obtain

$$U(g, t) = \int\limits_{\mathbf{S}}\int\limits_{0} \frac{\partial\eta(p, \tau)}{\partial\tau}(t-\tau)G_0(g, p, t-\tau) \times \frac{\cos(n_p, g-p)}{|g-p|}d\tau dp + G_0(g, g_0, t); \quad g \in \mathbf{Q}, \quad g_0 \in \mathbf{Q}_a; \tag{1.57}$$

$$\eta(g, t) + 2\int\limits_{\mathbf{S}}\int\limits_{0} \frac{\partial\eta(p, \tau)}{\partial\tau}(t-\tau)G_0(g, p, t-\tau)\frac{\cos(n_p, g-p)}{|g-p|}d\tau dp = -2G_0(g, g_0, t); \quad g \in \mathbf{S}, \tag{1.58}$$

where $(n_p, g-p)$ is an angle formed by an outer normal to the contour **S** at the point p and the vector connecting the points p and g.

In three-dimensional problems, for which the Huygens principle is true, the interference regions of different sections of the radiated surface are quite distinctly separated, making the integral equations look somewhat simpler. However, in case of two spatial variables for $g \neq p$, the operators in (1.56), (1.58) act upon potentials that are determined only by their values at times before the time of computation.

This essentially simplifies, at least for the variant (1.58), the numerical solution of integral equations, while reducing it in some cases to a sequential implementation of several direct formulas. The efficiency of computation schemes utilizing integral relations such as (1.55) to (1.58) and the accuracy of the results obtained are determined mostly by how precisely the individual details were studied in the course of constructing the algorithms for the problem (see, e.g., [105]). The principal problems of the method (the qualitative analysis of the convolution type equations) are solved in the framework of the classic mathematical disciplines and for the more general cases as considered above (presence of volume potentials, which corresponds to a partial inhomogeneous dielectric filling of an open resonator; piecewise smooth open and closed contours **S**, etc.). All this makes the integral equation method one of the most promising, universal, and reliable methods in computational electromagnetic theory.

Let us consider briefly, taking as an example the problem (1.27), (1.29) for a dielectric bounded body in free space (see Fig. 1.1C; $\overline{\text{int}\mathbf{S}} = \varnothing$), the possible schemes for transferring to integral equations directly in time domain. Assuming that $U(g,t) = 0$ and $F(g,t) = 0$ for $t < 0$, transfer first to the generalized statement of the corresponding initial boundary value problem [34]:

$$\begin{aligned} P_{1,0}[U(g,t)] &= F(g,t) - \delta^{(1)}(t)\varphi(g) - \delta(t)\psi(g) \\ &\quad + \left[(\varepsilon(g)-1)\frac{\partial^2}{\partial t^2} + \sigma(g)\frac{\partial}{\partial t}\right][U(g,t)] \\ &\equiv f(g,t); \quad -\infty < t < \infty, \quad g \in \mathbf{Q} \equiv \mathbf{R}^2. \end{aligned}$$

Then, using the definition and the properties of the fundamental solution $G(g,t) = -\chi(t-|g|)/2\pi\sqrt{t^2-|g|^2}$ of the wave operator $P_{1,0}[U(g,t)]$ (see Section 1.2.5), we obtain

$$\begin{aligned} U(g,t) &= [G(g,t) * f(g,t)] = U^i(g,t) \\ &\quad + \iint\limits_{\mathbf{Q}\ 0} G(g-p,t-\tau)\left[(\varepsilon(p)-1)\frac{\partial^2}{\partial \tau^2} + \sigma(p)\frac{\partial}{\partial \tau}\right][U(p,\tau)]dpd\tau; \\ &\quad g \in \mathbf{Q}, \quad t > 0. \end{aligned} \tag{1.59}$$

Here, $U^i(g,t)$ is the source field in free space (see the Poisson formula in Section 1.2.5). The second term in the integrodifferential equation with respect to the unknown total field $U(g,t)$ represents a correction to the source field, and it is conditioned by the presence of a compact semi-transparent scatterer. The reduction of (1.59) to the standard form for integral equations of the second kind is possible by using the basic properties of the convolution $(f * g)$ of the generalized functions f and g, particularly, the property $(\partial^\alpha f) * g = \partial^\alpha (f * g) = f * (\partial^\alpha g)$ [34]. The scheme is simple and universal. The sequence of steps required for an equivalent integral reformulation of differential initial boundary value problems (1.27) to (1.29) is practically the same for a wide class of objects with different material and geometric parameters.

1.4.3. The Directional Decomposition Method (Also Called the Invariant Imbedding or the Wave-Splitting Method)

The wave-splitting technique has been widely used for solving direct and inverse problems of wave scattering in layered inhomogeneous media, including random ones, in the frequency domain (the Riccati equation method). Nowadays, thanks to the studies by J.P. Corones, R.J. Krueger, V.H. Weston, L. Fishman, G. Kristensson, S. He, and many others (see, e.g., [107–109]), this technique is one of the most important ones in the theory of transient wave processes. It has been generalized to fairly general three-dimensional discontinuities, and extended to lossy inhomogeneous media. We note that it has recently been extended to anisotropic heterogeneous media [110]. This technique is distinguished by its efficiency for numerical solution of inverse problems (see, e.g., [107]).

We exemplify the main analytical concepts of this approach in the time domain by the simplest model problem and correspondingly adapt the results obtained in [107–109].

The field $U(g,t)$ of the source $F(g,t)=f(y,t)\delta(z_0-z)$ concentrated in the plane $z=z_0<0$ and exciting a regular plane-parallel waveguide with a longitudinal inhomogeneous dielectric filling (Fig. 1.1D: $\varepsilon(z<0)\equiv 1, \varepsilon\,(z>0)\to 1$ as $z\to 0$) satisfies the equation

$$\left[-\varepsilon(z)\frac{\partial^2}{\partial t^2}+\frac{\partial^2}{\partial z^2}+\frac{\partial^2}{\partial y^2}\right]U(g,t)=f(y,t)\delta(z_0-z);\quad t>0,\quad g\in\mathbf{Q},\tag{1.60}$$

the Dirichlet boundary conditions on the waveguide walls $y=0$ and $y=\theta$, and the zero initial data at time $t=0$. The separation of the transversal space variable y reduces (1.60) to the disconnected Cauchy problems (the boundary $z=0$ does not perform the mode transformation),

$$\left[-\varepsilon(z)\frac{\partial^2}{\partial t^2}+\frac{\partial^2}{\partial z^2}-\lambda_n^2\right]u_n(z,t)=a_n(t)\delta(z-z_0);$$
$$|z|<\infty,\quad t>0,\quad \lambda_n=\frac{n\pi}{\theta},\quad n=1,2,\ldots\tag{1.61}$$

for one-dimensional equations of the Klein–Gordon type and to the formula

$$U(g,t)=\sum_n u_n(z,t)\sin(\lambda_n y),\quad a_n(t)=\frac{2}{\theta}\int_0^\theta f(y,t)\sin(\lambda_n y)dy.$$

Omitting for the time being the inferior index n in $u_n(z,t)$ and λ_n, we split in the region $z>z_0$ the wave $u(z,t)$ into two parts: the incident wave (u^+) and reflected wave (u^-):

$$u^\pm=\frac{1}{2}\left(u\mp K\frac{\partial}{\partial z}u\right),\quad K^{-1}=\left(\varepsilon(z)\frac{\partial^2}{\partial t^2}+\lambda^2\right)K.\tag{1.62}$$

The wave u^+ in the domain $z_0 < z \leq 0$ is the wave of excitation, corresponding to the source from the right-hand side of (1.61). It may be easily determined using the definition of the fundamental solution of the operator (see Section 1.2.5). That is why farther on we consider it to be known. The choice of the splitting operator K is not unique and this fact can be used to construct individual optimal computation schemes for each specific case. The variant (1.62) corresponds to the physics of the problem under consideration [107].

The introduction of the split kernels $G^{\pm}(z, t)$ which allows us, according to

$$u^{\pm}(z, t + \eta(z)) = a(z) \begin{Bmatrix} u^+(0, t) \\ 0 \end{Bmatrix} + G^{\pm}(z, t) * u^+(0, t) \tag{1.63}$$

to calculate the field $u(z, t)$ in the region $z > 0$ from the incident field $u^+(0, t)$ in the plane $z = 0$, is called the Green function method [108]. Here,

$$\eta(z) = \int_0^z (\varepsilon(w))^{1/2} dw$$

is the time of excitation distribution from zero to z, $a(z) = [\varepsilon(z)]^{-1/4}$, and

$$f(z, t) * g(z, t) = \int_0^t f(z, t - \tau) g(z, \tau) d\tau.$$

Using (1.61) to (1.63) one obtains the following initial boundary value problem for the set of two connected linear differential equations of the first kind

$$\begin{aligned} \frac{\partial G^{\pm}}{\partial z} &= \begin{Bmatrix} -aA \\ 2\varepsilon^{1/2} \dfrac{\partial}{\partial t} G^- \end{Bmatrix} \\ &\pm \left[\frac{a\varepsilon' \lambda^2 C}{4\varepsilon} - A * G^{\pm} - \frac{\varepsilon'}{4\varepsilon}(G^+ - G^-) + \frac{\varepsilon' \lambda^2}{4\varepsilon} C * (G^+ - G^-) \right] \\ z &\geq 0, \quad G^+(0, t) = 0, \quad G^-(z, 0) = -\frac{1}{8} a \varepsilon' \varepsilon^{-3/2}. \end{aligned} \tag{1.64}$$

Here, $A(z, t) = \lambda J_1(\lambda \varepsilon^{-1/2} t) t^{-1}$, $C(z, t) = \lambda^{-1} \varepsilon^{-1/2} \sin(\lambda \varepsilon^{-1/2} t)$ and $\varepsilon'(z) = \partial \varepsilon(z)/\partial z$. On defining $G^+(z, t)$ and $G^-(z, t)$ from (1.64), the field $U(g, t)$ in the region $z > 0$ is calculated according to (1.63). The details of the numerical implementation of the algorithms based on (1.64) (explicit scheme) can be found, for example, in [107].

The proper imbedding equations for the kernels $R(z, t)$ of the reflection operator, which in the region $z \geq 0$ connects the fields $u^+(z, t)$ and $u^-(z, t)$ according to the expression

$$u^-(z, t + \eta(z)) = \int_0^t R(z, t - \tau) u^+(z, \tau + \eta(z)) d\tau, \tag{1.65}$$

can be obtained either by the traditional method, while substituting the expressions following from (1.65) into (1.61), or according to the formula $G^-(z, t) = a(z)R(z, t) + G^+(z, t) * R(z, t)$, that follows from (1.65) and (1.63). The latter variant after the substitution of $G^-(z, t)$ into (1.64) yields the following nonlinear problem with respect to the kernels $R(z, t)$.

$$\begin{aligned} &\frac{\partial}{\partial z}R = 2\varepsilon^{1/2}\frac{\partial}{\partial t}R - \frac{1}{4}\frac{\varepsilon'}{\varepsilon}\left[R * R + \lambda^2(C - C * R * R)\right] + 2A * R; \\ &R(z, 0) = -\frac{1}{8}a\varepsilon'\varepsilon^{-3/2}. \end{aligned} \tag{1.66}$$

Additionally, from the formula for $G^-(z, t)$ and the zero boundary conditions for $G^+(z, t)$ we derive $R(0, t) = G^-(0, t)$: the kernel of the reflection operator for the structure shown in Fig. 1.1D as a whole. Essentially, problem (1.64) is a linearized analogue to the nonlinear problem (1.66).

The technique that uses wave splitting can hardly as yet be characterized as a widely used one. Its main applications so far have been the numerical solution to the inverse problems where the complexity of the problem adequately motivates the efforts required in the application of this technique. As illustrated by the earlier sections of this chapter, for direct problems of transient wave scattering by local discontinuities in the propagation medium, there are alternative, and for the reader perhaps more familiar, methods available and we focus on those. Hence, here we limit ourselves to the above brief presentation of some key features of the emerging directional decomposition approach.

1.4.4. The Method of Separation of Variables and Other Analytic Methods

The method of mode basis [46,111], the method of partial separation of variables [32,45], and the method of evolutionary equations [111,112] that have much in common with classic analytic methods applied to the canonical problems of the theory of transient electromagnetic fields, confirm and vividly demonstrate by their results the value of a more detailed scrutiny of the old theoretical developments in electromagnetics, which have not been sufficiently exploited until the present time. Results of this kind (mainly the analytic ones, viz. the direct representations of electromagnetic fields generated by transient currents in regular domains) are extremely important for qualitative physical analysis of possible spatio–temporal variations both in the field as a whole and in its separate partial components (modes). Without such results, without preliminary nontrivial preparatory work, each stage of which is strictly oriented towards achieving the maximum possible progress in the analysis, no efficient numerical–analytical methods can be constructed; and it is scarcely possible to obtain reliable and unambiguous physical interpretation of the results of numerical experiments.

Let us consider one such result that gives an exact solution to some rather simple problems in the time domain. It utilizes an important characteristic feature of hyperbolic equations—the finiteness of the velocity of propagation of an excitation—and allows us, for relatively short time intervals $t \in (0; T)$, to construct a generalized solution to (1.27) to (1.29) with $\sigma(g) \equiv 0$, while formally following the classic scheme of the Fourier method.

Let T be specified. Then determine the domain $\mathbf{Q}_b \subset \mathbf{Q}$ such that the disturbance created by the sources $F(g,t)$, $\varphi(g)$, and $\psi(g)$ at the time $t = T$ has not reached the outer boundary yet. Then, formally, the solution $U(g,t)$ can be given as the series

$$U(g,t) = \sum_{n=1}^{\infty} \left[a_n \cos(\lambda_n t) + b_n \sin(\lambda_n t) - (\lambda_n)^{-1} \int_0^t F_n(\tau) \sin \lambda_n (t-\tau)\, d\tau \right] v_n(g). \tag{1.67}$$

Here, a_n, $b_n \lambda_n$, and $F_n(t)$ are the Fourier coefficients of the functions $\varphi(g)$, $\psi(g)$, and $F(g,t)$ in the orthonormal basis of the space $\mathbf{L}_2(\mathbf{Q}_b)$ consisting of the eigenfunctions of the Sturm–Liouville problem

$$\begin{cases} \left[\dfrac{\partial^2}{\partial z^2} + \dfrac{\partial^2}{\partial y^2} + \varepsilon(g)\lambda^2 \right] v(g) = 0; \quad g \in \mathbf{Q}_b. \\ v(g)\big|_{g \in \bar{\mathbf{Q}}_b \setminus \mathbf{Q}_b} = 0 \end{cases} \tag{1.68}$$

The validation of the Fourier method consists in studying the convergence of the series (1.67) in the norm of a certain function space and in determining the implication in which the sum of this series can be considered to be the solution of the original problem. The following statement is true [35].

Statement 1.5. *Let $\varphi(g) \in \mathring{\mathbf{W}}_2^1(\mathbf{Q}_b)$, $\psi(g) \in \mathbf{L}_2(\mathbf{Q}_b)$, and $F(g,t) \in \mathbf{L}_2(\mathbf{Q}_b^T)$. Then the series (1.67) and the series obtained by a single differentiation of it with respect to y, z, or t, converge in the norm $\mathbf{L}_2(\mathbf{Q}_b)$ uniformly over $t \in (0; T)$. The sum of the series (1.67) is the generalized solution $U(g,t)$ from the energy class of the original problem.*

A complete study of this method in all spaces $\mathbf{W}_2^l$, $l = 1, 2, \ldots$ at all three classical boundary conditions on $\mathbf{S}$ is presented in the book [75]. The practical computational aspect of the problem is also supplied with the necessary mathematical results, for example, with estimates of the residue after the reduction of the series (1.67). The characteristics, including the asymptotic ones, of the spectral sets and eigenvectors of the problems of type (1.68) have been studied in detail [113]. The dynamics of how they vary as functions of changes of the region $\mathbf{Q}_b$ and the potential $\varepsilon(g)$ has been analyzed. The relation between the behavior of $|a_n|$, $|b_n|$, $|F_n(t)|$ at $n \to \infty$ and the smoothness of the corresponding functions has been determined and universal, efficient, and reliable algorithms of numerical solutions of (1.68), the self-adjoint spectral problem in the frequency domain, have been developed and tested in many problems.

There are several topics not covered by this brief review, among them the method of finite elements. The basic schemes of these and other powerful direct numerical methods are very transparent, and the problems that arise are similar to those arising in the FDTD method.

In conclusion, a few comments have to be made about the problem of processing the results of numerical experiments and their reduction to physically well-established characteristics of the modeled processes. This problem deserves special attention and separate discussion, because without a satisfactory solution to it, a valid qualitative analysis, a study of the features, regularities, and implementation mechanisms of various resonance and irregular phenomena becomes difficult or impossible. Then the extraction of new information from the data can be unnecessarily complicated or even blocked, even if all the necessary reliable and efficient algorithms and solution methods for the corresponding model problems are available. In this review we only mention the publications [114,115], whose results clearly demonstrate the advantages of intelligent, problem-oriented processing of numerical data. The applied technique of the spatio–temporal representations [116] (superposition of windows differentiating the contribution of different signal components) divides the information into controllable doses, and this property can be naturally employed by a detailed electromagnetic analysis of structures embodying complex dispersion laws (see, e.g., [6–9] and Chapter 6 of this book).

2
Waveguides and Periodic Structures: Exact Absorbing Conditions on Virtual Boundaries in Cross-Section of Regular Waveguiding Structures

2.1. Introduction

In this and subsequent chapters, we discuss scalar and vector initial boundary-value model problems for two- and three-dimensional telegraphy and wave equations describing the scattering of electromagnetic waves by compact objects in unbounded domains $\mathbf{Q}$ of the spaces $\mathbf{R}^2$ and $\mathbf{R}^3$. The central problem arising in the course of the solution of such problems by finite-difference methods is caused by the progressively expanding support of the scalar or vector function $U(g,t)$, $g \in \mathbf{Q}$, $t \geq 0$, which determines the resulting field for the increasing observation times t. The analysis domain can be bounded by introducing anywhere in $\mathbf{Q}$ the artificial boundary $\mathbf{L}$ (see Fig. 1.1C) and supplementing the original initial boundary value problem with the condition

$$M\left[U(g,t)\right]\big|_{g\in\mathbf{L}} = 0; \qquad t \geq 0. \tag{2.1}$$

Here, $\boldsymbol{M}$ is an integrodifferential operator on $\mathbf{L}$. Such a modification of the problem should satisfy at least two requirements:

1. Condition (2.1) does not change the correctness classes either of the original problem or its discrete analogue; and
2. Condition (2.1) distorts the physical processes simulated by the mathematics only slightly.

About twenty-five years ago, B. Engquist and A. Majda [22] and a bit later G. Mur [23] suggested the so-called absorbing boundary conditions, the classical ABCs of the lowest orders of approximation, for use in (2.1). Requirement 1 for these conditions turns out to be satisfied, whereas requirement 2 is satisfied in one situation only, namely, when a complex plane wave is incident on the plane boundary $\mathbf{L}$ at near-normal angles. In all other cases, the incident wave $U(g,t)$ is reflected by the artificial boundary rather strongly. The simulated process of free propagation (in the corresponding domain) of $U(g,t)$ is distorted. The resultant calculation error cannot be estimated analytically. Too many factors affect its value: the fine structure of the field $U(g,t)$ near the boundary $\mathbf{L}$, the distance between

$\mathbf{L}$ and the region where the sources and the efficient scatterers are located, the observation time t, and so on.

Numerous attempts to improve the approximate classical ABCs (see a review of the literature in Chapter 1 and in [74]) have been made. Most of them are aimed at weakening the virtual (computational) effects caused by the presence of the imaginary boundary $\mathbf{L}$. However, as a rule, requirement (1) has been ignored. This may lead to possible negative consequences such as the violation of the stability of the finite-difference computational scheme, uncontrollable growth of total errors, and the onset of a computational catastrophe.

The approach suggested in this book focuses on the alternative technique clearly defined in [18–21]. The essence of our approach is as follows.

In the first step, a regular domain ${}_L\mathbf{Q} = \mathbf{Q}\backslash(\mathbf{Q}_L \cup \mathbf{L})$ is chosen, where the wave $U(g, t)$ propagates freely, moving away from the domain $\mathbf{Q}_L$ enclosing the sources and scattering objects.

In the second step, rigorous radiation conditions for the solution $U(g, t)$ of the original problem are formulated in ${}_L\mathbf{Q}$. These conditions reflect the general property of the solution $U(g, t)$ associated with an outgoing wave and, hence, their use as exact conditions on $\mathbf{L}$ changes nothing in the original problem and does not introduce additional model distortions in the process under study. The operator M is defined by the transport operator (see Chapter 5 and references [21,30,117,118], which describes spatio–temporal variations of the evolutionary basis of the signal $U(g, t)$ during its free propagation through the regular domain ${}_L\mathbf{Q}$. For different geometries of ${}_L\mathbf{Q}$, M is different. However, the algorithm of its derivation is based, in all cases, on the common sequence of operations widely used in the theory of hyperbolic equations [32,45]: incomplete separation of variables in the equations of telegraphy, integral transformations in the one-dimensional Klein–Gordon equations, solution of the auxiliary boundary value problems for ordinary differential equations, and inverse integral transforms. This technique results usually in nonlocal (in space and time) conditions. A change to local conditions can be performed by replacing a number of integral forms with differential ones, followed by the formulation of an additional (simple for computation) initial boundary value problem with respect to some auxiliary function (or vector function) of time and transversal coordinates.

In the third and final step, the exact radiation conditions referred to the coordinate boundary $\mathbf{L}$ are incorporated properly into the standard finite-difference scheme. The analysis domain (the domain where the initial boundary value problem is discretized) is restricted to $\mathbf{Q}_L$. The implementation of this algorithm results in the numerical solution of the problem for all observation times t, which is as simple and exact as in the case of physically closed regions $\mathbf{Q}_L$.

If the boundary $\mathbf{L}$ is a composite one (e.g., a boundary of a rectangular region in $\mathbf{R}^2$-space in Cartesian rectangular coordinates), the theoretical problem of corner points arises. For its rigorous resolution, a rather simple and universal mathematical technique has been suggested in [21,119].

This approach is detailed in the Section 2.2 for a problem concerning the transformation of E- and H-polarized transient waves by a compact waveguide unit whose geometry, parameters, and sources of excitation are uniform in the x-direction. The analytical results obtained in this section are then applied (in Section 2.3), without essential changes, to two-dimensional problems concerning an analysis of axially symmetric structures excited by TE_0-waves ($\partial/\partial\phi \equiv 0$, $E_\rho = E_z = H_\phi = j_\rho = j_z \equiv 0$) and TM_0-waves ($H_\rho = H_z = E_\phi = j_\phi \equiv 0$). The problems arising in the analysis of inhomogeneities of guided-wave structures in a three-dimensional vector case are discussed in Section 2.4. Section 2.5 is concerned with model problems of the electromagnetic theory of gratings, namely, the problems concerning compact inhomogeneities in the plane-parallel (scalar case) and rectangular in their cross-section (vector case) Floquet channels.

2.2. Compact Waveguide Units: 2-D Scalar Problems in Cartesian Coordinates

2.2.1. Transformation of the Evolutionary Basis for a Signal in a Regular Plane-Parallel Waveguide

A waveguide device (waveguide transformer or open waveguide resonator) is a classical electromagnetic model object possessing highly diversified features along with a wide range of various material and geometric parameters. This section discusses the waveguide units whose cross-sections remain constant in any plane $x = \text{const}$. The analysis of electromagnetic fields in such systems is reduced to the solution of the following 2-D (in the plane $y0z$) initial boundary value problems

$$\begin{cases} \left[-\varepsilon(g)\dfrac{\partial^2}{\partial t^2} - \sigma(g)\dfrac{\partial}{\partial t} + \dfrac{\partial^2}{\partial z^2} + \dfrac{\partial^2}{\partial y^2}\right] U(g,t) = F(g,t); \\ t > 0, \quad g = \{y,z\} \in \mathbf{Q} \\ U(g,t)\big|_{t=0} = \varphi(g), \quad \dfrac{\partial}{\partial t} U(g,t)\big|_{t=0} = \Psi(g); \quad g \in \overline{\mathbf{Q}} \\ E_{tg}(g,t)\big|_{g\in\mathbf{S}} = 0; \qquad t \geq 0 \end{cases} . \tag{2.2}$$

An example of the structure is given in Figure 2.1 (a waveguide T-junction). For $U(g,t) = E_x$ we have $E_{tg}(g,t)|_{g\in\mathbf{S}} = U(g,t)|_{g\in\mathbf{S}}$, and problems (2.2)

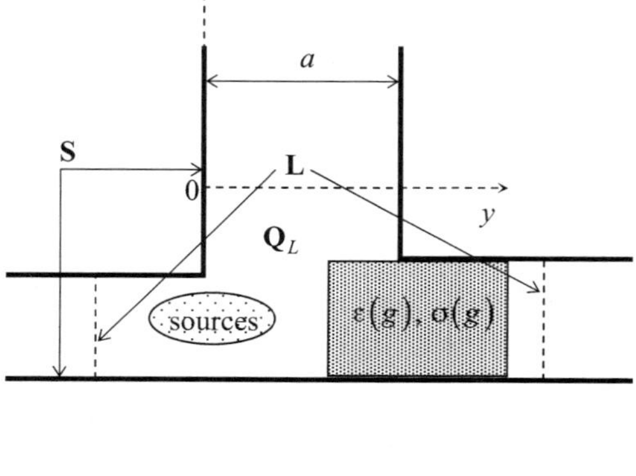

FIGURE 2.1. Waveguide T-junction.

correspond to the E-polarization (see Section 1.2.2). For $U(g,t) = H_x$ values $E_{tg}(g,t)|_{g\in \mathbf{S}}$ are given by derivatives $\partial U(g,t)/\partial \vec{n}|_{g\in \mathbf{S}}$ ($\vec{n}$ is a normal to the contour $\mathbf{S}$) and, for the piecewise constant functions $\varepsilon(g)$ and $\sigma(g)$, the solutions to (2.2) determine the H-polarized fields. The analysis domain $\mathbf{Q}$ is the region in the $\mathbf{R}^2$-plane that is bounded by the contour $\mathbf{S}$. $\mathbf{S} \times [|x| \leq \infty]$ is the surface of the perfectly conducting objects. It is assumed that the functions $F(g,t)$, $\varphi(g) = U^i(g,0)$, $\psi(g) = \partial U^i(g,t)/\partial t|_{t=0}$, $\sigma(g)$, and $\varepsilon(g) - 1$ ($U^i(g,t)$ is the incident wave), being finite in the closure of $\mathbf{Q}$, satisfy the theorem about a unique solvability of problems (2.2) in the energy class (the Sobolev space) $\mathbf{W}_2^1(\mathbf{Q}^T)$, $\mathbf{Q}^T = \mathbf{Q} \times (0;T)$, $T < \infty$ (see Statement 1.1 and reference [35]). The supports of these functions belong to the set $\overline{\mathbf{Q}_L} \backslash \mathbf{L}$. The regular waveguides (in the domain ${}_L\mathbf{Q} = \mathbf{Q} \backslash (\mathbf{Q}_L \cup \mathbf{L})$), along which the field formed by the unit may propagate infinitely far, are free from the sources and scatterers. The artificial boundary $\mathbf{L}$ (dashed lines in Fig. 2.1) coincides with the cross-sections of these waveguides.

Let us dwell (for definiteness' sake) on a vertical ($z > 0$) regular waveguide of the waveguide unit. Here $\varepsilon(g) \equiv 1$ and $\sigma(g) = \varphi(g) = \psi(g) = F(g,t) \equiv 0$. By assuming that the excitation $U(g,t)$ has not yet reached the boundary $z = 0$ by the time $t = 0$, we obtain via the separation of variables the following representation for the solutions $U(g,t)$ to (2.2).

$$U(g,t) = \sum_n u_n(z,t)\,\mu_n(y); \quad z \geq 0, \quad 0 \leq y \leq a, \qquad t \geq 0, \quad n \in \{n\}. \tag{2.3}$$

The orthonormal systems of the transversal functions $\{\mu_n(y)\}$, complete in the space $\mathbf{L}_2[(0;a)]$ of square-integrable functions $f(y): f(0) = f(a) = 0$ or $df(y)/dy|_{y=0,a} = 0$, are determined by nontrivial solutions to the homogeneous (spectral) problems

$$\left\{ \begin{array}{l} \left[\dfrac{d^2}{dy^2} + \lambda_n^2 \right] \mu_n(y) = 0, \quad 0 < y < a \\ \mu_n(0) = \mu_n(a) = 0\,(E - \text{case}) \quad \text{or} \quad d\mu_n(y)/dy|_{y=0,a} = 0\,(H - \text{case}) \end{array} \right. , \tag{2.4}$$

and the spatio–temporal amplitudes $\{u_n(z,t)\}$ (the evolutionary basis) of the signal $U(g,t)$ are given by the solutions of the initial boundary value problems

$$\left\{ \begin{array}{l} \left[-\dfrac{\partial^2}{\partial t^2} + \dfrac{\partial^2}{\partial z^2} - \lambda_n^2 \right] u_n(z,t) = 0, \qquad t > 0 \\ u_n(z,0) = 0, \qquad \dfrac{\partial}{\partial t} u_n(z,t)\Big|_{t=0} = 0 \end{array} \right. ; \quad z \geq 0, \quad n \in \{n\}. \tag{2.5}$$

In the case of the E-polarized field, $\{n\} = 1, 2, \ldots$, $\mu_n(y) = \sqrt{2/a}\,\sin(n\pi y/a)$, and $\lambda_n = n\pi/a$. In the H-case, $\{n\} = 0, 1, 2, \ldots$, $\mu_n(y) = \sqrt{(2 - \delta_0^n)/a}\,\cos(n\pi y/a)$, and $\lambda_n = n\pi/a$.

The cosine Fourier transform of (2.5) with respect to z on the semi-axis $z \geq 0$ (image $\leftrightarrow$ original)

$$\tilde{f}(\omega) = F_c[f](\omega) \equiv \sqrt{\frac{2}{\pi}} \int_0^{\infty} f(z)\cos(\omega z)dz$$

$$\leftrightarrow f(z) = F_c^{-1}[\tilde{f}](z) \equiv \sqrt{\frac{2}{\pi}} \int_0^{\infty} \tilde{f}(\omega)\cos(\omega z)d\omega \tag{2.6}$$

results in the following Cauchy problems for the images $\tilde{u}_n(\omega, t) \leftrightarrow u_n(z, t)$.

$$\begin{cases} D\left(\sqrt{\lambda_n^2 + \omega^2}\right)[\tilde{u}_n(\omega, t)] \equiv \left[\dfrac{\partial^2}{\partial t^2} + (\lambda_n^2 + \omega^2)\right]\tilde{u}_n(\omega, t) = -\sqrt{\dfrac{2}{\pi}}u_n'(0, t); \\ \omega > 0, \qquad t > 0 \\ \tilde{u}_n(\omega, 0) = 0, \qquad \dfrac{\partial}{\partial t}\tilde{u}_n(\omega, t)\Big|_{t=0} = 0; \qquad \omega \geq 0 \end{cases} . \tag{2.7}$$

Here,

$$u_n'(b, t) = \left.\frac{\partial u_n(z, t)}{\partial z}\right|_{z=b} = \int_0^a \left.\frac{\partial U(g, t)}{\partial z}\right|_{z=b} \mu_n(y)\,dy \tag{2.8}$$

($\{\mu_n(y)\}$ is the orthonormal basis). It has also been considered that

$$-\omega^2 \tilde{f}(\omega) - \sqrt{\frac{2}{\pi}}\left[\frac{d}{dz}f(z)\right]\Big|_{z=0} \leftrightarrow \frac{d^2}{dz^2}f(z)$$

and that the wave $U(z, t)$ in the region of interest of the ${}_L\mathbf{Q}$-domain does not contain components propagating in the sense of decreasing z. The outgoing components toward $z = \infty$ are equal to zero for sufficiently large z at any finite instant of time $t = T$.

By extending the functions $\tilde{u}_n(\omega, t)$ with zero on the semi-axis $t < 0$, let us pass on to the generalized statement of the Cauchy problems (2.7) [34]:

$$\begin{aligned} D\left(\sqrt{\lambda_n^2 + \omega^2}\right)[\tilde{u}_n(\omega, t)] &\equiv \left[\frac{\partial^2}{\partial t^2} + (\lambda_n^2 + \omega^2)\right]\tilde{u}_n(\omega, t) \\ &= -\sqrt{\frac{2}{\pi}}u_n'(0, t) + \delta^{(1)}(t)\,\tilde{u}_n(\omega, 0) + \delta(t)\left.\frac{\partial}{\partial t}\tilde{u}_n(\omega, t)\right|_{t=0} \\ &= -\sqrt{\frac{2}{\pi}}u_n'(0, t); \qquad \omega > 0, \quad -\infty < t < \infty. \end{aligned} \tag{2.9}$$

The convolution of the fundamental solution $G(\lambda, t) = \chi(t)\lambda^{-1}\sin\lambda t$ of the operator $D(\lambda)$ (see Section 1.2.5) with the right-hand side of equation (2.9) allows us to write $\tilde{u}_n(\omega, t)$ in the form

$$\tilde{u}_n(\omega, t) = -\sqrt{\frac{2}{\pi}} \int_0^t \sin\left[(t-\tau)\sqrt{\lambda_n^2 + \omega^2}\right] \frac{u_n'(0,\tau)}{\sqrt{\lambda_n^2+\omega^2}} d\tau; \quad \omega \geq 0, \quad t \geq 0. \tag{2.10}$$

By subjecting (2.10) to the inverse Fourier transform (2.6), we have for the originals $u_n(z, t)$:

$$u_n(z,t) = -\int_0^t J_0\left[\lambda_n\left((t-\tau)^2 - z^2\right)^{1/2}\right] \chi[(t-\tau) - z]\, u_n'(0,\tau) d\tau;$$
$$z \geq 0, \quad t \geq 0. \tag{2.11}$$

Expressions (2.11) reflect the general property of the solutions $U(g, t)$ to problems (2.2) in ${}_L\mathbf{Q}$, namely, the solutions satisfying zero initial conditions and being free of the components (modes) propagating toward the compact inhomogeneity (the unit). These expressions define the diagonal transport operator $Z_{0\to z}(t)$ (see Chapter 5 and references [21,30,117,118]), which operates according to the rule

$$u(z,t) = \{u_n(z,t)\} = Z_{0\to z}(t)\left[u'(0,\tau)\right]; \quad u'(b,\tau) = \left\{u_n'(b,\tau)\right\},$$
$$z \geq 0, \quad t \geq \tau \geq 0$$

and allows one to trace the field transformations during a free path of the transient wave in a finite section of a regular waveguide. It is evident that (2.11) can also be given as

$$u_n(z,t) = -\int_0^t J_0\left[\lambda_n\left((t-\tau)^2 - (z-z_0)^2\right)^{1/2}\right]\chi[(t-\tau) - (z-z_0)]\, u_n'(z_0,\tau) d\tau;$$
$$z \geq z_0 \geq 0, \quad t \geq 0. \tag{2.12}$$

2.2.2. *Nonlocal Absorbing Conditions*

By dropping the observation point in (2.11) onto the artificial boundary $\mathbf{L}$ ($z = 0$), we obtain

$$u_n(0,t) = -\int_0^t J_0\left[\lambda_n(t-\tau)\right]\chi(t-\tau) u_n'(0,\tau) d\tau; \quad t \geq 0. \tag{2.13}$$

Differentiating (2.13) with respect to t, we can write

$$\left[\frac{\partial}{\partial t} + \frac{\partial}{\partial z}\right] u_n(z,t)\Big|_{z=0} = \lambda_n \int_0^t J_1\left[\lambda_n(t-\tau)\right]\chi(t-\tau)\, u_n'(0,\tau) d\tau; \quad t \geq 0. \tag{2.14}$$

Here, the familiar relationships $dJ_0(x)/dx = -J_1(x)$, $J_0(0) = 1$, and $\chi^{(1)}(t-\tau) = \delta(t-\tau)$, where $\chi^{(1)}$ stands for the generalized derivative of χ, have been used.

Let us now take the Laplace transform in t of (2.14) (image $\leftrightarrow$ original)

$$\tilde{f}(s) = L[f](s) \equiv \int_0^\infty f(t)e^{-st}dt \;\leftrightarrow\; f(t) = L^{-1}\left[\tilde{f}\,\right](t) \equiv \frac{1}{2\pi i}\int_{\alpha-i\infty}^{\alpha+i\infty} \tilde{f}(s)e^{st}ds. \tag{2.15}$$

Taking into account the familiar formulas $\tilde{f}_1(s)\tilde{f}_2(s) \leftrightarrow \int_0^t f_1(t-\tau)f_2(\tau)d\tau$ (the convolution theorem), $\lambda^2[\sqrt{s^2+\lambda^2}(\sqrt{s^2+\lambda^2}+s)]^{-1} \leftrightarrow \lambda J_1(\lambda t)$ [120], and $s\tilde{f}(s) - f(0) \leftrightarrow df(t)/dt$, let us pass on to the following expression in the space of images $\tilde{u}_n(z, s)$,

$$\left[\frac{\partial}{\partial z} + s\right]\tilde{u}_n(z, s)\Big|_{z=0} = \frac{\lambda_n^2\tilde{u}_n'(0, s)}{\sqrt{s^2+\lambda^2}\left(\sqrt{s^2+\lambda^2}+s\right)}, \tag{2.16}$$

which may be rearranged to give

$$\tilde{u}_n'(0, s) = -\left(s + \frac{\lambda_n^2}{s+\sqrt{s^2+\lambda_n^2}}\right)\tilde{u}_n(0, s). \tag{2.17}$$

By taking the inverse Laplace transform of (2.17), we turn back to the originals:

$$\left[\frac{\partial}{\partial t} + \frac{\partial}{\partial z}\right]u_n(z, t)\Big|_{z=0} = -\lambda_n\int_0 J_1[\lambda_n(t-\tau)](t-\tau)^{-1}\chi(t-\tau)\,u_n(0, \tau)\,d\tau; \quad t \geq 0. \tag{2.18}$$

Here, the formula $(s+\sqrt{s^2+\lambda^2})^{-1} \leftrightarrow (\lambda t)^{-1}J_1(\lambda t)$ [121] has been used.

Rewriting (2.13), (2.14), and (2.18) in terms of (2.3) and (2.6), we obtain

$$U(y, 0, t) = -\sum_n\left\{\int_0^t J_0[\lambda_n(t-\tau)]\left[\int_0^a \frac{\partial U(\tilde{y}, z, \tau)}{\partial z}\Big|_{z=0}\mu_n(\tilde{y})d\tilde{y}\right]d\tau\right\}\mu_n(y) = V_1(y, t); \quad 0 \leq y \leq a, \quad t \geq 0, \tag{2.19}$$

$$\left[\frac{\partial}{\partial t} + \frac{\partial}{\partial z}\right]U(y, z, t)\Big|_{z=0} = \sum_n\left\{\int_0^t J_1[\lambda_n(t-\tau)]\left[\int_0^a \frac{\partial U(\tilde{y}, z, \tau)}{\partial z}\Big|_{z=0}\mu_n(\tilde{y})d\tilde{y}\right]d\tau\right\} \times \lambda_n\mu_n(y) = V_2(y, t); \quad 0 \leq y \leq a, \quad t \geq 0, \tag{2.20}$$

$$\left[\frac{\partial}{\partial t}+\frac{\partial}{\partial z}\right]U(y,z,t)\Bigg|_{z=0}$$
$$=-\sum_n\left\{\int_0^t J_1[\lambda_n(t-\tau)](t-\tau)^{-1}\left[\int_0^a U(\tilde{y},0,\tau)\mu_n(\tilde{y})d\tilde{y}\right]d\tau\right\}\lambda_n\mu_n(y)$$
$$=V_3(y,t);\quad 0\leq y\leq a,\quad t\geq 0. \tag{2.21}$$

Turning back to the main objective of the work, let us pre-estimate the possibility of using (2.19) to (2.21) as the conditions bounding the open propagation region. The following statement holds.

Statement 2.1. *Problems (2.2) and problems (2.2) supplemented with any one of conditions (2.19) to (2.21) are equivalent. The requirements that ensure their unique solvability (correctness classes) are identical.*

Formulas (2.19) to (2.21) are exact and, hence, their addition to the original problems does not actually increase the computation error or distort the simulated process.

When discretizing problems (2.2) supplemented by conditions (2.19) to (2.21), the instant (for $\tau = t$) impacts of the function $U(g,\tau)$, which is contained in the right-hand sides $V_j(y,t)$, can be entirely excluded [21,117]. Hence, during the calculation, when moving through time layers, the functions $V_j(y,t)$ may be considered as known, which were determined at the previous layers $\tau < t$.

Conditions (2.19) to (2.21) can be incorporated properly into the standard finite-difference computational scheme; they do not distort the stability of the scheme and do not complicate its numerical implementation.

The evidence for the retention of the stability of the computational scheme is too cumbersome to present here, however, it is based on the equivalence of the original and modified problems and on classical results [35].

Relations (2.13), (2.14), (2.18), and (2.19) to (2.21) are the exact radiation conditions for outgoing transient waves formed by the waveguide unit: formulas (2.13), (2.14), and (2.18) specify the behavior of spatio–temporal amplitudes of all partial components (modes) of the waves guided by the regular structure, whereas formulas (2.19) to (2.21) specify the behavior of the fields as a whole. These nonlocal (both in the spatial and temporal variables) conditions are related to the artificial boundaries that coincide with the waveguide cross-sections. Conditions (2.13) and (2.19) were first used as exact ABCs in [18].

Above, when formulating problems (2.2) and defining the domains $\mathbf{Q}$ and $\mathbf{Q}_L$, we have assumed that the functions associated with the sources exciting the waveguide unit are finite in the closure of $\mathbf{Q}$, whereas their supports belong to $\overline{\mathbf{Q}}_L\backslash\mathbf{L}$ for all observation times $0\leq t\leq T$. Similar assumptions are used in what follows. This allows the conditions on the artificial boundaries to be formulated in terms of the total field $U(g,t)$. Obviously, the restrictions that are associated with these assumptions can be removed, wholly or in part (see

Chapter 4), by assuming that the incident (primary) wave $U^i(g,t)$ exists in the region ${}_L\mathbf{Q}$. In this case, the function $U(g,t)$ should be replaced in all formulas with the function $U^s(g,t) = U(g,t) - U^i(g,t)$ describing the scattered (secondary) field. This simple procedure enables a nonrational enlargement of the computation domain to be prevented in the case of long-duration primary signals.

If the right-hand sides $V_j(y,t)$ in conditions (2.20) and (2.21) are replaced with zeros, these conditions will coincide with the simplest classical ABC of the first order of approximation in the angle of arrival of a plane complex wave (the absence of reflection is simulated) [22,23]. Thus, the functions $V_j(y,t)$ can serve as external estimators of the accuracy for the numerical methods that use the corresponding approximate absorbing condition. Similar estimates can also be obtained for other known ABCs (see the subsequent Sections and chapters).

2.2.3. *Local Absorbing Conditions*

The direct implementation of conditions (2.19) to (2.21) requires considerable computer memory capacity to store arrays of the $V_j(y,t)$ values. The array size grows at each time step. Moreover, at each time step, all of the array's elements take part in forming the values V_1, V_2, and V_3. (as a consequence of the nonlocality). The problem can be partially resolved (reduction to the nonlocal, in time, conditions) by analytically separating the variables in the arguments of the Bessel functions [117]: with a modest increase in the number of operations, the dimension of the arrays decreases by unity.

We can also pass on to the local conditions by applying the following easily realizable scheme. With the help of the representation [122]

$$J_0(x) = \frac{2}{\pi}\int\limits_0^{\pi/2} \cos\left[x\sin(\varphi)\right] d\varphi,$$

let us rewrite (2.13) as

$$u_n(0,t) = -\frac{2}{\pi}\int\limits_0^{\pi/2}\left\{\int\limits_0 \cos\left[\lambda_n(t-\tau)\sin(\varphi)\right]\chi(t-\tau)\,u_n'(0,\tau)\,d\tau\right\}d\varphi;\quad t\geq 0. \tag{2.22}$$

Denote

$$w_n(t,\varphi) = -\int\limits_0 \frac{\sin\left[\lambda_n(t-\tau)\sin(\varphi)\right]\chi(t-\tau)\,u_n'(0,\tau)}{\lambda_n\sin(\varphi)}d\tau;\quad t\geq 0,$$
$$0\leq\varphi\leq\pi/2. \tag{2.23}$$

Then

$$\frac{\partial w_n(t,\varphi)}{\partial t} = -\int\limits_0 \cos\left[\lambda_n(t-\tau)\sin(\varphi)\right]\chi(t-\tau)\,u_n'(0,\tau)\,d\tau,$$

and we have from (2.22) that

$$u_n(0,t) = \frac{2}{\pi}\int_0^{\pi/2} \frac{\partial w_n(t,\varphi)}{\partial t} d\varphi; \quad t \geq 0. \tag{2.24}$$

The integral form (2.23) is the equivalent to the following differential form.

$$\begin{cases} \left[\dfrac{\partial^2}{\partial t^2} + \lambda_n^2 \sin^2(\varphi)\right] w_n(t,\varphi) = -u_n'(0,t); \quad t > 0 \\ w_n(0,\varphi) = \left.\dfrac{\partial w_n(t,\varphi)}{\partial t}\right|_{t=0} = 0 \end{cases}. \tag{2.25}$$

Indeed, by passing on to the generalized formulation of the corresponding Cauchy problem and using the fundamental solution $G(\lambda, t) = \chi(t)\lambda^{-1}\sin\lambda t$ of the operator $D(\lambda) \equiv [d^2/dt^2 + \lambda^2]$ (see Section 1.2.5), we can easily verify that formulas (2.23) and (2.25) determine the same function $w_n(t,\varphi)$.

Let us now multiply (2.24) and (2.25) by $\mu_n(y)$ and sum over all $n \in \{n\}$. As a result, taking into consideration that

$$\sum_n \lambda_n^2 w_n(t,\varphi)\mu_n(y) = -\frac{\partial^2 W(y,t,\varphi)}{\partial y^2}$$

for

$$W(y,t,\varphi) = \sum_n w_n(t,\varphi)\mu_n(y)$$

(see problems (2.4)), we obtain

$$U(y,0,t) = \frac{2}{\pi}\int_0^{\pi/2} \frac{\partial W(y,t,\varphi)}{\partial t} d\varphi; \quad t \geq 0, \quad 0 \leq y \leq a,$$

$$\begin{cases} \left[\dfrac{\partial^2}{\partial t^2} - \sin^2(\varphi)\dfrac{\partial^2}{\partial y^2}\right] W(y,t,\varphi) = -\left.\dfrac{\partial U(y,z,t)}{\partial z}\right|_{z=0}; \quad 0 < y < a, \quad t > 0 \\ W(y,0,\varphi) = \left.\dfrac{\partial W(y,t,\varphi)}{\partial t}\right|_{t=0} = 0; \quad 0 \leq y \leq a \\ W(0,t,\varphi) = W(a,t,\varphi) = 0 \quad (E-\text{case}) \quad \text{or} \\ \partial W(y,t,\varphi)/\partial y\big|_{y=0,a} = 0 \quad (H-\text{case}); \quad t \geq 0 \end{cases}. \tag{2.26}$$

This is an exact local absorbing condition (both with respect to the spatial and the temporal variable), allowing one to truncate efficiently the computational domain when solving problems (2.2) numerically. From here on, $W(y,t,\varphi)$ is an auxiliary function, which can be determined by solving a separate initial boundary value problem being the inner problem with respect to the corresponding condition; $0 \leq \varphi \leq \pi/2$ is a parameter.

The transformation of (2.14) and (2.18) according to a similar scheme results in the following exact local ABCs, differing from (2.26).

$$\left[\frac{\partial}{\partial t}+\frac{\partial}{\partial z}\right]U(y,z,t)\Big|_{z=0}=\frac{2}{\pi}\int\limits_0^{\pi/2}W(y,t,\varphi)\cos^2\varphi d\varphi;\quad t\geq 0,\quad 0\leq y\leq a,$$

$$\begin{cases}\left[\dfrac{\partial^2}{\partial t^2}-\cos^2(\varphi)\dfrac{\partial^2}{\partial y^2}\right]W(y,t,\varphi)=-\dfrac{\partial^2}{\partial y^2}\left[\dfrac{\partial}{\partial z}U(y,z,t)\Big|_{z=0}\right]; \\ 0<y<a,\quad t>0 \\ W(y,0,\varphi)=\dfrac{\partial W(y,t,\varphi)}{\partial t}\Big|_{t=0}=0;\quad 0\leq y\leq a \\ W(0,t,\varphi)=W(a,t,\varphi)=0\quad (E-\text{case})\quad \text{or} \\ \partial W(y,t,\varphi)/\partial y\big|_{y=0,a}=0\quad (H-\text{case});\quad t\geq 0\end{cases};\tag{2.27}$$

$$\left[\frac{\partial}{\partial t}+\frac{\partial}{\partial z}\right]U(y,z,t)\Big|_{z=0}=\frac{2}{\pi}\int\limits_0^{\pi/2}\frac{\partial W(y,t,\varphi)}{\partial t}\sin^2\varphi d\varphi;\quad t\geq 0,\quad 0\leq y\leq a,$$

$$\begin{cases}\left[\dfrac{\partial^2}{\partial t^2}-\cos^2(\varphi)\dfrac{\partial^2}{\partial y^2}\right]W(y,t,\varphi)=\dfrac{\partial^2 U(y,0,t)}{\partial y^2};\quad 0<y<a,\quad t>0 \\ W(y,0,\varphi)=\dfrac{\partial W(y,t,\varphi)}{\partial t}\Big|_{t=0}=0;\quad 0\leq y\leq a \\ W(0,t,\varphi)=W(a,t,\varphi)=0\quad (E-\text{case})\quad \text{or} \\ \partial W(y,t,\varphi)/\partial y\big|_{y=0,a}=0\quad (H-\text{case});\quad t\geq 0\end{cases}.\tag{2.28}$$

Statement 2.2. *Problems (2.2) and problems (2.2) supplemented with any one of conditions (2.26) to (2.28) are equivalent. The inner initial boundary value problems in (2.26) to (2.28) are well posed with respect to the auxiliary functions* $W(y,z,\varphi)$.

In derivation of (2.27), the following formula has been used [123],

$$J_1(x)=\frac{2}{\pi}\int\limits_0^{\pi/2}\sin[x\cos(\varphi)]\cos(\varphi)\,d\varphi$$

along with the substitutions

$$w_n(t,\varphi)=\lambda_n\int\limits_0\frac{\sin[\lambda_n(t-\tau)\cos(\varphi)]\chi(t-\tau)u'_n(0,\tau)}{\cos(\varphi)}d\tau;\quad t\geq 0,$$

$$0\leq\varphi\leq\pi/2,$$

whereas in the derivation of (2.28) we have applied the integral Poisson formula [122]

$$J_1(x) = \frac{2x}{\pi} \int_0^{\pi/2} \cos[x \cos(\varphi)] \sin^2(\varphi)\, d\varphi$$

and

$$w_n(t, \varphi) = -\lambda_n \int_0 \frac{\sin[\lambda_n (t-\tau)\cos(\varphi)]\, \chi(t-\tau)\, u_n(0,\tau)}{\cos(\varphi)} d\tau; \quad t \geq 0,$$

$$0 \leq \varphi \leq \pi/2.$$

The assumption $W(y, t, \varphi) \equiv 0$ (which cannot be justified) reduces (2.27) and (2.28) to the classical ABC of the first order of approximation. With the finite sum of the trapezoidal method in place of the integral in (2.27), we arrive at the approximate conditions that coincide in form with certain of the conditions from [95].

2.3. Axially Symmetrical Waveguide Units: 2-D Scalar Problems in Cylindrical Coordinates

2.3.1. Statement of Model Initial Boundary Value Problems and Their General Solutions

The analysis of TE_0-waves ($\partial/\partial\phi \equiv 0$, ($E_\rho = E_z = H_\phi = j_\rho = j_z \equiv 0$); see Section 1.2.2) and TM_0-waves ($H_\rho = H_z = E_\phi = j_\phi \equiv 0$) in axially symmetrical structures (see Fig. 2.2 with the resonator in part of a circular coaxial waveguide as an example of an axially symmetrical geometry) reduces to the solution of the following 2-D (in the half-plane of the variable $g = \{\rho \geq 0, z\}$) scalar initial boundary value problems.

$$\begin{cases} \left[-\varepsilon(g) \frac{\partial^2}{\partial t^2} - \sigma(g) \frac{\partial}{\partial t} + \frac{\partial^2}{\partial z^2} + \frac{\partial}{\partial \rho}\left(\frac{1}{\rho}\frac{\partial}{\partial \rho}\rho\right) \right] U(g,t) = F(g,t); \\ t > 0; \quad g \in \mathbf{Q} \\ U(g,t)|_{t=0} = \varphi(g), \quad \frac{\partial}{\partial t} U(g,t)|_{t=0} = \Psi(g); \quad g \in \bar{\mathbf{Q}} \\ E_{tg}(g,t)]\,|_{g\in \mathbf{S}} = 0; \quad U(0,z,t) = 0; \quad t \geq 0 \end{cases} \tag{2.29}$$

The last condition in (2.29) is due to the symmetry of the problems: $\rho = 0$ coincides with the axis of circular symmetry and, hence, only E_z- and H_z-components can be different from zero. With $U(g,t) = E_\phi$ we have $E_{tg}(g,t)|_{g\in \mathbf{S}} = U(g,t)|_{g\in \mathbf{S}}$, and problems (2.29) describe spatio–temporal transformations of TE_0-waves, whereas with piecewise constant functions $\varepsilon(g)$, $\sigma(g)$, and $U(g,t) = H_\phi$, (2.29) are stated with respect to the TM_0-waves. The analysis

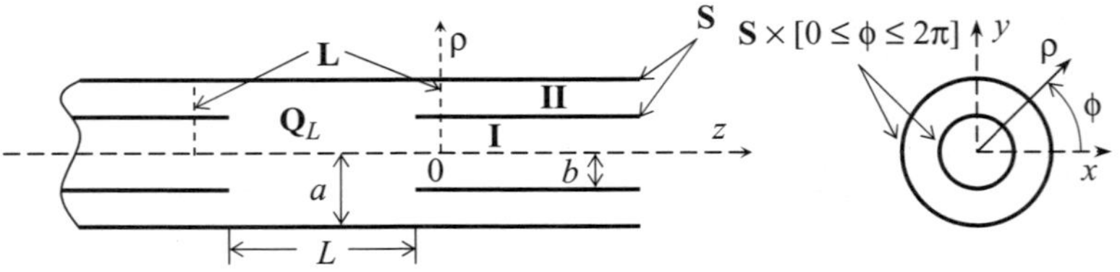

FIGURE 2.2. Discontinuity in a circular coaxial waveguide: $\phi = \pi/2$.

domain $\mathbf{Q}$ is the region in the $\phi =$ const half-plane that is bounded by the contour $\mathbf{S}$. $\mathbf{S} \times [0 \leq \phi \leq 2\pi]$ is the surface of perfectly conducting objects. It is assumed, as before, that the functions $F(g,t)$, $\varphi(g) = U^i(g,0)$, $\psi(g) = \partial U^i(g,t)/\partial t|_{t=0}$ ($U^i(g,t)$ stands for the incident wave), $\sigma(g)$, and $\varepsilon(g) - 1$ that are finite in the closure of $\mathbf{Q}$ satisfy the theorem about the unique solvability of problems (2.29) in the energy class (the Sobolev space) $\mathbf{W}_2^1(\mathbf{Q}^T)$ (see Statement 1.1 and Reference [35]). The supports of these functions belong to the set $\overline{\mathbf{Q}_L} \backslash \mathbf{L}$. The regular circular and coaxial circular waveguides (the region ${}_L\mathbf{Q} = \mathbf{Q} \backslash (\mathbf{Q}_L \cup \mathbf{L})$), along which the field formed by the waveguide unit may propagate infinitely far, are free from the sources and efficient scatterers. The artificial boundary $\mathbf{L}$ (dashed lines in Fig. 2.2) lies in their cross-section planes.

The geometry of the domain ${}_L\mathbf{Q}$ in problems (2.29) is such that the solutions to (2.29) in the relevant regular partial regions can be represented as (obviously, it is sufficient to restrict our consideration to the regions **I** and **II** with $z > 0$)

$$U(z,\rho,t) = \sum_n u_n(z,t)\mu_n(\rho); \quad t \geq 0, \tag{2.30}$$

where the orthonormal bases $\{\mu_n(\rho)\}$, $n \in \{n\}$ are specified by nontrivial solutions to the homogeneous (spectral) problems

$$\begin{cases} \left[\dfrac{d}{d\rho}\dfrac{1}{\rho}\dfrac{d}{d\rho}\rho + \lambda_n^2\right]\mu_n(\rho) = 0; \quad \rho \in (0;b) \\ \mu_n(0) = \mu_n(b) = 0 \quad (\mathrm{TE}_0 - \text{waves}) \quad \text{or} \\ \mu_n(0) = \left.\dfrac{d(\rho\,\mu_n(\rho))}{d\rho}\right|_{\rho=b} = 0 \quad (\mathrm{TM}_0 - \text{waves}) \end{cases} \tag{2.31}$$

(for region **I** relevant to the circular waveguide) and

$$\begin{cases} \left[\dfrac{d}{d\rho}\dfrac{1}{\rho}\dfrac{d}{d\rho}\rho + \lambda_n^2\right]\mu_n(\rho) = 0; \quad \rho \in (b;a) \\ \mu_n(b) = \mu_n(a) = 0 \quad (\mathrm{TE}_0 - \text{waves}) \quad \text{or} \\ \left.\dfrac{d(\rho\mu_n(\rho))}{d\rho}\right|_{\rho=b} = \left.\dfrac{d(\rho\,\mu_n(\rho))}{d\rho}\right|_{\rho=a} = 0 \quad (\mathrm{TM}_0 - \text{waves}) \end{cases} \tag{2.32}$$

(for region **II** relevant to the coaxial circular waveguide). The boundary conditions in problems (2.31) and (2.32) associated with electric waves (TM_0-waves), have been formulated in view of the following formulas (see Section 1.2.2), $U(g,t) = H_\phi$, $E_{tg}(g,t)|_{g\in \mathbf{S}} = E_z(g,t)|_{g\in \mathbf{S}}$, and

$$\frac{\partial E_z}{\partial t} = \frac{\eta_0}{\rho}\frac{\partial(\rho H_\phi)}{\partial \rho}.$$

The spatio–temporal amplitudes $u_n(z,t)$ (the elements of the evolutionary bases of the signals) are obtained by solving the initial boundary value problems

$$\left\{\begin{array}{l} \left[-\frac{\partial^2}{\partial t^2} + \frac{\partial^2}{\partial z^2} - \lambda_n^2\right] u_n(z,t) = 0; \quad t > 0 \\ u_n(z,0) = 0, \quad \frac{\partial}{\partial t} u_n(z,t)\Big|_{t=0} = 0 \end{array}\right. ; \quad z \geq 0, \quad n \in \{n\} \tag{2.33}$$

(it is assumed that the excitation $U(g,t)$, being generated by the sources $\varphi(g)$, $\psi(g)$, and $F(g,t)$ located in $\mathbf{Q}_L$, has not yet reached the boundary $z = 0$ of the regions **I** and **II** by the time $t = 0$).

For the axially symmetrical problems in question, we can usually distinguish two subregions of the region ${}_L\mathbf{Q}$: the interior parts of circular ($\rho < b$) and coaxial circular ($b < \rho < a$) semi-infinite waveguides. The sets $\{\mu_n\}$, $\{\lambda_n\}$, and $\{n\}$ of the solutions to the corresponding spectral problems (2.31) and (2.32) are known:

$$\left\{\begin{array}{l} \left\{\begin{array}{l} \mu_n(\rho) = J_1(\lambda_n\rho)\sqrt{2}\,[bJ_0(\lambda_n b)]^{-1}; \quad n = 1,2,\ldots \\ \lambda_n > 0 \text{ are the roots to the eq. } J_1(\lambda b) = 0 \end{array}\right. ; \quad \rho < b \\ \left\{\begin{array}{l} \mu_n(\rho) = G_1(\lambda_n,\rho)\sqrt{2}\left[a^2G_0^2(\lambda_n,a) - b^2G_0^2(\lambda_n,b)\right]^{-1/2}; \quad n = 1,2,\ldots \\ \lambda_n > 0 \text{ are the roots to the eq. } G_1(\lambda,a) = 0 \\ G_q(\lambda,\rho) = J_q(\lambda\rho)N_1(\lambda b) - N_q(\lambda\rho)J_1(\lambda b) \end{array}\right. ; \\ b < \rho < a \end{array}\right.$$

(for TE_0-waves);

$$\left\{\begin{array}{l} \left\{\begin{array}{l} \mu_n(\rho) = J_1(\lambda_n\rho)\sqrt{2}\,[bJ_1(\lambda_n b)]^{-1}, \quad n = 1,2,\ldots \\ \lambda_n > 0 \text{ are the roots to the eq. } J_0(\lambda b) = 0 \end{array}\right. ; \quad \rho < b \\ \left\{\begin{array}{l} \mu_n(\rho) = \tilde{G}_1(\lambda_n,\rho)\sqrt{2}\left[a^2\tilde{G}_1^2(\lambda_n,a) - b^2\tilde{G}_1^2(\lambda_n,b)\right]^{-1/2}; \quad n = 1,2,\ldots \\ \mu_0(\rho) = \left[\rho\sqrt{\ln(a/b)}\right]^{-1} \\ \lambda_n > 0 (n = 1,2,\ldots) \text{ are the roots to the eq. } \tilde{G}_0(\lambda,b) = 0, \quad \lambda_0 = 0 \\ \tilde{G}_q(\lambda,\rho) = J_q(\lambda\rho)N_0(\lambda a) - N_q(\lambda\rho)J_0(\lambda a) \end{array}\right. ; \\ b < \rho < a \end{array}\right.$$

(for TM_0-waves). Here, J_q и N_q are the Bessel and the Neumann functions, respectively.

In all the above cases, the systems of functions $\{\mu_n(\rho)\}$ being basis ones in the spaces whose elements meet the corresponding conditions at the ends of the intervals $0 < \rho < b$ and $b < \rho < a$ are orthonormal:

$$\int_0^b \mu_n(\rho)\mu_m(\rho)\,\rho\, d\rho = \begin{cases} 0, & n \neq m \\ 1, & n = m \end{cases}$$

or

$$\int_b^a \mu_n(\rho)\mu_m(\rho)\,\rho\, d\rho = \begin{cases} 0, & n \neq m \\ 1, & n = m \end{cases}.$$

2.3.2. *Exact Absorbing Conditions*

The evolutionary bases $u(z,t) = \{u_n(z,t)\}$ of the signals propagating along regular sections of closed waveguides and Floquet channels are qualitatively identical [21,30]. They are described by one-type initial boundary value problems for the one-dimensional Klein–Gordon equations (as exemplified by problems (2.5) and (2.33)). Therefore, in the situation considered here, we may simply rewrite (using proper notation and taking account of some obvious details) the nonlocal conditions (2.19) to (2.21), which were obtained in Section 2.2. All the stages in the construction of the exact ABCs

$$\begin{aligned} U(\rho,0,t) &= -\sum_n \left\{ \int_0^t J_0\left[\lambda_n(t-\tau)\right] \left[\int_{\rho_1}^{\rho_2} \left. \frac{\partial U(\tilde{\rho},z,\tau)}{\partial z} \right|_{z=0} \mu_n(\tilde{\rho})\,\tilde{\rho}\, d\tilde{\rho} \right] d\tau \right\} \mu_n(\rho) \\ &= V_1(\rho,t); \quad \rho_1 \le \rho \le \rho_2, \quad t \ge 0, \end{aligned} \tag{2.34}$$

$$\begin{aligned} &\left[\frac{\partial}{\partial t} + \frac{\partial}{\partial z} \right] U(\rho,z,t) \Big|_{z=0} \\ &\quad = \sum_n \left\{ \int_0^t J_1\left[\lambda_n(t-\tau)\right] \left[\int_{\rho_1}^{\rho_2} \left. \frac{\partial U(\tilde{\rho},z,\tau)}{\partial z} \right|_{z=0} \mu_n(\tilde{\rho})\,\tilde{\rho}\, d\tilde{\rho} \right] d\tau \right\} \\ &\quad\times \lambda_n \mu_n(\rho) = V_2(\rho,t); \quad \rho_1 \le \rho \le \rho_2, \quad t \ge 0, \end{aligned} \tag{2.35}$$

$$\begin{aligned} &\left[\frac{\partial}{\partial t} + \frac{\partial}{\partial z} \right] U(\rho,z,t) \Big|_{z=0} \\ &= -\sum_n \left\{ \int_0^t J_1\left[\lambda_n(t-\tau)\right](t-\tau)^{-1} \left[\int_{\rho_1}^{\rho_2} U(\tilde{\rho},0,\tau)\mu_n(\tilde{\rho})\,\tilde{\rho}\, d\tilde{\rho} \right] d\tau \right\} \lambda_n \mu_n(\rho) \\ &= V_3(\rho,t); \quad \rho_1 \le \rho \le \rho_2, \quad t \ge 0 \end{aligned} \tag{2.36}$$

for the channels of decaying energy in axially symmetrical waveguide units are almost identical to those realized for 2-D (in the $y0z$-plane) model problems (2.2). In (2.34) to (2.36), one should be set $\rho_1 = 0$ and $\rho_2 = b$ for a circular waveguide or $\rho_1 = b$ and $\rho_2 = a$ for a coaxial waveguide.

When going over from (2.34) to (2.36) to local conditions such as (2.26) to (2.28), we have to sum the corresponding exact local ABCs for partial components (modes) of the outgoing transient waves over all $n \in \{n\}$ (see Section 2.2.3) and then use the following representations for the basic ($u_n(z,t)$ and $U(\rho,z,t)$) and auxiliary ($w_n(t,\varphi)$ and $W(\rho,t,\varphi)$) functions, which follow from (2.30) to (2.32).

$$\sum_n \lambda_n^2 u_n(z,t)\mu_n(\rho) = -\frac{\partial}{\partial\rho}\frac{1}{\rho}\frac{\partial}{\partial\rho}\rho\, U(\rho,z,t),$$

$$\sum_n \lambda_n^2 w_n(t,\varphi)\mu_n(\rho) = -\frac{\partial}{\partial\rho}\frac{1}{\rho}\frac{\partial}{\partial\rho}\rho\, W(\rho,t,\varphi),$$

$$W(\rho,t,\varphi) = \sum_n w_n(t,\varphi)\mu_n(\rho).$$

As a result, we obtain

$$U(\rho,0,t) = \frac{2}{\pi}\int_0^{\pi/2} \frac{\partial W(\rho,t,\varphi)}{\partial t} d\varphi; \quad t \geq 0, \quad \rho_1 \leq \rho \leq \rho_2,$$

$$\begin{cases} \left[\dfrac{\partial^2}{\partial t^2} - \sin^2(\varphi)\dfrac{\partial}{\partial\rho}\dfrac{1}{\rho}\dfrac{\partial}{\partial\rho}\rho\right] W(\rho,t,\varphi) = -\left.\dfrac{\partial U(\rho,z,t)}{\partial z}\right|_{z=0}; \quad \rho_1 < \rho < \rho_2, \\ t > 0 \\ W(\rho,0,\varphi) = \left.\dfrac{\partial W(\rho,t,\varphi)}{\partial t}\right|_{t=0} = 0; \quad \rho_1 \leq \rho \leq \rho_2 \end{cases} ; \tag{2.37}$$

$$\left[\frac{\partial}{\partial t} + \frac{\partial}{\partial z}\right] U(\rho,z,t)\Big|_{z=0} = \frac{2}{\pi}\int_0^{\pi/2} W(\rho,t,\varphi)\cos^2\varphi d\varphi; \quad t \geq 0, \quad \rho_1 \leq \rho \leq \rho_2,$$

$$\begin{cases} \left[\dfrac{\partial^2}{\partial t^2} - \cos^2(\varphi)\dfrac{\partial}{\partial\rho}\dfrac{1}{\rho}\dfrac{\partial}{\partial\rho}\rho\right] W(\rho,t,\varphi) = -\dfrac{\partial}{\partial\rho}\dfrac{1}{\rho}\dfrac{\partial}{\partial\rho}\rho\left[\dfrac{\partial}{\partial z}U(\rho,z,t)\Big|_{z=0}\right]; \\ \rho_1 < \rho < \rho_2, \quad t > 0 \\ W(\rho,0,\varphi) = \left.\dfrac{\partial W(\rho,t,\varphi)}{\partial t}\right|_{t=0} = 0; \quad \rho_1 \leq \rho \leq \rho_2 \end{cases} ; \tag{2.38}$$

$$\left[\frac{\partial}{\partial t} + \frac{\partial}{\partial z}\right] U(\rho,z,t)\Big|_{z=0} = \frac{2}{\pi}\int_0^{\pi/2} \frac{\partial W(\rho,t,\varphi)}{\partial t}\sin^2\varphi d\varphi; \quad t \geq 0, \quad \rho_1 \leq \rho \leq \rho_2$$

$$\begin{cases} \left[\dfrac{\partial^2}{\partial t^2} - \cos^2(\varphi)\dfrac{\partial}{\partial\rho}\dfrac{1}{\rho}\dfrac{\partial}{\partial\rho}\rho\right] W(\rho,t,\varphi) = \dfrac{\partial}{\partial\rho}\dfrac{1}{\rho}\dfrac{\partial}{\partial\rho}\rho\, U(\rho,0,t); \quad \rho_1 < \rho < \rho_2, \\ t > 0 \\ W(\rho,0,\varphi) = \left.\dfrac{\partial W(\rho,t,\varphi)}{\partial t}\right|_{t=0} = 0; \quad \rho_1 \leq \rho \leq \rho_2 \end{cases} \tag{2.39}$$

Here, $\rho_1 = 0$ and $\rho_2 = b$ for the circular waveguide, and $\rho_1 = b$ and $\rho_2 = a$ for the coaxial waveguide. The inner initial boundary value problems in (2.37) to (2.39) with respect to the auxiliary functions $W(\rho, t, \varphi)$ should be supplemented with the following boundary conditions for all times $t \geq 0$.

$$\begin{cases} W(0,t,\varphi) = W(b,t,\varphi) = 0; & (\text{TE}_0 - \text{waves}) \\ W(0,t,\varphi) = \left.\dfrac{\partial(\rho W(\rho,t,\varphi))}{\partial\rho}\right|_{\rho=b} = 0; & (\text{TM}_0 - \text{waves}) \end{cases}$$

(for region **I** relevant to the circular waveguide) and

$$\begin{cases} W(b,t,\varphi) = W(a,t,\varphi) = 0; & (\text{TE}_0 - \text{waves}) \\ \left.\dfrac{\partial(\rho W(\rho,t,\varphi))}{\partial\rho}\right|_{\rho=b} = \left.\dfrac{\partial(\rho W(\rho,t,\varphi))}{\partial\rho}\right|_{\rho=a} = 0; & (\text{TM}_0 - \text{waves}) \end{cases}$$

(for region **II** relevant to the coaxial circular waveguide).

Statement 2.3. *Problems (2.29) and problems (2.29) supplemented with any one of nonlocal ((2.34) to (2.36)) or local ((2.37) to (2.39)) absorbing conditions are equivalent. The inner initial boundary value problems in conditions (2.37) to (2.39) are well posed with respect to the auxiliary functions* $W(\rho, t, \varphi)$.

2.4. Vector Problems of the Theory of Open Waveguide Resonators

In view of their moderate demands in terms of computer resources, model (scalar, two-dimensional) problems are well suited to be run through a large variety of situations and to visualize sufficiently accurately the real wave processes. Such exploratory operations with 3-D vector problems are as yet very costly in terms of computer resources and runtime. In this case, we most frequently deal with a standard problem of analysis of a particular electromagnetic object (unit, device) having a limited number of parameters. These parameters vary within a narrow range, whereas the requirements for accuracy and faithfulness of the numerical results are very stringent. One way to improve the accuracy of the widely used finite-difference algorithms is to apply the exact conditions on artificial boundaries limiting the domains where the original initial boundary value problems are formulated. In this section, the conditions of this type are derived for vector problems inherent in the theory of open waveguide resonators.

2.4.1. General Theoretical Questions

When considering an arbitrary waveguide unit, we neglect for a while the bounded region, where all the sources and inhomogeneities are located and where the standard, well-elaborated schemes [27,28] can be applied for solving problems (1.20) (see Section 1.2.3) with the corresponding boundary and initial conditions. We

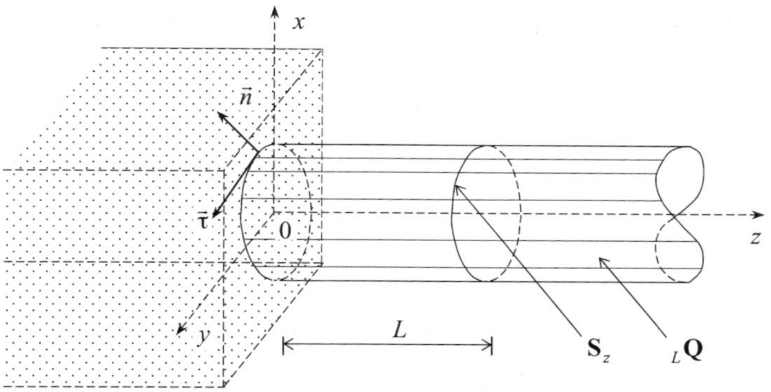

FIGURE 2.3. Model geometry for the vector problems.

focus our attention on the channels (closed semi-infinite regular waveguides) along which the signals formed by the waveguide unit are propagating.

Let us analyze one such channel (Fig. 2.3) and analyze it. In the region $_L\mathbf{Q} = \{g = \{x, y, z\} : \{x, y\} \in \operatorname{int}\mathbf{S}_z, z > L \geq 0\}$, the field $\vec{E}(g, t)$, which is nonzero only for times $t > 0$, represents an outgoing to infinity ($z = \infty$) wave and it satisfies the following vector initial boundary value problem.

$$\begin{cases} \left(-\dfrac{\partial^2}{\partial t^2} + \Delta\right) \vec{E}(g, t) = 0; \quad t > 0, \quad g \in {}_L\mathbf{Q} \\ \vec{E}(g, t)\Big|_{t=0} = \dfrac{\partial \vec{E}(g, t)}{\partial t}\Bigg|_{t=0} = 0; \quad g \in \overline{{}_L\mathbf{Q}} \\ E_z(g, t)\big|_{g\in\mathbf{S}} = \left(\vec{\tau} \cdot \vec{E}_\perp(g, t)\right)\Big|_{g\in\mathbf{S}} = 0; \quad t \geq 0 \end{cases}. \tag{2.40}$$

Here (see Fig. 2.3), $\vec{E} = E_z\vec{z} + \vec{E}_\perp$, $\vec{E}_\perp = E_x\vec{x} + E_y\vec{y}$; $\mathbf{S}_z$ is the boundary contour of the cross-section of the regular waveguide; $\vec{x}$, $\vec{y}$, $\vec{z}$, $\vec{\tau}$, and $\vec{n}$ are the unit vectors of the coordinate axes, the tangent and the normal to the contour $\mathbf{S}_z$; $\operatorname{int}\mathbf{S}_z$ is the region in the plane $z = \text{const}$ that is bounded by $\mathbf{S}_z$; and $\mathbf{S} = \mathbf{S}_z \times (0 < z < \infty)$.

When applying the scalar Borgnis functions $U^E(g, t)$ and $U^H(g, t)$ (see [32,33] and Section 1.2.3) such that

$$\begin{cases} \left(-\dfrac{\partial^2}{\partial t^2} + \Delta\right) \dfrac{\partial U^{E,H}(g, t)}{\partial t} = 0; \quad t > 0, \quad g \in {}_L\mathbf{Q} \\ U^E(g, t)\big|_{g\in\mathbf{S}} = \dfrac{\partial U^H(g, t)}{\partial \vec{n}}\Bigg|_{g\in\mathbf{S}} = 0; \quad t \geq 0 \end{cases}, \tag{2.41}$$

the general solution to the vector differential equation in problem (2.40) can be given as

$$E_x = \frac{\partial^2 U^E}{\partial x \partial z} - \frac{\partial^2 U^H}{\partial y \partial t}, \quad E_y = \frac{\partial^2 U^E}{\partial y \partial z} + \frac{\partial^2 U^H}{\partial x \partial t}, \quad E_z = \frac{\partial^2 U^E}{\partial z^2} - \frac{\partial^2 U^E}{\partial t^2}. \tag{2.42}$$

The field $\vec{E}(g,t)$ written in this form also satisfies the boundary conditions of problem (2.40):

$$\left(\vec{\tau}\cdot\vec{E}_{\perp}\right)\Big|_{\mathbf{S}_z} = \left(\frac{dx}{d\theta}\frac{\partial}{\partial x}+\frac{dy}{d\theta}\frac{\partial}{\partial y}\right)\frac{\partial}{\partial z}U^{E} - \left(\frac{dx}{d\theta}\frac{\partial}{\partial y}-\frac{dy}{d\theta}\frac{\partial}{\partial x}\right)\frac{\partial}{\partial t}U^{H}$$
$$= \frac{d}{d\theta}\left(\frac{\partial}{\partial z}U^{E}\right) - \frac{\partial}{\partial\vec{n}}\left(\frac{\partial}{\partial t}U^{H}\right) = 0.$$

Here, the contour $\mathbf{S}_z$ has been parametrized as

$$\mathbf{S}_z = \mathbf{S}_z(\theta) = \{x(\theta), y(\theta)\}_{\theta}$$

and the following representations for the vectors tangential and normal to $\mathbf{S}_z$ have been used [122],

$$\vec{\tau} = \frac{dx}{d\theta}\vec{x} + \frac{dy}{d\theta}\vec{y}, \quad \vec{n} = -\frac{dy}{d\theta}\vec{x} + \frac{dx}{d\theta}\vec{y}.$$

Separating the transverse variables (x, y) in problems (2.41), we obtain their solutions in the form

$$U^{E,H}(x,y,z,t) = \sum_{n=1,0}^{\infty} u_n^{E,H}(z,t)\,\mu_n^{E,H}(x,y), \tag{2.43}$$

where $\{\mu_n^E(x,y)\}_{n=1}^{\infty}$ and $\{\mu_n^H(x,y)\}_{n=0}^{\infty}$ are the complete orthonormal systems of the solutions to the Sturm–Liouville problems for the equation $(\partial^2/\partial x^2+\partial^2/\partial y^2+\lambda^2)\mu=0$ in the domain $\operatorname{int}\mathbf{S}_z$ and with the Dirichlet $(\mu^E(x,y)|_{\{x,y\}\in\mathbf{S}_z}=0)$ and Neumann $(\partial\mu^H(x,y)/\partial\vec{n}|_{\{x,y\}\in\mathbf{S}_z}=0)$ conditions on its boundary $\mathbf{S}_z$. λ_n^E and λ_n^H are the eigenvalues associated with these solutions.

When substituting (2.43) into (2.42), the field $\vec{E}(g,t)$ can be written as

$$E_z = \sum_{n=1}^{\infty} v_{n,z}(z,t)\xi_{n,z}(x,y), \quad \vec{E}_{\perp} = \sum_{n=-\infty}^{\infty} v_{n,\perp}(z,t)\vec{\xi}_{n,\perp}(x,y);$$
$$t\ge 0, \quad z\ge L. \tag{2.44}$$

Here, the scalar functions $v_{n,z}(z,t)$ and $v_{n,\perp}(z,t)$ are the solutions of the following initial boundary value problems.

$$\begin{cases}\left[-\dfrac{\partial^2}{\partial t^2}+\dfrac{\partial^2}{\partial z^2}-\alpha_{n,z}^2\right]v_{n,z}(z,t)=0; \quad t>0, \quad z>L \\ v_{n,z}(z,0) = \dfrac{\partial}{\partial t}v_{n,z}(z,t)\Big|_{t=0} = 0; \quad z\ge L, \quad n=1,2,\ldots\end{cases}, \tag{2.45}$$

$$\begin{cases}\left[-\dfrac{\partial^2}{\partial t^2}+\dfrac{\partial^2}{\partial z^2}-\alpha_{n,\perp}^2\right]v_{n,\perp}(z,t)=0; \quad t>0, \quad z>L \\ v_{n,\perp}(z,0) = \dfrac{\partial}{\partial t}v_{n,\perp}(z,t)\Big|_{t=0} = 0; \quad z\ge L, \quad n=0,\pm1,\pm2,\ldots\end{cases}. \tag{2.46}$$

Here, $\alpha_{n,z} = \lambda_n^E$; $\alpha_{n,\perp} = \lambda_n^E$ for $n = 1, 2, \ldots$ and $\alpha_{n,\perp} = \lambda_{-n}^H$ for $n = 0, -1, -2, \ldots$; $\xi_{n,z} = \mu_n^E$; $\vec{\xi}_{n,\perp} = (\partial\mu_n^E/\partial x)\vec{x} + (\partial\mu_n^E/\partial y)\vec{y}$ for $n = 1, 2, \ldots$ and $\vec{\xi}_{n,\perp} = -(\partial\mu_{-n}^H/\partial y)\vec{x} + (\partial\mu_{-n}^H/\partial x)\vec{y}$ for $n = 0, -1, -2, \ldots$.

The inversion formulas for (2.44)

$$v_{n,z}(z,t) = \int\limits_{\text{int}\,\mathbf{S}_z} E_z(x,y,z,t)\xi_{n,z}(x,y)\,dxdy,$$

$$v_{n,\perp}(z,t) = \left(\alpha_{n,\perp}\right)^{-2} \int\limits_{\text{int}\,\mathbf{S}_z} (\vec{E}_\perp(x,y,z,t)\cdot\vec{\xi}_{n,\perp}(x,y))dx\,dy \qquad (2.47)$$

can be obtained using the properties of the system of functions $\{\mu_n^{E,H}(x,y)\}$. The former formula is an evident consequence of the fact that the system $\{\mu_n^E(x,y)\}$ consists of the orthogonal functions of unit norm. To see that the latter formula is true, consider the integrals

$$\int\limits_{\text{int}\,\mathbf{S}_z} \left(\left(\frac{\partial\mu_n^E}{\partial x}\vec{x} + \frac{\partial\mu_n^E}{\partial y}\vec{y}\right)\cdot\left(\frac{\partial\mu_m^E}{\partial x}\vec{x} + \frac{\partial\mu_m^E}{\partial y}\vec{y}\right)\right)dx\,dy$$

$$= \int\limits_{\text{int}\,\mathbf{S}_z} (\text{grad}\mu_n^E \cdot \text{grad}\,\mu_m^E)dx\,dy$$

$$= -\int\limits_{\text{int}\,\mathbf{S}_z} \mu_m^E \Delta\mu_n^E dxdy + \int\limits_{\mathbf{S}_z} \mu_m^E \frac{\partial\mu_n^E}{\partial\vec{n}} ds = \begin{cases} 0, & m \neq n \\ (\lambda_n^E)^2, & m = n \end{cases},$$

$$\int\limits_{\text{int}\,\mathbf{S}_z} \left(\left(-\frac{\partial\mu_n^H}{\partial y}\vec{x} + \frac{\partial\mu_n^H}{\partial x}\vec{y}\right)\cdot\left(-\frac{\partial\mu_m^H}{\partial y}\vec{x} + \frac{\partial\mu_m^H}{\partial x}\vec{y}\right)\right)dx\,dy$$

$$= \int\limits_{\text{int}\,\mathbf{S}_z} ([\text{grad}\mu_n^H \times \vec{z}]\cdot[\text{grad}\mu_m^H \times \vec{z}])dx\,dy$$

$$= \int\limits_{\text{int}\,\mathbf{S}_z} (\text{grad}\,\mu_n^H \cdot \text{grad}\,\mu_m^H)dx\,dy = \begin{cases} 0, & m \neq n \\ (\lambda_n^H)^2, & m = n \end{cases},$$

$$\int\limits_{\text{int}\,\mathbf{S}_z} \left(\left(\frac{\partial\mu_n^E}{\partial x}\vec{x} + \frac{\partial\mu_n^E}{\partial y}\vec{y}\right)\cdot\left(-\frac{\partial\mu_m^H}{\partial y}\vec{x} + \frac{\partial\mu_m^H}{\partial x}\vec{y}\right)\right)dx\,dy$$

$$= \int\limits_{\mathbf{P}} (d\vec{\mathbf{P}}\cdot\text{rot}\vec{\mu}_m^H) = \int\limits_{\mathbf{S}_z} (d\vec{\mathbf{S}}_z\cdot\vec{\mu}_m^H) = 0.$$

Here, $\vec{\mu}_m^H = \mu_m^H\vec{z}$; $\mathbf{P}$ is the surface of the function $z = \mu_n^E(x,y)$ spanned on the contour $\mathbf{S}_z$;

$$d\vec{\mathbf{P}} = \left(-\frac{\partial\mu_n^E}{\partial x}\vec{x} - \frac{\partial\mu_n^E}{\partial y}\vec{y} + \vec{z}\right)dx\,dy$$

is the surface element vector of the surface $\mathbf{P}$; $d\vec{\mathbf{S}}_z = \vec{\tau}\, d\theta$ is the vector element of the contour $\mathbf{S}_z$.

2.4.2. Exact Absorbing Conditions in Vector Initial Boundary Value Problems

Problems (2.45) and (2.46) are identical to those ((2.5) and (2.33)) considered at length previously. Thus, without going into detail, we turn immediately to the analysis of their solutions:

$$v_n(L,t) = -\int_0^t J_0\left[\alpha_n(t-\tau)\right] v_n'(L,\tau)\,d\tau; \quad L \geq 0, \quad t \geq 0, \tag{2.48}$$

$$\left[\frac{\partial}{\partial z} + \frac{\partial}{\partial t}\right] v_n(z,t)\bigg|_{z=L} = \alpha_n \int_0^t J_1\left[\alpha_n(t-\tau)\right] v_n'(L,\tau)\,d\tau; \quad L \geq 0, \quad t \geq 0, \tag{2.49}$$

$$\left[\frac{\partial}{\partial z} + \frac{\partial}{\partial t}\right] v_n(z,t)\bigg|_{z=L} = -\alpha_n \int_0^t J_1\left[\alpha_n(t-\tau)\right](t-\tau)^{-1} v_n(L,\tau)\,d\tau; \quad L \geq 0, \quad t \geq 0. \tag{2.50}$$

On putting $\alpha_n = \alpha_{n,z}$, $n = 1, 2, \ldots$ in (2.48) to (2.50), we arrive at the exact conditions for the elements $v_{n,z}(z,t)$ of the evolutionary basis for the E_z-component of the outgoing wave $\vec{E}(g,t)$. With $\alpha_n = \alpha_{n,\perp}$, $n = 0, \pm 1, \pm 2, \ldots$, we have precisely the same expressions for $v_{n,\perp}(z,t)$, the amplitudes of the transverse electric field $\vec{E}_\perp(g,t)$ of the wave $\vec{E}(g,t)$. From here on, the symbols z, $\perp$, and $\rightarrow$ in the indices of the eigenvalues α_n and the functions v_n, ξ_n, and E are omitted; the way to extract any results following from (2.48) to (2.50), both in terms of longitudinal $E_z(g,t)$ and transverse $\vec{E}_\perp(g,t)$ fields should be clear.

Conditions (2.48) to (2.50) are nonlocal in time. Their use in the finite-difference algorithm gives rise to a spatial nonlocality when solving the general problem of determining $\vec{E}(g,t)$ at all points $g \in \mathbf{Q}_L$ of the waveguide unit: in each time step one should go over from the series (2.44) to the amplitudes (2.47) and vice versa. Following the procedure outlined in Section 2.2.3, let us go from (2.49), (2.50) to the conditions that are local both in the time and spatial coordinates:

$$\left[\frac{\partial}{\partial t} + \frac{\partial}{\partial z}\right] E(g,t)\bigg|_{z=L} = \frac{1}{\pi}\int_0^\pi W(x,y,t,\varphi)\,d\varphi; \quad \{x,y\} \in \overline{\mathrm{int}\mathbf{S}_z}, \quad t \geq 0,$$

$$\begin{cases} \left[\dfrac{\partial^2}{\partial t^2} - \sin^2\varphi\left(\dfrac{\partial^2}{\partial x^2} + \dfrac{\partial^2}{\partial y^2}\right)\right] W(x,y,t,\varphi) \\ \quad = -\sin^2\varphi\left(\dfrac{\partial^2}{\partial x^2} + \dfrac{\partial^2}{\partial y^2}\right)\left[\dfrac{\partial E(g,t)}{\partial z}\bigg|_{z=L}\right]; \quad \{x,y\} \in \mathrm{int}\mathbf{S}_z, \quad t > 0 \\ W(x,y,0,\varphi) = \dfrac{\partial W(x,y,t,\varphi)}{\partial t}\bigg|_{t=0} = 0; \quad \{x,y\} \in \overline{\mathrm{int}\mathbf{S}_z} \end{cases} \tag{2.51}$$

$$\left[\frac{\partial}{\partial t}+\frac{\partial}{\partial z}\right]E(g,t)\Big|_{z=L}$$
$$=\frac{2}{\pi}\int_0^{\pi/2}\frac{\partial W(x,y,t,\varphi)}{\partial t}\sin^2(\varphi)\,d\varphi;\quad \{x,y\}\in\overline{\mathrm{int}\mathbf{S}_z},\quad t\geq 0,$$

$$\begin{cases}\left[\frac{\partial^2}{\partial t^2}-\cos^2(\varphi)\left(\frac{\partial^2}{\partial x^2}+\frac{\partial^2}{\partial y^2}\right)\right]W(x,y,t,\varphi)=\left(\frac{\partial^2}{\partial x^2}+\frac{\partial^2}{\partial y^2}\right)E(x,y,L,t);\\ \{x,y\}\in\mathrm{int}\,\mathbf{S}_z,\quad t>0\\ W(x,y,0,\varphi)=\frac{\partial W(x,y,t,\varphi)}{\partial t}\Big|_{t=0}=0;\quad \{x,y\}\in\overline{\mathrm{int}\mathbf{S}_z}\end{cases}\tag{2.52}$$

(we suggest as a simple exercise rewriting the transformation (2.48) to a form similar to (2.26)). The auxiliary scalar and vector functions $W(x,y,t,\varphi)$ are presented in the same bases of the transversal functions $\xi_n(x,y)$ as the fields $E(g,t)$:

$$W(x,y,t,\varphi)=\sum_n w_n(t,\varphi)\xi_n(x,y).\tag{2.53}$$

The inner problems in (2.51), (2.52) should be supplemented with the boundary conditions (on the contour $\mathbf{S}_z$) for $W(x,y,t,\varphi)$ and $\vec{W}(x,y,t,\varphi)$, $t\geq 0$:

$$W(x,y,t,\varphi)|_{\{x,y\}\in\mathbf{S}_z}=0$$

in the case where the function $E(g,t)$ in (2.51), (2.52) corresponds to the E_z-component of the field $E(g,t)$ ($E(g,t)\to E_z(g,t)$) and

$$(\vec{\tau}\cdot\vec{W}(x,y,t,\varphi))|_{\{x,y\}\in\mathbf{S}_z}=0$$

in the case $E(g,t)\to\vec{E}_\perp(g,t)$. These conditions are the obvious consequence of (2.53).

In the case of (2.52) the technique for deriving the corresponding exact absorbing condition is just the same as the one used when obtaining the ABC (2.28), whereas in going from (2.49) to (2.51), some intermediate representations applied for the construction of (2.27) have been modified; namely, we have used [123]

$$J_1(x)=\frac{1}{\pi}\int_0^{\pi}\sin(x\sin(\varphi))\sin(\varphi)\,d\varphi$$

and

$$w_n(t,\varphi)=\alpha_n\sin(\varphi)\int_0^t\sin[\alpha_n(t-\tau)\sin(\varphi)]v_n'(L,\tau)\,d\tau;\quad t\geq 0,\quad 0\leq\varphi\leq\pi.$$

The result is changed correspondingly (compare (2.27) and (2.51)).

Formulas (2.48) to (2.52) do not cover all possible forms in which the exact conditions for the truncation of the analysis domain can be expressed as applied to the vector problems for open waveguide resonators. It is important to note also the following essential distinction between the nonlocal and local exact absorbing conditions. The former ones ((2.48) to (2.50)) require complete information on the systems of the eigenfunctions and eigenvalues of the Sturm–Liouville operator in the region int$\mathbf{S}_z$ with the Dirichlet or Neumann conditions on its boundary $\mathbf{S}_z$. The solution to the corresponding spectral problems may be too cumbersome and costly in terms of computer resources. The latter conditions ((2.51), (2.52)) are free from this shortcoming and, hence, they should be preferred when analyzing systems for which the transversal functions for the channels with decaying energy cannot be determined analytically.

2.5. Problems of the Electromagnetic Theory of Gratings

In this section, we extend the obtained results to initial boundary value problems of the electromagnetic theory of gratings. By passing on to the Fourier transforms of the functions describing physically realizable fields and sources, these problems can be reduced to those of scattering of transient waves by compact inhomogeneities in the Floquet channel. This channel has the form of a plane-parallel (scalar, 2-D case) or a rectangular in its cross-section (vector case) waveguide with specific (quasi-periodic) boundary conditions.

Diffraction gratings of various types (semi-transmitting and reflecting, metal and dielectric, with perfect and imperfect inclusions of different configurations) can be analyzed by the finite-difference method. In this section, we give the solution to the central problem for the efficient implementation of these algorithms, namely, the problem of the proper truncation of the computational domain. It should be mentioned that the reflecting structure with a perfect dielectric inclusion has been chosen (see Section 2.5.1) solely for demonstrating basic technical details of the approach being developed. A semi-transmitting grating (compact inhomogeneities in the infinite Floquet channel) and a grating with metal–dielectric inclusions (additional terms in the input equation and its discrete analogue) can be considered in the same way (with minor complications; see Section 2.5.3).

2.5.1. Scalar Problems for a Perfect Reflecting Grating

The scattering of E- and H-polarized transient waves by a grating (Fig. 2.4) is described by the initial boundary value problems

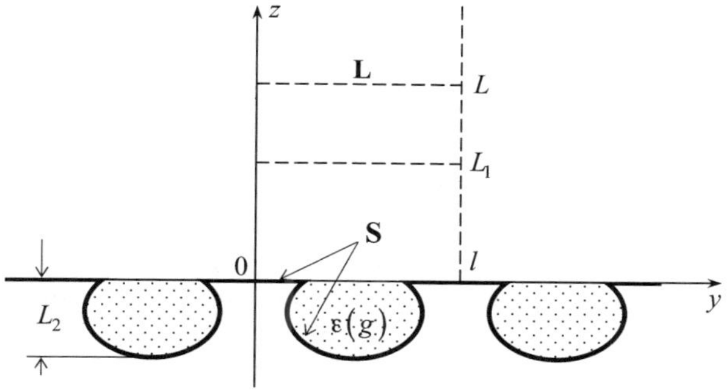

FIGURE 2.4. Reflecting periodic grating.

$$\left\{\begin{array}{l} \left[-\varepsilon(g)\frac{\partial^2}{\partial t^2}+\frac{\partial^2}{\partial z^2}+\frac{\partial^2}{\partial y^2}\right]U(g,t)=F(g,t);\quad t>0,\quad g\in\mathbf{Q} \\ U(g,t)|_{t=0}=\varphi(g)=U^i(g,t)\big|_{t=0}, \\ \quad \left.\frac{\partial}{\partial t}U(g,t)\right|_{t=0}=\psi(g)=\left.\frac{\partial}{\partial t}U^i(g,t)\right|_{t=0};\quad g\in\overline{\mathbf{Q}} \\ U(g,t)|_{g\in\mathbf{S}}=0\quad(E-\text{case})\quad\text{or}\quad\left.\frac{\partial U(g,t)}{\partial\vec{n}}\right|_{g\in\mathbf{S}}=0\quad(H-\text{case}); \\ t\geq 0 \\ U\left\{\frac{\partial U}{\partial y}\right\}(l,z,t)=e^{i2\pi\Phi}U\left\{\frac{\partial U}{\partial y}\right\}(0,z,t);\quad t\geq 0 \end{array}\right. \tag{2.54}$$

The analysis domain $\mathbf{Q}$ is the portion of the band $\mathbf{R}=\{g=\{y,z\}:0<y<l\}$ (the regular Floquet channel with l being the grating period) bounded below by $\mathbf{S}$. It is assumed that the functions $F(g,t)$, $\varphi(g)$, $\psi(g)$, and $\varepsilon(g)-1$ where $\operatorname{Im}\varepsilon(g)=0$, $\varepsilon(y,z>0)\equiv 1$ are finite in the closure of $\mathbf{Q}$, and satisfy the theorem on the unique solvability of problems (2.54) in the space $\mathbf{W}_2^1\left(\mathbf{Q}^T\right)$. $U(g,t)=E_x(g,t)$ in the case of the E-polarization. For the H-polarization $U(g,t)=H_x(g,t)$ and $\varepsilon(g)$ is a piecewise constant function. Objects, fields, and sources are uniform along the x-direction.

The complex-valued functions (as distinct from the ones discussed above) $F(g,t)$, $U^i(g,t)$ (the incident plane pulse wave), and $U(g,t)$ are the Fourier transforms $f(g,t,\Phi)$ of the real functions $f_{true}(g,t)$ describing the true sources are scattered fields:

$$\begin{aligned} f_{true}(g,t) &= \int_{-\infty}^{\infty}\tilde{f}(z,t,\Phi)e^{i\Phi 2\pi y/l}d\Phi \\ &= \int_{-0.5}^{0.5}\left[\sum_{n=-\infty}^{\infty}\tilde{f}(z,t,n+\Phi)\,e^{i(n+\Phi)2\pi y/l}\right]d\Phi=\int_{-0.5}^{0.5}f(g,t,\Phi)d\Phi. \end{aligned} \tag{2.55}$$

From (2.55), it follows that

$$f\left\{\frac{\partial f}{\partial y}\right\}(y+l, z, t, \Phi) = e^{i2\pi\Phi} f\left\{\frac{\partial f}{\partial y}\right\}(y, z, t, \Phi).$$

Using this condition in the original problems preceding (2.54) (in the problems with respect to the real functions $f_{true}(g,t)$) allows one to truncate the analysis domain down to the **R** band (see the last equation in (2.54) and Section 1.2.4) and thereby to properly simulate realistic physical situations in the framework of the traditional statement of initial boundary value problems in the electromagnetic theory of gratings [9]. The functions $f(g, t, \Phi)$ in this case represent the infinite series of complex pulse waves and make no clear physical sense; however, simple mathematical rearrangements of the result $(f(g, t, \Phi) \to f_{true}(g,t))$ restore all the values to a field of the usual physical notions [117].

2.5.2. The Transport Operator Specifying Spatio–Temporal Signal Transformations in the Floquet Channel and Exact Conditions for Outgoing Waves

Let $F(g,t) \equiv 0$. Represent the field formed by the grating in the reflection zone $z > 0$ as $U(g,t) = U^i(g,t) + U^s(g,t)$, where the functions

$$U^i(g,t) = \sum_n v_n(z,t)\,\mu_n(y), \qquad U^s(g,t) = \sum_n u_n(z,t)\,\mu_n(y) \tag{2.56}$$

describe the incident and the reflected transient waves, respectively. Here, $\{\mu_n(y)\}$: $\mu_n(y) = l^{-1/2}\exp(i\lambda_n y)$, $\lambda_n = \Phi_n 2\pi/l$, $\Phi_n = n + \Phi$, $n = 0, \pm 1, \ldots$ is the complete (in the $\mathbf{L}_2[(0;l)]$ space) orthonormal system of transverse functions, which arises when separating the y-variable in (2.54).

Let us assume that the wave $U^i(g,t)$ has not yet reached the grating by the time $t = 0$. Then we have from (2.54), (2.56) the following set of initial boundary value problems for the elements $u_n(z,t)$ of the evolutionary basis $u(z,t) = \{u_n(z,t)\}$ for the wave $U^s(g,t)$,

$$\begin{cases} \left[-\dfrac{\partial^2}{\partial t^2} + \dfrac{\partial^2}{\partial z^2} - \lambda_n^2\right] u_n(z,t) = 0; \quad z > 0, \quad t > 0 \\ u_n(z,0) = 0, \quad \dfrac{\partial}{\partial t} u_n(z,t)\Big|_{t=0} = 0; \quad z \ge 0 \end{cases}. \tag{2.57}$$

The solution $u(z,t) = \{u_n(z,t)\}$,

$$u_n(z,t) = -\int_0 J_0\left[\lambda_n\left((t-\tau)^2 - z^2\right)^{1/2}\right]\chi[(t-\tau) - z]\,u_n'(0,\tau)\,d\tau;$$

$$z \ge 0, \quad t \ge 0, \tag{2.58}$$

to these problems (obtained, e.g., by repeating the transfer (2.5) → (2.11)) specifies the diagonal transport operator $Z_{0\to z}(t)$, which acts according the rule

$$u(z,t) = \{u_n(z,t)\} = Z_{0\to z}(t)[u'(0,\tau)]; \quad u'(b,\tau) = \{u'_n(b,\tau)\},$$
$$z \geq 0, \quad t \geq \tau \geq 0$$

and allows one to trace the field transformations during the free propagation (from the plane $z=0$ to any plane $z \geq 0$) of the transient wave $U^s(g,t)$ in the finite section of the regular Floquet channel. Here,

$$u'_n(b,t) = \frac{\partial}{\partial z}u_n(z,t)\bigg|_{z=b} = \int_0^l \frac{\partial}{\partial z}U^S(y,z,t)\bigg|_{z=b}\mu_n^*(y)dy.$$

The symbol $*$ stands for a complex conjugation. It is clear that (2.58) can also be written in a form that determines the transport operator $Z_{z_0\to z}(t)$:

$$u_n(z,t) = -\int_0 J_0\left[\lambda_n\left((t-\tau)^2 - (z-z_0)^2\right)^{1/2}\right]\chi\left[(t-\tau)-(z-z_0)\right]u'_n(z_0,\tau)\,d\tau;$$
$$z \geq z_0 \geq 0, \quad t \geq 0. \tag{2.59}$$

By fixing the artificial boundary $z = L \geq 0$ somewhere in $\mathbf{Q}$ and using (2.56), we have, by simple rearrangements in (2.59),

$$U^s(y,L,t) = -\sum_n \left\{ \int_0^{t-(L-L_1)} J_0\left[\lambda_n\left((t-\tau)^2 - (L-L_1)^2\right)^{1/2}\right] \right.$$
$$\left. \times \left[\int_0^l \frac{\partial U^s(\tilde{y},z,\tau)}{\partial z}\bigg|_{z=L_1} \mu_n^*(\tilde{y})d\tilde{y}\right]d\tau \right\}$$
$$\times\, \mu_n(y) = V_1(y,t); \quad 0 \leq L_1 \leq L, \quad 0 \leq y \leq l, \quad t \geq 0 \tag{2.60}$$

and

$$\left[\frac{\partial}{\partial t} + \frac{\partial}{\partial z}\right]U^s(y,z,t)\bigg|_{z=L}$$
$$= \sum_n \left\{\int_0^t J_1\left[\lambda_n(t-\tau)\right]\left[\int_0^l \frac{\partial U^s(\tilde{y},z,\tau)}{\partial z}\bigg|_{z=L} \mu_n^*(\tilde{y})\,d\tilde{y}\right]d\tau\right\}\lambda_n\mu_n(y)$$
$$= V_2(y,t); \quad 0 \leq y \leq l, \quad t \geq 0. \tag{2.61}$$

We have taken into account the following formula, which is a direct consequence of (2.59),

$$\frac{\partial}{\partial t}u_n(L,t) = -u'_n(L,t) + \lambda_n\int_0 J_1\left[\lambda_n(t-\tau)\right]\chi(t-\tau)\,u'_n(L,\tau)\,d\tau.$$

The nonlocal ABCs (2.60) and (2.61) are exact. They describe analytically waves outgoing from the grating. In other words, relations (2.60) and (2.61) represent the exact radiation conditions for the solutions $U^s(g,t) = U(g,t) - U^i(g,t)$ to problems (2.54) referred to the artificial boundary $\mathbf{L}$ $(z = L)$ in the reflection zone of the periodic structure. As a consequence, problems (2.54) and problems (2.54) along with one of the conditions (2.60) or (2.61) are equivalent. The functions $V_j(y,t)$ are such that the momentary influence of the functions $U^s(g,\tau)$ on the left-hand (basic) sides of (2.60) and (2.61) is excluded completely. This property simplifies substantially the finite-difference computational scheme: at any time t, the functions $V_j(y,t)$ may be considered as given (determined earlier from times $\tau < t$).

By putting $L_1 = L$ in (2.60) and invoking the formulas [123]

$$J_0(x) = \frac{1}{2\pi}\int_{-\pi}^{\pi} \exp(ix\sin(\varphi))\,d\varphi, \quad J_1(x) = \frac{1}{\pi}\int_{0}^{\pi} \sin(x\sin(\varphi))\sin(\varphi)\,d\varphi,$$

we can pass on (as was done in Sections 2.2.3 and 2.4.2) to the exact local ABCs,

$$U^S(y,L,t) = -\frac{1}{2\pi}\int_{-\pi}^{\pi} W(y,t,\varphi)\,d\varphi; \quad 0 \le y \le l, \quad t \ge 0,$$

$$\begin{cases} \left[\dfrac{\partial}{\partial t} - \sin(\varphi)\dfrac{\partial}{\partial y}\right] W(y,t,\varphi) = \dfrac{\partial}{\partial z} U^S(y,z,t)\Big|_{z=L}; \quad 0 < y < l, \quad t > 0 \\ W(y,0,\varphi) = \dfrac{\partial W(y,t,\varphi)}{\partial t}\Big|_{t=0} = 0; \quad 0 \le y \le l \\ W\left\{\dfrac{\partial W}{\partial y}\right\}(l,t,\varphi) = e^{i2\pi\Phi} W\left\{\dfrac{\partial W}{\partial y}\right\}(0,t,\varphi); \quad t \ge 0 \end{cases}, \tag{2.62}$$

$$\left[\frac{\partial}{\partial t} + \frac{\partial}{\partial z}\right] U^s(g,t)\Big|_{z=L} = \frac{1}{\pi}\int_{0}^{\pi} W(y,t,\varphi)d\varphi; \quad 0 \le y \le l, \quad t \ge 0,$$

$$\begin{cases} \left[\dfrac{\partial^2}{\partial t^2} - \sin^2(\varphi)\dfrac{\partial^2}{\partial y^2}\right] W(y,t,\varphi) = -\sin^2(\varphi)\dfrac{\partial^2}{\partial y^2}\left[\dfrac{\partial U^s(g,t)}{\partial z}\Big|_{z=L}\right]; \\ 0 < y < l, \quad t > 0 \\ W(y,0,\varphi) = \dfrac{\partial W(y,t,\varphi)}{\partial t}\Big|_{t=0} = 0; \quad 0 \le y \le l \\ W\left\{\dfrac{\partial W}{\partial y}\right\}(l,t,\varphi) = e^{i2\pi\Phi} W\left\{\dfrac{\partial W}{\partial y}\right\}(0,t,\varphi); \quad t \ge 0 \end{cases}, \tag{2.63}$$

which are equivalent to (2.60), (2.61). We have applied the auxiliary functions $W(y,t,\varphi) = \sum_n w_n(t,\varphi)\,\mu_n(y)$ and made the substitutions

$$w_n(t,\varphi) = \int_0^t \exp\left[i\lambda_n(t-\tau)\sin(\varphi)\right]u_n'(L,\tau)\,d\tau; \quad |\varphi| \le \pi, \tag{2.64}$$

$$w_n(t,\varphi) = \lambda_n \sin(\varphi)\int_0^t \sin\left[\lambda_n(t-\tau)\sin(\varphi)\right]u_n'(L,\tau)\,d\tau; \quad 0 \le \varphi \le \pi$$

to derive conditions (2.62) and (2.63), respectively. Note also one novel technical detail: the differential form $[\partial/\partial t - i\lambda_n \sin(\varphi)]w_n(t,\varphi) = u_n'(L,\tau)$ (from which follows the equation with respect to $W(y,t,\varphi)$ in the inner initial boundary-value problem in (2.62)) which is equivalent to the integral form (2.64), has been constructed with the help of the fundamental solution $G(\lambda,t) = \chi(t)\exp(-\lambda t)$ of the operator $[d/dt + \lambda]$ (see Section 1.2.5).

2.5.3. The Conditions Truncating the Analysis Domain in Vector Problems of the Electromagnetic Theory of Gratings

Let a semi-transmitting or reflecting structure that is periodic both in the x-(period l_x) and y-directions (period l_y) (described by the functions $\mathbf{S}$, $\varepsilon(g)$, and $\sigma(g)$) be placed in the layer $-P \le z \le 0$, $P \ge 0$ of the $\mathbf{R}^3$-space. In addition, let us assume that the sources of an electromagnetic field are given by the functions $\vec{F}_{true}(g,t)$, $\vec{\varphi}_{true}(g)$, and $\vec{\psi}_{true}(g)$, finite in $\mathbf{R}^3$, whose supports are concentrated in the layer $-P-L < z < L$, $L \ge 0$. Then the real vector functions $\vec{E}_{true}(g,t)$ and $\vec{H}_{true}(g,t)$ $\left(\frac{\partial}{\partial t}\vec{H}_{true}(g,t) = -\frac{1}{\eta_0}\mathrm{rot}\vec{E}_{true}(g,t)\right)$ determined by the initial boundary value problems (see Sections 1.2.3 to 1.2.5)

$$\begin{cases} \left[\Delta - \ \mathrm{grad\,div} \ - \varepsilon(g)\dfrac{\partial^2}{\partial t^2} - \sigma(g)\dfrac{\partial}{\partial t}\right]\vec{E}_{true}(g,t) = \vec{F}_{true}(g,t); \\ g = \{x,y,z\} \in \mathbf{Q}_{true}, \quad t > 0 \\ \vec{E}_{true}(g,0) = \vec{\varphi}_{true}(g), \quad \left.\dfrac{\partial}{\partial t}\vec{E}_{true}(g,t)\right|_{t=0} = \vec{\psi}_{true}(g); \quad g \in \overline{\mathbf{Q}}_{true} \\ \left.(E_{true})_{tg}(g,t)\right|_{g\in\mathbf{S}} = 0; \quad t \ge 0 \end{cases}, \tag{2.65}$$

describe the true electromagnetic field formed by the grating in the region $\mathbf{Q}_{true}, = \mathbf{R}^3\backslash\overline{\mathrm{int}\mathbf{S}}$ (in the region of the $\mathbf{R}^3$-space that is restricted by the surfaces $\mathbf{S}$, i.e., the boundaries of the regions int$\mathbf{S}$ occupied by the perfectly conducting objects).

In the region ${}_L\mathbf{Q}_{true} = \{g \in \mathbf{Q}_{true} : z \notin [-P-L; L]\}$, the field $\vec{E}_{true}(g,t)$ is nonzero only for times $t > 0$ and represents a wave outgoing to infinity ($z = \pm\infty$). This field satisfies the following vector initial boundary value problem,

$$\begin{cases} \left(-\dfrac{\partial^2}{\partial t^2} + \Delta\right)\vec{E}_{true}(g,t) = 0; \quad t > 0, \quad g \in {}_L\mathbf{Q}_{true} \\ \left.\vec{E}_{true}(g,t)\right|_{t=0} = \left.\dfrac{\partial \vec{E}_{true}(g,t)}{\partial t}\right|_{t=0} = 0; \quad g \in \overline{{}_L\mathbf{Q}_{true}} \end{cases}. \tag{2.66}$$

Let us go in (2.65) and (2.66) to the Fourier transforms $f\left(g, t, \Phi_x, \Phi_y\right)$ of the real functions $f_{true}(g, t)$:

$$
\begin{aligned}
& f_{true}(g, t) \\
& = \int_{-\infty}^{\infty}\int_{-\infty}^{\infty} \tilde{f}(z, t, \Phi_x, \Phi_x) \exp\left(i\Phi_x \frac{2\pi x}{l_x}\right) \exp\left(i\Phi_y \frac{2\pi y}{l_y}\right) d\Phi_x d\Phi_y \\
& = \int_{-0.5}^{0.5}\int_{-0.5}^{0.5} \left[\sum_{n=-\infty}^{\infty}\sum_{n=-\infty}^{\infty} \tilde{f}\left(z, t, n+\Phi_x, m+\Phi_y\right) \exp\left(i(n+\Phi_x)\frac{2\pi x}{l_x}\right)\right. \\
& \quad \times \left.\exp\left(i\left(m+\Phi_y\right)\frac{2\pi y}{l_y}\right)\right] d\Phi_x d\Phi_y = \int_{-0.5}^{0.5}\int_{-0.5}^{0.5} f\left(g, t, \Phi_x, \Phi_y\right) d\Phi_x d\Phi_y .
\end{aligned}
\tag{2.67}
$$

From (2.67) it follows that

$$
\begin{aligned}
& f\left\{\frac{\partial f}{\partial x}\right\}(x + l_x, y, z, t, \Phi_x, \Phi_x) = e^{i2\pi\Phi_x} f\left\{\frac{\partial f}{\partial x}\right\}(x, y, z, t, \Phi_x, \Phi_x), \\
& f\left\{\frac{\partial f}{\partial y}\right\}\left(x, y + l_y, z, t, \Phi_x, \Phi_x\right) = e^{i2\pi\Phi_y} f\left\{\frac{\partial f}{\partial y}\right\}(x, y, z, t, \Phi_x, \Phi_x).
\end{aligned}
$$

The use of the foregoing conditions restricts the analysis domain up to the region $\mathbf{Q} = \{g \in \mathbf{Q}_{true} : 0 < x < l_x, 0 < y < l_y\}$, which is a part of the rectangular Floquet channel $\mathbf{R} = \{g \in \mathbf{R}^3 : 0 < x < l_x, 0 < y < l_y\}$.

We now turn to the problems that follow from (2.66) with the substitutions $\vec{E}_{true}(g, t) \to \vec{E}(g, t)$, ${}_L\mathbf{Q}_{true} \to {}_L\mathbf{Q} = \{g \in \mathbf{Q} : z \notin [-P - L; L]\}$ and the addition of the conditions

$$
\begin{aligned}
& \vec{E}\left\{\frac{\partial \vec{E}}{\partial x}\right\}(l_x, y, z, t) = e^{i2\pi\Phi_x} \vec{E}\left\{\frac{\partial \vec{E}}{\partial x}\right\}(0, y, z, t); \quad 0 \le y \le l_y \\
& \vec{E}\left\{\frac{\partial \vec{E}}{\partial y}\right\}\left(x, l_y, z, t\right) = e^{i2\pi\Phi_y} \vec{E}\left\{\frac{\partial \vec{E}}{\partial y}\right\}(x, 0, z, t); \quad 0 \le x \le l_x \\
& z \notin (-P - L; L), \quad t \ge 0
\end{aligned}
\tag{2.68}
$$

on the side surfaces of the semi-infinite regular Floquet channels in the reflection and transmission zones for the periodic structure. The solutions $\vec{E}(g, t)$ to these problems for the upper ($z \ge L$) channel can be written in the form

$$
\vec{E}(g, t) = \sum_{n,m=-\infty}^{\infty} \vec{u}_{nm}(z, t)\, \mu_{nm}(x, y); \quad \{x, y\} \in \overline{R_z}, \quad z \ge L, \quad t \ge 0
\tag{2.69}
$$

(the changes that should be made when considering the lower ($z \le -P - L$) channel are evident). Here $\mathbf{R}_z = (0 < x < l_x) \times (0 < y < l_y)$; $\{\mu_{nm}(x, y)\}$, $n, m = 0, \pm 1, \pm 2, \ldots$ is the complete in $\mathbf{L}_2(\mathbf{R}_z)$ and orthonormal system of functions $\mu_{nm}(x, y) = (l_x l_y)^{-1/2} \exp(i\alpha_n x)\exp(i\beta_m y)$, $\alpha_n = 2\pi(\Phi_x + n)/l_x$, $\beta_m = 2\pi(\Phi_y + m)/l_y$. The elements $\vec{u}_{nm}(z, t)$ of the evolutionary basis $\vec{u}(z, t) = \{\vec{u}_{nm}(z, t)\}$ of the wave $\vec{E}(g, t)$ can be obtained from a set of the initial boundary value problems such as

$$\begin{cases} \left[-\frac{\partial^2}{\partial t^2} + \frac{\partial^2}{\partial z^2} - \lambda_{nm}^2 \right] \vec{u}_{nm}(z, t) = 0; \quad z > L, \quad t > 0 \\ \vec{u}_{nm}(z, 0) = 0, \quad \left. \frac{\partial}{\partial t} \vec{u}_{nm}(z, t) \right|_{t=0} = 0; \quad z \geq L \end{cases},$$
$$\lambda_{nm}^2 = \alpha_n^2 + \beta_m^2. \tag{2.70}$$

In the further derivations for the vector version (2.70) of the scalar problems considered above for the one-dimensional Klein–Gordon equation, there is no principal distinction: we can write immediately (in the notation adopted) any one of the results (2.19) to (2.21), (2.26) to (2.28), (2.48) to (2.52), or (2.60) to (2.63), obtained previously when analyzing problems (2.5), (2.45), (2.46), and (2.57). Let us restrict our consideration to the relations

$$\vec{u}_{nm}(L, t) = -\int_0 J_0 [\lambda_{nm}(t - \tau)] \chi(t - \tau) \vec{u}'_{nm}(L, \tau) d\tau; \quad t \geq 0, \tag{2.71}$$

$$\left. \left[\frac{\partial}{\partial t} + \frac{\partial}{\partial z} \right] \vec{u}_{nm}(z, t) \right|_{z=L} = \lambda_{nm} \int_0 J_1 [\lambda_{nm}(t - \tau)] \chi(t - \tau) \vec{u}'_{nm}(L, \tau) d\tau; \quad t \geq 0, \tag{2.72}$$

$$\left. \left[\frac{\partial}{\partial t} + \frac{\partial}{\partial z} \right] \vec{u}_{nm}(z, t) \right|_{z=L} = -\lambda_{nm} \int_0 J_1 [\lambda_{nm}(t - \tau)] (t - \tau)^{-1} \chi(t - \tau) \vec{u}_{nm}(L, \tau) d\tau; \quad t \geq 0 \tag{2.73}$$

(the nonlocal ABCs for spatio–temporal amplitudes of the signal $\vec{E}(g, t)$), equivalent to (2.19) to (2.21) and

$$\vec{E}(x, y, L, t) = \frac{2}{\pi} \int_0^{\pi/2} \frac{\partial \vec{W}(x, y, t, \varphi)}{\partial t} d\varphi; \quad \{x, y\} \in \overline{\mathbf{R}_z}, \quad t \geq 0,$$

$$\left[\frac{\partial^2}{\partial t^2} - \sin^2(\varphi) \left(\frac{\partial^2}{\partial x^2} + \frac{\partial^2}{\partial y^2} \right) \right] \vec{W}(x, y, t, \varphi) = -\left. \frac{\partial \vec{E}(g, t)}{\partial z} \right|_{z=L}; \quad \{x, y\} \in \mathbf{R}_z, \quad t > 0; \tag{2.74}$$

$$\left[\frac{\partial}{\partial t}+\frac{\partial}{\partial z}\right]\vec{E}(g,t)\Big|_{z=L}=\frac{2}{\pi}\int\limits_0^{\pi/2}\vec{W}(x,y,t,\varphi)\cos^2\varphi d\varphi;\quad \{x,y\}\in\overline{\mathbf{R}_z},\quad t\geq 0,$$

$$\left[\frac{\partial^2}{\partial t^2}-\cos^2(\varphi)\left(\frac{\partial^2}{\partial x^2}+\frac{\partial^2}{\partial y^2}\right)\right]\vec{W}(x,y,t,\varphi)$$

$$=-\left(\frac{\partial^2}{\partial x^2}+\frac{\partial^2}{\partial y^2}\right)\left[\frac{\partial}{\partial z}\vec{E}(g,t)\Big|_{z=L}\right];\quad \{x,y\}\in\mathbf{R}_z,\quad t>0; \qquad (2.75)$$

$$\left[\frac{\partial}{\partial t}+\frac{\partial}{\partial z}\right]\vec{E}(g,t)\Big|_{z=L}=\frac{2}{\pi}\int\limits_0^{\pi/2}\frac{\partial\vec{W}(x,y,t,\varphi)}{\partial t}\sin^2\varphi d\varphi;\quad \{x,y\}\in\overline{\mathbf{R}_z}\quad t\geq 0,$$

$$\left[\frac{\partial^2}{\partial t^2}-\cos^2(\varphi)\left(\frac{\partial^2}{\partial x^2}+\frac{\partial^2}{\partial y^2}\right)\right]\vec{W}(x,y,t,\varphi)=\left(\frac{\partial^2}{\partial x^2}+\frac{\partial^2}{\partial y^2}\right)\vec{E}(x,y,L,t);$$

$$\{x,y\}\in\mathbf{R}_z,\quad t>0 \qquad (2.76)$$

(the local exact ABCs) "equivalent" to (2.26) to (2.28). The inner problems in (2.74) to (2.76) with respect to the auxiliary vector functions $\vec{W}(x,y,t,\varphi)$ should be supplemented with the initial and boundary conditions

$$\vec{W}(x,y,t,\varphi)\Big|_{t=0}=\frac{\partial\vec{W}(x,y,t,\varphi)}{\partial t}\Big|_{t=0}=0;\quad \{x,y\}\in\overline{\mathbf{R}_z},$$

$$\left\{\begin{array}{l}\vec{W}\left\{\dfrac{\partial\vec{W}}{\partial x}\right\}(l_x,y,t,\varphi)=e^{i2\pi\Phi_x}\vec{W}\left\{\dfrac{\partial\vec{W}}{\partial x}\right\}(0,y,t,\varphi);\quad 0\leq y\leq l_y\\[2ex] \vec{W}\left\{\dfrac{\partial\vec{W}}{\partial y}\right\}(x,l_y,t,\varphi)=e^{i2\pi\Phi_y}\vec{W}\left\{\dfrac{\partial\vec{W}}{\partial y}\right\}(x,0,t,\varphi);\quad 0\leq x\leq l_x\end{array}\right.;\quad t\geq 0.$$

3
Compact Inhomogeneities in Free Space: Virtual Coordinate Boundaries in Scalar and Vector Problems of Wave Scattering Theory

3.1. Introduction

Proceeding with the discussion started in the previous chapters, let us consider model scalar and vector initial boundary value problems typical for the theory of the scattering of transient waves from compact (bounded) inhomogeneities in $\mathbf{R}^2$-space (E- and H-polarization) and $\mathbf{R}^3$-space. The geometry of the problems (the shape of metal and dielectric inclusions, the supports of momentary and current sources) may be of a rather general nature. However, it is essential that for all of these geometries there exists a regular region ${}_L\mathbf{Q}$, where the outgoing signal $U(g,t)$ propagates freely. This point is of utmost importance in the algorithms suggested below for truncating the analysis domain.

In Section 3.2, exact absorbing conditions for artificial coordinate boundaries $\mathbf{L}$ (in cylindrical coordinates ρ, ϕ, x) are constructed for the problems

$$\begin{cases} \left[-\varepsilon(g)\dfrac{\partial^2}{\partial t^2} - \sigma(g)\dfrac{\partial}{\partial t} + \dfrac{1}{\rho}\dfrac{\partial}{\partial \rho}\rho\dfrac{\partial}{\partial \rho} + \dfrac{1}{\rho^2}\dfrac{\partial^2}{\partial \phi^2}\right]U(g,t) = F(g,t); \\ g = \{\rho,\phi\} \in \mathbf{Q}, \quad t > 0 \\ U(g,t)\big|_{t=0} = \varphi(g), \quad \dfrac{\partial}{\partial t}U(g,t)\Big|_{t=0} = \psi(g); \quad g \in \bar{\mathbf{Q}} \\ E_{tg}(g,t)\big|_{g\in\mathbf{S}} = 0, \quad U(\rho,\phi,t) = U(\rho,\phi+2\pi,t); \quad t \geq 0 \end{cases} \quad (3.1)$$

These problems describe transient states of an E-polarized ($E_\rho = E_\phi = H_x = 0$, $U(g,t) = E_x(g,t)$) and an H-polarized ($H_\rho = H_\phi = E_x = 0$, $U(g,t) = H_x(g,t)$, $\varepsilon(g)$, and $\sigma(g)$ are piecewise constant functions) fields that are generated by the sources and objects concentrated in the bounded region $\mathbf{Q}_L$ of the $\mathbf{R}^2$-plane of the variables ρ and ϕ (see Sections 1.2.2, 1.2.4, and Fig. 1.1C; $\partial/\partial x \equiv 0$). Here, as before, $\mathbf{Q} = \mathbf{R}^2 \backslash \overline{\text{int}\mathbf{S}}$, $\mathbf{S}$ stands for the boundaries of the regions int$\mathbf{S}$ occupied by perfectly conducting objects, and $\mathbf{Q} = \mathbf{Q}_L \cup {}_L\mathbf{Q} \cup \mathbf{L}$.

The solutions $U(g,t)$ to problems (3.1) should be bounded in the vicinity of $\rho = 0$. In the case where the origin of coordinates is not covered by a "perfect metal", this constraint, normally expressed as $\lim_{\rho\to 0}(\rho\,\partial U/\partial\rho) = 0$ [124], requires that the nonzero value $\rho_{\min}$ (a lower limit of the discrete values of the variable ρ in the $\mathbf{Q}$ region) be introduced while passing on to the finite-difference analogues of (3.1).

If we make use of the fundamental results by O.A. Ladyzhenskaya [35] in the equations of problems (3.1), the stability conditions for the finite-difference algorithm as applied to (3.1) (they are similar to (1.50)) can be written in the form

$$\frac{\eta\sqrt{2}}{\sqrt{\xi}}\frac{l}{h} < 1 \quad \text{or} \quad 2\sqrt{\eta}\frac{l}{h} < 1;$$

$$0 < \xi \le \left(\frac{b_1^2}{\varepsilon(g)} + \frac{b_2^2}{\varepsilon(g)\rho^2}\right)\frac{1}{b_1^2 + b_2^2} \le \eta, \quad g = \{\rho, \phi\} \in \mathbf{Q}(h). \tag{3.2}$$

Here, $\mathbf{Q}(h)$ stands for the discretization region; b_1 and b_2 are any given real numbers; $h = h_\rho = h_\phi$; h_ρ and h_ϕ are the grid space increments in ρ and ϕ, and l is the grid time increment. From the estimates

$$\eta < 1 + \rho_{\min}^{-2} \quad \text{and} \quad \xi > a^{-1}, \quad a = \max_{g \in \mathbf{Q}(h)} \left\{\varepsilon(g); \rho^2\varepsilon(g)\right\}.$$

it follows that the time-step l largely depends on $\rho_{\min}$ in stable algorithms. In the absence of metal objects covering the origin of coordinates, the order of magnitude of l should be equal to that of h^2. Therein lies a main reason for poor efficiency of discretization schemes in cylindrical (polar) coordinates: under this approach the computer resources are not utilized in the best way. Another reason, which aggravates the negative effect of the former one, is associated with the following physically justified requirement [9]: the maximal mesh size should be at most half of (the recommended value is at most one-fifth of) the characteristic dimensions of the analyzed objects (waves of highest frequencies in the medium, geometric details of the scatterers, etc.). With large $\rho_{\max}$—the upper limit of the ρ values in $\mathbf{Q}_L$—the grid space increment in ϕ must be substantially reduced.

In Section 3.3, the possibilities for a correct numerical solution of the problems equivalent to (3.1) in Cartesian coordinates are analyzed. The main intent is to improve the efficiency of the algorithms that allow one to extract complete and exact information on the transients in complex electrodynamic objects from reasonably simple models. The visualization of the fields for these model objects, possessing realistic material and geometric parameters, may simplify substantially and speed up the solution of relevant fundamental and applied electromagnetic problems.

In Chapter 2, the basis for the construction of the exact ABCs was the radiation conditions for the elements of the evolutionary bases of the signals propagating in regular guided-wave structures, namely, closed waveguides and Floquet channels. These evolutionary bases do not change qualitatively under changes of the type of guiding structure, which has allowed us to close the analysis domain by the transverse coordinate boundaries (with respect to the direction of wave propagation): one boundary for one propagation channel of an outgoing wave. When solving problems (3.1) in a Cartesian rectangular grid, one radiation channel should be closed by four coordinate boundaries. The problem of corner points (the intersection points of the boundaries) arises [125,126]. It is evident that the successful resolution of this and other related problems may improve the efficiency of the numerical analysis of key problems modeled by (3.1), as well as of many other 2-D problems.

Section 3.4 extends the results of the study of 2-D scalar problems to 3-D vector problems. In Section 3.5, using axially symmetric initial boundary-value problems as an example, the potential advantages of joining two different coordinate systems in the construction of reliable computational schemes are demonstrated: absorbing conditions are constructed for a spherical artificial boundary **L**, while being intended for use in a rectangular grid of the cylindrical coordinates ρ and z.

3.2. Exact Conditions for Artificial Boundaries in Cylindrical (Polar) Coordinates

3.2.1. Transformation of the Evolutionary Basis for a Diverging Cylindrical Wave

In the region ${}_L\mathbf{Q}$ (see Fig. 1.1C), the solutions $U(g,t)$ to the initial boundary value problems (3.1) are transient waves, outgoing from the region where the sources and the efficient scatterers are concentrated. From (3.1) it follows

$$\left\{\begin{array}{l}\left[-\dfrac{\partial^2}{\partial t^2}+\dfrac{1}{\rho}\dfrac{\partial}{\partial\rho}\rho\dfrac{\partial}{\partial\rho}+\dfrac{1}{\rho^2}\dfrac{\partial^2}{\partial\phi^2}\right]U(g,t)=0;\quad g\in{}_L\mathbf{Q},\quad t>0\\ U(g,t)\big|_{t=0}=0,\quad \dfrac{\partial}{\partial t}U(g,t)\Big|_{t=0}=0;\quad g\in{}_L\bar{\mathbf{Q}}\\ U(\rho,\phi,t)=U(\rho,\phi+2\pi,t);\quad t\geq 0\end{array}\right. . \tag{3.3}$$

The separation of the ϕ-variable in (3.3) results in the formula

$$U(\rho,\phi,t)=\sum_n u_n(\rho,t)\mu_n(\phi);\quad g\in{}_L\bar{\mathbf{Q}},\quad t\geq 0 \tag{3.4}$$

where $\mu_n(\phi)=(2\pi)^{-1/2}\exp(in\phi)$, $n=0,\pm1,\pm2,\ldots$, is the orthonormal system of transversal functions (complete in $\mathbf{L}_2\,[(0<\phi<2\pi)]$), whereas the spatio–temporal amplitudes $u_n(\rho,t)$ (the evolutionary basis $u(\rho,t)=\{u_n(\rho,t)\}$) of the wave $U(\rho,\phi,t)$ are determined by the solutions to the initial boundary value problems

$$\left\{\begin{array}{l}\left[-\dfrac{\partial^2}{\partial t^2}+\dfrac{1}{\rho}\dfrac{\partial}{\partial\rho}\rho\dfrac{\partial}{\partial\rho}-\dfrac{n^2}{\rho^2}\right]u_n(\rho,t)=0;\quad \rho>L,\quad t>0\\ u_n(\rho,0)=0,\quad \dfrac{\partial}{\partial t}u_n(\rho,t)\Big|_{t=0}=0;\quad \rho\geq L\end{array}\right. . \tag{3.5}$$

The expansion (3.4) transfers the analysis to a space of complex functions $u_n(\rho,t)$. We could (through minor analytic complications) pursue the analysis in a real space, thereby saving computer resources. However, for the moment, an important point is that the problem of constructing the conditions that truncate the analysis domain is solved almost identically within both approaches.

Multiply (3.5) by $\chi(\rho - L)$ and then apply a Hankel transformation in ρ on the semi-axis $\rho \geq 0$ (image $\leftrightarrow$ original):

$$\tilde{f}_n(\omega) = H[f_n](\omega) \equiv \int_0^\infty f_n(\rho)\rho J_{|n|}(\rho\omega)\,d\rho$$

$$\leftrightarrow f_n(\rho) = H^{-1}\left[\tilde{f}_n\right](\rho) \equiv \int_0^\infty \tilde{f}_n(\omega)\omega J_{|n|}(\rho\omega)\,d\omega. \qquad (3.6)$$

We have for the images $\tilde{Z}_n(\omega, t)$ of the functions $Z_n(\rho, t) = u_n(\rho, t)\chi(\rho - L)$:

$$\begin{cases} \left(\dfrac{\partial^2}{\partial t^2} + \omega^2\right)\tilde{Z}_n(\omega, t) = L\left[u_n(L, t)\,J'_{|n|}(\omega L) - u'_n(L, t)\,J_{|n|}(\omega L)\right]; \\ \omega > 0, \quad t > 0 \\ \tilde{Z}_n(\omega, 0) = \dfrac{\partial}{\partial t}\tilde{Z}_n(\omega, t)\Big|_{t=0} = 0; \quad \omega \geq 0 \end{cases}. \qquad (3.7)$$

Here,

$$u'_n(L, t) = \frac{\partial}{\partial\rho}u_n(\rho, t)\Big|_{\rho=L}, \qquad J'_{|n|}(\omega L) = \frac{\partial}{\partial\rho}J_{|n|}(\omega\rho)\Big|_{\rho=L}.$$

In the course of the derivation of (3.7), we have used the familiar formula [122]

$$-\omega^2\tilde{f}_n(\omega) \leftrightarrow \left[\frac{d^2}{d\rho^2} + \frac{d}{\rho\,d\rho} - \frac{n^2}{\rho^2}\right]f_n(\rho),$$

and a chain of the evident equalities

$$\begin{aligned} \chi(\rho - L)\left[\frac{1}{\rho}\frac{\partial}{\partial\rho}\rho\frac{\partial}{\partial\rho}\right]u_n(\rho, t) &= \chi(\rho - L)\left[\frac{1}{\rho}\frac{\partial}{\partial\rho} + \frac{\partial^2}{\partial\rho^2}\right]u_n(\rho, t) \\ &= \left[\frac{1}{\rho}\frac{\partial}{\partial\rho} + \frac{\partial^2}{\partial\rho^2}\right]Z_n(\rho, t) - \delta(\rho - L)\left[\frac{1}{\rho} + \frac{\partial}{\partial\rho}\right] \\ &\quad \times u_n(\rho, t) - \frac{\partial}{\partial\rho}\left[\delta(\rho - L)\,u_n(\rho, t)\right] \end{aligned}$$

as well as the equality $(\partial^\alpha f, \gamma) = (-1)^{|\alpha|}(f, \partial^\alpha\gamma)$, which defines the generalized derivative $\partial^\alpha f$ of the generalized function $f \in \tilde{\mathbf{D}}(\mathbf{R}^n)$ (see Section 1.2.5).

Problems (3.7) are similar to those in (2.7) from Section 2.2.1. Their solutions

$$\tilde{Z}_n(\omega, t) = \frac{L}{\omega}\int_0^t \sin[\omega(t - \tau)]\left[u_n(L, \tau)\,J'_{|n|}(\omega L) - u'_n(L, \tau)\,J_{|n|}(\omega L)\right]d\tau,$$

when subjected to the inverse Fourier transform (3.6), result in the following expression,

$$u_n(\rho, t) = L\int_0^t \left[u_n(L, \tau)\,f'_n(L, \rho, t - \tau) - u'_n(L, \tau)\,f_n(L, \rho, t - \tau)\right]d\tau;$$
$$\rho \geq L, \quad t \geq 0. \qquad (3.8)$$

Formula (3.8) describes variations in the spatio–temporal amplitudes of the outgoing pulsed cylindrical waves as they propagate from the circle $\mathbf{L}$ to any circle of radius $\rho \geq L$. Here,

$$f_n(r, \rho, t-\tau) = \int\limits_0^\infty \sin[\omega(t-\tau)]\, J_{|n|}(\omega r) J_{|n|}(\omega\rho)\, d\omega \qquad (3.9)$$

and

$$f_n'(L, \rho, t-\tau) = \frac{\partial}{\partial r} f_n(r, \rho, t-\tau)\bigg|_{r=L}.$$

The integration in (3.9) is reduced to the calculation of the Legendre functions of the first ($P_{|n|-1/2}(a)$) and the second kind ($Q_{|n|-1/2}(-a)$) for the argument values $a_{r,\rho} = [r^2 + \rho^2 - (t-\tau)^2]/(2\rho r)$ [127]:

$$f_n(r, \rho, t-\tau) = \begin{cases} 0; & 0 < t-\tau < \rho - r \\ P_{|n|-1/2}\left(a_{r,\rho}\right) \big/ \left[2(r\rho)^{1/2}\right]; & \rho - r < t-\tau < \rho + r \\ Q_{|n|-1/2}\left(-a_{r,\rho}\right) \cos(n\pi) \big/ \left[\pi (r\rho)^{1/2}\right]; & \rho + r < t-\tau \end{cases}$$

$$= \chi\,[(t-\tau) - (\rho - r)]\, Q_{|n|-1/2}(-a_{r,\rho}) \cos(n\pi) \big/ \left[\pi (r\rho)^{1/2}\right];$$

$0 < t-\tau$.

The latest step in the above formula has been performed with the use of the well-known property of the Legendre functions [123]:

$$P_v(x) = \cos(v\pi)\, P_v(-x) - \frac{2}{\pi} \sin(v\pi)\, Q_v(-x).$$

3.2.2. Radiation Conditions and Nonlocal Absorbing Conditions

Formula (3.8) plays the same role in problems (3.1) as (2.12) does in problems (2.2) for waveguide units. Let us continue the rearrangements as we have done repeatedly and write down the exact radiation conditions for the total field $U(g, t)$ in the region ${}_L\mathbf{Q}$. Considering [34,123] that the value of $Q_{|n|-1/2}(-a_{r,\rho})$ at $t - \tau = \rho - r$ is $Q_{|n|-1/2}(-1) = \pi P_{|n|-1/2}(1)/[2\cos(n\pi)] = \pi/[2\cos(n\pi)]$, and $\partial\chi[(t-\tau)-(\rho-r)]/\partial r = \delta[(t-\tau)-(\rho-r)]$, upon differentiation in (3.8) and summation of the results in accordance with (3.4), we obtain for $\rho \geq L$ and $t \geq 0$:

$$U(\rho, \phi, t)$$

$$= \frac{1}{2}\sqrt{\frac{L}{\rho}}\, U(L, \phi, t - \rho + L) + \frac{1}{\pi}\sqrt{\frac{L}{\rho}} \sum_n (-1)^n \mu_n(\phi) \int\limits_0^{t-(\rho-L)} \left\{ \frac{u_n(L,\tau)}{2L} \right.$$
$$\times \left[Q'_{|n|-1/2}(-a_{L,\rho}) \left(\frac{\rho^2 - L^2 - (t-\tau)^2}{L\rho} \right) - Q_{|n|-1/2}(-a_{L,\rho}) \right]$$
$$\left. - u_n'(L,\tau) Q_{|n|-1/2}(-a_{L,\rho}) \right\} d\tau. \qquad (3.10)$$

On the artificial boundary $\rho = L$, the exact absorbing condition resulting from (3.10) takes the form

$$U(L,\phi,t) = \frac{2}{\pi}\sum_n (-1)^n \mu_n(\phi)\left[\int_0^t \left[u_n(L,\tau)\,\xi_n(t-\tau) - u_n'(L,\tau)\,\eta_n(t-\tau)\right]d\tau\right];$$
$$0 \le \phi \le 2\pi, \quad t \ge 0. \tag{3.11}$$

In (3.10) and (3.11) the following notations have been introduced:

$$Q'_{|n|-1/2}(-a) = \left.\frac{\partial}{\partial x} Q_{|n|-1/2}(x)\right|_{x=-a},$$
$$\xi_n(t-\tau) = \left[2Q'_{|n|-1/2}(-a_{L,L})\,(a_{L,L}-1) - Q_{|n|-1/2}(-a_{L,L})\right](2L)^{-1},$$

and $\eta_n(t-\tau) = Q_{|n|-1/2}(-a_{L,L})$.

3.3. Exact Conditions for Artificial Boundaries in Cartesian Coordinates: The Problem of Corner Points and Its Solution

3.3.1. Truncation of the Analysis Domain Down to a Band in the Plane of the Variables $g = \{y, z\}$

Let us rewrite problems (3.1) in Cartesian coordinates:

$$\begin{cases} \left[-\varepsilon(g)\dfrac{\partial^2}{\partial t^2} - \sigma(g)\dfrac{\partial}{\partial t} + \dfrac{\partial^2}{\partial z^2} + \dfrac{\partial^2}{\partial y^2}\right]U(g,t) = F(g,t); \quad g = \{y,z\} \in \mathbf{Q} \\ t > 0 \\ U(g,t)|_{t=0} = \varphi(g), \quad \left.\dfrac{\partial}{\partial t}U(g,t)\right|_{t=0} = \psi(g); \quad g \in \overline{\mathbf{Q}} \\ E_{tg}(g,t)\big|_{g\in\mathbf{S}} = 0; \quad t \ge 0 \end{cases} \tag{3.12}$$

The functions $F(g,t)$, $\varphi(g)$, $\psi(g)$, $\sigma(g)$, and $\varepsilon(g) - 1$ are finite in $\mathbf{Q}$. Their supports for all observation times $0 \le t \le T$ belong to the closure of the region $\mathbf{Q}_L = \{g \in \mathbf{Q} : L_4 < y < L_3; L_2 < z < L_1\}$ (Fig. 3.1). Above (below, to the right, to the left) of the artificial boundary $z = L_1$ ($z = L_2$, $y = L_3$, $y = L_4$, respectively) there are neither sources nor scatterers. The function $U(g,t)$ here is associated with the outgoing wave intersecting the boundary in one direction only and satisfies the homogeneous problems (3.12) with $\varepsilon(g) - 1 = \sigma(g) \equiv 0$. By subjecting $U(g,t)$

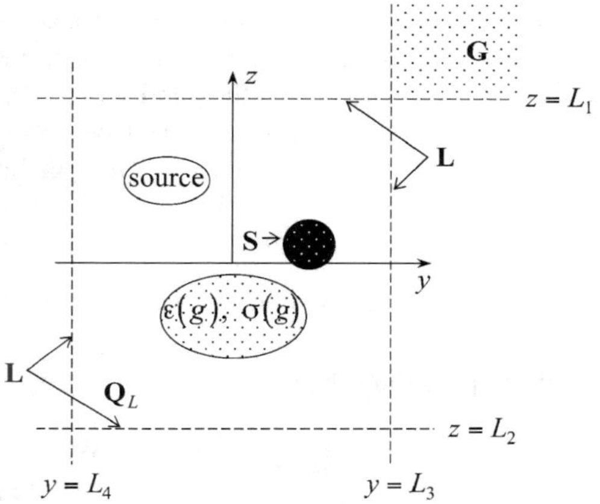

FIGURE 3.1. Geometry of problems (3.12).

to the Fourier transform

$$u_y(\lambda, z, t) = \frac{1}{2\pi}\int_{-\infty}^{\infty} U(y, z, t)e^{i\lambda y}dy \leftrightarrow U(y, z, t) = \int_{-\infty}^{\infty} u_y(\lambda, z, t)e^{-i\lambda y}d\lambda,$$

$$u_z(y, \mu, t) = \frac{1}{2\pi}\int_{-\infty}^{\infty} U(y, z, t)e^{i\mu z}dz \leftrightarrow U(y, z, t) = \int_{-\infty}^{\infty} u_z(y, \mu, t)e^{-i\mu z}d\mu$$

and applying the conventional technique for studying homogeneous problems for one-dimensional Klein–Gordon equations, we obtain

$$\left[\frac{\partial}{\partial t} \pm \frac{\partial}{\partial z}\right] u_y(\lambda, z, t) = \pm\lambda \int_0^t J_1(\lambda(t-\tau))\frac{\partial u_y(\lambda, z, \tau)}{\partial z}d\tau; \quad \begin{cases} z \geq L_1 \\ z \leq L_2 \end{cases}, \tag{3.13}$$

$$\left[\frac{\partial}{\partial t} \pm \frac{\partial}{\partial y}\right] u_z(y, \mu, t) = \pm\mu \int_0^t J_1(\mu(t-\tau))\frac{\partial u_z(y, \mu, \tau)}{\partial y}d\tau; \quad \begin{cases} y \geq L_3 \\ y \leq L_4 \end{cases}. \tag{3.14}$$

With a sequence of simple operations (see Section 2.2.2: Laplace transform in t $\rightarrow$ solution of the operational equations with respect to the derivatives in the spatial coordinates of the images of u_y and u_z $\rightarrow$ inverse Laplace transform), equations (3.13) and (3.14) can be written in the equivalent form:

$$\left[\frac{\partial}{\partial t} \pm \frac{\partial}{\partial z}\right] u_y(\lambda, z, t) = -\lambda \int_0^t \frac{J_1(\lambda(t-\tau))}{t-\tau}u_y(\lambda, z, \tau)\,d\tau; \quad \begin{cases} z \geq L_1 \\ z \leq L_2 \end{cases}, \tag{3.15}$$

$$\left[\frac{\partial}{\partial t} \pm \frac{\partial}{\partial y}\right] u_z(y, \mu, t) = -\mu \int_0^t \frac{J_1(\mu(t-\tau))}{t-\tau}u_z(y, \mu, \tau)\,d\tau; \quad \begin{cases} y \geq L_3 \\ y \leq L_4 \end{cases}. \tag{3.16}$$

Let us pass from the nonlocal conditions (3.13) to (3.16) (written in terms of the Fourier amplitudes of the field $U(g, t)$ and truncating the analysis domain in problems (3.12) down to the band $L_2 < z < L_1$ or $L_4 < y < L_3$), to relations that are local in time and space. Next we work with formulas (3.15) and (3.16), following the scheme originally applied in Section 2.2.3. First we invoke the Poisson integral formula [122],

$$J_1(x) = \frac{2x}{\pi}\int_0^{\pi/2} \cos\left[x\cos(\varphi)\right]\sin^2(\varphi)\,d\varphi,$$

and then apply the substitutions

$$v_y = -\lambda\int_0^t \frac{\sin\left[\lambda(t-\tau)\cos(\varphi)\right]}{\cos(\varphi)}u_y\,d\tau,$$

$$\frac{\partial v_y}{\partial t} = -\lambda^2\int_0^t \cos\left[\lambda(t-\tau)\cos(\varphi)\right]u_y\,d\tau \tag{3.17}$$

along with the fundamental solution $G(a, t) = \chi(t)\sin(at)a^{-1}$ of the operator $D(a) \equiv [d^2/dt^2 + a^2]$, in order to go from the integral forms (3.17) to the equivalent differential forms

$$\begin{cases} \left[\dfrac{\partial^2}{\partial t^2} + \lambda^2\cos^2(\varphi)\right]v_y = -\lambda^2 u_y; \quad t > 0 \\ \left.\dfrac{\partial v_y}{\partial t}\right|_{t=0} = \left.v_y\right|_{t=0} = 0 \end{cases}.$$

As a result, we obtain

$$\left[\frac{\partial}{\partial t} \pm \frac{\partial}{\partial z}\right]u_y(\lambda, z, t) = \frac{2}{\pi}\int_0^{\pi/2}\frac{\partial v_y(\lambda, z, t, \varphi)}{\partial t}\sin^2(\varphi)\,d\varphi; \quad t \geq 0,$$

$$\begin{cases} \left[\dfrac{\partial^2}{\partial t^2} + \lambda^2\cos^2(\varphi)\right]v_y(\lambda, z, t, \varphi) = -\lambda^2 u_y(\lambda, z, t); \quad t > 0 \\ \left.\dfrac{\partial v_y(\lambda, z, t, \varphi)}{\partial t}\right|_{t=0} = \left.v_y(\lambda, z, t, \varphi)\right|_{t=0} = 0 \end{cases}, \quad \begin{cases} z \geq L_1 \\ z \leq L_2 \end{cases}, \tag{3.18}$$

$$\left[\frac{\partial}{\partial t} \pm \frac{\partial}{\partial y}\right]u_z(y, \mu, t) = \frac{2}{\pi}\int_0^{\pi/2}\frac{\partial v_z(y, \mu, t, \varphi)}{\partial t}\sin^2(\varphi)\,d\varphi; \quad t \geq 0,$$

$$\begin{cases} \left[\dfrac{\partial^2}{\partial t^2} + \mu^2\cos^2(\varphi)\right]v_z(y, \mu, t, \varphi) = -\mu^2 u_z(y, \mu, t); \quad t > 0 \\ \left.\dfrac{\partial v_z(y, \mu, t, \varphi)}{\partial t}\right|_{t=0} = \left.v_z(y, \mu, t, \varphi)\right|_{t=0} = 0 \end{cases}, \quad \begin{cases} y \geq L_3 \\ y \leq L_4 \end{cases}. \tag{3.19}$$

Let

$$V_1(g,t,\varphi)=\int\limits_{-\infty}^{\infty} v_y(\lambda,z,t,\varphi)e^{-i\lambda y}d\lambda,\quad V_2(g,t,\varphi)=\int\limits_{-\infty}^{\infty} v_z(y,\mu,t,\varphi)e^{-i\mu z}d\mu.$$

Then, upon subjecting (3.18) and (3.19) to the inverse Fourier transform, the local ABCs truncating the analysis domain in the original problems (3.12) down to a band in $\mathbf{R}^2$, take the form

$$\left[\frac{\partial}{\partial t}\pm\frac{\partial}{\partial z}\right]U(g,t)=\frac{2}{\pi}\int\limits_0^{\pi/2}\frac{\partial V_1(g,t,\varphi)}{\partial t}\sin^2(\varphi)\,d\varphi;\quad |y|\le\infty,\quad t\ge 0,$$
$$\left\{\begin{array}{l}\left[\dfrac{\partial^2}{\partial t^2}-\cos^2(\varphi)\dfrac{\partial^2}{\partial y^2}\right]V_1(g,t,\varphi)=\dfrac{\partial^2}{\partial y^2}U(g,t);\quad |y|<\infty,\quad t>0\\ \left.\dfrac{\partial V_1(g,t,\varphi)}{\partial t}\right|_{t=0}=V_1(g,t,\varphi)|_{t=0}=0;\quad |y|\le\infty\end{array}\right.;$$
$$\left\{\begin{array}{l}z\ge L_1\\ z\le L_2\end{array}\right. \tag{3.20}$$

and

$$\left[\frac{\partial}{\partial t}\pm\frac{\partial}{\partial y}\right]U(g,t)=\frac{2}{\pi}\int\limits_0^{\pi/2}\frac{\partial V_2(g,t,\varphi)}{\partial t}\sin^2(\varphi)\,d\varphi;\quad |z|\le\infty,\quad t\ge 0,$$
$$\left\{\begin{array}{l}\left[\dfrac{\partial^2}{\partial t^2}-\cos^2(\varphi)\dfrac{\partial^2}{\partial z^2}\right]V_2(g,t,\varphi)=\dfrac{\partial^2}{\partial z^2}U(g,t),\quad |z|<\infty,\quad t>0\\ \left.\dfrac{\partial V_2(g,t,\varphi)}{\partial t}\right|_{t=0}=V_2(g,t,\varphi)|_{t=0}=0,\quad |z|\le\infty\end{array}\right.;$$
$$\left\{\begin{array}{l}y\ge L_3\\ y\le L_4\end{array}\right.. \tag{3.21}$$

3.3.2. The Corner Points: Correct Formulation of the Inner Initial Boundary Value Problems in the Exact Local Absorbing Conditions

Each of the four equations (3.20), (3.21) yields the exact absorbing condition, which narrows the analysis domain down to the half-plane $z<L_1, z>L_2, y<L_3$, or $y>L_4$. The involved inner differential problems (the Cauchy problems) formulated with respect to the functions $V_1(g,t,\varphi)$ (z is a parameter) and $V_2(g,t,\varphi)$ (y is a parameter) are well posed.

When truncating the domain down to a rectangular one, all four equations in (3.20), (3.21) should be allowed for, and the inner differential problems must be supplemented with the conditions at the ends of the intervals, at the intersection points of the boundaries $z = \text{const}$ and $y = \text{const}$. Several ways to resolve the problem of the corner points can be proposed, which differ by the analytic technique involved. Let us dwell on one such procedure, supposedly the clearest one.

Initially, consider the first equations (with the plus sign) in conditions (3.20) and (3.21). In Figure 3.1, the quadrant of the $\mathbf{R}^2$-plane, in which these equations are valid simultaneously, is dotted. Let us isolate the region

$$\mathbf{G} = \{g = \{y, z\} : L_3 < y < L_3 + 2\pi; \quad L_1 < z < L_1 + 2\pi\}$$

in this quadrant and use the following representation for the functions

$$f(g,t) \in \mathbf{W}_2^1(\mathbf{G}^\infty, \beta) = \left\{ f(g,t) : f(g,t)\exp(-\beta t) \in \mathbf{W}_2^1(\mathbf{G}^\infty) \right\}.$$

$$f(g,t) = \frac{1}{2\pi i} \int\limits_{\alpha - i\infty}^{\alpha + i\infty} \sum_{n,m=-\infty}^{\infty} \tilde{f}(n,m,s) e^{i(ny+mz)+st} ds; \quad \operatorname{Re} s \geq \beta \geq 0$$

$$\leftrightarrow \tilde{f}(n,m,s) = \frac{1}{4\pi^2} \int\limits_0^\infty \int\limits_{L_1}^{L_1+2\pi} \int\limits_{L_3}^{L_3+2\pi} f(g,t) e^{-i(ny+mz)-st} dy dz dt. \tag{3.22}$$

For the amplitudes $\tilde{u} = \tilde{u}(n,m,s)$, $\tilde{v}_j(\varphi) = \tilde{v}_j(n,m,s,\varphi)$, and $\tilde{w}_j(\varphi) = \tilde{w}_j(n,m,s,\varphi)$, $j = 1, 2$, of the functions $U(g,t)$, $V_j(g,t,\varphi)$, and $W_j(g,t,\varphi) = V_j(g,t,\varphi)\cos^2\varphi + U(g,t)$, we have

$$(s + im)\tilde{u} = \frac{2s}{\pi} \int\limits_0^{\pi/2} \sin^2(\varphi)\tilde{v}_1 d\varphi, \quad (s + in)\tilde{u} = \frac{2s}{\pi} \int\limits_0^{\pi/2} \sin^2(\varphi)\tilde{v}_2 d\varphi \tag{3.23}$$

$$\tilde{w}_1 = \frac{s^2}{s^2 + n^2\cos^2(\varphi)}\tilde{u}, \quad \tilde{w}_2 = \frac{s^2}{s^2 + m^2\cos^2(\varphi)}\tilde{u}, \tag{3.24}$$

$$\tilde{v}_1 = -\frac{n^2}{s^2 + n^2\cos^2(\varphi)}\tilde{u}, \quad \tilde{v}_2 = -\frac{m^2}{s^2 + m^2\cos^2(\varphi)}\tilde{u}, \tag{3.25}$$

$$s^2 + m^2 + n^2 = 0. \tag{3.26}$$

Now, consider the function

$$in\tilde{w}_1(\varphi)$$
$$= in\tilde{u}\frac{s^2}{s^2 + n^2\cos^2\varphi} = \frac{s^2}{s^2 + n^2\cos^2\varphi}\left[-s\tilde{u} + \frac{2s}{\pi}\int\limits_0^{\pi/2} \sin^2(\gamma)\tilde{v}_2(\gamma)d\gamma\right]$$

$$= -s\tilde{u}\frac{s^2}{s^2 + n^2\cos^2\varphi}\left[1 + \frac{2\sin^2\varphi}{\pi}\int_0^{\pi/2}\frac{\sin^2\gamma}{\cos^2\varphi + \sin^2(\varphi)\cos^2(\gamma)}d\gamma\right]$$

$$+\frac{2}{\pi}\int_0^{\pi/2} s\tilde{u}\frac{s^2}{s^2 + m^2\cos^2\gamma} \times \frac{\sin^2\gamma}{\cos^2\varphi + \sin^2(\varphi)\cos^2(\gamma)}d\gamma$$

$$= -\frac{s}{\cos\varphi}\tilde{w}_1(\varphi) + \frac{2}{\pi}\int_0^{\pi/2} s\tilde{w}_2(\gamma)\frac{\sin^2\gamma}{\cos^2\varphi + \sin^2(\varphi)\cos^2(\gamma)}d\gamma. \tag{3.27}$$

We have successively applied equations (3.24), (3.23), (3.25), the equality

$$\frac{s^2}{s^2 + an^2} \times \frac{s^2}{s^2 + bm^2} = \frac{a}{a + (1-a)b} \times \frac{s^2}{s^2 + an^2} + \frac{b}{a + (1-a)b} \times \frac{s^2}{s^2 + bm^2}$$

(which is true if (3.26) holds), and, once again, equations (3.24).

By subjecting (3.27) to the inverse transform, we arrive at

$$\left[\frac{\partial}{\partial t} + \cos\varphi\frac{\partial}{\partial y}\right] W_1(g, t, \varphi)$$

$$= \frac{2\cos\varphi}{\pi}\int_0^{\pi/2}\frac{\sin^2\gamma}{\cos^2\varphi + \sin^2(\varphi)\cos^2(\gamma)}\frac{\partial W_2(g, t, \gamma)}{\partial t}d\gamma; \quad z \geq L_1, \quad y \geq L_3.$$

We have not identified the upper boundaries $z = L_1 + 2\pi$ and $y = L_3 + 2\pi$ of the region **G**, where the above equation holds. It is obvious that there is no need to do so, inasmuch as the region **G** may be of arbitrary size.

By performing the operations described above for the functions $im\tilde{w}_2(\varphi)$ as the object to be investigated, we obtain

$$\left[\frac{\partial}{\partial t} + \cos\varphi\frac{\partial}{\partial z}\right] W_2(g, t, \varphi)$$

$$= \frac{2\cos\varphi}{\pi}\int_0^{\pi/2}\frac{\sin^2\gamma}{\cos^2\varphi + \sin^2(\varphi)\cos^2(\gamma)}\frac{\partial W_1(g, t, \gamma)}{\partial t}d\gamma; \quad z \geq L_1, \quad y \geq L_3.$$

The expressions relating the auxiliary functions $W_1(g, t, \varphi)$ and $W_2(g, t, \varphi)$ in all four regions such as **G** resolve the problem of corner points and allow the inner initial boundary value problems in conditions (3.20), (3.21) to be well posed on the finite sections of the outer boundary **L** enclosing the rectangular domain $\mathbf{Q}_L$. The relevant complete system of equations truncating the analysis domain **Q** down to the region $\mathbf{Q}_L$ is

$$\left[\frac{\partial}{\partial t} \pm \frac{\partial}{\partial z}\right] U(g,t) = \frac{2}{\pi} \int_0^{\pi/2} \frac{\partial V_1(g,t,\varphi)}{\partial t} \sin^2(\varphi)\, d\varphi; \quad L_4 \leq y \leq L_3, \quad t \geq 0,$$

$$\left\{ \begin{array}{l} \left[\dfrac{\partial^2 V_1(g,t,\varphi)}{\partial t^2} - \dfrac{\partial^2 W_1(g,t,\varphi)}{\partial y^2} \right] = 0; \quad L_4 < y < L_3, \quad t > 0 \\ \left. \dfrac{\partial V_1(g,t,\varphi)}{\partial t} \right|_{t=0} = V_1(g,t,\varphi)|_{t=0} = 0; \quad L_4 \leq y \leq L_3 \end{array} \right. ;$$

$$\left\{ \begin{array}{l} z = L_1 \\ z = L_2 \end{array} \right. , \tag{3.28}$$

$$\left[\frac{\partial}{\partial t} \pm \frac{\partial}{\partial y}\right] U(g,t) = \frac{2}{\pi} \int_0^{\pi/2} \frac{\partial V_2(g,t,\varphi)}{\partial t} \sin^2(\varphi)\, d\varphi; \quad L_2 \leq z \leq L_1, \quad t \geq 0,$$

$$\left\{ \begin{array}{l} \left[\dfrac{\partial^2 V_2(g,t,\varphi)}{\partial t^2} - \dfrac{\partial^2 W_2(g,t,\varphi)}{\partial z^2} \right] = 0; \quad L_2 < z < L_1, \quad t > 0 \\ \left. \dfrac{\partial V_2(g,t,\varphi)}{\partial t} \right|_{t=0} = V_2(g,t,\varphi)|_{t=0} = 0; \quad L_2 \leq z \leq L_1 \end{array} \right. ;$$

$$\left\{ \begin{array}{l} y = L_3 \\ y = L_4 \end{array} \right. , \tag{3.29}$$

$$\left\{ \begin{array}{l} \left[\dfrac{\partial}{\partial t} \pm \cos\varphi \dfrac{\partial}{\partial y}\right] W_1(g,t,\varphi) = \dfrac{2\cos\varphi}{\pi} \displaystyle\int_0^{\pi/2} \dfrac{\sin^2\gamma}{\cos^2\varphi + \sin^2(\varphi)\cos^2\gamma} \dfrac{\partial W_2(g,t,\gamma)}{\partial t} d\gamma \\ \left[\dfrac{\partial}{\partial t} \pm \cos\varphi \dfrac{\partial}{\partial z}\right] W_2(g,t,\varphi) = \dfrac{2\cos\varphi}{\pi} \displaystyle\int_0^{\pi/2} \dfrac{\sin^2\gamma}{\cos^2\varphi + \sin^2(\varphi)\cos^2\gamma} \dfrac{\partial W_1(g,t,\gamma)}{\partial t} d\gamma \end{array} \right.$$

$$\left\{ \begin{array}{c} + \\ + \end{array} \right\} \to g = \{L_3, L_1\}, \quad \left\{ \begin{array}{c} + \\ - \end{array} \right\} \to \{L_3, L_2\}, \quad \left\{ \begin{array}{c} - \\ + \end{array} \right\} \to \{L_4, L_1\}$$

$$\left\{ \begin{array}{c} - \\ - \end{array} \right\} \to \{L_4, L_2\}. \tag{3.30}$$

Actually, the three formulas (3.28) to (3.30) should have been combined into one formula, because only taken together do they determine the exact local absorbing condition for the entire artificial coordinate boundary **L**. Equations (3.30) here act as boundary conditions in the inner initial boundary value problems of (3.28) and (3.29). Symbols such as

$$\left\{ \begin{array}{c} + \\ + \end{array} \right\} \to g = \{L_3, L_1\}$$

specify the rules for the choice of signs in the upper and lower equations for different corner points $g = \{y, z\}$.

Statement 3.1. *Problems (3.12) with the analysis domain* $\mathbf{Q}$ *and problems (3.12) with the analysis domain* $\mathbf{Q}_L = \{g \in \mathbf{Q} : L_4 < y < L_3; L_2 < z < L_1\}$ *and conditions (3.28) to (3.30) on its outer rectangular boundary* $\mathbf{L}$ *are equivalent.*

The inner initial boundary value problems in (3.28) to (3.30) with respect to the functions $W_1(g, t, \varphi)$ *and* $W_2(g, t, \varphi)$ *are well posed.*

3.4. Vector Problems: Spherical Coordinates

3.4.1. Statement of the Problems and Preliminary Derivations

The initial boundary value problems

$$\begin{cases} \left[\Delta - \text{ grad div } - \varepsilon(g)\dfrac{\partial^2}{\partial t^2} - \sigma(g)\dfrac{\partial}{\partial t}\right]\vec{E}(g, t) = \vec{F}(g, t); \quad g \in \mathbf{Q} \subset \mathbf{R}^3, \\ t > 0 \\ \vec{E}(g, 0) = \vec{\varphi}(g), \quad \left.\dfrac{\partial \vec{E}(g, t)}{\partial t}\right|_{t=0} = \vec{\psi}(g); \quad g \in \overline{\mathbf{Q}} \\ \left.E_{tg}(g, t)\right|_{g \in \mathbf{S}} = 0; \quad t \geq 0 \end{cases} \tag{3.31}$$

describe the transient states of the electromagnetic field generated by the current sources ($\vec{F}(g, t)$) and the momentary sources ($\vec{\varphi}(g), \vec{\psi}(g)$) in the region of $\mathbf{R}^3$-space that contains compact metal and metal–dielectric inhomogeneities (the functions $\varepsilon(g) - 1 \geq 0$ and $\sigma(g) \geq 0$ are finite and the region int$\mathbf{S}$ is bounded). Only in spherical coordinates $g = \{r, \vartheta, \phi\}$, the analysis domain $\mathbf{Q}$ for these open problems can be truncated down to the bounded region $\mathbf{Q}_L = \{g \in \mathbf{Q} : r < L\}$ through the introduction of the unitary artificial coordinate boundary $\mathbf{L} = \{g \in \mathbf{R}^3 : r = L\}$ (Fig. 3.2). All the sources and scatterers are taken to be inside the solid sphere $\mathbf{S}(0, L)$. We must formulate conditions on $\mathbf{L}$ that do not distort the simulated physical processes. The strong solution of this problem consists in replacing the original open problem with an equivalent closed one. This Section is devoted to the construction of this solution.

The expression for $\Delta\vec{E}(g, t)$ in spherical coordinates is too cumbersome (see, for example, [128]), which practically rules out analytically working directly with equations (3.31). Some intermediate derivations are needed to simplify the situation. Below we use for this purpose the so-called Borgnis functions [32,33]. These are two scalar functions $U^E(g, t)$ and $U^H(g, t)$ through the use of which all six components of the electromagnetic field vectors $\vec{E}(g, t)$ and $\vec{H}(g, t)$ are determined (see Section 1.2.3). Clearly we have to do with the regions ${}_L\mathbf{Q} = \{g \in \mathbf{R}^3 : r > L\}$, where the relevant representations are valid and where

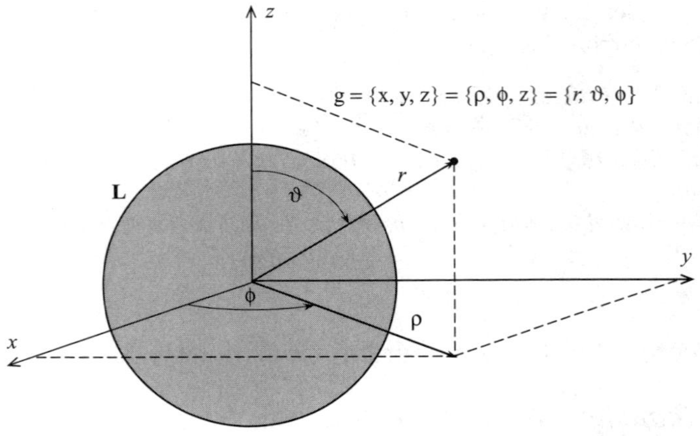

FIGURE 3.2. Vector initial boundary value problem: all sources and scatterers are located inside the sphere **L**.

all the analytical rearrangements will be performed in the course of constructing the exact absorbing conditions on the boundary **L**.

Following [32], we obtain for the solutions $\vec{E}(g,t)$, $g \in {}_L\overline{\mathbf{Q}}$, $t \geq 0$ of equations (3.31):

$$E_r = \frac{\partial^2 U^E}{\partial r^2} - \frac{\partial^2 U^E}{\partial t^2}, \qquad E_\vartheta = \frac{1}{r}\left[\frac{\partial^2 U^E}{\partial r \partial \vartheta} - \frac{1}{\sin\vartheta}\frac{\partial^2 U^H}{\partial \phi \partial t}\right],$$

$$E_\phi = \frac{1}{r}\left[\frac{1}{\sin\vartheta}\frac{\partial^2 U^E}{\partial r \partial \phi} + \frac{\partial^2 U^H}{\partial \vartheta \partial t}\right]. \tag{3.32}$$

The functions $U^E(g,t)$ and $U^H(g,t)$, specifying the TM- and TE-waves (with respect to the r-axis), respectively, satisfy the equivalent equations (with $r \neq 0$)

$$\left[-\frac{\partial^2}{\partial t^2} + \frac{1}{r^2 \sin\vartheta}\frac{\partial}{\partial \vartheta}\left(\sin\vartheta\frac{\partial}{\partial\vartheta}\right) + \frac{1}{r^2\sin^2\vartheta}\frac{\partial^2}{\partial\phi^2} + \frac{\partial^2}{\partial r^2}\right]\frac{\partial U^{E,H}(g,t)}{\partial t} = 0 \tag{3.33}$$

and

$$\left[-\frac{\partial^2}{\partial t^2} + \frac{1}{r^2 \sin\vartheta}\frac{\partial}{\partial \vartheta}\left(\sin\vartheta\frac{\partial}{\partial\vartheta}\right) + \frac{1}{r^2\sin^2\vartheta}\frac{\partial^2}{\partial\phi^2} + \frac{1}{r^2}\frac{\partial}{\partial r}\left(r^2\frac{\partial}{\partial r}\right)\right]\frac{\partial U^{E,H}(g,t)}{r\partial t}$$

$$= \left[-\frac{\partial^2}{\partial t^2} + \Delta\right]\frac{\partial U^{E,H}(g,t)}{r\partial t}; \quad g \in {}_L\overline{\mathbf{Q}}, \quad t \geq 0. \tag{3.34}$$

It is obvious that in the basis $\{\mu_{nm}(\vartheta,\phi)\}$ of the tesseral spherical harmonics [122],

$$\mu_{nm}(\vartheta,\phi) = \frac{1}{2}\sqrt{\frac{(2n+1)}{\pi}\frac{(n-|m|)!}{(n+|m|)!}}P_n^{|m|}(\cos\vartheta)\,\mathrm{e}^{im\phi};$$

$$n = 0, 1, 2, \ldots, \qquad m = 0, \pm 1, \pm 2, \ldots, \pm n$$

($P_n^{|m|}(x)$ are the associated Legendre functions of the first kind of degree n and of order $|m|$) they can be given by the expansions

$$U^{E,H}(g,t) = \sum_{n,m} u_{nm}^{E,H}(r,t)\mu_{nm}(\vartheta,\phi). \tag{3.35}$$

The system $\mu(\vartheta,\phi) = \{\mu_{nm}(\vartheta,\phi)\}$ is complete and orthonormal on the sphere $0 \le \vartheta \le \pi$, $0 \le \phi \le 2\pi$: for any two functions f_1 and f_2 such that they belong to $\mu(\vartheta,\phi)$ we have

$$\left[f_1 f_2^*\right]_\perp \equiv \int_0^{2\pi} d\phi \int_0^{\pi} f_1(\theta,\phi) f_2^*(\theta,\phi)\sin\vartheta d\vartheta = \begin{cases} 0, & f_1 \ne f_2 \\ 1, & f_1 = f_2 \end{cases}. \tag{3.36}$$

The functions $\mu_{nm}(\vartheta,\phi)$ satisfy the equations [122],

$$\left[\frac{1}{\sin\vartheta}\frac{\partial}{\partial\vartheta}\left(\sin\vartheta\frac{\partial}{\partial\vartheta}\right) + \frac{1}{\sin^2\vartheta}\frac{\partial^2}{\partial\phi^2} + n(n+1)\right]\mu_{nm}(\vartheta,\phi)$$
$$= \left[\Delta_\perp + n(n+1)\right]\mu_{nm}(\vartheta,\phi) = 0. \tag{3.37}$$

From (3.33) to (3.35) and (3.37), we have the following equivalent equations to determine in the closure of ${}_L\mathbf{Q}$, $t \ge 0$, the spatio–temporal amplitudes $u_{nm}^{E,H}(r,t)$ of the functions $U^{E,H}(g,t)$.

$$\left[-\frac{\partial^2}{\partial t^2} + \frac{\partial^2}{\partial r^2} - \frac{n(n+1)}{r^2}\right]\frac{\partial u_{nm}^{E,H}(r,t)}{\partial t} = 0, \tag{3.38}$$

$$\left[-\frac{\partial^2}{\partial t^2} + \frac{1}{r^2}\frac{\partial}{\partial r}\left(r^2\frac{\partial}{\partial r}\right) - \frac{n(n+1)}{r^2}\right]\frac{\partial u_{nm}^{E,H}(r,t)}{r\partial t} = 0. \tag{3.39}$$

Let us now rewrite the vector $\vec{E}(g,t)$ in the form

$$\vec{E}(g,t) = -\frac{\partial^2\vec{U}^E}{\partial t^2} + \text{grad}\frac{\partial U^E}{\partial r} - \text{rot}\frac{\partial\vec{U}^H}{\partial t}, \quad \vec{U}^{E,H}(g,t) = U^{E,H}(g,t)\vec{r}, \tag{3.40}$$

which is equivalent to (3.32). By substituting (3.35) into (3.40), we obtain

$$E_r(g,t) = \sum_{n,m} v_{nm}^r(r,t)\,\mu_{nm}(\vartheta,\phi), \tag{3.41}$$

$$\vec{E}_\perp(g,t) = E_\vartheta(g,t)\vec{\vartheta} + E_\phi(g,t)\vec{\phi} = \vec{E}_\perp^E(g,t) + \vec{E}_\perp^H(g,t)$$
$$= \sum_{n,m}\left[v_{nm}^{\perp,E}(r,t)\,\text{grad}_\perp\mu_{nm}(\vartheta,\phi) + v_{nm}^{\perp,H}(r,t)\,\text{rot}_\perp\mu_{nm}(\vartheta,\phi)\right]. \tag{3.42}$$

Here,

$$\text{grad}_\perp \equiv \vec{\vartheta}\frac{\partial}{\partial\vartheta} + \vec{\phi}\frac{1}{\sin\vartheta}\frac{\partial}{\partial\phi}, \quad \text{rot}_\perp \equiv \vec{\vartheta}\frac{1}{\sin\vartheta}\frac{\partial}{\partial\phi} - \vec{\phi}\frac{\partial}{\partial\vartheta},$$

and

$$v_{nm}^{r}(r,t) = \left(-\frac{\partial^2}{\partial t^2} + \frac{\partial^2}{\partial r^2}\right) u_{nm}^{E}(r,t) = \frac{n(n+1)}{r^2} u_{nm}^{E}(r,t), \tag{3.43}$$

$$v_{nm}^{\perp,E}(r,t) = \frac{1}{r}\frac{\partial u_{nm}^{E}(r,t)}{\partial r}, \tag{3.44}$$

$$v_{nm}^{\perp,H}(r,t) = -\frac{1}{r}\frac{\partial u_{nm}^{H}(r,t)}{\partial t}. \tag{3.45}$$

On the other hand,

$$v_{nm}^{r}(r,t) = \left[E_r(g,t)\,\mu_{nm}^{*}(\vartheta,\phi)\right]_{\perp}, \tag{3.46}$$

$$v_{nm}^{\perp,E}(r,t) = [n(n+1)]^{-1}\left[\left(\vec{E}_{\perp}(g,t)\cdot \mathrm{grad}_{\perp}\mu_{nm}^{*}(\vartheta,\phi)\right)\right]_{\perp}, \tag{3.47}$$

$$v_{nm}^{\perp,H}(r,t) = [n(n+1)]^{-1}\left[\left(\vec{E}_{\perp}(g,t)\cdot \mathrm{rot}_{\perp}\mu_{nm}^{*}(\vartheta,\phi)\right)\right]_{\perp}. \tag{3.48}$$

Equation (3.46) follows immediately from (3.36). To show that (3.47) and (3.48) are true, consider the integrals

$$\begin{aligned}
&\left[\left(\mathrm{grad}_{\perp}\mu_{ps}(\vartheta,\phi)\cdot \mathrm{grad}_{\perp}\mu_{nm}^{*}(\vartheta,\phi)\right)\right]_{\perp}\\
&= \int_0^{2\pi}\int_0^{\pi}\int_0^{1}\left(\mathrm{grad}\mu_{ps}(\vartheta,\phi)\cdot \mathrm{grad}\mu_{nm}^{*}(\vartheta,\phi)\right) r^2\sin(\vartheta)\,dr\,d\vartheta\,d\phi\\
&= \int_{\mathbf{S}(0,1)}\left(\mathrm{grad}\mu_{ps}(\vartheta,\phi)\cdot \mathrm{grad}\mu_{nm}^{*}(\vartheta,\phi)\right)d\mathbf{v}\\
&= -\int_{\mathbf{S}(0,1)}\mu_{ps}(\vartheta,\phi)\,\Delta\mu_{nm}^{*}(\vartheta,\phi)\,d\mathbf{v}\\
&\quad + \int_{\partial\mathbf{S}(0,1)}\mu_{ps}(\vartheta,\phi)\,\frac{\partial\mu_{nm}^{*}(\vartheta,\phi)}{\partial\vec{n}}\,d\mathbf{s}\\
&= -\left[\mu_{ps}(\vartheta,\phi)\,\Delta_{\perp}\mu_{nm}^{*}(\vartheta,\phi)\right]_{\perp}\\
&= \begin{cases} 0; & p\neq n \quad \text{or} \quad s\neq m\\ n(n+1); & p=n \quad \text{and} \quad s=m \end{cases},
\end{aligned} \tag{3.49}$$

$$\begin{aligned}
&\left[\left(\mathrm{rot}_{\perp}\mu_{ps}(\vartheta,\phi)\cdot \mathrm{rot}_{\perp}\mu_{nm}^{*}(\vartheta,\phi)\right)\right]_{\perp}\\
&\quad = \left[\left(\left[\mathrm{grad}_{\perp}\mu_{ps}(\vartheta,\phi)\times\vec{r}\right]\cdot\left[\mathrm{grad}_{\perp}\mu_{nm}^{*}(\vartheta,\phi)\times\vec{r}\right]\right)\right]_{\perp}\\
&\quad = \left[\left(\mathrm{grad}_{\perp}\mu_{ps}(\vartheta,\phi)\cdot \mathrm{grad}_{\perp}\mu_{nm}^{*}(\vartheta,\phi)\right)\right]_{\perp},
\end{aligned}$$

$$\left[\left(\mathrm{grad}_\perp \mu_{ps}(\vartheta,\phi)\cdot \mathrm{rot}_\perp \mu^*_{nm}(\vartheta,\phi)\right)\right]_\perp$$

$$= \int_0^{2\pi}\int_0^{\pi}\left[\frac{\partial \mu_{ps}(\vartheta,\phi)}{\partial\theta}\frac{\partial \mu^*_{nm}(\vartheta,\phi)}{\partial\phi} - \frac{\partial \mu_{ps}(\vartheta,\phi)}{\partial\phi}\frac{\partial \mu^*_{nm}(\vartheta,\phi)}{\partial\vartheta}\right]d\vartheta d\phi$$

$$\overset{\{\vartheta,\phi\}\Rightarrow\{x,y\}}{=} -\int_0^{2\pi}\int_0^{\pi}\left(\mathrm{rot}\left[\mu^*_{nm}(x,y)\vec{z}\right]\cdot\left[-\frac{\partial \mu_{ps}(x,y)}{\partial x}\vec{x} - \frac{\partial \mu_{ps}(x,y)}{\partial y}\vec{y} + \vec{z}\right]dxdy\right)$$

$$= -\int_{\mathbf{M}}\left(\mathrm{rot}\left[\mu^*_{nm}(x,y)\vec{z}\right]\cdot\overrightarrow{d\mathbf{m}}\right) = -\oint_{\mathbf{S}}\left(\overrightarrow{d\mathbf{r}}\cdot\mu^*_{nm}(x,y)\vec{z}\right) = 0. \tag{3.50}$$

Here, $\mathbf{S}(0,1)$ is a solid sphere of the unit radius with the origin of coordinates as a center; $\partial\mathbf{S}(0,1)$ is its spherical surface; $\vec{n}$ is the outer normal to $\partial\mathbf{S}(0,1)$; $dv = r^2\sin(\vartheta)\,dr d\vartheta d\phi$ is the element of volume; $\mathbf{M}$ is the complex "surface" of the function $z = \mu_{ps}(x,y)$, $\{x,y\}\in[0;\pi]\times[0;2\pi]$, spanned on the complex "contour" $\mathbf{S}$; $\overrightarrow{d\mathbf{m}}$ is the vector element of the area of this surface; and $\overrightarrow{d\mathbf{r}}$ is the vector element of the contour $\mathbf{S}$. In the course of deriving (3.49), the Green theorem and equations (3.37) have been used, whereas in the derivation of (3.50) we have relied on the Stokes theorem, the periodicity in ϕ of the spherical harmonics (the period equals 2π), and the particular values $P_n^{|m|}(\pm1) = 0$ for $n = 0,1,2,\ldots, m = \pm1,\pm2,\ldots,\pm n$, and $P_n(\pm1) = (\pm1)^n$, $\partial\mu_{n0}(\vartheta,\phi)/\partial\phi = 0$, for $n = 0,1,2,\ldots$ [123,129].

3.4.2. Nonlocal Radiation Conditions for the Borgnis Functions

Rewrite equation (3.38) in the form

$$\left[-\frac{\partial^2}{\partial t^2} + \frac{\partial^2}{\partial r^2} - \frac{n(n+1)}{r^2}\right]u_n(r,t) = 0;\quad r\ge L,\quad t>0,$$

and subject it to the integral transform

$$\tilde{f}(\omega) = \int_L^{\infty} f(r) Z_\gamma(\omega,r)\,dr;\quad \omega\ge 0, \tag{3.51}$$

where the kernel $Z_\gamma(\omega,r) = r^a\left[\alpha(\omega)J_\gamma(\omega r) + \beta(\omega)N_\gamma(\omega r)\right]$ (involving the unknown parameters a, γ, $\alpha(\omega)$, and $\beta(\omega)$) satisfies the equation [122],

$$\left[\frac{\partial^2}{\partial r^2} + \frac{1-2a}{r}\frac{\partial}{\partial r} + \omega^2 + \frac{a^2-\gamma^2}{r^2}\right]Z_\gamma(\omega,r) = 0. \tag{3.52}$$

Here, $u_n(r,t) = \partial u^{E,H}_{nm}(r,t)/\partial t$ or $u_n(r,t) = u^{E,H}_{nm}(r,t)$: because at the initial instant of time $t=0$ the excitation is absent in ${}_L\mathbf{Q}$, the functions $u^{E,H}_{nm}(r,t)$ in this region satisfy the same equations as the functions $\partial u^{E,H}_{nm}(r,t)/\partial t$ do.

Considering that

$$\int_L^\infty \frac{\partial^2 u_n(r,t)}{\partial r^2} Z_\gamma(\omega,r)\,dr = \frac{\partial u_n(r,t)}{\partial r} Z_\gamma(\omega,r)\Big|_L^\infty - \int_L^\infty \frac{\partial u_n(r,t)}{\partial r}\frac{\partial Z_\gamma(\omega,r)}{\partial r}dr$$

$$= \frac{\partial u_n(r,t)}{\partial r} Z_\gamma(\omega,r)\Big|_L^\infty - u_n(r,t)\frac{\partial Z_\gamma(\omega,r)}{\partial r}\Big|_L^\infty + \int_L^\infty u_n(r,t)\frac{\partial^2 Z_\gamma(\omega,r)}{\partial^2 r}dr,$$

$$\int_L^\infty u_n(r,t)\frac{\partial^2 Z_\gamma(\omega,r)}{\partial^2 r}dr$$

$$= -\int_L^\infty u_n(r,t)\left[\frac{1-2a}{r}\frac{\partial}{\partial r} + \omega^2 + \frac{a^2-\gamma^2}{r^2}\right] Z_\gamma(\omega,r)\,dr,$$

and that the functions $u_n(r,t)$ are zero for arbitrary finite t and sufficiently large r, we obtain for $a = 1/2$ and $\gamma = n + 1/2$.

$$\int_L^\infty \left[-\frac{\partial^2}{\partial t^2} - \omega^2\right] u_n(r,t)\,Z_\gamma(\omega,r)\,dr - \frac{\partial u_n(r,t)}{\partial r}\Big|_{r=L} Z_\gamma(\omega,L)$$
$$+ u_n(L,t)\frac{\partial Z_\gamma(\omega,r)}{\partial r}\Big|_{r=L} = 0. \tag{3.53}$$

From (3.53) the simple differential equation for the images $\tilde{u}_n(\omega,t)$ of the functions $u_n(r,t)$ follows.

$$\left[\frac{\partial^2}{\partial t^2} + \omega^2\right]\tilde{u}_n(\omega,t) = u_n(L,t)\frac{\partial Z_\gamma(\omega,r)}{\partial r}\Big|_{r=L} - \frac{\partial u_n(r,t)}{\partial r}\Big|_{r=L} Z_\gamma(\omega,L);$$
$$\omega \geq 0, \quad t > 0. \tag{3.54}$$

Similar equations have been examined before. Now all one needs to do is to define the parameters $\alpha(\omega)$ and $\beta(\omega)$ along with the kernels $Z_\gamma(\omega,r)$ (the values $a = 1/2$ and $\gamma = n + 1/2$ have already been fixed) so that, on the one hand, transformation (3.51) has a known inverse transform; and that furthermore, the result (the exact absorbing conditions or the exact radiation conditions for outgoing waves) is as simple and convenient to implement as possible.

To this end, we try (at this stage) to eliminate the radial derivatives of the $u_n(r,t)$ functions in the corresponding conditions. These derivatives are hard to approximate accurately on curved boundaries when solving problems in a rectangular Cartesian grid; on the other hand, the schemes that include the discretization procedure in angular coordinates are costly in terms of computer resources and runtime. Thus, let the coefficient $Z_\gamma(\omega,L)$ of $\partial w_n(r,t)/\partial r|_{r=L}$ in (3.54) be identically equal to zero:

$$Z_\gamma(\omega,L) = \sqrt{L}\left[\alpha(\omega)J_\gamma(\omega L) + \beta(\omega)N_\gamma(\omega L)\right] = 0; \quad \omega \geq 0. \tag{3.55}$$

From (3.55), we have

$$\alpha(\omega) = -N_\gamma(\omega L), \quad \beta(\omega) = J_\gamma(\omega L).$$

For these values of the parameters the coefficient of the remaining term in the right-hand side of (3.54) has the form

$$\left.\frac{\partial Z_\gamma(\omega, r)}{\partial r}\right|_{r=L} = \frac{1}{2\sqrt{L}}\left[J_\gamma(\omega L) N_\gamma(\omega L) - N_\gamma(\omega L) J_\gamma(\omega L)\right] + \omega\sqrt{L}\left[J_\gamma(\omega L) N_\gamma'(\omega L) - N_\gamma(\omega L) J_\gamma'(\omega L)\right] = \frac{2}{\pi\sqrt{L}}, \tag{3.56}$$

whereas (3.51) becomes the Weber–Orr transform having the known inverse [129]:

$$\tilde{f}(\omega) = \int_L^\infty \left[J_\gamma(\omega L) N_\gamma(\omega r) - N_\gamma(\omega L) J_\gamma(\omega r)\right] f(r)\sqrt{r}\,dr$$
$$\leftrightarrow f(r) = \sqrt{r}\int_0^\infty \frac{J_\gamma(\omega L) N_\gamma(\omega r) - N_\gamma(\omega L) J_\gamma(\omega r)}{J_\gamma^2(\omega L) + N_\gamma^2(\omega L)}\tilde{f}(\omega)\,\omega d\omega. \tag{3.57}$$

Here, $W\{J_\gamma(\omega L), N_\gamma(\omega L)\} = 2/(\pi\omega L)$ is the Wronskian [122], whereas $N_\gamma'(\omega L)$ and $J_\gamma'(\omega L)$ are the derivatives of the functions $N_\gamma(\omega L)$ and $J_\gamma(\omega L)$ with respect to ωL.

Let us construct the solutions $\tilde{u}_n(\omega, t)$ to the initial boundary value problems

$$\begin{cases} \left[\dfrac{\partial^2}{\partial t^2} + \omega^2\right]\tilde{u}_n(\omega, t) = \dfrac{2}{\pi\sqrt{L}} u_n(L, t); \quad t > 0, \quad \omega \ge 0 \\ \left.\dfrac{\partial \tilde{u}_n(\omega, t)}{\partial t}\right|_{t=0} = \tilde{u}_n(\omega, 0) = 0; \quad \omega \ge 0 \end{cases}$$

(which follow immediately from (3.54) to (3.56) and the zero initial values for $u_n(r, t)$), using the fundamental solution (see Section 1.2.5)

$$G(\lambda, t) = \chi(t)\lambda^{-1}\sin(\lambda t) = \lambda^{-2}\left[\delta(t)\cos(\lambda t) - d\left[\chi(t)\cos(\lambda t)\right]/dt\right]$$

of the operator $D(\lambda) \equiv [d^2/dt^2 + \lambda^2]$ and the convolution properties [34] $(\partial^\alpha f * g) = \partial^\alpha(f * g) = (f * \partial^\alpha g)$:

$$\tilde{u}_n(\omega, t) = \frac{2}{\pi\omega^2\sqrt{L}}\left[u_n(L, t) - \int_0^t \cos[\omega(t - \tau)]\frac{\partial u_n(L, \tau)}{\partial \tau} d\tau\right]; \quad \omega \ge 0, \quad t \ge 0. \tag{3.58}$$

By subjecting (3.58) to the inverse transform (3.57) and taking into account that [127],

$$\int_0^\infty \frac{J_\gamma(\omega L)\, N_\gamma(\omega r) - N_\gamma(\omega L)\, J_\gamma(\omega r)}{J_\gamma^2(\omega L) + N_\gamma^2(\omega L)} \frac{d\omega}{\omega} = \frac{\pi}{2}\left(\frac{L}{r}\right)^\gamma; \quad L < r,$$

we arrive at the following expression for the amplitudes $u_n(r, t)$,

$$u_n(r,t) = \left(\frac{L}{r}\right)^{\gamma-1/2} u_n(L,t) - \frac{2}{\pi}\sqrt{\frac{r}{L}} \int_0^t F_\gamma(r, L, t-\tau) \frac{\partial u_n(L,\tau)}{\partial \tau} d\tau; \quad r > L. \tag{3.59}$$

To evaluate the function

$$F_\gamma(r, L, t-\tau) = \int_0^\infty \frac{J_\gamma(\omega L) N_\gamma(\omega r) - N_\gamma(\omega L) J_\gamma(\omega r)}{J_\gamma^2(\omega L) + N_\gamma^2(\omega L)} \frac{\cos[\omega(t-\tau)]}{\omega} d\omega, \tag{3.60}$$

consider the auxiliary integral

$$\tilde{F}_\gamma(r, L, t-\tau) = \int_{-\infty}^\infty \frac{H_\gamma^{(1)}(zr)}{H_\gamma^{(1)}(zL)} \frac{\cos[z(t-\tau)]}{z} dz \tag{3.61}$$

along the real axis of the complex z-plane. The singular point $z = 0$ is avoided by passing along a semicircle $\mathbf{C}_\varepsilon$ of infinitely small radius ε in the upper half-plane; $H_\gamma^{(1)}$ is the Hankel function of the first kind. The integral (3.61) and the function $F_\gamma(r, L, t-\tau)$ are related by the apparent equality

$$\frac{H_\gamma^{(1)}(zr)}{H_\gamma^{(1)}(zL)} - \frac{H_\gamma^{(1)}(e^{i\pi} zr)}{H_\gamma^{(1)}(e^{i\pi} zL)} = 2i \frac{J_\gamma(zL) N_\gamma(zr) - N_\gamma(zL) J_\gamma(zr)}{J_\gamma^2(zL) + N_\gamma^2(zL)},$$

by virtue of which

$$\begin{aligned}
&\tilde{F}_\gamma(r, L, t-\tau) \\
&= \int_{-\infty}^{-\varepsilon} \frac{H_\gamma^{(1)}(zr)}{H_\gamma^{(1)}(zL)} \frac{\cos[z(t-\tau)]}{z} dz + \int_\varepsilon^\infty \frac{H_\gamma^{(1)}(zr)}{H_\gamma^{(1)}(zL)} \frac{\cos[z(t-\tau)]}{z} dz \\
&\quad + \int_{\mathbf{C}_\varepsilon} \frac{H_\gamma^{(1)}(zr)}{H_\gamma^{(1)}(zL)} \frac{\cos[z(t-\tau)]}{z} dz = -\int_\varepsilon^\infty \frac{H_\gamma^{(1)}(e^{i\pi} zr)}{H_\gamma^{(1)}(e^{i\pi} zL)} \frac{\cos[z(t-\tau)]}{z} dz \\
&\quad + \int_\varepsilon^\infty \frac{H_\gamma^{(1)}(zr)}{H_\gamma^{(1)}(zL)} \frac{\cos[z(t-\tau)]}{z} dz - \pi i \left(\frac{L}{r}\right)^\gamma \\
&= 2i F_\gamma(r, L, t-\tau) - \pi i \left(\frac{L}{r}\right)^\gamma.
\end{aligned} \tag{3.62}$$

Now wrie the function $\tilde{F}_\gamma(r, L, t-\tau)$ as

$$\tilde{F}_\gamma(r, L, t-\tau) = \frac{1}{2}\left\{\int\limits_{-\infty}^{\infty} \frac{H_\gamma^{(1)}(zr)}{H_\gamma^{(1)}(zL)} \frac{\exp[iz(t-\tau)]}{z} dz + \int\limits_{-\infty}^{\infty} \frac{H_\gamma^{(1)}(zr)}{H_\gamma^{(1)}(zL)} \frac{\exp[-iz(t-\tau)]}{z} dz\right\} \tag{3.63}$$

and for evaluating integrals in brackets, invoke the standard technique, which is based on the Cauchy theorem and the Jordan lemma [130]. Because $H_\gamma^{(1)}(zr)/H_\gamma^{(1)}(zL) \approx (L/r)^{1/2}\exp[iz(r-L)]$ for $|zL| \geq \gamma$ as $z \to \infty$ [122], the integration contours in (3.63) for $r - L > t - \tau$ can be closed by a circular arc of infinitely large radius in the upper half-plane of the complex variable z. For $r - L < t - \tau$, the contour of the first integral in (3.63) is closed in the upper half-plane, whereas the contour of the second integral is closed in the lower half-plane. Taking into consideration that all the singularities of the function $H_\gamma^{(1)}(zr)/H_\gamma^{(1)}(zL)$ are a finite number of poles at the points $z = z_s$: $\operatorname{Im} z_s < 0$, $s = 1, 2, \ldots, n$, coinciding with the zeros of the function $H_\gamma^{(1)}(zL)$ [120], and which as $z \to 0$ satisfies $H_\gamma^{(1)}(zr)/H_\gamma^{(1)}(zL) \approx (L/r)^\gamma$ along with the equality $dH_\gamma^{(1)}(zL)/dz = L[H_{\gamma-1}^{(1)}(zL) - \gamma H_\gamma^{(1)}(zL)/(zL)]$, we obtain

$$\tilde{F}_\gamma(r, L, t-\tau) = \begin{cases} -\pi i\left[\left(\dfrac{L}{r}\right)^\gamma + \displaystyle\sum_s \frac{H_\gamma^{(1)}(z_s r)}{H_{\gamma-1}^{(1)}(z_s L)} \frac{\exp[-iz_s(t-\tau)]}{z_s L}\right]; & \tau < t-(r-L) \\ 0; \quad \tau > t-(r-L) & \end{cases}. \tag{3.64}$$

By substituting (3.64) into (3.62), we have finally that

$$F_\gamma(r, L, t-\tau) = \begin{cases} -\dfrac{\pi}{2}\displaystyle\sum_s \frac{H_\gamma^{(1)}(z_s r)}{H_{\gamma-1}^{(1)}(z_s L)} \frac{\exp[-iz_s(t-\tau)]}{z_s L} \\ = -\dfrac{\pi}{2} S_\gamma(r, L, t-\tau); \quad \tau < t-(r-L). \\ \dfrac{\pi}{2}\left(\dfrac{L}{r}\right)^\gamma; \quad \tau > t-(r-L) \end{cases} \tag{3.65}$$

The summation in (3.65) may also be taken over half of the roots z_s, $s = 1, 2, \ldots, n$ of the equation $H_\gamma^{(1)}(zL) = 0$. For example, it may be taken only over those that lie in the fourth quadrant of the complex variable z-plane [21]. This step is justified by the fact that the function $H_\gamma^{(1)}(zL)$ has n complex zeros z_s : $\operatorname{Im} z_s < 0$, which are symmetric about the imaginary axis and are located approximately along the finite circular arc joining points $zL = -n$ and $zL = n$ [120].

Having calculated the function $S_\gamma(r, L, t-\tau)$, through integration by parts in (3.59), we arrive at the following exact radiation conditions for the spatio–temporal amplitudes $u_n(r, t)$ of the Borgnis functions, which determine the field $\vec{E}(g, t)$ in the region ${}_L\mathbf{Q}$ (see formulas (3.40) to (3.45); $u_n(r, t) = \partial u_{nm}^{E,H}(r, t)/\partial t$

or $u_n(r,t) = u_{nm}^{E,H}(r,t)$).

$$u_n(r,t) = \left(\frac{L}{r}\right)^n u_n(L, t-(r-L)) + \sqrt{\frac{r}{L}} \int_0^{t-(r-L)} S_{n+1/2}(r, L, t-\tau) \frac{\partial u_n(L,\tau)}{\partial \tau} d\tau$$

$$= \left[\left(\frac{L}{r}\right)^n + \sqrt{\frac{r}{L}} S_{n+1/2}(r, L, r-L)\right] u_n(L, t-(r-L))$$

$$-\sqrt{\frac{r}{L}} \int_0^{t-(r-L)} u_n(L,\tau) \frac{\partial S_{n+1/2}(r, L, t-\tau)}{\partial \tau} d\tau;$$

$$r > L, \quad t \geq 0, \quad n = 0, 1, 2, \ldots. \tag{3.66}$$

When placing the observation point r onto a sphere of radius L, the equalities (3.66) turn into identities. The conditions for $U^{E,H}(g,t)$ (see below), resulting from (3.66) and (3.35), (3.36), are nonlocal. They do not contain directional derivatives of $U^{E,H}(g,t)$ and, hence, can be realized in the rectangular Cartesian grid with minimum errors. An important point is that formula (3.66) efficiently resolves the problem of determining the far-zone field from the near-zone field: the transport operator $Z_{L\to r}(t)$, given by (3.66) and operating in space of the amplitudes $u(r,t) = \{u_n(r,t)\}$ according to the rule

$$u(r,t) = Z_{L\to r}(t)[u(L,\tau)]; \quad r > L, \quad t \geq \tau \geq 0,$$

allows us to follow accurately the changes of the electromagnetic field during the free propagation of a transient wave moving away from the region $\mathbf{Q}_L$.

To evaluate the amplitudes $v_{nm}^{\perp,E}(r,t)$ of the electric field of the TM-wave (see formula (3.44)), the radiation conditions for radial derivatives of the functions $u_n(r,t)$ should also be constructed. Through differentiation with respect to r in the first equality in (3.66), we obtain

$$\frac{\partial u_n(r,t)}{\partial r} = -\frac{n}{r}\left(\frac{L}{r}\right)^n u_n(L, t-(r-L))$$

$$-\left[\left(\frac{L}{r}\right)^n + \sqrt{\frac{r}{L}} S_{n+1/2}(r, L, r-L)\right] \left.\frac{\partial u_n(L,\tau)}{\partial \tau}\right|_{\tau=t-(r-L)}$$

$$+\sqrt{\frac{r}{L}} \int_0^{t-(r-L)} \left[\frac{S_{n+1/2}(r, L, t-\tau)}{2r} + \frac{\partial S_{n+1/2}(r, L, t-\tau)}{\partial r}\right]$$

$$\times \frac{\partial u_n(L,\tau)}{\partial \tau} d\tau; \quad r > L, \quad t \geq 0, \quad n = 0, 1, 2, \ldots. \tag{3.67}$$

The radial derivative of $S_\gamma(r, L, t-\tau)$ can be evaluated as easily as the function $S_\gamma(r, L, t-\tau)$ itself (see formula (3.65)):

$$\frac{\partial S_\gamma(r, L, t-\tau)}{\partial r} = \frac{1}{L} \sum_{s=1}^{n} \frac{\left[H_\gamma^{(1)}\right]'(z_s r)}{H_{\gamma-1}^{(1)}(z_s L)} e^{-iz_s(t-\tau)}; \quad \tau < t-(r-L).$$

Now let us pass, by using expansions (3.35), (3.36), from conditions (3.66), (3.67) for the amplitudes to the exact radiation conditions for the Borgnis functions $U^{E,H}(g,t)$:

$$U(g,t) = \sum_{n,m} \left\{ \left(\frac{L}{r}\right)^n \left[U\left(L,\tilde{\vartheta},\tilde{\phi},t-(r-L)\right)\mu^*_{nm}\left(\tilde{\vartheta},\tilde{\phi}\right)\right]_\perp + \sqrt{\frac{r}{L}} \int\limits_0^{t-(r-L)} \right.$$

$$\left. \times\, S_{n+1/2}(r,L,t-\tau)\left[\frac{\partial U\left(L,\tilde{\vartheta},\tilde{\phi},\tau\right)}{\partial \tau}\mu^*_{nm}\left(\tilde{\vartheta},\tilde{\phi}\right)\right]_\perp d\tau \right\} \mu_{nm}(\vartheta,\phi)$$

$$= V_1(g,t); \quad g=\{r,\vartheta,\phi\}\in {}_L\mathbf{Q}, \quad t\geq 0, \qquad (3.68)$$

$$\frac{\partial U(g,t)}{\partial r} = -\sum_{n,m}\left\{\frac{n}{r}\left(\frac{L}{r}\right)^n \left[U\left(L,\tilde{\vartheta},\tilde{\phi},t-(r-L)\right)\mu^*_{nm}\left(\tilde{\vartheta},\tilde{\phi}\right)\right]_\perp\right.$$

$$+\left[\left(\frac{L}{r}\right)^n + \sqrt{\frac{r}{L}} S_{n+1/2}(r,L,t-\tau)\right]$$

$$\times \left[\left.\frac{\partial U\left(L,\tilde{\vartheta},\tilde{\phi},\tau\right)}{\partial \tau}\right|_{\tau=t-(r-L)} \mu^*_{nm}\left(\tilde{\vartheta},\tilde{\phi}\right)\right]_\perp$$

$$-\sqrt{\frac{r}{L}} \int\limits_0^{t-(r-L)} \left[\frac{S_{n+1/2}(r,L,t-\tau)}{2r} + \frac{\partial S_{n+1/2}(r,L,t-\tau)}{\partial r}\right]$$

$$\left.\times \left[\frac{\partial U\left(L,\tilde{\vartheta},\tilde{\phi},\tau\right)}{\partial \tau}\mu^*_{nm}\left(\tilde{\vartheta},\tilde{\phi}\right)\right]_\perp d\tau\right\} \mu_{nm}(\vartheta,\phi) = V_2(g,t)$$

$$g\in {}_L\mathbf{Q}, \quad t\geq 0. \qquad (3.69)$$

Here, the function $U(g,t)=\sum_{n,m} u_n(r,t)\,\mu_{nm}(\vartheta,\phi)$, $g=\{r,\vartheta,\phi\}$, acts as a symbol that may be replaced either with the functions $U^{E,H}(g,t)$, or their time derivatives $\partial U^{E,H}(g,t)/\partial t$.

3.4.3. Exact Radiation Conditions for the Components of the Electric Field Vector

Formulas (3.66), (3.67), stemming from (3.43) to (3.45) through simple rearrangements, are transformed into the following exact radiation conditions for the spatio–temporal amplitudes of the $\vec{E}(g,t)$ field components.

$$v^r_{nm}(r,t) = \left(\frac{L}{r}\right)^{n+2} v^r_{nm}\left(L,t-(r-L)\right)$$

$$+\left(\frac{L}{r}\right)^{3/2} \int\limits_0^{t-(r-L)} S_{n+1/2}(r,L,t-\tau)\frac{\partial v^r_{nm}(L,\tau)}{\partial \tau}d\tau;$$

$$r>L, \quad t\geq 0, \quad n=0,1,2,\ldots, \quad m=0,\pm 1,\pm 2,\ldots,\pm n, \qquad (3.70)$$

$$v_{nm}^{\perp,E}(r,t) = -\frac{1}{n+1}\left(\frac{L}{r}\right)^{n+2} v_{nm}^{r}(L, t-(r-L)) - \frac{L}{n(n+1)}$$
$$\times \left[\left(\frac{L}{r}\right)^{n+1} + \sqrt{\frac{L}{r}}S_{n+1/2}(r, L, r-L)\right] \frac{\partial v_{nm}^{r}(L,\tau)}{\partial \tau}\bigg|_{\tau=t-(r-L)}$$
$$+ \frac{L}{n(n+1)}\sqrt{\frac{L}{r}} \int_{0}^{t-(r-L)} \left[\frac{S_{n+1/2}(r, L, t-\tau)}{2r} + \frac{\partial S_{n+1/2}(r, L, t-\tau)}{\partial r}\right]$$
$$\times \frac{\partial v_{nm}^{r}(L,\tau)}{\partial \tau} d\tau; \quad r > L, \quad t \geq 0,$$
$$n = 0, 1, 2, \ldots, \quad m = 0, \pm 1, \pm 2, \ldots, \pm n, \tag{3.71}$$

$$v_{nm}^{\perp,H}(r,t) = \left(\frac{L}{r}\right)^{n+1} v_{nm}^{\perp,H}(L, t-(r-L))$$
$$+ \sqrt{\frac{L}{r}} \int_{0}^{t-(r-L)} S_{n+1/2}(r, L, t-\tau) \frac{\partial v_{nm}^{\perp,H}(L,\tau)}{\partial \tau} d\tau;$$
$$r > L, \quad t \geq 0, \quad n = 0, 1, 2, \ldots, \quad m = 0, \pm 1, \pm 2, \ldots, \pm n. \tag{3.72}$$

For the components $E_r(g,t)\vec{r}$ and $\vec{E}_\perp(g,t)$ we obtain from (3.70) to (3.72) and (3.41), (3.42), (3.46) to (3.48):

$$E_r(g,t) = \sum_{n,m}\left\{\left(\frac{L}{r}\right)^{n+2} \left[E_r\left(L, \tilde{\vartheta}, \tilde{\phi}, t-(r-L)\right)\mu_{nm}^{*}\left(\tilde{\vartheta}, \tilde{\phi}\right)\right]_\perp + \left(\frac{L}{r}\right)^{3/2}\right.$$
$$\left. \times \int_{0}^{t-(r-L)} S_{n+1/2}(r, L, t-\tau) \left[\frac{\partial E_r\left(L, \tilde{\vartheta}, \tilde{\phi}, \tau\right)}{\partial \tau}\mu_{nm}^{*}\left(\tilde{\vartheta}, \tilde{\phi}\right)\right]_\perp d\tau\right\}$$
$$\times \mu_{nm}(\vartheta, \phi); \quad g \in {}_{L}\mathbf{Q}, \quad t \geq 0 \tag{3.73}$$

and

$$\vec{E}_\perp(g,t)$$
$$= \sum_{n,m}\left\{-\frac{1}{n+1}\left\{\left(\frac{L}{r}\right)^{n+2} \left[E_r\left(L, \tilde{\vartheta}, \tilde{\phi}, t-(r-L)\right)\mu_{nm}^{*}\left(\tilde{\vartheta}, \tilde{\phi}\right)\right]_\perp\right.\right.$$
$$+ \frac{L}{n}\left[\left(\frac{L}{r}\right)^{n+1} + \sqrt{\frac{L}{r}}S_{n+1/2}(r, L, r-L)\right]\left[\frac{\partial E_r\left(L, \tilde{\vartheta}, \tilde{\phi}, \tau\right)}{\partial \tau}\bigg|_{\tau=t-(r-L)}\right.$$
$$\left. \times \mu_{nm}^{*}\left(\tilde{\vartheta}, \tilde{\phi}\right)\right]_\perp - \frac{L}{n}\sqrt{\frac{L}{r}} \int_{0}^{t-(r-L)} \left[\frac{S_{n+1/2}(r, L, t-\tau)}{2r} + \frac{\partial S_{n+1/2}(r, L, t-\tau)}{\partial r}\right]$$

$$\times\left[\frac{\partial E_r\left(L,\tilde{\vartheta},\tilde{\phi},\tau\right)}{\partial\tau}\mu^*_{nm}\left(\tilde{\vartheta},\tilde{\phi}\right)\right]_\perp d\tau\Bigg\}\operatorname{grad}_\perp\mu_{nm}(\vartheta,\phi)$$

$$+\frac{1}{n(n+1)}\left\{\left(\frac{L}{r}\right)^{n+1}\left[\left(\vec{E}_\perp\left(L,\tilde{\vartheta},\tilde{\phi},t-(r-L)\right)\cdot\operatorname{rot}_\perp\mu^*_{nm}\left(\tilde{\vartheta},\tilde{\phi}\right)\right)\right]_\perp\right.$$

$$\left.+\sqrt{\frac{L}{r}}\int\limits_0^{t-(r-L)}S_{n+1/2}(r,L,t-\tau)\left[\left(\frac{\partial\vec{E}_\perp\left(L,\tilde{\vartheta},\tilde{\phi},\tau\right)}{\partial\tau}\cdot\operatorname{rot}_\perp\mu^*_{nm}\left(\tilde{\vartheta},\tilde{\phi}\right)\right)\right]_\perp d\tau\right\}$$

$$\times\operatorname{rot}_\perp\mu_{nm}(\vartheta,\phi)\Bigg\};\quad g\in{}_L\mathbf{Q},\quad t\ge 0. \tag{3.74}$$

Statement 3.2. *The open problems (3.31) with the unbounded analysis domain* $\mathbf{Q}=\mathbf{R}^3\backslash\mathrm{int}\mathbf{S}$ *and the closed problems (3.31) with the analysis domain* $\mathbf{Q}_{\tilde{L}}=\{g\in\mathbf{Q}:r<\tilde{L}\}$ *and supplemented by conditions (3.73), (3.74) on the outer boundary* $\tilde{\mathbf{L}}=\{g:r=\tilde{L},\tilde{L}>L\}$ *are equivalent.*

The radiation conditions constructed in this section admit of various schemes for their numerical implementation. The analysis of the advantages and disadvantages of such schemes is the subject of a special study. Here, we confine ourselves to noting that the results presented above are sufficient to solve numerically, with reasonable efficiency, rather complex model problems arising in transient electromagnetics. It should be emphasized once again that the numerical implementation of the obtained conditions (in contrast, for example, to the result from [19]) will not require numerical differentiation with respect to the normal to the artificial spherical boundary **L**. This feature, which might appear (at first glance) to be inessential, is actually very important in cases where the absorbing conditions derived in spherical coordinates close the analysis domain of the original initial boundary value problem that is approximated in some rectangular Cartesian grid.

Evidently, the approaches that are oriented from the outset to gain the most analytic and computational benefit from the work in different coordinates are very promising. For example, the open problems discussed in this section are closed by the unitary coordinate boundary exclusively in the spherical coordinates, whereas the corresponding finite-difference schemes are best constructed in the rectangular Cartesian grid. Obviously, interfacing the different systems of coordinates presents difficulties. However, they may be obviated as done, for example, in the next section dealing with axially symmetric compact inhomogeneities in free space.

3.5. Axially Symmetric Problems: Spherical and Cylindrical Coordinates

3.5.1. Formulation of the Initial Boundary Value Problems and Some General Statements

The analysis of pulsed TE_0- and TM_0-waves in axially symmetric structures ($\partial/\partial\phi\equiv 0$; the geometry is exemplified in Fig. 3.3A) is reduced to the solution of

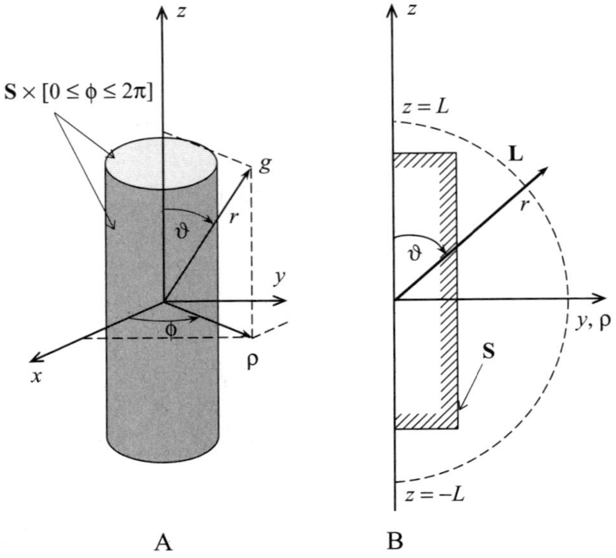

FIGURE 3.3. (A) Example of the geometry for the axially symmetric problem and (B) its analysis domain in the half-plane $\phi = \pi/2$.

one of the following 2-D scalar initial boundary value problems (see Section 1.2.2 and Fig. 3.3B; $\phi = \pi/2$).

$$\begin{cases} \left[-\varepsilon(g)\dfrac{\partial^2}{\partial t^2} - \sigma(g)\dfrac{\partial}{\partial t} + \dfrac{\partial^2}{\partial z^2} + \dfrac{\partial}{\partial \rho}\left(\dfrac{1}{\rho}\dfrac{\partial}{\partial \rho}\rho\right)\right] U(g,t) = F(g,t); \\ t > 0, \quad g \in \mathbf{Q} \\ U(g,t)\big|_{t=0} = \varphi(g), \quad \dfrac{\partial}{\partial t} U(g,t)\Big|_{t=0} = \psi(g); \quad g \in \overline{\mathbf{Q}} \\ E_{tg}(g,t)\big|_{g\in \mathbf{S}} = 0, \quad U(0,z,t) = 0; \quad t \geq 0 \end{cases} \tag{3.75}$$

(in cylindrical coordinates; $\mathbf{Q} \subset \{g = \{\rho, z\} : \rho \geq 0, |z| < \infty\}$) and

$$\begin{cases} \left[-\varepsilon(g)\dfrac{\partial^2}{\partial t^2} - \sigma(g)\dfrac{\partial}{\partial t} + \dfrac{1}{r}\dfrac{\partial^2}{\partial r^2}r + \dfrac{1}{r^2}\dfrac{\partial}{\partial \vartheta}\left(\dfrac{1}{\sin\vartheta}\dfrac{\partial}{\partial \vartheta}\sin\vartheta\right)\right] \\ \times U(g,t) = F(g,t); \quad t > 0, \quad g \in \mathbf{Q} \\ U(g,t)\big|_{t=0} = \varphi(g), \quad \dfrac{\partial}{\partial t} U(g,t)\Big|_{t=0} = \psi(g); \quad g \in \overline{\mathbf{Q}} \\ E_{tg}(g,t)\big|_{g\in \mathbf{S}} = 0; \quad t \geq 0 \\ U(r,0,t) = U(r,\pi,t) = 0; \quad r \geq 0, \quad t \geq 0 \end{cases} \tag{3.76}$$

(in spherical coordinates; $\mathbf{Q} \subset \{g = \{r, \vartheta\} : r \geq 0, 0 \leq \vartheta \leq \pi\}$).

The last-named conditions in (3.75) and (3.76) result from the symmetry of the problems: the axis $\rho = 0$ coincides with the axis of circular symmetry and, hence, only E_z- and H_z-components (E_r- and H_r-components) can be different from zero. With $U(g,t) = E_\phi$, problems (3.75) and (3.76) describe spatio–temporal transformations of TE_0-waves, whereas with piecewise constant functions $\varepsilon(g)$, $\sigma(g)$, and $U(g,t) = H_\phi$, they are stated with respect to the TM_0-waves. The analysis domain $\mathbf{Q}$ is the region in the half-plane $\phi = \pi/2$ that is bounded by the contour $\mathbf{S}$. $\mathbf{S} \times [0 \le \phi \le 2\pi]$ is the surface of perfectly conducting objects. It is assumed, as before, that the functions $F(g,t)$, $\varphi(g) = U^i(g,0)$, $\psi(g) = \partial U^i(g,t)/\partial t|_{t=0}$ ($U^i(g,t)$ stands for the incident wave), $\sigma(g)$, and $\varepsilon(g) - 1$ are finite in the closure of $\mathbf{Q}$ and satisfy the theorem about the unique solvability of problems (3.75) and (3.76) in the energetic class $\mathbf{W}_2^1(\mathbf{Q}^T)$.

The region $\mathbf{Q}_L \subset \mathbf{Q}$, which contains all the primary and secondary (the scatterers) sources of the field $U(g,t)$, can be closed by the unitary artificial boundary $\mathbf{L}$ only in spherical coordinates ($\mathbf{L} = \{g = \{r, \vartheta\} \in \mathbf{Q} : r = L\}$). On the other hand, the finite-difference scheme for the problems in question is optimal if constructed in the rectangular grid of the coordinates $g = \{\rho, z\}$. Let us try to give due consideration, when solving the problem, to both of these circumstances: the exact absorbing conditions are written for the spherical boundary $\mathbf{L}$, however, they are such that their implementation in the rectangular grid $g = \{\rho, z\}$ will not cause a growth in the total computational error. We stress that two different systems of coordinates can be integrated in such a way only in the event that the exact (for the spherical boundary $\mathbf{L}$) absorbing conditions contain no radial derivatives of the unknown function $U(g,t)$. This is a possibility as evidenced in the previous section.

3.5.2. *Exact Radiation Conditions for the Artificial Spherical Boundary*

In the region ${}_L\mathbf{Q} = \mathbf{Q} \backslash (\mathbf{Q}_L \cup \mathbf{L})$, where the field $U(g,t)$ propagates freely and infinitely far as $t \to \infty$, problems (3.76) transform to

$$\begin{cases} \left[-\dfrac{\partial^2}{\partial t^2} + \dfrac{1}{r}\dfrac{\partial^2}{\partial r^2} r + \dfrac{1}{r^2}\dfrac{\partial}{\partial \vartheta}\left(\dfrac{1}{\sin\vartheta}\dfrac{\partial}{\partial \vartheta}\sin\vartheta \right) \right] U(g,t) = 0; \\ t > 0, \quad g \in {}_L\mathbf{Q} \\ U(g,t)|_{t=0} = 0, \quad \dfrac{\partial}{\partial t} U(g,t)\Big|_{t=0} = 0; \quad g \in {}_L\overline{\mathbf{Q}} \\ U(r,0,t) = U(r,\pi,t) = 0; \quad r \ge L, \quad t \ge 0 \end{cases} . \tag{3.77}$$

By separating the variable ϑ in (3.77), we can represent the solution $U(g,t)$ as

$$U(r,\vartheta,t) = \sum_n u_n(r,t)\,\mu_n(\cos(\vartheta)); \quad r \ge L,$$

$$u_n(r,t) = \int_0^\pi U(r,\vartheta,t)\mu_n(\cos\vartheta)\sin\vartheta d\vartheta, \tag{3.78}$$

where $\{\mu_n(\cos\vartheta)\}$ is the orthonormal system of functions (complete in $\mathbf{L}_2[(0 < \vartheta < \pi)]$), which is defined by the nontrivial solutions of the homogeneous

Sturm–Liouville problem [113]

$$\begin{cases} \left[\dfrac{d}{d\vartheta}\left(\dfrac{1}{\sin\vartheta}\dfrac{d}{d\vartheta}\sin\vartheta\right)+\lambda^2\right]\mu(\cos\vartheta) \\ = \left[\dfrac{d^2}{d\vartheta^2}+\mathrm{ctg}\dfrac{d}{d\vartheta}-\dfrac{1}{\sin^2\vartheta}+\lambda^2\right]\mu(\cos\vartheta)=0; \quad 0<\vartheta<\pi, \\ \mu(\cos\vartheta)|_{\vartheta=0,\pi}=0 \end{cases} \tag{3.79}$$

whereas the spatio-temporal amplitudes $u_n(\rho, t)$ of the partial components of the spherical wave $U(g, t)$ are determined by the solutions to the following initial boundary value problems.

$$\begin{cases} \left[-\dfrac{\partial^2}{\partial t^2}+\dfrac{1}{r}\dfrac{\partial^2}{\partial r^2}r-\dfrac{\lambda_n^2}{r^2}\right]u_n(r,t)=\left[-\dfrac{\partial^2}{\partial t^2}+\dfrac{\partial^2}{\partial r^2}-\dfrac{\lambda_n^2}{r^2}\right]ru_n(r,t)=0; \\ r\geq L, \quad t>0 \\ u_n(r,0)=\dfrac{\partial}{\partial t}u_n(r,t)\Big|_{t=0}=0; \quad r\geq L \end{cases}. \tag{3.80}$$

Here, λ_n are the eigenvalues associated with the eigenfunctions $\mu_n(\cos\vartheta)$. The substitution of variables $x=\cos\vartheta$ ($\tilde{\mu}_n(x)=\mu_n(\cos\vartheta)$) reduces problem (3.79) to the form:

$$\begin{cases} \left[\left(1-x^2\right)\dfrac{d^2}{dx^2}-2x\dfrac{d}{dx}+\left(\lambda^2-\dfrac{1}{1-x^2}\right)\right]\tilde{\mu}(x)=0; \quad |x|<1 \\ \tilde{\mu}(-1)=\tilde{\mu}(1)=0 \end{cases}. \tag{3.81}$$

From (3.81), by employing the properties [123,129] of the linear independent solutions to the relevant differential equation (the behavior of the functions $P_n^1(x)$ and $Q_n^1(x)$ as $x\to\pm 1$, the values of the integral $\int_{-1}^{1}P_n^1(x)P_s^1(x)\,dx$, etc.), we obtain

$$\tilde{\mu}_n(x)=\mu_n(\cos\vartheta)=\sqrt{(2n+1)/(2n(n+1))}P_n^1(\cos\vartheta)$$

($P_n^1(x)$ are the associated Legendre functions of the first kind of degree n); $\lambda_n^2=n(n+1)$; $n=1,2,3,\ldots$.

The exact radiation conditions for the solutions $ru_n(r,t)$ of initial boundary value problems (3.80) have been constructed in Section 3.4.2. Rewriting them (see formula (3.66)) in terms of the spatio-temporal amplitudes $u_n(r,t)$, we arrive at

$$\begin{aligned} u_n(r,t) &= \left(\frac{L}{r}\right)^{n+1}u_n(L,t-(r-L))+\sqrt{\frac{L}{r}}\int_0^{t-(r-L)}S_{n+1/2}(r,L,t-\tau)\frac{\partial u_n(L,\tau)}{\partial\tau}d\tau \\ &= \left[\left(\frac{L}{r}\right)^{n+1}+\sqrt{\frac{L}{r}}S_{n+1/2}(r,L,r-L)\right]u_n(L,t-(r-L)) \end{aligned}$$

$$-\sqrt{\frac{L}{r}}\int\limits_{0}^{t-(r-L)} u_n(L,\tau)\frac{\partial S_{n+1/2}(r,L,t-\tau)}{\partial\tau}d\tau;\quad r>L,$$
$$t\geq 0,\quad n=1,2,3,\ldots. \tag{3.82}$$

In view of the expansions (3.78), we can go from conditions (3.82) to the exact radiation conditions for the functions $U(g,t)$, which determine the axially symmetric fields $\vec{E}(g,t)$ and $\vec{H}(g,t)$:

$$U(g,t)=\sum_{n=1}^{\infty}\left\{\left(\frac{L}{r}\right)^{n+1}\int\limits_{0}^{\pi}U\left(L,\tilde{\vartheta},t-(r-L)\right)\mu_n\left(\cos\tilde{\vartheta}\right)\sin\tilde{\vartheta}d\tilde{\vartheta}+\sqrt{\frac{L}{r}}\right.$$
$$\left.\times\int\limits_{0}^{t-(r-L)}S_{n+1/2}(r,L,t-\tau)\left[\int\limits_{0}^{\pi}\frac{\partial U\left(L,\tilde{\vartheta},\tau\right)}{\partial\tau}\mu_n\left(\cos\tilde{\vartheta}\right)\sin\tilde{\vartheta}d\tilde{\vartheta}\right]d\tau\right\}$$
$$\times\,\mu_n(\cos\vartheta);\quad g=\{r,\vartheta\}\in{}_L\mathbf{Q},\quad t\geq 0. \tag{3.83}$$

3.5.3. Some Peculiarities Arising in Implementation of Exact Absorbing Conditions in a Rectangular Grid of Coordinates $g=\{\rho,z\}$

When placing the observation point r onto a sphere of radius L, the equalities (3.82) and (3.83) rearrange to identities. Although the conditions (3.83) for $U(r,\vartheta,t)$ are nonlocal, they can be reduced, by following the scheme given in [21,117], to local conditions without any marked increase in the number of computational operations. These conditions contain no directional derivatives of the functions $U(r,\vartheta,t)$, and hence, they are easily approximated in the rectangular grid $\{\rho_j,z_k,t_m\}$ of the cylindrical coordinates ρ and z ($\rho_j=h_\rho j$, $z_k=h_z k$, $t_m=lm$; h_ρ, h_z, and l are grid space and time increments; $j=0,1,2,\ldots,J$, $k=0,\pm1,\ldots,\pm K$, $m=0,1,2,\ldots,M$). Note also that formulas (3.82) and (3.83) make it possible to determine the far-zone field from the field in the near zone of sources and inhomogeneities.

Let us consider briefly one possible numerical implementation of the algorithm for solving problems (3.75), (3.82), (3.83) [21].

The problems are discretized in the rectangular grid of cylindrical coordinates ρ and z, $h_\rho=h_z=h$. The boundary of the computational domain, which is associated with the arc $r=L$ (Fig. 3.3B), coincides with the broken line $\mathbf{P}_L$ whose vertices are situated at the grid nodes and are determined in the following way:

1. The ends of the broken line coincide with the points $\rho=0,z=L$ and $\rho=0,z=-L$;
2. To select the vertex of the broken line from four vertices of the mesh, which are crossed by the arc $r=L$, we should choose the vertex closest to the arc belonging to the region $\mathbf{Q}_L$.

In addition to the standard grid nodes and those that have already been determined, let us introduce the nodes that lie on the arc $r = L$. They are situated at the intersection points of vertical straight lines $\rho_j = jh$ with the upper $(0 < \vartheta < \pi/4)$ and lower $(3\pi/4 < \vartheta < \pi)$ parts of the arc as well as of horizontal straight lines $z_k = kh$ with the central part of the arc $(\pi/4 < \vartheta < 3\pi/4)$. Let us similarly introduce additional nodes, which lie on the inner arc $r = L_1$, $L_1 < L$. For this arc, we define a broken line $\mathbf{P}_{L_1}$ in the same manner as for the arc $r = L$, with the only difference being that the line vertices are located strictly on the other side of the arc. This requirement imposed on the nodes of the broken line is caused by the subsequent field interpolation from the broken line onto the arc $(r = L_1)$ and conversely $(r = L)$. The algorithm for the numerical implementation of conditions (3.82) and (3.83) involves the following steps:

1. Interpolation of the field $U(g, t)$ from the nodes of the broken line $\mathbf{P}_{L_1}$ onto the nodes of the arc $r = L_1$;
2. Computation of the amplitudes $u_n(L_1, t)$ of the angular field harmonics through the integration with respect to the nodes of the inner arc $r = L_1$ (see formula (3.78));
3. Computation of the amplitudes $u_n(L, t)$ by integral formula (3.82);
4. Computation of the field $U(g, t)$ at the nodes on the arc $r = L$ through the summation of the series (3.78); and
5. Interpolation of the field $U(g, t)$ from the nodes of the arc $r = L$ onto the nodes of the outer broken line $\mathbf{P}_L$.

The efficiency of this algorithm strongly depends upon how the Hankel functions $H_v^{(1)}(z)$ of half-integer index $v = n \pm 1/2$ as well as their complex zeros are calculated (see the formulas defining the functions $S_\gamma(r, L, t - \tau)$ along with their radial derivatives $\partial S_\gamma(r, L, t - \tau)/\partial r$). On this point, in the range of small n $(n < 9)$, the recommendations of paper [131] may be of use. As for large n and the values of the argument z comparable with n, the uniform asymptotic expansion 9.3.37 from [120] could be used.

4 The Simplest Modifications of the Exact ABCs Approach and the Associated Numerical Tests

4.1. Introduction

The analysis of pattern-forming structures (PFSs) with compact elements (whose parameters are given by the functions $\varepsilon(g)$, $\sigma(g)$, and the boundaries $\mathbf{S}$) excited by compact current and momentary sources (the functions $F(g,t)$, $\varphi(g)$, and $\psi(g)$ are finite) is reduced to the solution of the initial boundary value problems considered in Chapter 3. The antennas' near-zone fields are determined directly in the computational domain $\mathbf{Q}_L$ of the finite-difference method. The far-zone fields can be evaluated by converting the fields $\vec{U}(g,t)$ $(U(g,t))$ from the artificial boundary $\mathbf{L}$ onto the artificial boundary $\mathbf{P}$, which is removed from $\mathbf{L}$ at a required distance: the exact radiation conditions allow one to construct the transport operator $Z_{p\in\mathbf{L}\to g\in\mathbf{P}}(t)$ such that $\vec{U}(g,t) = Z_{p\in\mathbf{L}\to g\in\mathbf{P}}(t)[\vec{U}(p,\tau)]$.

Yet another class of pulse radiators can be analyzed rigorously with the results obtained in the Chapters 2 and 3. These are model PFSs, in which the signal propagation region is separated by infinite perfectly conducting flanges from the region containing the feeding waveguides (see, e.g., Fig. 4.1). The analysis domain $\mathbf{Q}_L$ of the corresponding initial boundary value problems is closed by the sections of the antenna elements' surface $\mathbf{S}$ along with two artificial boundaries $\mathbf{L}_1$ and $\mathbf{L}_2$: $\mathbf{L}_1$ is the unitary or compound coordinate boundary in the radiation zone $z > 0$, whereas $\mathbf{L}_2$ coincides with the cross-section $z = -L_2$, $L_2 > 0$, of the semi-infinite feeding waveguide. The region ${}_L\mathbf{Q} = \mathbf{Q}\backslash(\mathbf{Q}_L \cup \mathbf{L}_1 \cup \mathbf{L}_2)$, as before, contains neither sources nor scatterers. The absorbing conditions for the boundary $\mathbf{L}_2$ have been obtained in Chapter 2. All one needs to do in the relevant analytical representations is to account for changes in the direction of free propagation of the outgoing transient waves. On the $\mathbf{L}_1$ boundary we apply the conditions from Chapter 3: the systems of functions $\{\mu_n(\phi)\}$, $\{\mu_n(\cos\vartheta)\}$, $\{\mu_{nm}(\vartheta,\phi)\}$, and so on are being modified according to variations in the geometry of ${}_L\mathbf{Q}$ in the radiation zone of the PFS (see Section 4.2).

In Section 4.3, we elaborate the approach presented in section 3.3: the rectangular boundary $\mathbf{L}$ (Fig. 4.2) closes the analysis domain in the region of free propagation of pulsed E- or H-polarized waves, which are radiated from a plane-parallel waveguide with rather arbitrary aperture geometry. The associated plane

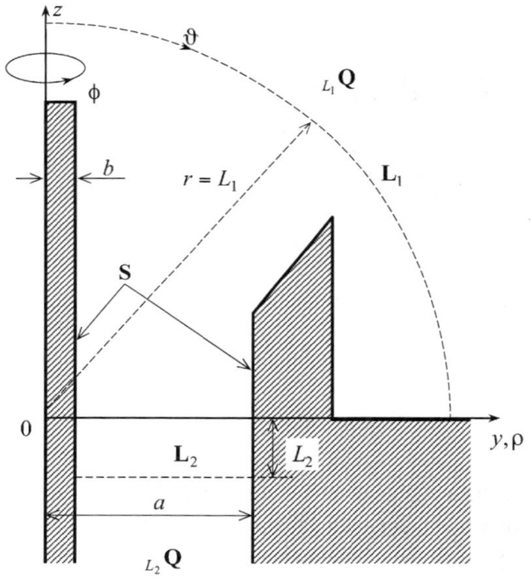

FIGURE 4.1. An example of an axially symmetric radiator with infinite flanges.

initial boundary value problem is considered in the Cartesian coordinates x, y, and z. The boundary $\mathbf{L}$ is a compound one. Hence, when replacing the original open problem with the equivalent closed one, it is necessary to go through several stages: the truncation of the computational domain $\mathbf{Q} = \mathbf{R}^2 \backslash \overline{\text{int}\,\mathbf{S}}$ down to a

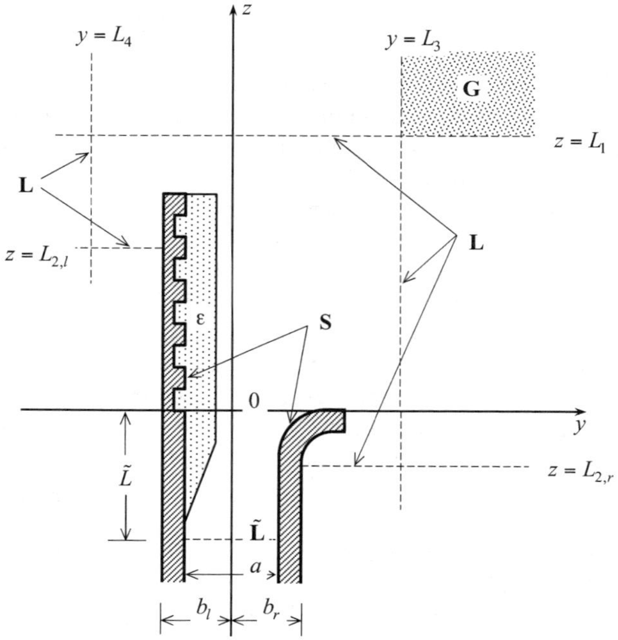

FIGURE 4.2. Plane model of the antenna with a grating as a dispersive element.

half-plane and a band in $\mathbf{R}^2$, the resolution of the problem of the corner points, and so on.

A similar technique may be applied in constructing the strong solution to the problem of the proper and efficient truncation of the analysis domain in related axially symmetric initial boundary value problems: the exact absorbing conditions are formulated on the coordinate (in the half-plane of the cylindrical coordinates ρ and z) compound boundary $\mathbf{L}$. Moreover, the implementation of this approach is possible for vector problems for compact discontinuities in the $\mathbf{R}^3$ space. The vector problems in the cylindrical coordinate system may be reduced by using the convenient field component combination leading to independent scalar problems [33]. The compound boundary $\mathbf{L}$—the surface of circular cylinder finite segment—has only two wedges, generating a problem that is similar to the problem of edge points in planar problems. The difficulties appearing in this way are connected in the main with the absence of "already done" results in the relatively rarely used theory of integral transformations of the Weber–Orr type (see Section 3.4.2).

Having removed several model limitations that had been introduced in the formulation of initial boundary value problems, we can make the result of previous chapters more profound. The subject of Sections 4.4 and 4.5 is the revision that focuses on the nearest possibilities for such advancement.

In Section 4.4, we have considered problems where solutions require only limited additional analytical efforts. And we can avoid unreasonable increase of the computational domain for initial signals of long duration, and for huge remote field sources. The approaches suggested are rather clear and versatile. A complete enough description of their implementation is supplied with several examples, presented below.

Section 4.5 is focused on the problems appearing while filling the domains $_L\mathbf{Q}$ with a medium different from vacuum. We focus on the resolution of one aspect of the problems. Several others are waiting to be solved in the future.

The basic results of the Chapters 2 to 4 consist in formulation of exact absorbing conditions. These conditions provide the possibility of carrying out the analysis within bounded space domains, comprising all efficient sources, for any interval of variation of the variable t. In essence, the well-known radiation condition

$$U(g,t)|_{g\in\mathbf{G},t\in[0;T]} = 0$$

for outgoing waves (for $t \leq T$ the excitation $U(g,t)$ is not able to reach the points g, belonging to the domain $\mathbf{G}$) is transferred onto the virtual boundaries $\mathbf{L}$ (see representation (2.1)) into regions where the spatio–temporal transformation intensity may be either big or small. The explicit analytical relations are derived for the integrodifferential operator M. Boundaries $\mathbf{L}$ divide infinite analysis domain $\mathbf{Q}$ of the original problem into two: $\mathbf{Q}_L$ and $_L\mathbf{Q}$; $\mathbf{Q} = \mathbf{Q}_L \cup {}_L\mathbf{Q} \cup \mathbf{L}$. In the first one (the bounded one) while seeking the value of function $U(g,t)$ (that is solving the initial boundary value problem with condition (2.1)) the conventional computational schemes are implemented. In the second domain, the field $U(g,t)$ is calculated over its values on $\mathbf{L}$.

The original open problems are equivalent to the closed ones. Relevant statements are formulated in Chapters 2 to 4. The following three facts provide the basis for their validity: the original problems are uniquely solvable [35], the solution to the original problem is also the solution to the modified problem (according the derivations) and the solution to the modified problem is unique. The last statement may be proved relying on the so-called "energy" estimates of the real function $U(g,t)$ (estimates for U^2, $(\partial U/\partial t)^2$, and $|\text{grad}U|^2$; see Section 1.2.6 and the books [34,35,132]).

The results obtained hold for practically all changes in $\mathbf{Q}_L$ that are not affecting the existence of regular domain ${}_L\mathbf{Q}$ of free outgoing propagation of the pulsed waves. Such changes, in particular, may concern the geometry of the domain $\mathbf{Q}_L$, and the electrical parameters and also geometry of the scattering objects. In the domain $\mathbf{Q}_L$ so far, the presence of dielectric objects (parameters $\varepsilon(g)$ and $\sigma(g)$) and perfectly conducting (metal) discontinuities have been considered. The inclusion into $\mathbf{Q}_L$ of magnetic and plasma-type discontinuities and even nonlinear objects [133] does not bring any problem into the construction of the solution. All depends, in fact, on the numerical schema used for solving the modified problem. The standard schemata of the finite-difference methods are rather flexible and versatile. The exact absorbing conditions make them rather reliable also; the scope of the open problems of transient electromagnetic theory that may be solved in a rigorous way becomes considerably larger. It follows from the equivalence of the original and modified problems that exact ABCs do not increase the computational errors; they remain level, determined by the discretization of the problem. Section 4.6 of this chapter is devoted to the verification of this statement.

4.2. Radiators with Infinite Flanges

The results obtained in Chapters 2 and 3 are easily adaptable to antennas with infinite flanges. Thus we consider here only one example, namely, the excitation of an axially symmetric structure by pulsed TE_0- or TM_0-waves of a circular or coaxial circular waveguide (Fig. 4.1).

4.2.1. Statement of the Initial Boundary Value Problems

TE_0-waves ($E_\rho = E_z = H_\phi = j_\rho = j_z \equiv 0$, see Section 1.2.2) and TM_0-waves ($H_\rho = H_z = E_\phi = j_\phi \equiv 0$) in axially symmetric structures ($\partial/\partial\phi \equiv 0$) are determined by the solutions $U(g,t)$ to the following 2-D (e.g., in the half-plane $\phi = \pi/2$) scalar initial boundary value problems

$$\begin{cases} \left[-\varepsilon(g)\dfrac{\partial^2}{\partial t^2} - \sigma(g)\dfrac{\partial}{\partial t} + \dfrac{\partial^2}{\partial z^2} + \dfrac{\partial}{\partial \rho}\left(\dfrac{1}{\rho}\dfrac{\partial}{\partial \rho}\rho\right)\right] U(g,t) = F(g,t); \\ t > 0, \quad g = \{\rho, z\} \in \mathbf{Q} \\ U(g,t)\big|_{t=0} = \varphi(g), \quad \dfrac{\partial}{\partial t}U(g,t)\Big|_{t=0} = \psi(g); \quad g \in \overline{\mathbf{Q}} \\ E_{tg}(g,t)\big|_{g\in\mathbf{S}} = 0, \quad U(0,z,t) = 0; \quad t \geq 0 \end{cases} \tag{4.1}$$

In (4.1), the condition $U(0, z, t) = 0, t \geq 0$, is a consequence of the symmetry of the problems: the axis $\rho = 0$ coincides with the axis of circular symmetry and, hence, only E_z- and H_z-components of the field vectors can be different from zero. With $U(g, t) = E_\phi$ we have $E_{tg}(g, t)|_{g \in \mathbf{S}} = U(g, t)|_{g \in \mathbf{S}}$ and problems (4.1) describe spato–temporal transformations of TE_0-waves, whereas with piecewise constant functions $\varepsilon(g)$, $\sigma(g)$, and $U(g, t) = H_\phi$, they are stated with respect to the TM_0-waves. The analysis domain $\mathbf{Q}$ is the region in the half-plane $\phi = \text{const}$ that is bounded by the contour $\mathbf{S}$. $\mathbf{S} \times [0 \leq \phi \leq 2\pi]$ is the surface of perfectly conducting objects. It is assumed, as before, that the functions describing the sources as well as the geometry and material parameters of the metal–dielectric inclusions, being finite in the closure of $\mathbf{Q}$, satisfy the theorem about the unique solvability of problems (4.1) in the energy class. Their supports belong to the set $\overline{\mathbf{Q}_L} \backslash (\mathbf{L}_1 \cup \mathbf{L}_2)$. In the region ${}_L\mathbf{Q} = \mathbf{Q} \backslash (\mathbf{Q}_L \cup \mathbf{L})$, $\mathbf{L} = \mathbf{L}_1 \cup \mathbf{L}_2$ (in the region ${}_{L_2}\mathbf{Q}$: $z < -L_2$, $L_2 > 0$ of regular circular and coaxial circular waveguides and in the region ${}_{L_1}\mathbf{Q}$: $r > L_1 > 0$ of the structure radiation zone; r, ϑ, ϕ are spherical coordinates), there are neither sources nor scatterers. The artificial boundaries $\mathbf{L}_1$ ($r = L_1, 0 \leq \vartheta \leq \pi/2$) and $\mathbf{L}_2$ (it coincides with the waveguide cross-sections by the plane $z = -L_2$) are given in Figure 4.1 by the dashed lines.

4.2.2. Exact Absorbing Conditions in the Radiation Zone of an Axially Symmetric Structure

For the part ${}_{L_1}\mathbf{Q}$ of the region ${}_L\mathbf{Q}$, let us rewrite problems (4.1) in spherical coordinates:

$$\left\{\begin{array}{l}
\left[-\dfrac{\partial^2}{\partial t^2} + \dfrac{1}{r}\dfrac{\partial^2}{\partial r^2} r + \dfrac{1}{r^2}\dfrac{\partial}{\partial \vartheta}\left(\dfrac{1}{\sin\vartheta}\dfrac{\partial}{\partial \vartheta}\sin\vartheta\right)\right] U(g, t) = 0; \\
t > 0, \quad g = \{r, \vartheta\} \in {}_{L_1}\mathbf{Q} \\
U(g, t)|_{t=0} = 0, \quad \left.\dfrac{\partial}{\partial t} U(g, t)\right|_{t=0} = 0; \quad g \in {}_{L_1}\overline{\mathbf{Q}} \\
\left\{\begin{array}{ll}
U(r, 0, t) = U\left(r, \pi/2, t\right) = 0; & TE_0 - \text{waves} \\
U(r, 0, t) = \dfrac{1}{\sin\vartheta} \left.\dfrac{\partial\left[U(r, \vartheta, t)\sin\vartheta\right]}{\partial \vartheta}\right|_{\vartheta=\pi/2} = 0; & TM_0 - \text{waves}
\end{array}\right. , \\
r \geq L_1, \quad t \geq 0
\end{array}\right. . \tag{4.2}$$

Here, it has been taken into account that the tangential component of the electric field vector on the flange $\vartheta = \pi/2$ coincides with $U(r, \vartheta, t)$ in the case of TE_0-waves, whereas for TM_0-waves we have $E_{tg} = E_r$ and

$$\frac{\partial E_r}{\partial t} = \frac{\eta_0}{r\sin\vartheta}\frac{\partial\left(H_\phi \sin\vartheta\right)}{\partial \vartheta} = \frac{\eta_0}{r\sin\vartheta}\frac{\partial (U\sin\vartheta)}{\partial \vartheta}$$

(see formulas (1.19) in Section 1.2.2).

Next, we follow the scheme given in Section 3.5.2. Namely, let us separate the variable ϑ in problems (4.2) and represent their solutions $U(g,t)$ as

$$\begin{cases} U(r,\vartheta,t) = \sum_n u_n(r,t)\,\mu_n(\cos(\vartheta)); \quad r \geq L_1, \quad 0 \leq \vartheta \leq \pi/2, \\ u_n(r,t) = \int_0^{\pi/2} U(r,\vartheta,t)\mu_n(\cos\vartheta)\sin\vartheta d\vartheta \end{cases} . \tag{4.3}$$

When constructing the complete systems (on the interval $0 < \vartheta < \pi/2$) of the orthonormal functions

$$\mu_n(\vartheta) = \sqrt{\frac{2n+1}{n(n+1)}} P_n^1(\cos\vartheta); \quad n = \begin{cases} 2k, & \mathrm{TE}_0 - \text{waves} \\ 2k-1, & \mathrm{TM}_0 - \text{waves} \end{cases}, \\ k = 1, 2, 3, \ldots, \tag{4.4}$$

we are using the known [122,129] particular values of the associated Legendre functions

$$P_n^1(0) = \frac{2}{\sqrt{\pi}}\cos\left(\frac{\pi}{2}(n+1)\right)\frac{\Gamma\left(1+\frac{n}{2}\right)}{\Gamma\left(\frac{n}{2}\right)},$$

$$\left.\frac{dP_n^1(x)}{dx}\right|_{x=0} = \frac{4}{\sqrt{\pi}}\sin\left(\frac{\pi}{2}(n+1)\right)\frac{\Gamma\left(\frac{3}{2}+\frac{n}{2}\right)}{\Gamma\left(\frac{n}{2}\right)}$$

along with the values of the tabulated integral

$$\int_0^1 P_n^1(x)P_s^1(x)\,dx = \delta_n^s\frac{n(n+1)}{2n+1},$$

where $n, s = 2k$ or $n, s = 2k-1$, $k = 1, 2, 3, \ldots$ (Γ stands for the gamma function and δ_n^s is the Kronecker delta).

The spatio–temporal amplitudes $u_n(r,t)$ of the partial components of the pulsed spherical waves $U(r,\vartheta,t)$ (see (4.3)) are determined by the solutions to the following initial boundary value problems.

$$\begin{cases} \left[-\frac{\partial^2}{\partial t^2} + \frac{1}{r}\frac{\partial^2}{\partial r^2}r - \frac{n(n+1)}{r^2}\right]u_n(r,t) = \left[-\frac{\partial^2}{\partial t^2} + \frac{\partial^2}{\partial r^2} - \frac{n(n+1)}{r^2}\right] \\ \times r u_n(r,t) = 0; \quad r \geq L_1, t > 0 \\ u_n(r,0) = \left.\frac{\partial}{\partial t}u_n(r,t)\right|_{t=0} = 0; \quad r \geq L_1 \end{cases},$$

$$n = \begin{cases} 2k, & \mathrm{TE}_0 - \text{waves} \\ 2k-1, & \mathrm{TM}_0 - \text{waves} \end{cases}, \quad k = 1, 2, 3, \ldots.$$

Their behavior is subject to the radiation conditions

$$u_n(r,t) = \left(\frac{L_1}{r}\right)^{n+1} u_n(L_1, t-(r-L_1))$$

$$+\sqrt{\frac{L_1}{r}} \int\limits_0^{t-(r-L_1)} S_{n+1/2}(r, L_1, t-\tau)\frac{\partial u_n(L_1,\tau)}{\partial \tau} d\tau$$

$$= \left[\left(\frac{L_1}{r}\right)^{n+1} + \sqrt{\frac{L_1}{r}} S_{n+1/2}(r, L_1, r-L_1)\right] u_n(L_1, t-(r-L_1))$$

$$-\sqrt{\frac{L_1}{r}} \int\limits_0^{t-(r-L_1)} u_n(L_1,\tau)\frac{\partial S_{n+1/2}(r, L_1, t-\tau)}{\partial \tau} d\tau;$$

$$n = \begin{cases} 2k, & \text{TE}_0 - \text{waves} \\ 2k-1, & \text{TM}_0 - \text{waves} \end{cases}, \quad k = 1, 2, 3, \ldots,$$

$$r > L_1, \quad t \geq 0, \tag{4.5}$$

which are identical in form to the exact nonlocal conditions (3.82) from Section 3.5.2.

By using expansions (4.3), let us go from conditions (4.5) for the amplitudes to the exact conditions for the functions $U(g,t)$, which determine the axially-symmetric fields $\vec{E}(g,t)$ and $\vec{H}(g,t)$ in the radiation zone of the structure:

$$U(g,t) = \sum_n \left\{ \left(\frac{L_1}{r}\right)^{n+1} \int\limits_0^{\pi/2} U\left(L_1, \tilde{\vartheta}, t-(r-L_1)\right) \mu_n\left(\cos\tilde{\vartheta}\right) \sin\tilde{\vartheta} d\tilde{\vartheta} \right.$$

$$+\sqrt{\frac{L_1}{r}} \int\limits_0^{t-(r-L_1)} S_{n+1/2}(r, L_1, t-\tau)$$

$$\left. \times \left[\int\limits_0^{\pi/2} \frac{\partial U\left(L_1, \tilde{\vartheta}, \tau\right)}{\partial \tau} \mu_n\left(\cos\tilde{\vartheta}\right) \sin\tilde{\vartheta} d\tilde{\vartheta} \right] d\tau \right\}$$

$$\times \mu_n(\cos\vartheta); \quad g = \{r, \vartheta\} \in {}_{L_1}\mathbf{Q}, \quad t \geq 0,$$

$$U(g,t) = \begin{cases} E_\phi(g,t) & (\text{TE}_0 - \text{waves}) \quad \text{for} \quad n = 2k \\ H_\phi(g,t) & (\text{TM}_0 - \text{waves}) \quad \text{for} \quad n = 2k-1 \end{cases};$$

$$k = 1, 2, 3, \ldots. \tag{4.6}$$

4.2.3. Exact Absorbing Conditions in Cross-Sections of the Feeding Waveguides

In solving this part of the problem, we are guided by the results obtained in Section 2.3. In the region ${}_{L_2}\mathbf{Q}$ (Fig. 4.1), the functions $U(\rho, z, t)$ satisfy the system of equations

$$\begin{cases} \left[-\dfrac{\partial^2}{\partial t^2} + \dfrac{\partial^2}{\partial z^2} + \dfrac{\partial}{\partial \rho}\left(\dfrac{1}{\rho}\dfrac{\partial}{\partial \rho}\rho \right) \right] U(g,t) = 0; \quad t > 0, \quad g = \{\rho, z\} \in {}_{L_2}\mathbf{Q} \\ U(g,t)\big|_{t=0} = 0, \quad \dfrac{\partial}{\partial t} U(g,t)\Big|_{t=0} = 0; \quad g \in {}_{L_2}\bar{\mathbf{Q}} \\ E_{tg}(g,t)\big|_{g\in\mathbf{S}} = 0; \quad t \geq 0 \end{cases},$$

which is herein equivalent to initial boundary value problems (4.1). Let a circular coaxial waveguide act as a feeding channel (as is shown in Fig. 4.1). Then,

$$\begin{cases} U(z,\rho,t) = \sum\limits_n \tilde{u}_n(z,t)\,\tilde{\mu}_n(\rho); \quad z \leq -L_2, \quad t \geq 0 \\ \tilde{u}_n(z,t) = \int\limits_b^a U(z,\rho,t)\,\tilde{\mu}_n(\rho)\,\rho\,d\rho \end{cases}, \tag{4.7}$$

where (see Section 2.3.1)

$$\begin{cases} \tilde{\mu}_n(\rho) = G_1(\lambda_n,\rho)\sqrt{2}\left[a^2 G_0^2(\lambda_n,a) - b^2 G_0^2(\lambda_n,b)\right]^{-1/2}; \quad n = 1,2,\ldots \\ \lambda_n > 0 \text{ are the roots of eq. } G_1(\lambda,a) = 0 \\ G_q(\lambda,\rho) = J_q(\lambda\rho)\,N_1(\lambda b) - N_q(\lambda\rho)\,J_1(\lambda b) \end{cases};$$
$$b < \rho < a$$

in the case of TE_0-waves and

$$\begin{cases} \tilde{\mu}_n(\rho) = \tilde{G}_1(\lambda_n,\rho)\sqrt{2}\left[a^2 \tilde{G}_1^2(\lambda_n,a) - b^2 \tilde{G}_1^2(\lambda_n,b)\right]^{-1/2}; \quad n = 1,2,\ldots \\ \tilde{\mu}_0(\rho) = \left[\rho\sqrt{\ln(a/b)}\right]^{-1} \\ \lambda_n > 0\,(n = 1,2,\ldots) \text{ are the roots of eq. } \tilde{G}_0(\lambda,b) = 0, \quad \lambda_0 = 0 \\ \tilde{G}_q(\lambda,\rho) = J_q(\lambda\rho)\,N_0(\lambda a) - N_q(\lambda\rho)\,J_0(\lambda a) \end{cases};$$
$$b < \rho < a$$

for TM_0-waves. Evidently, choosing definite waveguide geometry, we reduce the quantity of the formulas duplicated from Section 2.3.

The elements $\tilde{u}_n(z,t)$ of the evolutionary bases for the signals $U(z,\rho,t)$ can be obtained by solving the following initial boundary value problems.

$$\begin{cases} \left[-\dfrac{\partial^2}{\partial t^2} + \dfrac{\partial^2}{\partial z^2} - \lambda_n^2 \right] \tilde{u}_n(z,t) = 0, \quad t > 0 \\ \tilde{u}_n(z,0) = 0, \quad \dfrac{\partial}{\partial t}\tilde{u}_n(z,t)\Big|_{t=0} = 0 \end{cases}; \quad z \leq -L_2,$$
$$n = \begin{cases} 1,2,\ldots; & TE_0 - \text{waves} \\ 0,1,2,\ldots; & TM_0 - \text{waves} \end{cases} \tag{4.8}$$

(it is assumed that the excitation $U(g,t)$, being generated by the sources $\varphi(g)$, $\psi(g)$, and $F(g,t)$ located in $\mathbf{Q}_L$, has not yet reached the boundary $z = -L_2$ of the region ${}_{L_2}\mathbf{Q}$ by the instant of time $t = 0$).

From (2.33) (the problems similar to (4.8)), three types of exact nonlocal (formulas (2.34) to (2.36)) and local (formulas (2.37) to (2.39)) absorbing conditions have been derived. Below we dwell, for example, on conditions (2.35) and (2.38). Taking into account the change both in the direction of free propagation of nonsinusoidal waves (toward $z \to -\infty$ instead of $z \to +\infty$) and in the position of the artificial boundary $\mathbf{L}_2$ ($z = -L_2$ instead of $z = 0$), we have

$$\left[\frac{\partial}{\partial t} - \frac{\partial}{\partial z}\right] U(\rho, z, t)\Big|_{z=-L_2} = -\sum_n \left\{ \int_0^t J_1[\lambda_n (t-\tau)] \left[\int_b^a \frac{\partial U(\tilde{\rho}, z, \tau)}{\partial z}\Big|_{z=-L_2} \tilde{\mu}_n(\tilde{\rho})\tilde{\rho}d\tilde{\rho} \right] d\tau \right\} \times \lambda_n \tilde{\mu}_n(\rho); \quad b \le \rho \le a, \quad t \ge 0 \tag{4.9}$$

and

$$\left[\frac{\partial}{\partial t} - \frac{\partial}{\partial z}\right] U(\rho, z, t)\Big|_{z=-L_2} = \frac{2}{\pi}\int_0^{\pi/2} W(\rho, t, \varphi)\cos^2\varphi d\varphi; \quad t \ge 0, \quad b \le \rho \le a,$$

$$\begin{cases} \left[\dfrac{\partial^2}{\partial t^2} - \cos^2(\varphi)\dfrac{\partial}{\partial \rho}\dfrac{1}{\rho}\dfrac{\partial}{\partial \rho}\rho\right] W(\rho, t, \varphi) = \dfrac{\partial}{\partial \rho}\dfrac{1}{\rho}\dfrac{\partial}{\partial \rho}\rho\left[\dfrac{\partial}{\partial z} U(\rho, z, t)\Big|_{z=-L_2}\right]; \\ b < \rho < a, \quad t > 0 \\ W(\rho, 0, \varphi) = \dfrac{\partial W(\rho, t, \varphi)}{\partial t}\Big|_{t=0} = 0; \quad b \le \rho \le a \end{cases}. \tag{4.10}$$

Here, $n = 1, 2, \ldots$, for TE_0-waves, and the inner problems with respect to the auxiliary functions $W(\rho, t, \varphi)$ in this case should be supplemented with the boundary conditions $W(b, t, \varphi) = W(a, t, \varphi) = 0$. In the case of TM_0-waves, $n = 0, 1, 2, \ldots$, and

$$\frac{\partial(\rho W(\rho, t, \varphi))}{\partial \rho}\Big|_{\rho=b} = \frac{\partial(\rho W(\rho, t, \varphi))}{\partial \rho}\Big|_{\rho=a} = 0.$$

Statement 4.1. *Open problems (4.1) with the analysis domain* $\mathbf{Q}$ *are equivalent to closed problems (4.1) with the analysis domain* $\mathbf{Q}_L$ *and conditions (4.6), (4.9) (or (4.6), (4.10)) on its outer boundary* $\mathbf{L} = \mathbf{L}_1 \cup \mathbf{L}_2$.

4.3. Wave Radiation from a Plane-Parallel Waveguide of Arbitrary Aperture

4.3.1. Truncation of the Analysis Domain Down to a Half-Plane and a Band

The original initial boundary value problems for plane pulse antennas (see an example of the geometry of such PFS in Fig. 4.2) can be written in the rectangular

Cartesian coordinates as

$$\begin{cases} \left[-\varepsilon(g) \dfrac{\partial^2}{\partial t^2} - \sigma(g) \dfrac{\partial}{\partial t} + \dfrac{\partial^2}{\partial z^2} + \dfrac{\partial^2}{\partial y^2} \right] U(g,t) = F(g,t); \\ g = \{y, z\} \in \mathbf{Q}, \quad t > 0 \\ U(g,t)\big|_{t=0} = \varphi(g), \quad \dfrac{\partial}{\partial t} U(g,t)\Big|_{t=0} = \psi(g); \quad g \in \overline{\mathbf{Q}} \\ E_{tg}(g,t)\big|_{g\in\mathbf{S}} = 0; \quad t \geq 0 \end{cases} . \tag{4.11}$$

The functions $F(g,t), \varphi(g), \psi(g), \sigma(g)$, and $\varepsilon(g) - 1$ are finite in $\mathbf{Q} = \mathbf{R}^2 \backslash \overline{\mathbf{intS}}$. Their supports belong to the closure of the region

$$\mathbf{Q}_L = \left\{ g \in \mathbf{Q} : L_4 < y < L_3; \begin{Bmatrix} L_{2,l}, & y < -b_l \\ -\tilde{L}, & |y| < a/2 \\ L_{2,r}, & y > b_r \end{Bmatrix} < z < L_1 \right\}$$

over all observation times $0 \leq t \leq T$.

Above (below, to the right, to the left) of the artificial boundary $\mathbf{L} \cup \tilde{\mathbf{L}}$ there are neither sources nor scatterers. The functions $U(g,t)$ ($U(g,t) = E_x(g,t)$ in the case of E-polarization, and in the H-case $U(g,t) = H_x(g,t)$ and $\sigma(g)$, $\varepsilon(g)$ are the piecewise constant functions) are associated with the outgoing pulsed waves intersecting the corresponding boundary in one direction only and satisfy homogeneous problems (4.11) with $\varepsilon(g) - 1 = \sigma(g) \equiv 0$.

Let us subject these functions to the Fourier transform $\tilde{f}(\omega) = F[f](z) \leftrightarrow f(z) = F^{-1}[\tilde{f}](z)$ (image $\leftrightarrow$ original) and to the modified cosine-Fourier transforms in the case of E-polarization or the modified sine-Fourier transforms in the H-case:

$$u_z(y,\mu,t) = \frac{1}{2\pi} \int_{-\infty}^{\infty} U(y,z,t) e^{i\mu z} dz \leftrightarrow U(y,z,t) = \int_{-\infty}^{\infty} u_z(y,\mu,t) e^{-i\mu z} d\mu;$$

$$\begin{Bmatrix} y \geq L_3 \\ y \leq L_4 \end{Bmatrix},$$

$$u_y(\lambda,z,t) = \frac{1}{2\pi} \int_{-\infty}^{\infty} U(y,z,t) e^{i\lambda y} dy \leftrightarrow U(y,z,t) = \int_{-\infty}^{\infty} u_y(\lambda,z,t) e^{-i\lambda y} d\lambda;$$

$$z \geq L_1$$

and

$$u_{y,r}(\lambda_r,z,t) = \sqrt{\frac{2}{\pi}} \int_{b_r}^{\infty} U(y,z,t) \begin{Bmatrix} \cos[\lambda_r(y - b_r)] \\ \sin[\lambda_r(y - b_r)] \end{Bmatrix} dy$$

$$\leftrightarrow U(y,z,t) = \sqrt{\frac{2}{\pi}} \int_0^{\infty} u_{y,r}(\lambda_r,z,t) \begin{Bmatrix} \cos[\lambda_r(y - b_r)] \\ \sin[\lambda_r(y - b_r)] \end{Bmatrix} d\lambda_r;$$

$$z \leq L_{2,r},$$

$$u_{y,l}(\lambda_l, z, t) = \mp\sqrt{\frac{2}{\pi}} \int\limits_{-b_l}^{-\infty} U(y, z, t) \left\{ \begin{matrix} \cos[\lambda_l(y + b_l)] \\ \sin[\lambda_l(y + b_l)] \end{matrix} \right\} dy$$

$$\leftrightarrow U(y, z, t) = \pm\sqrt{\frac{2}{\pi}} \int\limits_{0}^{\infty} u_{y,l}(\lambda_l, z, t) \left\{ \begin{matrix} \cos[\lambda_l(y + b_l)] \\ \sin[\lambda_l(y + b_l)] \end{matrix} \right\} d\lambda_l;$$

$$z \leq L_{2,l}.$$

Then, for the images $u_z(y, \mu, t)$, $u_y(\lambda, z, t)$, $u_{y,r}(\lambda_r, z, t)$, and $u_{y,l}(\lambda_l, z, t)$, by applying the familiar properties

$$\left\{ \begin{matrix} F \\ F_c \\ F_s \end{matrix} \right\} \left[\frac{d^2 f(z)}{dz^2} \right](\omega) = -\omega^2 \tilde{f}(\omega) - \sqrt{\frac{2}{\pi}} \left\{ \begin{matrix} 0 \\ df(z)/dz|_{z=0} \\ \omega f(0) \end{matrix} \right\}$$

of the Fourier transform $\tilde{f}(\omega) = F[f](z)$, the standard cosine- and sine-transforms $\tilde{f}(\omega) = F_s[f](z)$ and $\tilde{f}(\omega) = F_c[f](z)$ (see, e.g., formula (2.6)) and allowing for the boundary condition $E_{tg}(g, t)\big|_{g \in \mathbf{S}} = 0$, we have from (4.11):

$$\begin{cases} \left[-\frac{\partial^2}{\partial t^2} + \frac{\partial^2}{\partial y^2} - \mu^2 \right] u_z(y, \mu, t) = 0; \quad t > 0 \\ u_z(y, \mu, 0) = \left. \frac{\partial u_z(y, \mu, t)}{\partial t} \right|_{t=0} = 0 \end{cases}; \quad \begin{cases} y \geq L_3 \\ y \leq L_4 \end{cases}$$

and

$$\begin{cases} \left[-\frac{\partial^2}{\partial t^2} + \frac{\partial^2}{\partial z^2} - \tilde{\lambda}^2 \right] \tilde{u}_y\left(\tilde{\lambda}, z, t\right) = 0; \\ t > 0 \\ \tilde{u}_y\left(\tilde{\lambda}, z, 0\right) = \left. \frac{\partial \tilde{u}_y\left(\tilde{\lambda}, z, t\right)}{\partial t} \right|_{t=0} = 0 \end{cases};$$

$$\tilde{u}_y\left(\tilde{\lambda}, z, t\right) = \begin{cases} u_y(\lambda, z, t); & z \geq L_1 \\ u_{y,r}(\lambda_r, z, t); & z \leq L_{2,r} \\ u_{y,l}(\lambda_l, z, t); & z \leq L_{2,l} \end{cases}.$$

Problems of this type have been considered repeatedly in previous chapters (see, e.g., Sections 2.2.1 and 3.3.1). Their solutions are subject to the following conditions.

$$\left[\frac{\partial}{\partial t} \pm \frac{\partial}{\partial y} \right] u_z(y, \mu, t) = \pm\mu \int\limits_{0}^{t} J_1(\mu(t - \tau)) \frac{\partial u_z(y, \mu, \tau)}{\partial y} d\tau; \quad \begin{cases} y \geq L_3 \\ y \leq L_4 \end{cases}, \tag{4.12}$$

$$\left[\frac{\partial}{\partial t} + \frac{\partial}{\partial z} \right] u_y(\lambda, z, t) = \lambda \int\limits_{0}^{t} J_1(\lambda(t - \tau)) \frac{\partial u_y(\lambda, z, \tau)}{\partial z} d\tau; \quad z \geq L_1, \tag{4.13}$$

$$\left[\frac{\partial}{\partial t}-\frac{\partial}{\partial z}\right]\tilde{u}_y\left(\tilde{\lambda},z,t\right)=-\tilde{\lambda}\int_0^t J_1\left(\tilde{\lambda}(t-\tau)\right)\frac{\partial \tilde{u}_y\left(\tilde{\lambda},z,\tau\right)}{\partial z}d\tau;$$

$$\tilde{u}_y\left(\tilde{\lambda},z,t\right)=\begin{cases}u_{y,r}\left(\lambda_r,z,t\right), & z\leq L_{2,r}\\ u_{y,l}\left(\lambda_l,z,t\right), & z\leq L_{2,l}\end{cases}. \tag{4.14}$$

The conditions

$$\left[\frac{\partial}{\partial t}\pm\frac{\partial}{\partial y}\right]u_z\left(y,\mu,t\right)=-\mu\int_0^t\frac{J_1\left(\mu(t-\tau)\right)}{t-\tau}u_z\left(y,\mu,\tau\right)d\tau;\quad \begin{cases}y\geq L_3\\ y\leq L_4\end{cases}, \tag{4.15}$$

$$\left[\frac{\partial}{\partial t}+\frac{\partial}{\partial z}\right]u_y\left(\lambda,z,t\right)=-\lambda\int_0^t\frac{J_1\left(\lambda(t-\tau)\right)}{t-\tau}u_y\left(\lambda,z,\tau\right)d\tau;\quad z\geq L_1, \tag{4.16}$$

$$\left[\frac{\partial}{\partial t}-\frac{\partial}{\partial z}\right]\tilde{u}_y\left(\tilde{\lambda},z,t\right)=-\tilde{\lambda}\int_0^t\frac{J_1\left(\tilde{\lambda}(t-\tau)\right)}{t-\tau}\tilde{u}_y\left(\tilde{\lambda},z,\tau\right)d\tau;$$

$$\tilde{u}_y\left(\tilde{\lambda},z,t\right)=\begin{cases}u_{y,r}\left(\lambda_r,z,t\right), & z\leq L_{2,r}\\ u_{y,l}\left(\lambda_l,z,t\right), & z\leq L_{2,l}\end{cases} \tag{4.17}$$

are derived from (4.12) to (4.14) as a result of simple operations (see Section 2.2.2): the Laplace transform in t $\rightarrow$ the solution of the operator equations with respect to the derivatives of the images of the functions u_z, u_y, and $\tilde{u}_y$ in spatial variables $\rightarrow$ the inverse Laplace transform.

Let us go from nonlocal conditions (4.15) to (4.17) to the local ones, which truncate the analysis domain $\mathbf{Q}=\mathbf{R}^2\backslash\overline{\text{int}\mathbf{S}}$ down to a vertical or horizontal band in the plane $\mathbf{R}^2$ of the variables y and z, by repeating in detail all steps of the approach from Section 3.3. As a result, we obtain

$$\left[\frac{\partial}{\partial t}+\frac{\partial}{\partial z}\right]U(g,t)=\frac{2}{\pi}\int_0^{\pi/2}\frac{\partial V_1(g,t,\varphi)}{\partial t}\sin^2(\varphi)\,d\varphi;\quad t\geq 0,$$

$$\begin{cases}\left[\dfrac{\partial^2}{\partial t^2}-\cos^2(\varphi)\dfrac{\partial^2}{\partial y^2}\right]V_1(g,t,\varphi)=\dfrac{\partial^2}{\partial y^2}U(g,t); & t>0\\ \left.\dfrac{\partial V_1(g,t,\varphi)}{\partial t}\right|_{t=0}=\left.V_1(g,t,\varphi)\right|_{t=0}=0 & \end{cases};$$

$$|y|<\infty,\quad z\geq L_1, \tag{4.18}$$

$$\left[\frac{\partial}{\partial t}-\frac{\partial}{\partial z}\right]U(g,t)=\frac{2}{\pi}\int_0^{\pi/2}\frac{\partial \tilde{V}_1(g,t,\varphi)}{\partial t}\sin^2(\varphi)\,d\varphi;\quad t\geq 0,$$

$$\begin{cases} \left[\dfrac{\partial^2}{\partial t^2} - \cos^2(\varphi)\dfrac{\partial^2}{\partial y^2}\right]\tilde{V}_1(g,t,\varphi) = \dfrac{\partial^2}{\partial y^2}U(g,t); \quad t>0 \\ \left.\dfrac{\partial \tilde{V}_1(g,t,\varphi)}{\partial t}\right|_{t=0} = \left.\tilde{V}_1(g,t,\varphi)\right|_{t=0} = 0 \end{cases},$$

$$\tilde{V}_1(g,t,\varphi) = \begin{cases} V_{1,r}(g,t,\varphi); & y \geq b_r, \quad z \leq L_{2,r} \\ V_{1,l}(g,t,\varphi); & y \leq -b_l, \quad z \leq L_{2,l} \end{cases}, \tag{4.19}$$

and

$$\left[\frac{\partial}{\partial t} \pm \frac{\partial}{\partial y}\right]U(g,t) = \frac{2}{\pi}\int_0^{\pi/2} \frac{\partial V_2(g,t,\varphi)}{\partial t}\sin^2(\varphi)\,d\varphi; \quad t \geq 0,$$

$$\begin{cases} \left[\dfrac{\partial^2}{\partial t^2} - \cos^2(\varphi)\dfrac{\partial^2}{\partial z^2}\right]V_2(g,t,\varphi) = \dfrac{\partial^2}{\partial z^2}U(g,t); \quad t>0 \\ \left.\dfrac{\partial V_2(g,t,\varphi)}{\partial t}\right|_{t=0} = \left.V_2(g,t,\varphi)\right|_{t=0} = 0 \end{cases},$$

$$\begin{cases} y \geq L_3 \\ y \leq L_4 \end{cases}, \quad |z| < \infty. \tag{4.20}$$

The inner problems in (4.19) with respect to the auxiliary functions $\tilde{V}_1(y,z,t,\varphi)$ must be supplemented with the boundary conditions (for $t \geq 0$)

$$\begin{cases} \tilde{V}_1\left(\tilde{b},z,t,\varphi\right) = 0; & E-\text{polarization} \\ \left.\dfrac{\partial \tilde{V}_1(y,z,t,\varphi)}{\partial y}\right|_{y=\tilde{b}} = 0; & H-\text{polarization} \end{cases};$$

$$\tilde{V}_1(y,z,t,\varphi) = \begin{cases} V_{1,r}(y,z,t,\varphi); & y \geq \tilde{b} = b_r, \quad z \leq L_{2,r} \\ V_{1,l}(y,z,t,\varphi); & y \leq \tilde{b} = -b_l, \quad z \leq L_{2,l} \end{cases}.$$

The Cauchy problems in (4.18) and (4.20) with respect to the functions $V_1(g,t,\varphi)$ and $V_2(g,t,\varphi)$ are well posed.

4.3.2. The Problem of the Corner Points and the Exact Absorbing Conditions on a Rectangular Coordinate Boundary

When truncating the analysis domain down to a rectangular one in $\mathbf{R}^2$, all the inner problems in (4.18) to (4.20) should be supplemented with boundary conditions at points $\{y = L_3, z = L_1\}$, $\{L_3, L_{2,r}\}$, $\{L_4, L_{2,l}\}$, and $\{L_4, L_1\}$, where the rectilinear artificial boundaries intersect. This problem, which is called the problem of the corner points, is now solved according the algorithm from Section 3.3.2. The principal steps are as follows: the separation of the quadrants $\mathbf{G}$ in the $\mathbf{R}^2$-plane, where two of four conditions (4.18) to (4.20) are valid simultaneously → the subjection of these conditions to global transformations (3.22) → the solution of the systems of operator equations with respect to the amplitudes (images) of the functions $W_j(g,t,\varphi) = V_j(g,t,\varphi)\cos^2(\varphi) + U(g,t)$, $j = 1, 2$, and $\tilde{W}_1(g,t,\varphi) = \tilde{V}_1(g,t,\varphi)\cos^2(\varphi) + U(g,t)$ ($W_{1,r}(g,t,\varphi) = V_{1,r}(g,t,\varphi)\cos^2(\varphi) + U(g,t)$ and $W_{1,l}(g,t,\varphi) = V_{1,l}(g,t,\varphi)\cos^2(\varphi) + U(g,t)$) → application of inverse transformations (3.22) and determination of analytical relations between the functions $W(g,t,\varphi)$ (W_j, $W_{1,r}$, and $W_{1,l}$) for the points $g \in \overline{\mathbf{G}}$.

As a result, we have (see also formulas (3.28) to (3.30)):

$$\left[\frac{\partial}{\partial t} + \frac{\partial}{\partial z}\right] U(g,t) = \frac{2}{\pi}\int_0^{\pi/2} \frac{\partial V_1(g,t,\varphi)}{\partial t}\sin^2(\varphi)\,d\varphi; \quad t \geq 0,$$

$$\left\{\begin{array}{l} \left[\dfrac{\partial^2 V_1(g,t,\varphi)}{\partial t^2} - \dfrac{\partial^2 W_1(g,t,\varphi)}{\partial y^2}\right] = 0; \quad t > 0 \\ \left.\dfrac{\partial V_1(g,t,\varphi)}{\partial t}\right|_{t=0} = V_1(g,t,\varphi)|_{t=0} = 0 \end{array}\right., \quad L_4 \leq y \leq L_3, \quad z = L_1, \tag{4.21}$$

$$\left[\frac{\partial}{\partial t} - \frac{\partial}{\partial z}\right] U(g,t) = \frac{2}{\pi}\int_0^{\pi/2} \frac{\partial \tilde{V}_1(g,t,\varphi)}{\partial t}\sin^2(\varphi)\,d\varphi; \quad t \geq 0,$$

$$\left\{\begin{array}{l} \left[\dfrac{\partial^2 \tilde{V}_1(g,t,\varphi)}{\partial t^2} - \dfrac{\partial^2 \tilde{W}_1(g,t,\varphi)}{\partial y^2}\right] = 0; \quad t > 0 \\ \tilde{V}_1\left(\tilde{b},z,t,\varphi\right) = 0 \quad (E - \text{polarization}) \quad \text{or} \quad \left.\dfrac{\partial \tilde{V}_1(y,z,t,\varphi)}{\partial y}\right|_{y=\tilde{b}} = 0 \; ; \\ (H - \text{polarization}); \quad t \geq 0 \\ \left.\dfrac{\partial \tilde{V}_1(g,t,\varphi)}{\partial t}\right|_{t=0} = \tilde{V}_1(g,t,\varphi)\Big|_{t=0} = 0 \end{array}\right.$$

$$\tilde{V}_1(y,z,t,\varphi) = \left\{\begin{array}{lll} V_{1,r}(y,z,t,\varphi), & \tilde{b} = b_r \leq y \leq L_3, & z = L_{2,r} \\ V_{1,l}(y,z,t,\varphi), & L_4 \leq y \leq \tilde{b} = -b_l, & z = L_{2,l} \end{array}\right., \tag{4.22}$$

$$\left[\frac{\partial}{\partial t} \pm \frac{\partial}{\partial y}\right] U(g,t) = \frac{2}{\pi}\int_0^{\pi/2} \frac{\partial V_2(g,t,\varphi)}{\partial t}\sin^2(\varphi)\,d\varphi; \quad t \geq 0,$$

$$\left\{\begin{array}{l} \left[\dfrac{\partial^2 V_2(g,t,\varphi)}{\partial t^2} - \dfrac{\partial^2 W_2(g,t,\varphi)}{\partial z^2}\right] = 0; \quad t > 0 \\ \left.\dfrac{\partial V_2(g,t,\varphi)}{\partial t}\right|_{t=0} = V_2(g,t,\varphi)|_{t=0} = 0 \end{array}\right.,$$

$$\left\{\begin{array}{ll} y = L_3, & L_{2,r} \leq z \leq L_1 \\ y = L_4, & L_{2,l} \leq z \leq L_1 \end{array}\right., \tag{4.23}$$

$$\begin{cases} \left[\dfrac{\partial}{\partial t} \pm \cos\varphi \dfrac{\partial}{\partial y}\right] Z_1(g,t,\varphi) \\ \quad = \dfrac{2\cos\varphi}{\pi} \displaystyle\int_0^{\pi/2} \dfrac{\sin^2\gamma}{\cos^2\varphi + \sin^2\varphi\cos^2\gamma} \dfrac{\partial W_2(g,t,\gamma)}{\partial t} d\gamma \\ \left[\dfrac{\partial}{\partial t} \pm \cos\varphi \dfrac{\partial}{\partial z}\right] W_2(g,t,\varphi) \\ \quad = \dfrac{2\cos\varphi}{\pi} \displaystyle\int_0^{\pi/2} \dfrac{\sin^2\gamma}{\cos^2\varphi + \sin^2\varphi\cos^2\gamma} \dfrac{\partial Z_1(g,t,\gamma)}{\partial t} d\gamma \\ t \geq 0 \end{cases};,$$

$$\begin{Bmatrix} + \\ + \end{Bmatrix} \to g = \{L_3, L_1\} \quad \text{and} \quad Z_1 = W_1$$

$$\begin{Bmatrix} + \\ - \end{Bmatrix} \to g = \left\{L_3, L_{2,r}\right\} \quad \text{and} \quad Z_1 = W_{1,r} \quad .$$

$$\begin{Bmatrix} - \\ + \end{Bmatrix} \to g = \{L_4, L_1\} \quad \text{and} \quad Z_1 = W_1$$

$$\begin{Bmatrix} - \\ - \end{Bmatrix} \to g = \left\{L_4, L_{2,l}\right\} \quad \text{and} \quad Z_1 = W_{1,l} \qquad (4.24)$$

The exact local absorbing conditions for the entire artificial coordinate boundary **L** are determined exclusively by four formulas (4.21) to (4.24) taken together. Equations (4.24) in these ABCs act as boundary conditions in the inner initial boundary value problems of (4.21) to (4.23). Symbols such as

$$\begin{Bmatrix} + \\ + \end{Bmatrix} \to g = \{L_3, L_1\}$$

and $Z_1 = W_1$ specify here the rules for the choice of signs in the upper and bottom equations (4.24) and for the values of $Z_1(g,t,\varphi)$ functions for different points $g = \{y, z\}$.

4.3.3. Conditions for the Artificial Boundary $\tilde{\mathbf{L}}$ in the Cross-Section of a Plane-Parallel Waveguide

$\tilde{\mathbf{L}} = \{g : |y| \leq a/2; z = -\tilde{L}\}$ is a section of the entire artificial boundary of the truncated analysis domain $\mathbf{Q}_L$ in initial boundary value problems (4.11). Let us choose the local or nonlocal absorbing conditions for this section from those constructed in Section 2.2. For definiteness' sake, we dwell on the related ABCs (2.21) and (2.28). With allowance made for the change both in the direction of free propagation of transient waves (toward $z \to -\infty$ instead of $z \to +\infty$) and in the position of the artificial boundary $\tilde{\mathbf{L}}$ ($z = -\tilde{L}$ instead of $z = 0$), these conditions can be written as

$$\left[\frac{\partial}{\partial t} - \frac{\partial}{\partial z}\right] U(y,z,t)\Big|_{z=-\tilde{L}} = -\sum_{n}\left\{\int_{0}^{t} J_1[\lambda_n(t-\tau)](t-\tau)^{-1}\left[\int_{-a/2}^{a/2} U\left(\tilde{y},-\tilde{L},\tau\right)\mu_n(\tilde{y})\,d\tilde{y}\right]d\tau\right\} \times \lambda_n \mu_n(y), \quad |y| \le a/2, \quad t \ge 0, \tag{4.25}$$

$$\left[\frac{\partial}{\partial t} - \frac{\partial}{\partial z}\right] U(y,z,t)\Big|_{z=-\tilde{L}} = \frac{2}{\pi}\int_{0}^{\pi/2} \frac{\partial W(y,t,\varphi)}{\partial t}\sin^2(\varphi)\,d\varphi;$$

$$t \ge 0, \quad |y| \le a/2,$$

$$\begin{cases} \left[\dfrac{\partial^2}{\partial t^2} - \cos^2\varphi\dfrac{\partial^2}{\partial y^2}\right] W(y,t,\varphi) = \dfrac{\partial^2 U\left(y,-\tilde{L},t\right)}{\partial y^2}; \quad |y| < a/2, \quad t > 0 \\ W(y,0,\varphi) = \dfrac{\partial W(y,t,\varphi)}{\partial t}\Big|_{t=0} = 0; \quad |y| \le a/2 \\ W\left(\pm a/2, t, \varphi\right) = 0 \quad (E - \text{case}) \quad \text{or} \\ \partial W(y,t,\varphi)/\partial y\big|_{y=\pm a/2} = 0 \quad (H - \text{case}); \quad t \ge 0 \end{cases} . \tag{4.26}$$

Here, as before (see Section 2.2.1), $\lambda_n = n\pi/a$, whereas $\mu_n(y) = \sqrt{2/a} \times \sin[n\pi(y + a/2)/a]$, $\{n\} = 1, 2, \ldots$, in the case of the E-polarized field and $\mu_n(y) = \sqrt{(2 - \delta_0^n)/a} \times \cos[n\pi(y + a/2)/a]$, $\{n\} = 0, 1, 2, \ldots$, in the H-case.

Statement 4.2. *Problems (4.11) with the analysis domain* $\mathbf{Q}$ *and problems (4.11) with the analysis domain* $\mathbf{Q}_L$ *and conditions (4.21) to (4.24), (4.25) (conditions (4.21) to (4.24), (4.26)) on its outer boundary* $\mathbf{L} \cup \tilde{\mathbf{L}}$ *are equivalent.*

The inner initial boundary value problems in (4.21) to (4.24) with respect to the auxiliary functions W *and* V *are well posed.*

4.4. The Problems of Strong and Remote Field Sources

Above, in the course of the formulation of the initial boundary value problems and the determination of the domains $\mathbf{Q}$ and $\mathbf{Q}_L$ we have assumed that functions of sources, which are exciting the waveguide junctions (Sections 2.2 to 2.4), compact discontinuities in open space (Chapter 3), or pattern-forming structures (Sections 4.2 and 4.3), are finite in the closure of complete analysis domains $\mathbf{Q}$, and their supports belong to $\bar{\mathbf{Q}}_L \backslash \mathbf{L}$ for all times considered; that is, $0 \le t \le T$. This assumption has allowed us to formulate conditions on virtual boundaries $\mathbf{L}$ in terms of total field $U(g,t)$. The limitations connected with these assumptions may be partially or completely removed if we assume that a certain part of the current and (or) momentary sources is located in the domain ${}_L\mathbf{Q}$. It is only necessary to annihilate the input of the incoming (primary) wave $U^i(g,t)$, generated by these sources, in the field $U(g,t)$ in $\mathbf{L}$, by introducing $U^s(g,t) = U(g,t) - U^i(g,t)$, describing a scattered (secondary) field, in the way, for example, it has been done in the Section 2.5.1 for scalar problems for reflective gratings. The final equations for the modified problem can be formulated either in terms of the total field $U(g,t)$, or in terms of the secondary field $U^s(g,t)$. The first way is preferable, as a formally correct separation of the field $U(g,t)$ into the components $U^s(g,t)$ and $U^i(g,t)$ may be physically misleading with respect to partial subdomains of the domain $\mathbf{Q}$.

4.4.1. *Waveguide Open Resonators: 2-D Scalar Problems*

Consider the problems (2.2) for waveguide junctions of constant cross-section in any plane $x = \text{const}$. An example of a simple junction of this type is depicted in Figure 4.3 (open waveguide resonator). Let us rewrite the problems (2.2) in the form

$$\begin{cases} \left[-\varepsilon(g)\dfrac{\partial^2}{\partial t^2} - \sigma(g)\dfrac{\partial}{\partial t} + \dfrac{\partial^2}{\partial z^2} + \dfrac{\partial^2}{\partial y^2}\right] U(g,t) = F(g,t) + \tilde{F}(g,t); \\ t > 0, \quad g = \{y,z\} \in \mathbf{Q} \\ U(g,t)\big|_{t=0} = \varphi(g) + \tilde{\varphi}(g), \quad \dfrac{\partial}{\partial t} U(g,t)\Big|_{t=0} = \psi(g) + \tilde{\psi}(g); \quad g \in \bar{\mathbf{Q}} \\ E_{tg}(g,t)\big|_{g\in\mathbf{S}} = 0; \quad t \geq 0 \end{cases} \tag{4.27}$$

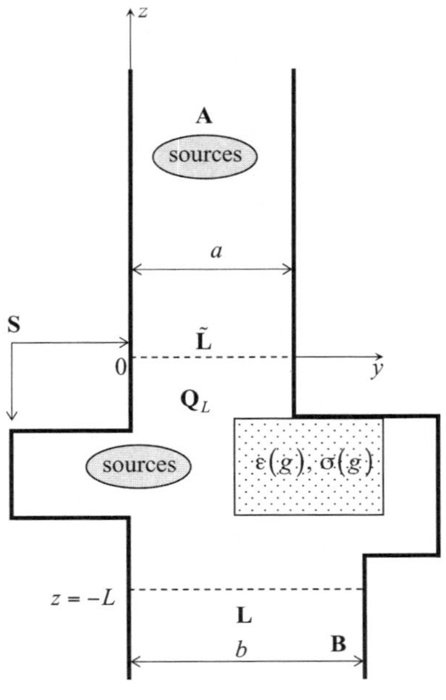

FIGURE 4.3. Waveguide open resonator.

When $U(g,t) = E_x$ we have $E_{tg}(g,t)|_{g\in\mathbf{S}} = U(g,t)|_{g\in\mathbf{S}}$, and problems (4.27) correspond to the case of an E-polarized field. For $U(g,t) = H_x$ the values of $E_{tg}(g,t)|_{g\in\mathbf{S}}$ are defined by derivatives $\partial U(g,t)/\partial\vec{n}|_{g\in\mathbf{S}}$, $\vec{n}$ is the normal vector to the contour $\mathbf{S}$, and for piece-wise constant $\varepsilon(g)$, $\sigma(g)$ the solutions to the problems (4.27) determine H-polarized fields. The analysis domain $\mathbf{Q}$ which is part of the plane $\mathbf{R}^2$, bounded with contour $\mathbf{S}$. $\mathbf{S} \times [|x| \le \infty]$ is the surface of perfect contactors. It is assumed that the finite function supports $\varphi(g)$, $\psi(g)$, $\sigma(g)$, and $\varepsilon(g) - 1$ belong to the set $\overline{\mathbf{Q}_L} \backslash \mathbf{L} \cup \tilde{\mathbf{L}}$. The virtual boundaries $\tilde{\mathbf{L}}$ and $\mathbf{L}$ of the domain $\mathbf{Q}_L$ coincide with the cross-section of regular semi-infinite waveguides $\mathbf{A}$ and $\mathbf{B}$. Sources, corresponding to functions $\tilde{F}(g,t)$, $\tilde{\varphi}(g)$, and $\tilde{\psi}(g)$, are situated beyond the boundary $\tilde{\mathbf{L}}$ in waveguide (in the domain) $\mathbf{A}$. In waveguide $\mathbf{B}$ along which the field formed by scattering from the waveguide junction can propagate infinitely far away, there are no sources and efficient scatterers.

Represent the total field $U(g,t)$ in waveguide $\mathbf{A}$ in the form of $U(g,t) = U^i(g,t) + U^s(g,t)$, where $U^i(g,t)$ is the field, produced by sources $\tilde{F}(g,t)$, $\tilde{\varphi}(g)$, and $\tilde{\psi}(g)$. Then, for determining $U^s(g,t)$ in $\mathbf{A}$ and function $U(g,t)$ in $\mathbf{B}$ we obtain the following homogeneous initial boundary value problems.

$$\left\{ \begin{array}{l} \left[-\dfrac{\partial^2}{\partial t^2} + \dfrac{\partial^2}{\partial z^2} + \dfrac{\partial^2}{\partial y^2} \right] \left\{ \begin{array}{l} U^s(g,t) \\ U(g,t) \end{array} \right\} = 0; \quad t > 0, \quad g = \{y,z\} \in \left\{ \begin{array}{l} \mathbf{A} \\ \mathbf{B} \end{array} \right\} \\ \left\{ \begin{array}{l} U^s(g,t) \\ U(g,t) \end{array} \right\} \Bigg|_{t=0} = 0, \quad \dfrac{\partial}{\partial t} \left\{ \begin{array}{l} U^s(g,t) \\ U(g,t) \end{array} \right\} \Bigg|_{t=0} = 0; \quad g \in \left\{ \begin{array}{l} \bar{\mathbf{A}} \\ \bar{\mathbf{B}} \end{array} \right\} \\ \left\{ \begin{array}{l} E^s_{tg}(g,t) \\ E_{tg}(g,t) \end{array} \right\} \Bigg|_{g\in\mathbf{S}} = 0; \quad t \ge 0 \end{array} \right. \tag{4.28}$$

(it is assumed that at the time $t = 0$ the perturbation, produced by sources, concentrated in $\mathbf{Q}_L$, does not reach boundaries $\tilde{\mathbf{L}}$ and $\mathbf{L}$ of domains $\mathbf{A}$ and $\mathbf{B}$). Here $E^s_{tg}(g,t)$, $g \in \mathbf{S}$ is the tangential component of electric field density, corresponding to the function $U^s(g,t)$. Solutions to the problems (4.28)—function $U(g,t)$ in the domain $\mathbf{B}$ and function $U^s(g,t)$ in the domain $\mathbf{A}$—determine outgoing waves into the directions $z \to -\infty$ and $z \to +\infty$. That is why we can consider the fact that

$$\begin{aligned} &U(y,0,t) - U^i(y,0,t) \\ &= -\sum_n \left\{ \int_0^t J_0\left[\lambda_n(t-\tau)\right] \left[\int_0^a \frac{\partial\left(U(\tilde{y},z,\tau) - U^i(\tilde{y},z,\tau)\right)}{\partial z} \Bigg|_{z=0} \mu_n(\tilde{y})\, d\tilde{y} \right] d\tau \right\} \\ &\quad \times \mu_n(y); \quad 0 \le y \le a, \quad t \ge 0, \end{aligned} \tag{4.29}$$

$$\begin{aligned} U(y,-L,t) = \sum_n \left\{ \int_0^t J_0\left[\tilde{\lambda}_n(t-\tau)\right] \left[\int_0^b \frac{\partial U(\tilde{y},z,\tau)}{\partial z} \Bigg|_{z=-L} \tilde{\mu}_n(\tilde{y})\, d\tilde{y} \right] d\tau \right\} \\ \times \tilde{\mu}_n(y); \quad 0 \le y \le b, \quad t \ge 0 \end{aligned} \tag{4.30}$$

and

$$U(y,0,t) - U^i(y,0,t) = \frac{2}{\pi}\int\limits_0^{\pi/2} \frac{\partial W(y,t,\varphi)}{\partial t} d\varphi; \quad t \geq 0, \quad 0 \leq y \leq a,$$

$$\begin{cases} \left[\dfrac{\partial^2}{\partial t^2} - \sin^2(\varphi)\dfrac{\partial^2}{\partial y^2}\right] W(y,t,\varphi) = -\left.\dfrac{\partial\left(U(y,z,t) - U^i(y,z,t)\right)}{\partial z}\right|_{z=0}; \\ 0 < y < a, \quad t > 0 \\ W(y,0,\varphi) = \left.\dfrac{\partial W(y,t,\varphi)}{\partial t}\right|_{t=0} = 0; \quad 0 \leq y \leq a \\ W(0,t,\varphi) = W(a,t,\varphi) = 0 \quad (E-\text{ case }) \quad \text{or} \\ \left.\dfrac{\partial W(y,t,\varphi)}{\partial y}\right|_{y=0,a} = 0 \quad (H-\text{ case }); \quad t \geq 0 \end{cases}, \tag{4.31}$$

$$U(y,-L,t) = \frac{2}{\pi}\int\limits_0^{\pi/2} \frac{\partial \tilde{W}(y,t,\varphi)}{\partial t} d\varphi; \quad t \geq 0, \quad 0 \leq y \leq b,$$

$$\begin{cases} \left[\dfrac{\partial^2}{\partial t^2} - \sin^2(\varphi)\dfrac{\partial^2}{\partial y^2}\right] \tilde{W}(y,t,\varphi) = \left.\dfrac{\partial U(y,z,t)}{\partial z}\right|_{z=-L}; \quad 0 < y < b, \quad t > 0 \\ \tilde{W}(y,0,\varphi) = \left.\dfrac{\partial \tilde{W}(y,t,\varphi)}{\partial t}\right|_{t=0} = 0; \quad 0 \leq y \leq b \\ \tilde{W}(0,t,\varphi) = \tilde{W}(b,t,\varphi) = 0 \quad (E-\text{ case }) \quad \text{or} \\ \left.\dfrac{\partial \tilde{W}(y,t,\varphi)}{\partial y}\right|_{y=0,b} = 0 \quad (H-\text{ case }); \quad t \geq 0 \end{cases} \tag{4.32}$$

to be proved (see Section 2.2). Here, as before, $W(y,t,\varphi)$ and $\tilde{W}(y,t,\varphi)$ are certain auxiliary functions. For E-polarized fields $\{n\} = 1, 2, \ldots,$ $\mu_n(y) = \sqrt{2/a}\sin(n\pi y/a)$ and $\lambda_n = n\pi/a$, $\tilde{\mu}_n(y) = \sqrt{2/b}\sin(n\pi y/b)$, $\tilde{\lambda}_n = n\pi/b$. In the case of H-polarization, $\{n\} = 0, 1, 2, \ldots,$ $\mu_n(y) = \sqrt{(2-\delta_0^n)/a}\cos(n\pi y/a)$ and $\lambda_n = n\pi/a$, $\tilde{\mu}_n(y) = \sqrt{(2-\delta_0^n)/b}\cos(n\pi y/b)$, $\tilde{\lambda}_n = n\pi/b$.

The couple of relations (4.29), (4.31) are the exact (nonlocal and local) absorbing conditions on the boundary $\tilde{\mathbf{L}}$ in the waveguide **A** cross-section at $z = 0$. The couple of relations (4.30), (4.32) are the same conditions for the boundary **L** in the waveguide **B** cross-section $z = -L$. They are the direct analogues of the conditions (2.19), (2.26), constructed in Section 2.2. It is evident that the other nonlocal and local conditions from Section 2.2 may be adjusted for the situation considered herein.

In the domain $\mathbf{Q}_L$ the function $U(g,t)$ is defined by equations

$$\begin{cases} \left[-\varepsilon(g)\dfrac{\partial^2}{\partial t^2} - \sigma(g)\dfrac{\partial}{\partial t} + \dfrac{\partial^2}{\partial z^2} + \dfrac{\partial^2}{\partial y^2}\right] U(g,t) = F(g,t); \\ t > 0, \quad g = \{y,z\} \in \mathbf{Q}_L \\ U(g,t)\big|_{t=0} = \varphi(g), \quad \dfrac{\partial}{\partial t} U(g,t)\Big|_{t=0} = \psi(g); \quad g \in \bar{\mathbf{Q}}_L \\ E_{tg}(g,t)\big|_{g\in\mathbf{S}} = 0; \quad t \geq 0 \end{cases} . \tag{4.33}$$

The following statement is valid.

Statement 4.3. *The problems (4.27) and problems (4.33), supplied with conditions (4.29), (4.30) or conditions (4.31), (4.32), in the domain* $\mathbf{Q}_L$ *result in the same solutions* $U(g,t)$ *for any observation time* $t \in [0;T]$. *In modified problems the functions* $U^i(g,t)$, *contained in the exact absorbing conditions (4.29), (4.31) for virtual boundary* $\tilde{\mathbf{L}}$ *play the role of real sources, situated out of bounded analysis domain* $\mathbf{Q}_L$.

4.4.2. *Compact Discontinuities in* $\mathbf{R}^2$ *Space: Formulation of Modified Problems in Terms of Secondary Field* $U^s(g,t)$

The problems

$$\begin{cases} \left[-\varepsilon(g)\dfrac{\partial^2}{\partial t^2} - \sigma(g)\dfrac{\partial}{\partial t} + \dfrac{1}{\rho}\dfrac{\partial}{\partial \rho}\rho\dfrac{\partial}{\partial \rho} + \dfrac{1}{\rho^2}\dfrac{\partial^2}{\partial \phi^2}\right] U(g,t) = F(g,t) + \tilde{F}(g,t); \\ g = \{\rho,\phi\} \in \mathbf{Q} = \mathbf{R}^2 \backslash \overline{\mathrm{int}\mathbf{S}}, \quad t > 0 \\ U(g,t)\big|_{t=0} = \varphi(g) + \tilde{\varphi}(g), \quad \dfrac{\partial}{\partial t} U(g,t)\Big|_{t=0} = \psi(g) + \tilde{\psi}(g); \quad g \in \bar{\mathbf{Q}} \\ E_{tg}(g,t)\big|_{g\in\mathbf{S}} = 0, \quad U(\rho,\phi,t) = U(\rho,\phi+2\pi,t); \quad t \geq 0 \end{cases} \tag{4.34}$$

differ from the problems (3.1), considered in Section 3.2, only by the fact that part of the sources is concentrated out of the domain $\mathbf{Q}_L$, including all efficient scatterers (see Fig. 4.4; cylindrical coordinate system ρ, ϕ, x; all sources and scatterers are homogeneous along the x-direction). Functions $\tilde{F}(g,t)$, $\tilde{\varphi}(g)$, and $\tilde{\psi}(g)$ correspond to these sources. The supports of functions $\tilde{\varphi}(g)$ and $\tilde{\psi}(g)$ may be unbounded as, for example, in the case of illumination of the domain $\mathbf{Q}_L$ by a plane pulsed wave.

Let the sources $\tilde{F}(g,t)$, $\tilde{\varphi}(g)$, and $\tilde{\psi}(g)$ generate the field $U^i(g,t)$ in the space $\mathbf{R}^2$. In other words, let the function $U^i(g,t)$ be the solution to following Cauchy

FIGURE 4.4. Compact discontinuities of $\mathbf{R}^2$ space: Part of sources is located outside the domain $\mathbf{Q}_L$.

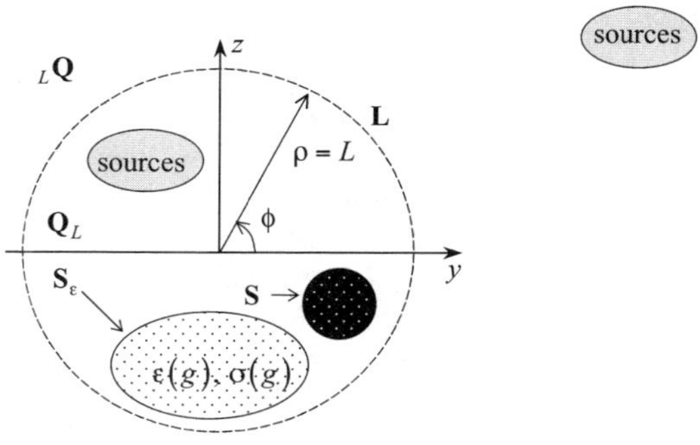

problem,

$$\begin{cases} \left[-\frac{\partial^2}{\partial t^2} + \frac{1}{\rho}\frac{\partial}{\partial \rho}\rho\frac{\partial}{\partial \rho} + \frac{1}{\rho^2}\frac{\partial^2}{\partial \phi^2} \right] U^i(g,t) = \tilde{F}(g,t); \\ g = \{\rho,\phi\} \in \mathbf{R}^2, \quad t > 0 \\ U^i(g,t)\big|_{t=0} = \tilde{\varphi}(g), \quad \frac{\partial}{\partial t}U^i(g,t)\Big|_{t=0} = \tilde{\psi}(g); \\ U^i(\rho,\phi,t) = U^i(\rho,\phi+2\pi,t) \end{cases} . \tag{4.35}$$

Then, we derive from (4.34), (4.35) for the function $U^s(g,t) = U(g,t) - U^i(g,t)$, $g \in \mathbf{Q} = \mathbf{Q}_L \cup {}_L\mathbf{Q} \cup \mathbf{L}$ the following initial boundary value problems in the domains $\mathbf{Q}_L$ and ${}_L\mathbf{Q}$.

$$\begin{cases} \left[-\varepsilon(g)\frac{\partial^2}{\partial t^2} - \sigma(g)\frac{\partial}{\partial t} + \frac{1}{\rho}\frac{\partial}{\partial \rho}\rho\frac{\partial}{\partial \rho} + \frac{1}{\rho^2}\frac{\partial^2}{\partial \phi^2} \right] U^s(g,t) \\ = F(g,t) - \left[[1-\varepsilon(g)]\frac{\partial^2}{\partial t^2} - \sigma(g)\frac{\partial}{\partial t} \right] U^i(g,t) = \Phi(g,t); \\ g \in \mathbf{Q}_L, \quad t > 0 \\ U^s(g,t)|_{t=0} = \varphi(g), \quad \frac{\partial}{\partial t}U^s(g,t)\Big|_{t=0} = \psi(g); \quad g \in \bar{\mathbf{Q}}_L \\ E^s_{tg}(g,t)\big|_{g\in\mathbf{S}} = -\,E^i_{tg}(g,t)\big|_{g\in\mathbf{S}}, \quad U^s(\rho,\phi,t) = U^s(\rho,\phi+2\pi,t); \quad t \geq 0 \end{cases} , \tag{4.36}$$

$$\begin{cases} \left[-\frac{\partial^2}{\partial t^2} + \frac{1}{\rho}\frac{\partial}{\partial \rho}\rho\frac{\partial}{\partial \rho} + \frac{1}{\rho^2}\frac{\partial^2}{\partial \phi^2} \right] U^s(g,t) = 0; \quad g \in_L \mathbf{Q}, \quad t > 0 \\ U^s(g,t)\Big|_{t=0} = 0, \quad \frac{\partial}{\partial t}U^s(g,t)\Big|_{t=0} = 0; \quad g \in_L \bar{\mathbf{Q}} \\ U^s(\rho,\phi,t) = U^s(\rho,\phi+2\pi,t); \quad t \geq 0 \end{cases} . \tag{4.37}$$

Here, $E^i_{tg}(g,t)|_{g\in\mathbf{S}}$ is the tangential component of electric field density vector, corresponding to function $U^i(g,t)$, and $E^s_{tg}(g,t)|_{g\in\mathbf{S}} + E^i_{tg}(g,t)|_{g\in\mathbf{S}} = E_{tg}(g,t)|_{g\in\mathbf{S}}$. The current source $\Phi(g,t)$ on the right-hand side of telegraph equation (4.36) contains true source $F(g,t)$ and so called "equivalent currents", induced by the field $U^i(g,t)$ at dielectric discontinuities of $\mathbf{R}^2$-space.

Problem (4.37) coincides with problem (3.3). Thus, repeating derivations carried out in Section 3.2.1, one can obtain

$$U^s(L,\phi,t) = \frac{2}{\pi}\sum_n (-1)^n \mu_n(\phi)\left[\int_0^t \left[u_n(L,\tau)\,\xi_n(t-\tau) - u'_n(L,\tau)\,\eta_n(t-\tau)\right]d\tau\right];$$
$$0 \le \phi \le 2\pi, \quad t \ge 0. \tag{4.38}$$

Here (see Section 3.2), $\mu_n(\phi) = (2\pi)^{-1/2}\exp(in\phi)$, $n = 0, \pm 1, \pm 2, \dots$ is a complete (in $\mathbf{L}_2[(0<\phi<2\pi)]$) orthonormal system of transverse functions; $Q_{|n|-1/2}$ are Legendre functions of the second order; $a_{r,\rho} = [r^2+\rho^2-(t-\tau)^2]/(2\rho r)$; $\eta_n(t-\tau) = Q_{|n|-1/2}(-a_{L,L})$, and

$$U^s(\rho,\phi,t) = \sum_n u_n(\rho,t)\mu_n(\phi) \leftrightarrow u_n(\rho,t) = \int_0^{2\pi} U^s(\rho,\phi,t)\,\mu_n^*(\phi)\,d\phi;$$

$$u'_n(L,t) = \left.\frac{\partial}{\partial\rho}u_n(\rho,t)\right|_{\rho=L}; \quad Q'_{|n|-1/2}(-a) = \left.\frac{\partial}{\partial x}Q_{|n|-1/2}(x)\right|_{x=-a};$$
$$\xi_n(t-\tau) = \left[2Q'_{|n|-1/2}\left(-a_{L,L}\right)\left(a_{L,L}-1\right) - Q_{|n|-1/2}\left(-a_{L,L}\right)\right](2L)^{-1}.$$

Statement 4.4. *Problems (4.34) and problems (4.36), (4.38) are equivalent: in the domain $\mathbf{Q}_L$ they result in the same solutions $U(g,t)$ for all observation times $t \in [0;T]$.*

4.4.3. Determination of the Given Sources Field

For the implementation of the schemas suggested above, it is necessary to know the values of the function $U^i(g,t)$ in the closure of the domains $\mathbf{Q}_L \times (0;T)$ (see Section 4.4.1); and sometimes in the whole domain of the original initial boundary problem formulation (Section 4.4.2) in any observation instant t. The range of real problems posing such requirements is rather wide. In essence, all these problems may be reduced to well-studied Cauchy problems about the disturbance propagation in regular domains. There are already many results concerning this issue and decent "recipes" for dealing with such problems may be found in [32,34,36,45,134].

The most complicated problem of the determination of the field $\{\vec{E}(g,t), \vec{H}(g,t)\}$ generated by given current and momentary ($\vec{E}(g,0)$ and $\vec{H}(g,0)$) sources in hollow ($\varepsilon(g)\equiv 1$, $\sigma(g)\equiv 0$) waveguides with perfectly conducting walls is considered in Chapter 5. In this section, we present only routine for function $U^i(g,t)$ calculation in the problems of compact discontinuities in open space.

Calculation of the spatio–temporal characteristics of electromagnetic fields produced by compact sources in $\mathbf{R}^3$ and freely propagating in vacuum may be reduced to the solving of the following vector Cauchy problems.

$$\begin{cases} \left[-\frac{\partial^2}{\partial t^2}+\Delta\right]\begin{Bmatrix}\vec{E}\\ \vec{H}\end{Bmatrix}=\begin{Bmatrix}\partial\vec{J}/\partial t+\text{ grad }\rho_2\\ -\text{rot}\vec{j}\end{Bmatrix}=\begin{Bmatrix}\vec{F}^E\\ \vec{F}^H\end{Bmatrix}; \\ g=\{x,y,z\}\in\mathbf{R}^3,\quad t>0 \\ \begin{Bmatrix}\vec{E}(g,0)\\ \vec{H}(g,0)\end{Bmatrix}=\begin{Bmatrix}\vec{\varphi}^E\\ \vec{\varphi}^H\end{Bmatrix},\ \begin{Bmatrix}\partial\vec{E}(g,t)/\partial t\big|_{t=0}=\eta_0\text{rot}\vec{H}(g,0)\\ \partial\vec{H}(g,t)/\partial t\big|_{t=0}=-\eta_0^{-1}\text{rot}\vec{E}(g,0)\end{Bmatrix}=\begin{Bmatrix}\vec{\psi}^E\\ \vec{\psi}^H\end{Bmatrix}; \\ g\in\mathbf{R}^3 \end{cases} \tag{4.39}$$

Here, $\vec{J}=\eta_0\vec{j}$, $\vec{j}\equiv\vec{j}(g,t)$ is the vector of external current density (see Section 1.2.1). Function $\rho_2(g,t)$ describes the volume density of external electric charges. Hence the term grad ρ_2 is moved to the right-hand side of wave equations defining the current sources of electric and magnetic fields. The operator grad div$\vec{E}=$ grad ρ_1/ε (here the function $\rho_1(g,t)$ denotes the volume density of induced changes) in vector equations for the field $\vec{E}$ can be omitted from consideration [31]. In homogeneous media ($\varepsilon=\text{const}\geq 1$ and $\sigma=\text{const}\geq 0$)

$$\rho_1(g,t)=\rho_1(g,0)\,e^{-t\sigma/\varepsilon}, \tag{4.40}$$

and if $\rho_1(g,0)=0$, then $\rho_1(g,t)=0$ for any $t>0$. Relation (4.40) follows from the equations (1.2) and (1.3).

In fact, (4.39) is simply the convenient (vector) form of the formulation of six scalar Cauchy problems of the same type for 3-D wave equations:

$$\begin{cases} \left[-\frac{\partial^2}{\partial t^2}+\Delta\right]U(g,t)=F(g,t);\quad g=\{x,y,z\}\in\mathbf{R}^3,\quad t>0 \\ U(g,0)=\varphi(g),\quad \frac{\partial U(g,t)}{\partial t}\Big|_{t=0}=\psi(g);\quad g\in\mathbf{R}^3 \end{cases}; \tag{4.41}$$

$U(g,t)$ is one of the components of the field vectors $\{\vec{E},\vec{H}\}$, and $F(g,t)$, $\varphi(g)$ and $\psi(g)$ are the corresponding current and momentary sources. In the generalized formulation the problems (4.41) are stated in the following way [34].

$$P_{1,0}[U]\equiv\left[-\frac{\partial^2}{\partial t^2}+\Delta\right]U(g,t)=F(g,t)-\delta^{(1)}(t)\varphi(g)-\delta(t)\psi(g)=f(g,t);$$
$$g\in\mathbf{R}^3,\quad -\infty<t<\infty \tag{4.42}$$

(the functions $U(g,t)$ and $F(g,t)$ are continued with zero on semi-axis $t<0$). The solutions $U(g,t)$ are obtained from the convolution $U(g,t)=G(g,t)*f(g,t)$; $g\in\mathbf{R}^3$, $t>0$ of the fundamental solution

$$G(g,t)=-\frac{\chi(t)}{2\pi}\delta\left(t^2-|g|^2\right)$$

of wave operator $P_{1,0}[U]$ with right-hand side parts from equations (4.42).

The Poisson formula (see Section 1.2.5) may be used in the course of the derivation of either classic or generalized (locally integrable) solutions to scalar Cauchy problems for 2-D wave equations, describing E- or H-polarized electromagnetic fields, generated by sources homogeneous along the x-direction. Sometimes it is more convenient to operate with representations of the corresponding functions $U^i(g,t)$ in a fixed coordinated system in the form of a series over transverse functions. One of the problems arising in such consideration is treated below. Namely, we consider the problem (4.35) from Section 4.4.2. Let us separate here the transverse coordinate ϕ and for spatio–temporal amplitudes $u_n(\rho,t)$ of waves

$$U^i(\rho,\phi,t)=\sum_n u_n(\rho,t)\mu_n(\phi); \quad g=\{\rho,\phi\}\in\mathbf{R}^2, \quad t>0 \tag{4.43}$$

we obtain following initial problems.

$$\left\{\begin{array}{l}\left[-\dfrac{\partial^2}{\partial t^2}+\dfrac{1}{\rho}\dfrac{\partial}{\partial\rho}\rho\dfrac{\partial}{\partial\rho}-\dfrac{n^2}{\rho^2}\right]u_n(\rho,t)=F_n(\rho,t); \quad t>0 \\ u_n(\rho,0)=\varphi_n(\rho), \quad \dfrac{\partial}{\partial t}u_n(\rho,t)\Big|_{t=0}=\psi_n(\rho)\end{array}\right. ;$$
$$\rho\geq 0, \quad n=0,\pm1,\pm2,\ldots. \tag{4.44}$$

Here, $F_n(\rho,t)$, $\varphi_n(\rho)$, and $\psi_n(\rho)$ are Fourier coefficients of functions $\tilde{F}(g,t)$, $\tilde{\varphi}(g)$, and $\tilde{\psi}(g)$ in the $\{\mu_n(\phi)\}$ basis.

The use of the Hankel transform (3.6) in (4.44) results in the problems

$$\left\{\begin{array}{l}\left[\dfrac{\partial^2}{\partial t^2}+\omega^2\right]\tilde{u}_n(\omega,t)=-\tilde{F}_n(\omega,t); \quad t>0 \\ \tilde{u}_n(\omega,0)=\tilde{\varphi}_n(\omega), \quad \dfrac{\partial}{\partial t}\tilde{u}_n(\omega,t)\Big|_{t=0}=\tilde{\psi}_n(\omega)\end{array}\right. ;$$
$$\omega\geq 0, \quad n=0,\pm1,\pm2,\ldots \tag{4.45}$$

for the transforms $\tilde{u}_n(\omega,t)\leftrightarrow u_n(\rho,t)$, where $\tilde{F}_n(\omega,t)\leftrightarrow F_n(\rho,t)$, $\tilde{\varphi}_n(\omega)\leftrightarrow\varphi_n(\rho)$, and $\tilde{\psi}_n(\omega)\leftrightarrow\psi_n(\rho)$.

We obtain the solutions

$$\tilde{u}_n(\omega,t)=\frac{1}{\omega}\int_{-\infty}^{t}\sin[(t-\tau)\omega]\left[\delta^{(1)}(\tau)\tilde{\varphi}_n(\omega)+\delta(\tau)\tilde{\psi}_n(\omega)-\tilde{F}_n(\omega,\tau)\right]d\tau;$$
$$\omega\geq 0, \quad t\geq 0$$

and

$$u_n(\rho, t) = \int_0^\infty \tilde{u}_n(\omega, t)\omega J_{|n|}(\rho\omega)\,d\omega; \quad \rho \geq 0, \quad t > 0$$

after going over to generalized formulation of the problems (4.45) (functions $\tilde{u}_n(\omega, t)$ and $\tilde{F}_n(\omega, \tau)$ are continued with zero on the semi-axis $t < 0$), taking into consideration the properties of the fundamental solution for the operator $D(\omega) = [d^2/dt^2 + \omega^2]$ and application of the inverse transforms (3.6).

Complex plane waves are widely used in mathematical modeling of electromagnetic processes in $\mathbf{R}^2$- and $\mathbf{R}^3$-spaces as excitation signals $U^i(g, t)$ ($U^i(g, 0) = \varphi(g)$, $\partial U^i(g, t)/\partial t|_{t=0} = \psi(g)$) or as partial components of more complicated excitation signals. The field $\{\vec{E}, \vec{H}\}$ of a plane wave is transverse, vectors $\vec{E}$, $\vec{H}$, and $\vec{\alpha}$ (vector $\vec{\alpha}$ gives the propagation direction; $|\vec{\alpha}| = 1$) make up a right-handed vector trio. The Cartesian components of the vectors $\vec{E}$ and $\vec{H}$ satisfy the wave equation $P_{1,0}[U^i] = 0$ (the operator $P_{1,0}[U^i]$ is defined in (4.42)) and may be given by an arbitrary function $U^i(g, t) = U^i(\alpha_x x + \alpha_y y + \alpha_z z - t)$.

For example, the function

$$U^i(z, t) = \mathrm{e}^{-(z+t-\tilde{T})^2/4\tilde{\alpha}^2} \cos\left[\tilde{k}\left(z + t - \tilde{T}\right)\right] \tag{4.46}$$

in 2-D scalar problems ($\partial/\partial x \equiv 0$) may describe E_x- or H_x-field components of a primary plane wave normally incident at the plane $z = 0$ from the positive z-direction. (4.46) is a Gaussian pulse. The parameter $\tilde{\alpha}$ defines its space duration and frequency range band; $\tilde{k}$ defines the spectral domain aptitude center of the signal and $\tilde{T}$ is the time of retard (the time during which the signal passes the plane $z = 0$).

4.5. Evolutionary Basis of Outgoing Waves in the Domains $_L\mathbf{Q}$ with Homogeneous and Inhomogeneous Filling

In the formulation of the problems considered above, we have always assumed that in the complete domain of analysis $\mathbf{Q}$, there is a regular subdomain $_L\mathbf{Q}$ that is effectively free from scatterers and sources. The field $\{\vec{E}, \vec{H}\}$ (or $\{\vec{E}^s, \vec{H}^s\}$) of outgoing waves propagates freely in this domain, without experiencing appreciable transformations. Only the dispersive properties of the hollow wave-guiding domain $_L\mathbf{Q}$ influence the level of transformations.

How will the results change if the constitutive parameters of the material in the domain $_L\mathbf{Q}$ material are different from those of vacuum? In which situations are the approaches developed above valid? The answers to these questions make up the content of the present section.

Let us come back to the problem (2.2) for a waveguide junction, assuming now that the regular waveguide $0 < y < a$ (see Fig. 2.1) is filled with a medium, characterized by $\varepsilon = \text{const} \geq 1$ and $\sigma = \text{const} \geq 0$. The problem is not a very complicated one, but judging from the results in the previous chapters, it can be expected to be a key problem and its solution can be taken as a basis for the solution of more complicated vector and scalar problems.

Assuming that the excitation $U(g,t)$ at the time $t=0$ has not reached the boundary $z=0$, and using the representation

$$U(g,t)=\sum_n u_n(z,t)\,\mu_n(y);\quad z\geq 0,\quad 0\leq y\leq a,\quad t\geq 0,\quad n\in\{n\} \tag{4.47}$$

(see Section 2.2), we obtain for the determination of the spatio–temporal amplitudes $u_n(z,t)$ (evolutionary basis elements) of the signal $U(g,t)$ the following initial boundary value problems.

$$\left\{\begin{array}{l}\left[-\varepsilon\dfrac{\partial^2}{\partial t^2}-\sigma\dfrac{\partial}{\partial t}+\dfrac{\partial^2}{\partial z^2}-\lambda_n^2\right]u_n(z,t)=0;\quad t>0\\[2ex] u_n(z,0)=0,\quad \dfrac{\partial}{\partial t}u_n(z,t)\Big|_{t=0}=0\end{array}\right.;\quad z\geq 0,\quad n\in\{n\}. \tag{4.48}$$

Transformation (2.6) transfers (4.48) into the Cauchy problems

$$\left\{\begin{array}{l}D\left(\dfrac{\sigma}{\varepsilon},\dfrac{\lambda_n^2+\omega^2}{\varepsilon}\right)[\tilde{u}_n(\omega,t)]\equiv\left[\dfrac{\partial^2}{\partial t^2}+\dfrac{\sigma}{\varepsilon}\dfrac{\partial}{\partial t}+\dfrac{\lambda_n^2+\omega^2}{\varepsilon}\right]\tilde{u}_n(\omega,t)\\[2ex] =-\sqrt{\dfrac{2}{\pi}}\dfrac{u_n'(0,t)}{\varepsilon};\quad \omega>0,\quad t>0\\[2ex] \tilde{u}_n(\omega,0)=0,\quad \dfrac{\partial}{\partial t}\tilde{u}_n(\omega,t)\Big|_{t=0}=0;\quad \omega\geq 0\end{array}\right. \tag{4.49}$$

for the transforms $\tilde{u}_n(\omega,t)\leftrightarrow u_n(z,t)$. The derivatives $u_n'(0,t)$ are defined by (2.8).

We construct the fundamental solution

$$G(a,b,t)=\chi(t)\frac{2}{\sqrt{4b-a^2}}e^{-at/2}\sin\left(\frac{\sqrt{4b-a^2}}{2}t\right) \tag{4.50}$$

of the operator $D(a,b)\equiv\partial^2/\partial t^2+a\partial/\partial t+b$, following prescriptions from [34]: $G(a,b,t)=\chi(t)V(t)$, where $V(t)$is a solution to equation $G(a,b,t)[V]=0$, such that $V(0)=0$ and $\partial V(t)/\partial t|_{t=0}=1$. Substituting the values $a=\sigma/\varepsilon$, $b=(\lambda_n^2+\omega^2)/\varepsilon$ into (4.50) and inverting the generalized Cauchy problems (4.49), we arrive at

$$\tilde{u}_n(\omega,t)=-\sqrt{\frac{2}{\pi\varepsilon}}\int_0^t\sin\left[\frac{(t-\tau)}{\sqrt{\varepsilon}}\sqrt{\lambda_n^2-\frac{\sigma^2}{4\varepsilon}+\omega^2}\right]e^{-\sigma(t-\tau)/2\varepsilon}\frac{u_n'(0,\tau)}{\sqrt{\lambda_n^2-\dfrac{\sigma^2}{4\varepsilon}+\omega^2}}d\tau;$$
$$\omega\geq 0,\quad t\geq 0, \tag{4.51}$$

and after passing to the originals $u_n(z,t)$ in (4.51) we have

$$u_n(z,t) = -\frac{1}{\sqrt{\varepsilon}}\int_0 J_0\left[\sqrt{\lambda_n^2 - \frac{\sigma^2}{4\varepsilon}}\left(\frac{(t-\tau)^2}{\varepsilon} - z^2\right)^{1/2}\right] e^{-\sigma(t-\tau)/2\varepsilon}\chi\left[\frac{(t-\tau)}{\sqrt{\varepsilon}} - z\right] \times u_n'(0,\tau)\,d\tau; \quad z \geq 0, \quad t \geq 0. \tag{4.52}$$

Relations (4.52) give a diagonal-type transport operator $Z_{0\to z}(t)$, which acts according to the rule

$$u(z,t) = \{u_n(z,t)\} = Z_{0\to z}(t)\left[u'(0,\tau)\right]; \quad u'(b,\tau) = \left\{u_n'(b,\tau)\right\}, \quad z \geq 0, \quad t \geq \tau \geq 0$$

and determines the signal's spatio–temporal transformations in the plane-parallel waveguide, filled with an absorptive medium.

Lowering the observation point onto the virtual boundary $\mathbf{L}$ $(z = 0)$ in (4.52), and using (4.47), (2.8) we arrive at

$$U(y,0,t) = -\frac{1}{\sqrt{\varepsilon}}\sum_{n\in\{n\}}\left\{\int_0 J_0\left[(t-\tau)\eta_n\right]e^{-\sigma(t-\tau)/2\varepsilon}\chi(t-\tau) \times \left[\int_0^a \left.\frac{\partial U(\tilde{y},z,\tau)}{\partial z}\right|_{z=0}\mu_n(\tilde{y})\,d\tilde{y}\right]d\tau\right\}\mu_n(y); \quad 0 \leq y \leq a, \quad t \geq 0 \tag{4.53}$$

which is the exact nonlocal absorbing condition for transient waves formed by the waveguide junction and outgoing into the direction of increasing z. Here,

$$\eta_n = \varepsilon^{-1/2}\sqrt{\lambda_n^2 - \sigma^2/4\varepsilon}.$$

By setting $z = 0$, and taking into consideration the representation [122]

$$J_0(s) = \frac{2}{\pi}\int_0^{\pi/2}\cos\left[s\sin(\varphi)\right]d\varphi$$

we rewrite relations (4.52) in the form

$$u_n(0,t) = -\frac{2}{\pi\sqrt{\varepsilon}}\int_0^{\pi/2}\left\{\int_0 \cos\left[(t-\tau)\eta_n\sin\varphi\right]\chi(t-\tau)e^{-\sigma(t-\tau)/2\varepsilon}u_n'(0,\tau)\,d\tau\right\}d\varphi; \quad t \geq 0 \tag{4.54}$$

and introduce the auxiliary functions

$$w_n(t,\varphi) = -\int_0 \frac{\sin\left[(t-\tau)\eta_n\sin\varphi\right]\chi(t-\tau)e^{-\sigma(t-\tau)/2\varepsilon}u_n'(0,\tau)}{\eta_n\sin\varphi}d\tau; \quad t \geq 0, \quad 0 \leq \varphi \leq \pi/2. \tag{4.55}$$

Then,

$$\frac{\partial w_n(t,\varphi)}{\partial t} = -\int_0 \cos\left[(t-\tau)\eta_n \sin\varphi\right]\chi(t-\tau)\,e^{-\sigma(t-\tau)/2\varepsilon} \times u_n'(0,\tau)\,d\tau - \frac{\sigma}{2\varepsilon}w_n(t,\varphi),$$

and from (4.54), we derive

$$u_n(0,t) = \frac{2}{\pi\sqrt{\varepsilon}}\int_0^{\pi/2}\left[\frac{\partial w_n(t,\varphi)}{\partial t} + \frac{\sigma}{2\varepsilon}w_n(t,\varphi)\right]d\varphi; \quad t \geq 0. \tag{4.56}$$

The integral forms (4.55) are equivalent to the differential forms

$$\begin{cases} \left[\dfrac{\partial^2}{\partial t^2} + \dfrac{\sigma}{\varepsilon}\dfrac{\partial}{\partial t} + \dfrac{\lambda_n^2}{\varepsilon}\sin^2\varphi + \dfrac{\sigma^2}{4\varepsilon^2}\cos^2\varphi\right]w_n(t,\varphi) = -u_n'(0,t); \quad t > 0 \\ w_n(0,\varphi) = \left.\dfrac{\partial w_n(t,\varphi)}{\partial t}\right|_{t=0} = 0 \end{cases},$$

$$0 \leq \varphi \leq \pi/2. \tag{4.57}$$

Indeed, if we transfer in (4.57) to the generalized statement of Cauchy problems and recall the fundamental solution (4.50) of the operator $D(a,b) \equiv \partial^2/\partial t^2 + a\partial/\partial t + b$, then we can clearly see that the representations (4.55) and (4.57) define the same function $w_n(t,\varphi)$.

Now, let us multiply (4.56) and (4.57) by $\mu_n(y)$ and sum terms for all $n \in \{n\}$. Taking into account that

$$\sum_n \lambda_n^2 w_n(t,\varphi)\mu_n(y) = -\frac{\partial^2 W(y,t,\varphi)}{\partial y^2}$$

when

$$W(y,t,\varphi) = \sum_n w_n(t,\varphi)\mu_n(y),$$

as a result we derive

$$U(y,0,t) = \frac{2}{\pi\sqrt{\varepsilon}}\int_0^{\pi/2}\left[\frac{\partial W(y,t,\varphi)}{\partial t} + \frac{\sigma}{2\varepsilon}W(y,t,\varphi)\right]d\varphi;$$

$$t \geq 0, \quad 0 \leq y \leq a,$$

$$\begin{cases} \left[\dfrac{\partial^2}{\partial t^2} + \dfrac{\sigma}{\varepsilon}\dfrac{\partial}{\partial t} - \dfrac{\sin^2\varphi}{\varepsilon}\dfrac{\partial^2}{\partial y^2} + \dfrac{\sigma^2\cos^2\varphi}{4\varepsilon^2}\right]W(y,t,\varphi) = -\left.\dfrac{\partial U(y,z,t)}{\partial z}\right|_{z=0}; \\ 0 < y < a, \quad t > 0 \\ W(y,0,\varphi) = \left.\dfrac{\partial W(y,t,\varphi)}{\partial t}\right|_{t=0} = 0; \quad 0 \leq y \leq a \\ W(0,t,\varphi) = W(a,t,\varphi) = 0 \quad (E-\text{ case }) \quad \text{or} \\ \left.\dfrac{\partial W(y,t,\varphi)}{\partial y}\right|_{y=0,a} = 0 \quad (H-\text{ case }); \quad t \geq 0 \end{cases}. \tag{4.58}$$

This is the exact local absorbing condition on the virtual boundary **L** located in the cross-section of the waveguide $0 < y < a$ (see Fig. 2.1), filled with a homogeneous medium with parameters $\varepsilon = \text{const} \geq 1$ and $\sigma = \text{const} \geq 0$.

Let us compare the couples of conditions (2.19), (2.26) for hollow waveguides and (4.53), (4.58) for waveguides filled with a homogeneous medium. The distinction is rather slight. The technique of their derivation also differs slightly. This statement is valid for several other problems and approaches considered in this book. Namely, if the domain ${}_L\mathbf{Q}$ is filled with a medium with constitutive parameters ε and σ, this does not bring any new problems to the construction of the routine for correct limitation of the computational space with absorbing conditions at the virtual boundaries **L**, or when studying the processes of spatio–temporal transformations of transient electromagnetic fields. Only one result is impossible to generalize within the framework of common techniques, and that is the problem of edge points (see Section 3.3.2) for the medium with losses which is yet unsolved.

A parallel-plane waveguide $0 < y < a$ (see Fig. 2.1), filled with a medium with parameters $\varepsilon = \varepsilon(y)$ and $\sigma = 0$, is the simplest model problem that may serve for tests and verification of mathematical models and algorithms for the study of peculiarities of signal propagation in transversely irregular guiding structures. Transient electromagnetic processes are rather different in waveguides which are regular and irregular, respectively, in the transverse direction. Apparently, these differences considerably reduce the efficiency of the techniques applied above. Presently, there is no analytic solution to the problem of limiting the computational domain for structures loaded on the not uniform in cross-section domains ${}_L\mathbf{Q}$, although such structures are of considerable interest in various applications.

4.6. Numerical Tests of the New Exact Conditions

4.6.1. A Finite-Difference Scheme with Exact ABCs at the Coordinate Boundary in the Floquet Channel

In the simplest finite-difference versions of the problems (2.54) that have been considered in Section 2.5 (in this section we consider only the case of an E-polarized field, assuming that of the grating period is equal to 2π, and $0 \leq t \leq T$), it is necessary to determine the mesh functions $U(j,k,m)$, satisfying the difference equations

$$\begin{aligned}
&U(j,k,m+1) \\
&\quad = 2U(j,k,m) - U(j,k,m-1) + \frac{l^2}{\varepsilon(j,k)\,h^2}\,[U(j,k+1,m) \\
&\qquad + U(j,k-1,m) + U(j+1,k,m) + U(j-1,k,m) - 4U(j,k,m)]
\end{aligned} \tag{4.59}$$

at the spatio–temporal mesh points $\{y_j, z_k, t_m\}$:

$$\begin{cases} y_j = h(j-1); \quad h = 2\pi/(J-1), \quad j = 1, 2, \ldots, J \\ z_k = -L_2 + h(k-1); \quad (L_2 + L)/(K-1) = h, \quad k = 1, 2, \ldots, K. \\ t_m = lm; \quad m = 1, 2, \ldots, M = T/l \end{cases}$$

Equations (4.59) should be complemented by the equations

$$\begin{cases} U(j,k,0) = \varphi(j,k) \\ U(j,k,1) = \varphi(j,k) + l\psi(j,k) + \dfrac{l^2}{2h^2}\left[\varphi(j+1,k) + \varphi(j,k+1)\right., \\ \left.+\varphi(j-1,k) + \varphi(j,k-1) - 4\varphi(j,k)\right] \end{cases} \tag{4.60}$$

$$U(1,k,m) = e^{-i2\pi\Phi}U(J,k,m), \tag{4.61}$$

as well as by

$$U^s(j,K,m+1) = V_1(j,m+1) \tag{4.62}$$

if the conditions (2.60) are chosen; otherwise,

$$\frac{3(h+l)}{2hl}U^s(j,K,m+1) - 2\left[\frac{U^s(j,K,m)}{l} + \frac{U^s(j,K-1,m+1)}{h}\right] + \frac{1}{2}\left[\frac{U^s(j,K,m-1)}{l} + \frac{U^s(j,K-2,m+1)}{h}\right] = V_2(j,m+1), \tag{4.63}$$

in case of conditions (2.61). Equations (4.59) to (4.61) approximate the wave equations of the problems (2.54) (central-difference approximation), their initial conditions (the Taylor formula), and the quasi-periodicity conditions of the Floquet channel **R**. Equations (4.62) and (4.63) approximate the conditions (2.60) and (2.61) (difference approximation for the backward interpolation) at the imaginary boundary $z = L$. The approximation is everywhere of the order $O(h^2 + l^2)$. In (4.59) to (4.63) the conventional notations for the mesh approximations of the functions $U(y_j, z_k, t_m)$, $U^s(y_j, z_k, t_m)$, $\varepsilon(y_j, z_k)$, $V_1(y_j, t_m)$, and $V_2(y_j, t_m)$ are used, that is, $U(j,k,m)$, respectively, and so on. The bottom boundary **S** of the region $\mathbf{Q}_L = \{g \in \mathbf{Q} : z < L\}$ is determined by the matrix of the values $\{\varepsilon^{-1}(j,k)\}_{j,k}$: if the point z_j, y_k is on or below the boundary **S**, then $\varepsilon^{-1}(j,k)$ is assumed to be zero. It is clear that by such an approach, the condition $U(g,t) = 0$, $g \in \mathbf{S}$, is approximated by equations (4.59). It is also obvious that the two possible sets of equations (4.59) to (4.62) or (4.59) to (4.61), (4.63) uniquely determine $U(j,k,m)$. Then $U(j,k,m)$ can be computed without any inversions of matrix operators, and the explicit three-layer scheme is implemented.

Let us now consider more precisely some details in the technique for the implementation of the scheme of computation. First, in the case of conditions (2.60), the values of L_1, h, and l should be matched in such a way that the value of $z = L_1$ falls exactly on the corresponding space mesh ($L_1 = z_N$, $N < K$), and the interval $L - L_1$ contains a whole number of time steps l ($L - L_1 = lp$, p is an integer).

Then

$$V_1(j,m+1) = -\frac{l}{2}B\left[U^s(j,N,m+1-p)\right]$$
$$-l\sum_n \mu_n(j)\sum_{r=1}^{m-p} J_{0,n}(K-N,m+1-r)B\left[u_n(N,r)\right]. \quad (4.64)$$

Here, the quadrature formulas of the trapezoidal method are used that do not violate the accepted order of approximation accuracy, as well as the following notation for the mesh approximations of the functions and the operator $B\left[f(N)\right]$,

$$J_\xi\left[\Phi_n\left(t_m^2 - z_k^2\right)^{1/2}\right] = J_{\xi,n}(k,m), \quad u_n\left(z_k, t_m\right) = u_n(k,m),$$
$$\mu_n\left(y_j\right) = \mu_n(j); \quad B\left[f(N)\right] = \left[3f(N) - 4f(N-1) + f(N-2)\right]/2h.$$

Similarly, with the same notations, derive

$$V_2(j,m+1) = l\sum_n \Phi_n\mu_n(j)\sum_{r=1}^{m} J_{1,n}(1,m+1-r)\,B\left[u_n(K,r)\right]. \quad (4.65)$$

Note also that to obtain the values of $U(1,k,m+1)$ and $U(J,k,m+1)$ (see (4.59)) the values of $U(0,k,m)$ and $U(J+1,k,m)$ at the points beyond the computation area should be known. To return to the standard mesh we use the quasi-periodicity conditions (4.61) in the form

$$U(0,k,m) = U(J-1,k,m)\,\mathrm{e}^{-i2\pi\Phi}$$

and

$$U(J+1,k,m) = U(2,k,m)\,\mathrm{e}^{i2\pi\Phi}.$$

The discretization of the relation

$$U^s(y,L,t) = -\sum_n\left\{\int_0^t J_0\left[\Phi_n(t-\tau)\right]u_n'(L,\tau)\,d\tau\right\}\mu_n(y), \quad (4.66)$$

that is obtained from (2.60) for $L = L_1$, yields

$$U^s(j,K,m+1) + \frac{l}{2}B\left[U^s(j,K,m+1)\right] = V_3(j,m+1), \quad (4.67)$$

where

$$V_3(j,m+1) = -l\sum_n \mu_n(j)\sum_{r=1}^{m} J_{0,n}(1,m+1-r)\,B\left[u_n(K,r)\right].$$

The only qualitative difference among (4.66) and (2.60) and (2.61) is that the first does not exclude the momentary effect (at $\tau = t$) of the function $u_n'(L,\tau)$ on the boundary value of $U^s(y,L,t)$ in the left-hand side. Therefore one cannot use (4.66) directly while closing the analysis area by the artificial boundary $z = L$. Still, in (4.67) this restriction is already completely eliminated. The condition (4.67) can complement the series of conditions (4.62), (4.63) that do not influence the

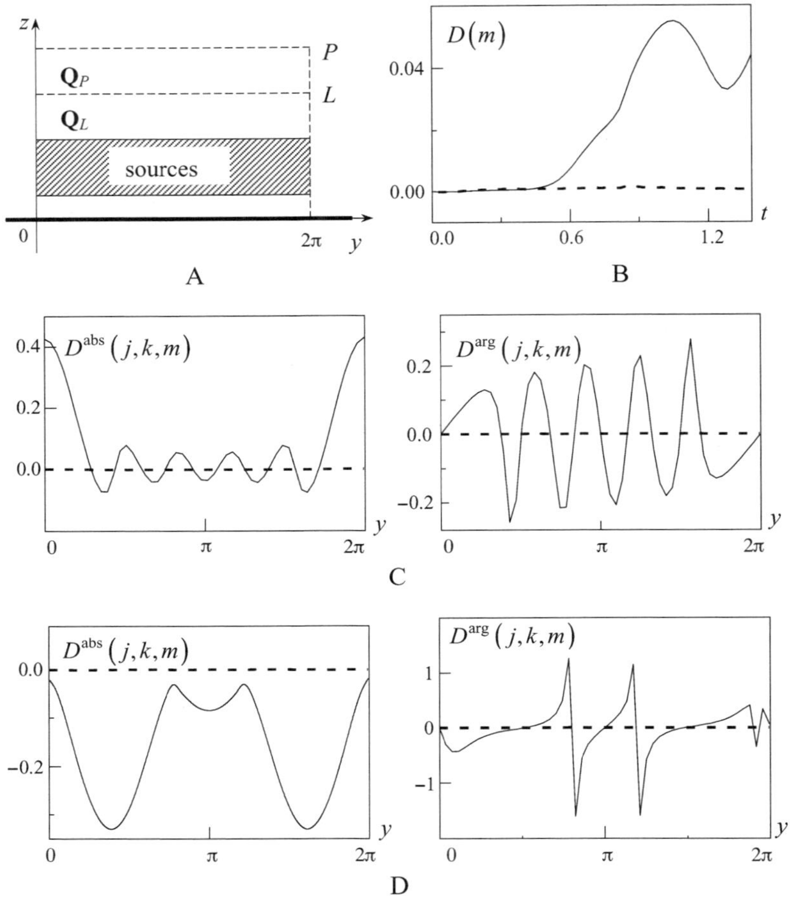

FIGURE 4.5. (A) Configuration of the test problem. (B) Global and (C, D) local errors, caused by applying ABCs of second order of approximation (solid lines) and conditions (2.61) (dashed lines). C: $t = 0.6$; D: $t = 1.4$.

modeled process and are fit for the simple explicit scheme of the finite-difference method.

The numerical tests of the new exact conditions have been done for the scheme, so as to allow an evaluation of the real error of the results. The exact (rigorous) solution to the model problems has been compared with the solution obtained by using the new strict conditions and conventional heuristic ABCs.

In the first series of experiments, the local (individual points of the mesh) and global (overall in the mesh for the region $\mathbf{Q}_L$) errors have been determined that occur due to the absorbing conditions (2.1) of different types that have been introduced into the problem. In Figure 4.5A, two regions $\mathbf{Q}_L$ and $\mathbf{Q}_P \supset \mathbf{Q}_L$ over the perfect conduction plane $z = 0$ are shown, in which the algorithm of the finite-difference method has been simultaneously applied to solve two initial boundary problems (2.54) with $\varepsilon(g) \equiv 1$ and the functions of the E-polarized field sources $F(g, t)$, $\varphi(g)$, and $\psi(g)$ such that their supports at any moment of time $t \in [0; T]$ belong to $\mathbf{Q}_L$. At the boundary $z = L$ of the region $\mathbf{Q}_L$ the tested boundary conditions have been established (the first problem; its solution is denoted by U_L). The boundary $z = P$ of the region $\mathbf{Q}_P$ has been chosen so that the field $U(g, t)$ belonging to it was zero for all $t \in [0; T]$ (the second problem with solution U_P, whose error is conditioned only by the discretization of the problem (2.54)).

Any field distortion at the boundary $\mathbf{L}$ caused by the absorbing conditions, results in a difference in the values of solutions U_L and U_P, found in the regions $\mathbf{Q}_L$ and $\mathbf{Q}_P$ in the corresponding points within $\mathbf{Q}_L$. To evaluate these differences, we determine at each time step m for all y_j and z_k belonging to $\mathbf{Q}_L$ the values $D(j, k, m)$ and $R(j, k, m)$ with

$$D^{\text{abs}}(j, k, m) = |U_L(j, k, m)| - |U_P(j, k, m)|,$$
$$D^{\text{arg}}(j, k, m) = \arg U_L(j, k, m) - \arg U_P(j, k, m),$$

and

$$R^{\text{abs}}(j, k, m) = \frac{D^{\text{abs}}(j, k, m)}{|U_P(j, k, m)|}, \qquad R^{\text{arg}}(j, k, m) = \frac{D^{\text{arg}}(j, k, m)}{\arg U_P(j, k, m)}.$$

It is obvious that $D(j, k, m)$ and $R(j, k, m)$ cause the local absolute and relative computation errors due to the introduction of the imaginary boundary $z = L$. The values

$$D(m) = \frac{1}{N} \sum_{j,k} \left| D^{\text{abs}}(j, k, m) \right|, \qquad R(m) = \frac{1}{N} \sum_{j,k} \left| R^{\text{abs}}(j, k, m) \right|$$

determine the average global error at the mth time step overall within the mesh for the region $\mathbf{Q}_L$ (N is the total number of the mesh points).

In the experiment, we have analyzed the results obtained by using the strict boundary conditions suggested in Section 2.5, by comparing them with the results obtained under the conventional ABCs [22] of the first

$$\left(\frac{\partial}{\partial z} + \frac{\partial}{\partial t} \right) U(g, t) \bigg|_{z=L} = 0,$$

the second

$$\left(-\frac{\partial^2}{\partial z \partial t} - \frac{\partial^2}{\partial t^2} + \frac{1}{2} \frac{\partial^2}{\partial y^2} \right) U(g, t) \bigg|_{z=L} = 0,$$

and the third

$$\left(-\frac{\partial^3}{\partial z \partial t^2} + \frac{1}{4} \frac{\partial^3}{\partial y^2 \partial z} - \frac{\partial^3}{\partial t^3} + \frac{3}{4} \frac{\partial^3}{\partial y^2 \partial t} \right) U(g, t) \bigg|_{z=L} = 0$$

approximation orders. The excitation conditions allowed a comparison of all the results with the analytical solution to the problem and to follow the differences in the errors depending on the changes in the structure of the field arriving at the boundary $z = L$.

Here we stress that the classic ABCs are rather sensitive to such changes, whereas the error occurring when using the strict boundary conditions remains within the error limit of the proper difference scheme; that is, this limit is not violated.

Figure 4.5 illustrates the global and local errors in the amplitude and phase caused by using the classic ABC of the second approximation order (solid lines) and the strict condition (2.61) (dashed lines). The discretization step for the space variables is $h = h_1 = 6.1359 \cdot 10^{-3}$; for the time variable it is $l = l_1 = 8.7656 \cdot 10^{-4}$; $\varphi(g) = F(g, t) \equiv 0$, $\psi(g) = \sum_{n=0}^{N} \psi_n(z) e^{i\Phi_n y}$, $\Phi = 0.3$, $N = 5$, $\psi_n(z) = 10$ at $z \in [0.1; 0.3]$ and $\psi_n(z) = 0$ at $z \notin [0.1; 0.3]$. The local error was calculated for mesh points on the line $z = L = 0.4$.

Figure 4.6 shows, for various observation times t, similar results, that is, the results of a comparative test of the strict condition (4.66) and the classic ABCs of the first, second, and third approximation orders. In the numerical experiments, only one parameter has been changed, namely, N; in Figure 4.6 it is equal to ten. The rapidly reducing strength of the field $U(g, t)$ in the region $\mathbf{Q}_L$ with growing t (the effect of an instantaneous source) makes the test somewhat less sensitive. The errors, whose values are presented in Figure 4.7, are determined by solving the problem (2.54) with $\varphi(g) = \psi(g) \equiv 0$ and $F(g, t) = e^{-i\tilde{\kappa}t} \sum_{n=0}^{11} \psi_n(z) e^{i\Phi_n y}$, $\tilde{\kappa} = 0.5$ (steadily acting current source).

The results obtained clearly demonstrate that the strict conditions are quite efficient for solving the problem of limiting the computation area by discretizing of the initial boundary problem formulated in an unlimited area. They do not distort the mathematically modeled physical processes and do not increase the computation errors, keeping them on a level determined by the chosen finite difference scheme. The tendency for the errors to be depressed by higher approximation orders of the classic ABCs can appear quite reassuring and optimistic at the first sight. Still, it is a delusive impression: in the resonance case (not yet considered) any virtual effect caused by using the imaginary boundaries will be much more strongly pronounced.

The sinusoidal plane E-polarized wave $\tilde{U}^i(g, \kappa) = \exp[i(\Phi y - \Gamma_0 z)]$ incident on the grating (see Fig. 2.4) induces in the reflection zone the secondary monochromatic field [6,9],

$$\tilde{U}^s(g, \kappa) = \sum_{n=-\infty}^{\infty} a_n \exp[i(\Phi_n y + \Gamma_n z)]; \quad z \geq 0,$$

where $\Gamma_n = \left(\kappa^2 - \Phi_n^2\right)^{1/2}$, $\mathrm{Re}\Gamma_n \geq 0$, $\mathrm{Im}\Gamma_n \geq 0$, κ is a real frequency parameter, and the time dependence in the space–time coordinates used is given by the factor $\exp(-i\kappa t)$. The complex amplitudes $\{a_n\}$ are complicated functions of Φ, κ, and of the geometric and material parameters of the grating.

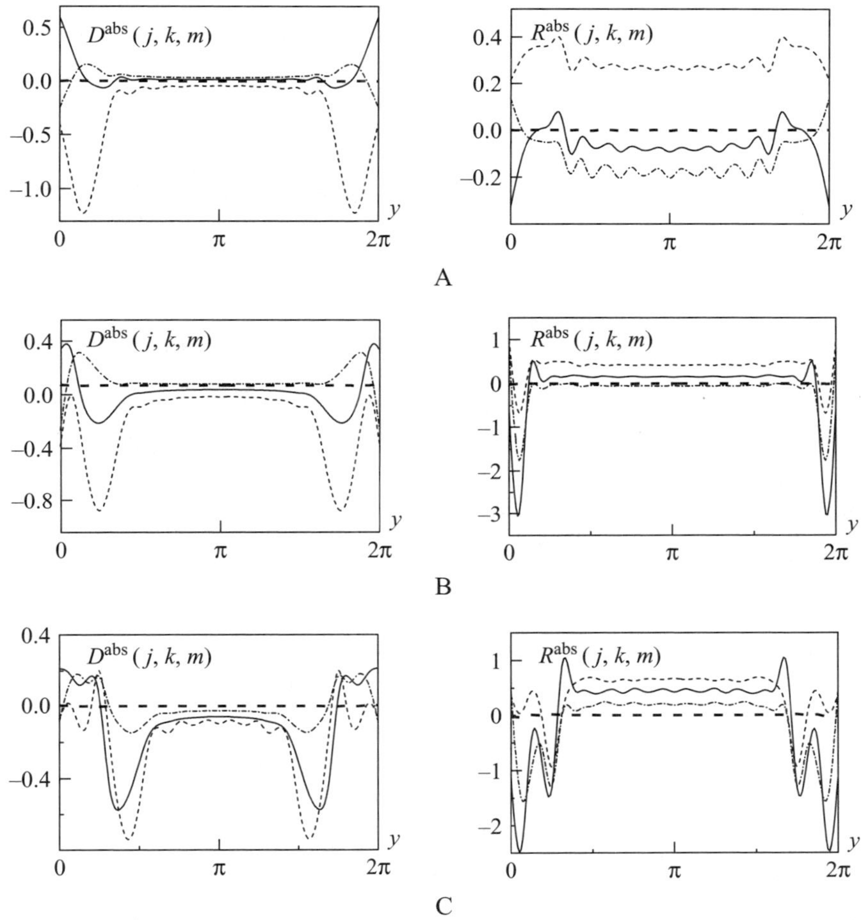

FIGURE 4.6. Local (in the points of the boundary $z = L = 0.4$) errors, caused by applying 'absorbing' conditions of various types. A: $t = 0.7$; B: $t = 0.9$; C: $t = 1.4$. Exact condition (4.66): ‐ ‐ ‐ ‐ ; ABC of the first order: ‐‐‐‐‐ ; ABC of the second order: —— ; ABC of the third order: ‐·‐·‐·· .

Their exact numerical values that have been used to test the finite-difference method algorithm with the new absorbing conditions, have been obtained from solving the corresponding elliptic boundary problem by the analytical regularization method [9,10].

The scheme of the computation experiments is as follows. The field $U^i(g, t)$ of the transient wave exciting the grating (see formula (2.56) and Fig. 2.4), has been chosen in the form

$$U^i(g, t) = e^{i\Phi y} \int_b^c J_0\left[\Phi\left(t^2 - |z-\omega|^2\right)^{1/2}\right] \chi(t - |z-\omega|)d\omega, \quad 0 \le z < b < c.$$

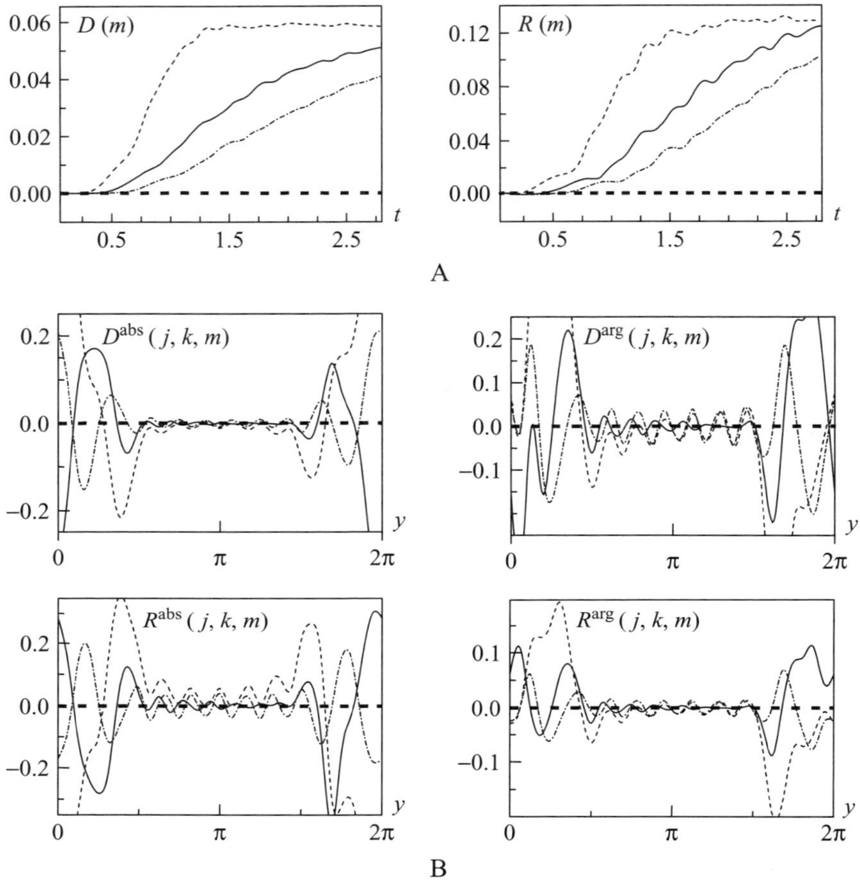

FIGURE 4.7. (A) Global and (B) local errors (B: $t = 1.4$), caused by using absorbing conditions of various types. Exact condition (4.66): ---- ; ABC of the first order: ----- ; ABC of the second order: —— ; ABC of the third order: -·-·-·- .

Because

$$\tilde{U}^i(g,\kappa)\,A(\kappa) = \frac{1}{2\pi}\int_{-\infty}^{\infty} U^i(g,t)\mathrm{e}^{i\kappa t}dt, \quad A(\kappa) = \frac{i}{2\pi\Gamma_0}\int_b^c \mathrm{e}^{i\Gamma_0\omega}d\omega, \quad (4.68)$$

the fields generated by the grating in the frequency ($\tilde{U}^s(g,\kappa)$) and time ($U^s(g,t)$) domain can be connected by the same integral Fourier transform. A relation similar to (4.68) was applied to transfer the results, obtained in the time domain, into the frequency domain, and to derive the corresponding results for $\{\tilde{a}_n(\kappa)\}$.

Figure 4.8 presents the frequency dependences of the exact values $\{a_n(\kappa)\}$: the modules and phases of the complex amplitudes of the three fundamental spatial

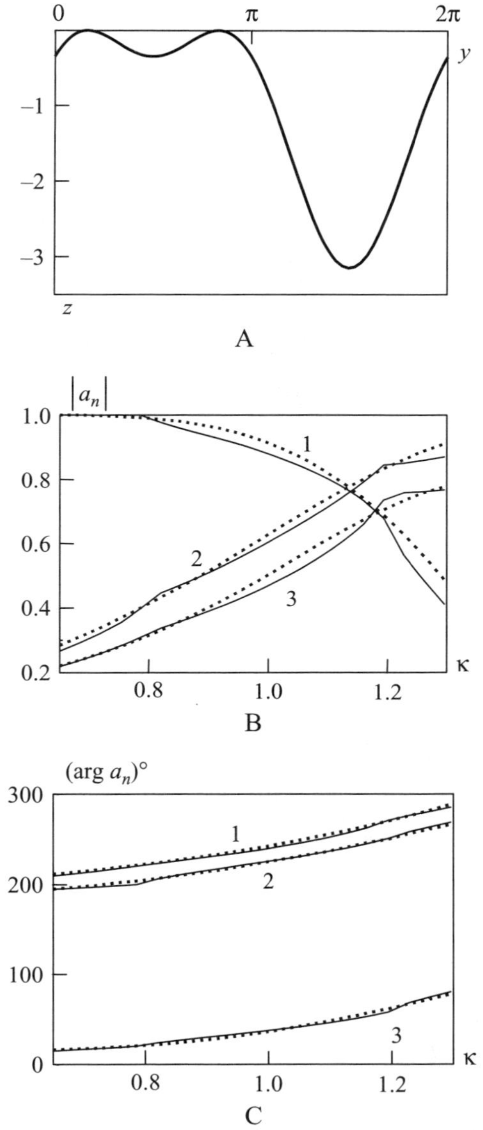

FIGURE 4.8. (A) Configuration of the test problem and (B, C) results of comparison of frequency responses: accurate (solid lines) and obtained in TD (dashed lines). 1: $n = 0$; 2: $n = -1$; 3: $n = 1$.

harmonics of the grating ($n = 0, \pm 1$, solid lines). The dashed lines show the dependences of $\{\tilde{a}_n(\kappa)\}$ obtained through a transfer from the time domain, on implementing the finite-difference method algorithm with the condition (2.61) at the virtual boundary $z = 0.4$. The profile $\mathbf{S}$ of the periodic structure (see Fig. 4.8A) is described by the function $z = 1.4(\sin y + \cos^2 y - 1.25)$, $\varepsilon(g) \equiv 1$, $\mathbf{\Phi} = 0.2$. The space discretization step is $h = h_1/2$, and the time step is $l = l_1/2$; the number M of time layers for the transform (4.68) during the implementation is 5705. The

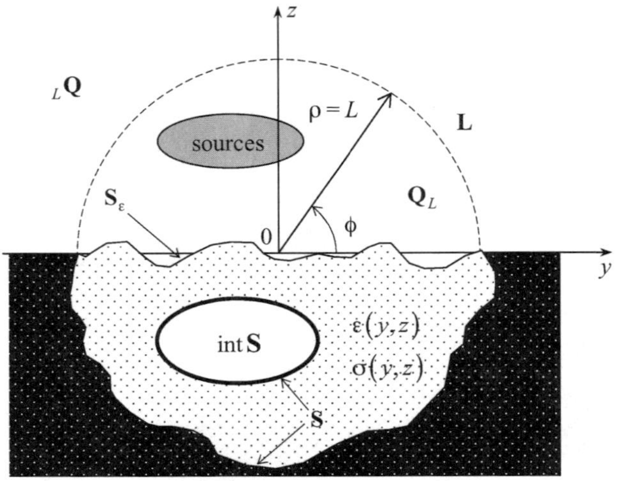

FIGURE 4.9. Geometry of the model problem of underground sounding.

agreement of the exact results with those obtained through the transfer from the time domain can be considered as satisfactory. As expected, the computational accuracy increases when l and h decrease, and as the higher limit $t = T$ increases in the transformation (4.68).

4.6.2. A Finite-Difference Scheme with Exact ABCs for 2-D Problems in Polar Coordinate Systems

The model problems of underground sounding (see Fig. 4.9)

$$\begin{cases} \left[-\varepsilon(g) \dfrac{\partial^2}{\partial t^2} - \sigma(g) \dfrac{\partial}{\partial t} + \dfrac{1}{\rho} \dfrac{\partial}{\partial \rho} \rho \dfrac{\partial}{\partial \rho} + \dfrac{1}{\rho^2} \dfrac{\partial^2}{\partial \phi^2} \right] U(g, t) = F(g, t); \\ g = \{\rho, \phi\} \in \mathbf{Q}, \quad t > 0 \\ U(g, t)|_{t=0} = \varphi(g), \quad \dfrac{\partial}{\partial t} U(g, t)\Big|_{t=0} = \psi(g); \quad g \in \bar{\mathbf{Q}} \\ E_{tg}(g, t)\big|_{g \in \mathbf{S}} = 0; \quad t \geq 0 \end{cases} \tag{4.69}$$

differ from the problems (3.1), considered in Section 3.2, only by their transverse dimensions $0 \leq \phi \leq \pi$ in the domain $_L\mathbf{Q}$ of freely propagating outgoing waves

$$U(\rho, \phi, t) = \sum_n u_n(\rho, t) \mu_n(\phi); \quad g \in {}_L\bar{\mathbf{Q}}, \quad t \geq 0.$$

Here, in the case of an E-polarized field $U(g, t) = E_x(g, t)$, $\mu_n(\phi) = \sin(n\phi)$, and $n = 1, 2, \ldots$. For an H-polarized field $U(g, t) = H_x(g, t)$, $\varepsilon(g)$ and $\sigma(g)$ are piecewise constant functions, $\mu_n(\phi) = \cos(n\phi)$, and $n = 0, 1, 2, \ldots$. We have pointed out the distinctions of this kind in Sections 4.1 and 4.2. In fact, they do not bear any complications in the construction of absorbing conditions for coordinate type boundaries **L**. The open problems (4.69) turn into the equivalent closed problems by adding to (4.69) the conditions

$$U(L,\phi,t) = \frac{2}{\pi}\sum_{n=\left\{{1 \atop 0}\right\}}^{\infty}(-1)^n\left\{\begin{matrix}\sin(n\phi)\\ \cos(n\phi)\end{matrix}\right\}\left[\int_0^t\left[u(L,\tau)\xi_n(t-\tau)-u'_n(L,\tau)\eta_n(t-\tau)\right]d\tau\right];$$

$$\left\{\begin{matrix}E-\text{polarization}\\ H-\text{polarization}\end{matrix}\right\},\quad 0\le\phi\le\pi,\quad t\ge 0. \tag{4.70}$$

Here (see also Section 3.2.2), $\eta_n(t-\tau)=Q_{n-1/2}\left(-a_{L,L}\right)$ and

$$\xi_n(t-\tau)=\left[2Q'_{n-1/2}\left(-a_{L,L}\right)\left(a_{L,L}-1\right)-Q_{n-1/2}\left(-a_{L,L}\right)\right](2L)^{-1}.$$

Conditions (3.11) and (4.70) are referred to the coordinate boundary $\rho = L$. Hence, they can be implemented most efficiently by discretization of the problems (3.1) and (4.69) in the cylindrical (polar) coordinates [124] (the case of an E-polarized field):

$$\begin{cases}\left[-\varepsilon(j,k)D_+^tD_-^t-\sigma(j,k)D_0^t+L^{\rho,\phi}\right][U(j,k,m)]=F(j,k,m), \quad m\neq 0\\ U(j,k,0)=\varphi(j,k)\\ U(j,k,1)=\varphi(j,k)+l\psi(j,k)\\ \quad +l^2\left[L^{\rho,\phi}[\varphi(j,k)]-\sigma(j,k)\psi(j,k)\right]/2\varepsilon(j,k)\\ U(j,k,m)=0;\quad \left\{j,k: g_{j,k}=\left\{\rho_j,\phi_k\right\}\in\text{int}\,\mathbf{S}\right\}\end{cases};$$

$$\begin{cases}j=1,2,\ldots,J & \left(\rho_j=h_\rho(j-0.5),h_\rho=L/(J-0.5)\right)\\ k=1,2,\ldots,K & \left(\phi_k=h_\phi(k-1),h_\phi=2\pi/(K-1)\right)\\ m=0,1,2,\ldots,M & (t_m=lm,l=T/M)\end{cases}. \tag{4.71}$$

Here, $f(j,k,m)$ are the mesh approximations of the functions $f\left(\rho_j,\phi_k,t_m\right)$;

$$D_+^t[U(j,k,m)]=[U(j,k,m+1)-U(j,k,m)]l^{-1},$$
$$D_-^t[U(j,k,m)]=[U(j,k,m)-U(j,k,m-1)]l^{-1},$$
$$D_0^t[U(j,k,m)]=[U(j,k,m+1)-U(j,k,m-1)](2l)^{-1}$$

are the operators of the right, left, and central difference derivatives. The interval of variation of the independent variable ρ is bounded above by the value of $\rho = L$; it is assumed that all compact discontinuities of the wave propagation medium in problems (4.69) (see Fig. 4.9) lie within a circle of radius L. The difference operator

$$L^{\rho,\phi}[U(j,k,m)] = \frac{1}{\rho_j^2} D_+^{\phi} D_-^{\phi} [U(j,k,m)] + \begin{cases} \frac{1}{\rho_j} D_+^{\rho}\left(\bar{\rho}_j D_-^{\rho}[U(j,k,m)]\right), & j \neq 1 \\ \frac{1}{\rho_1 h_\rho}\left(\bar{\rho}_2 D_-^{\rho}[U(2,k,m)]\right), & j = 1 \end{cases};$$

$$\bar{\rho}_j = 0.5\left(\rho_j + \rho_{j-1}\right), \quad j = 2, \ldots, J, \quad \bar{\rho}_0 = 0$$

provides (in the form of the condition $\lim_{\rho\to 0}(\rho\,\partial U/\partial\rho) = 0$) the requirement of boundedness of the solutions $U(\rho,\phi,t)$ near the point $\rho = 0$. In the case of problem (3.1), the difference equations (4.71) should be complemented by the equation $U(j,1,m) = U(j,K,m)$ that approximates the repeatability condition $U(\rho,\phi,t) = U(\rho,\phi+2\pi,t)\,;t \geq 0$. Using the quadrature formulas of the trapezoidal method, we obtain the discrete analogues of conditions (3.11), (4.70) (they differ only in the form of the transverse, $\mu_n(\phi)$, and auxiliary, $\xi_n(t-\tau)$ and $\eta_n(t-\tau)$, functions):

$$U(J,k,m+1) + \frac{l}{2}\left(\frac{U(J,k,m+1)}{2L} + B[U(J,k,m+1)]\right)$$
$$= \frac{2l}{\pi}\sum_n \mu_n(k)(-1)^n \sum_{r=1}^{m}[u_n(J,r)\,\xi_n(m+1-r) - B[u_n(J,r)]\,\eta_n(m+1-r)]. \tag{4.72}$$

Here, $\xi_n(m+1-r) = \xi_n(t_{m+1}-\tau_r)$, $\eta_n(m+1-r) = \eta_n(t_{m+1}-\tau_r)$, and

$$B[f(J)] = [3f(J) - 4f(J-1) + f(J-2)]\left(2h_\rho\right)^{-1}.$$

It is obvious that the possible sets of equations, such as (4.71), (4.72), uniquely determine the mesh functions $U(j,k,m)$. An explicit three-layer scheme enables one to evaluate relatively easily the series $\{U(j,k,m)\}_{h_\rho,h_\phi,l}$, which for $h_\rho, h_\phi, l \to 0$ converges to the exact values of solutions to the original problems $U\left(\rho_j,\phi_k,t_m\right)$. The only requirement is that the conditions for the stability of the scheme should be satisfied (see Section 3.1).

The experiments, aimed at testing the strict conditions (3.11), (4.70) are basically of the same kind as those implemented to verify the conditions (2.60), (2.61), and (4.66). Omitting the technical details, which are similar to those discussed in the Section 4.6.1, we cite only one result that is characteristic for the whole situation.

Figure 4.10 allows us to compare the errors caused by the application of the strict condition (4.70) (shown by dashed lines) and the classic heuristic ABC of the first approximation order [22],

$$\left(\frac{\partial}{\partial\rho} + \frac{\partial}{\partial t} + \frac{1}{2L}\right) U(\rho,\phi,t)\Big|_{\rho=L} = 0$$

(solid lines). Two problems of the type (4.69) for simple geometries are considered (see Fig. 4.10A). The supports of the source functions $\psi(g)$ are closed by an

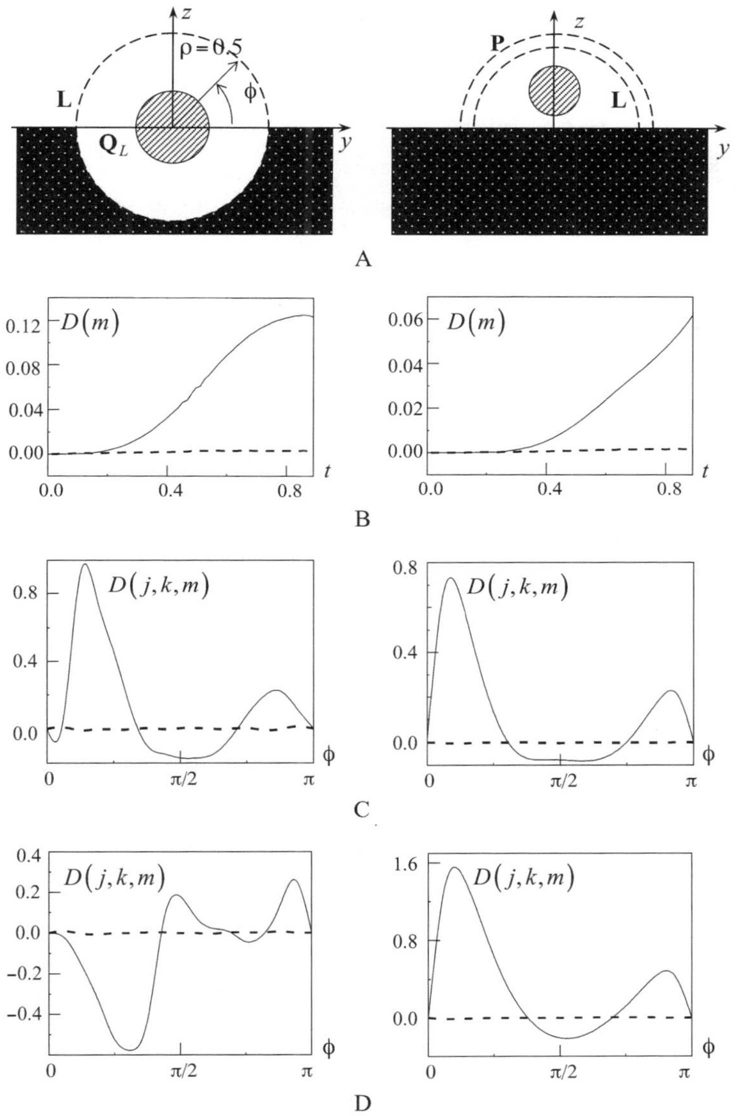

FIGURE 4.10. Results of the comparative tests of exact conditions (4.70) (dashed lines) and ABCs of the first order of approximation (solid lines) for two different geometries (A) of model problems (4.69). B: global errors; C ($t = 0.6$) and D ($t = 0.9$): local errors in the knots on virtual boundary $\rho = L = 0.5$.

oblique hatching, $\varphi = F = \sigma = \varepsilon - 1 \equiv 0$. For the data from the left column

$$\psi(\rho, \phi) = \begin{cases} 10\,(1 + \cos\phi + \cos 2\phi)\,; & \rho \le 0.45 \\ 0; & \rho > 0.45 \end{cases}, \qquad 0 \le \phi \le 2\pi,$$

and for the data from the right column

$$\psi(\rho, \phi) = \begin{cases} 10\,(1 + \cos\phi + \cos 2\phi)\,; & |\rho - 0.2| \le 0.15 \\ 0; & |\rho - 0.2| > 0.15 \end{cases}, \qquad 0 \le \phi \le \pi.$$

The local errors $D(j, k, m) = U_L(j, k, m) - U_P(j, k, m)$ are calculated at the mesh points at the imaginary boundary $\rho = L = 0.5$, $0 \le \phi \le \pi$. The value of

$$D(m) = \frac{1}{N} \sum_{j,k} |D(j, k, m)|$$

determines, as previously, the averaged global error at the mth time step overall within the whole mesh for the region $\mathbf{Q}_L$ (N is the total number of the mesh points). The problems have been discretized by a mesh of polar space coordinates ρ and ϕ [124]. The discretization step for ρ and ϕ (indices j and k) was $h = h_2 = 1.22 \cdot 10^{-2}$, and for the time $l = l_2 = 3.53 \cdot 10^{-5}$. The approximation order was $O(h^2 + l^2)$.

4.6.3. Absorbing Conditions on the Boundaries with Corner Points

An example of a numerical implementation of the algorithm based on the solution of (3.12), (3.28) to (3.30) is illustrated by Figure 4.11. A teflon radiator

$$\varepsilon(g) = 2.1\,(\chi(y)\,\chi(10.1 - y)\,\chi(1.35 - |z|) + \chi(y - 10.1)\,\chi(12.1 - y)\,\chi(0.75 - |z|))$$

(for the definition of the step functions of the kind $\chi[f_1(g)]\chi[f_2(g)]\ldots\chi[f_m(g)]$ see the appendix) is inserted into an aluminum cup (shell)

$$\sigma(g) = 13.3 \cdot 10^7\,(\chi(y + 0.25)\,\chi(2.5 - y)\,\chi(1.6 - |z|) - \chi(y)\,\chi(2.5 - y)\,\chi(1.35 - |z|))$$

and is excited by a "soft" source (a term taken from [27])

$$F(g, t) = 10\chi\left[0.25^2 - z^2 - (y - 1.25)^2\right]\cos(1.2t), \qquad \varphi(g) = \psi(g) \equiv 0$$

of E-polarized waves. The closing of the analysis area is $\overline{\mathbf{Q}_L} = [-0.5 \le y \le 15] \times [-2.5 \le z \le 2.5]$, the time step of the mesh is $l = 0.01$, and the value of $\sigma(g)$ is chosen on the assumption that all the dimensions are given in centimeters. The two-dimensional continuous-tone images show the distribution of the electric field strength $U(g, t) = E_x(g, t)$ overall in $\mathbf{Q}_L$ for different observation times t. A real antenna is modeled that is a part of a subsurface sounding radar (a development

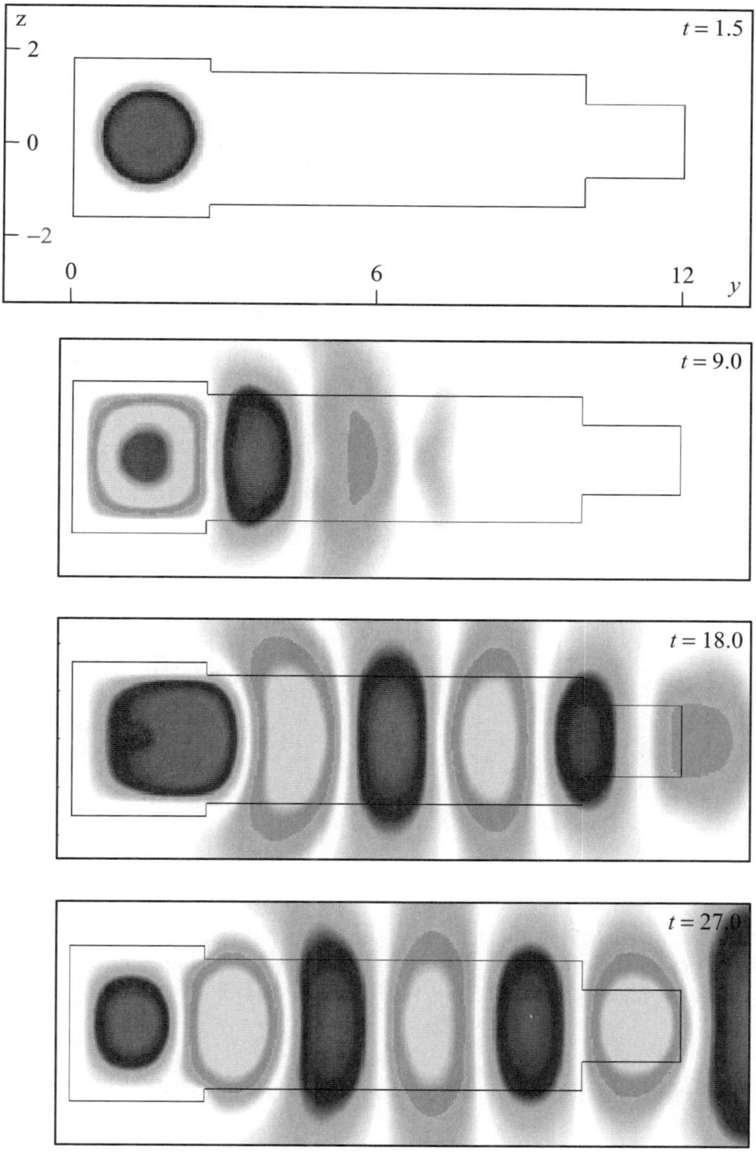

FIGURE 4.11. Modeling of the antenna for the underground sounding locator.

of the Turkish–Ukrainian Joint Research Laboratory, TUBITAK-MRC, Turkey, headed by Professor A.A. Vertiy).

The numerical experiments have been preceded by a comprehensive testing of the algorithm. One of the results of such a routine, but highly important, preparatory

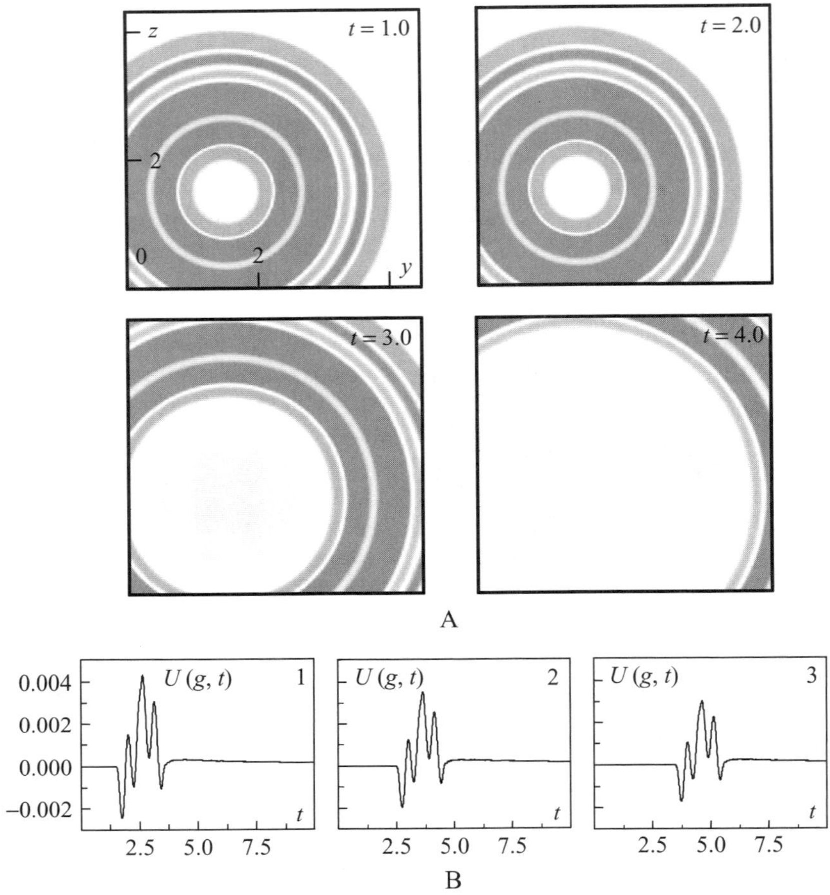

FIGURE 4.12. Results of the tests absorbing conditions, following from equations (3.28) to (3.30). A: field pattern $U(g,t) = E_x$ in domain $\mathbf{Q}_L$ in different instants of observation t; B: dynamics of function $U(g,t)$ changing in points $y = z = \sqrt{8}$ (1), $y = z = \sqrt{12.5}$ (2), and $y = z = \sqrt{18}$ (3) of the domain $\mathbf{Q}_L$.

study is presented in Figure 4.12. A current pulsed source of E-polarized waves

$$F(g,t) = \chi\left[0.25 - \left(y - \sqrt{2}\right)^2 - \left(z - \sqrt{2}\right)^2\right]\chi(1-t)\cos(4\pi(t-0.5))$$

is placed in the section $\mathbf{Q}_L = [0 < y < \sqrt{18}] \times [0 < z < \sqrt{18}]$ of free space $\mathbf{R}^2$. The central frequency of these pulses corresponds to the wavelength $\lambda = 0.5$. The time step of the mesh is $l = 0.005$; the space coordinate step is $h = 0.011$. The two-dimensional Figures 4.12A give us an idea of the field strength distribution $U(g,t) = E_x(g,t)$ overall in $\mathbf{Q}_L$ at various observation times t. The time when the excitation reaches the corner point $y = z = 0$ of the boundary $\mathbf{L}$ is $t = 1.5$. This is the corner point of the boundary that is nearest to the source. The most remote corner point ($y = z = \sqrt{18}$) is reached at $t = 3.5$. The boundaries of region $\mathbf{Q}_L$ are completely transparent for the cylindrical pulsed wave outgoing toward the region ${}_L\mathbf{Q}$. They are transparent not only at their smooth sections, but also at the broken ones. This fact is confirmed by Figures 4.12B, showing the dynamics of the changes in the field strength $U(g,t); 0 < t < 10$ at the points $y = z = \sqrt{8}$, $y = z = \sqrt{12.5}$, $y = z = \sqrt{18}$ that are separated by equal spacing in the diagonally situated region $\mathbf{Q}_L$. None of the partial components of the pulse shows any deviations from the main direction of the process.

4.6.4. Spherical Conditions in Axially Symmetric Problems

The analysis of an axially symmetric pulsed antenna based on the application of approximate absorbing boundary conditions to FDTD schemas are the subject of numerous papers published during the last couple of decades (see, e.g., [135,136]). We almost repeat the result from [136] in Figure 4.13 (the parameters of radiator

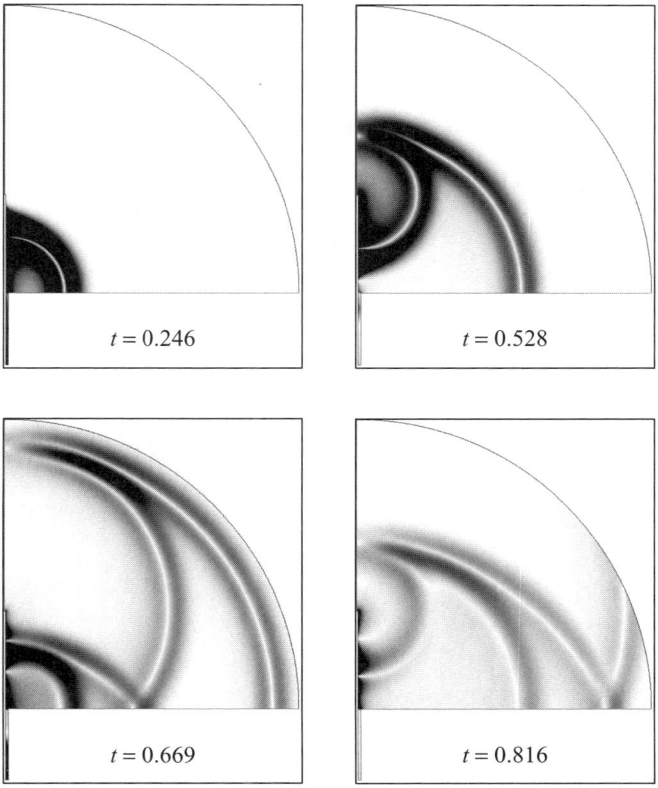

FIGURE 4.13. The pulsed TEM wave is exciting the monopole placed in the aperture of a circular coaxial waveguide. The pattern of $|\vec{E}(g,t)|$ in computational space $\mathbf{Q}_L$ for various time of observation t.

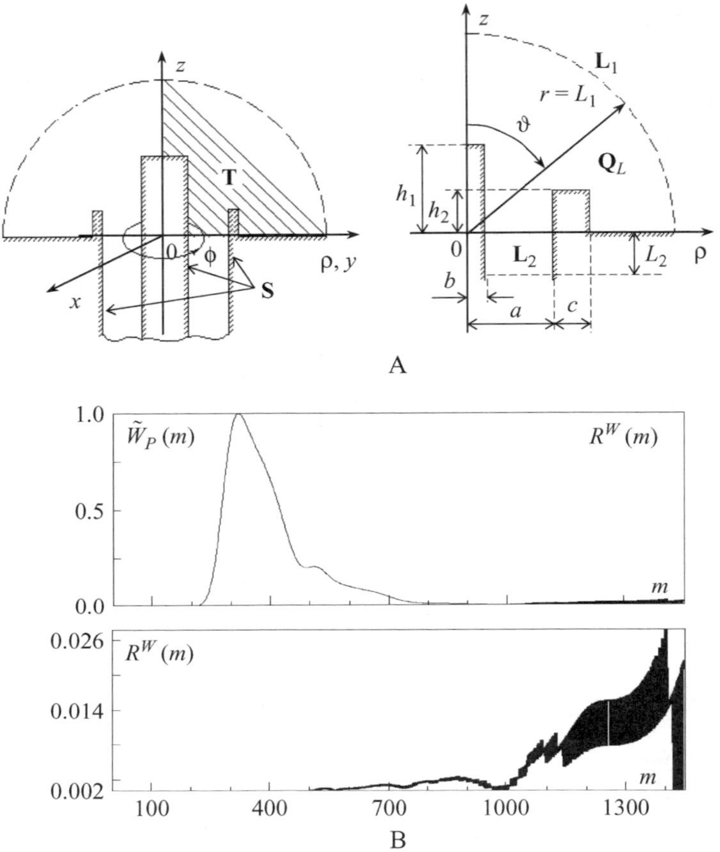

FIGURE 4.14. (A) Problem geometry and (B) results of the tests of conditions (4.5), (4.6) when antenna is excited with TEM-pulsed wave of coaxial waveguide.

and primary wave are somewhat changed): a monopole that is the continuation of the inner rod of a coaxial waveguide is excited by a pulsed TEM-wave with

$$E_{\rho}^{i}(g,t)=\frac{73}{\rho}\exp\left[-\left(\frac{z-t+\tilde{T}+L_2}{\tilde{T}/4}\right)^2\right];\quad \tilde{T}=0.0618.$$

Here (see Fig. 4.14A), $a=0.0048$, $b=0.0016$, $h_1=0.136$, $L_1=0.4$, $L_2=0.1$, and $h_2=0$; the grid dimensions are $h=h_\rho=h_z=2l=0.0008$ (l is the grid size for t). The distribution of values $|\vec{E}(g,t)|$ in the computational domain $\mathbf{Q}_L$ for different times t is computed by solving the closed problem (4.1), (4.6), (4.9). In the wave equation of problem (4.1) we have a zero right-hand side (there are no current sources). Condition (4.9) has been formulated in terms of the secondary field $U^s(g,t)$ (the supports of the functions $\varphi(g)=U^i(g,t)|_{t=0}$ and

$\varphi(g) = \partial U^i(g,t)/\partial t|_{t=0}$ lie in the domain ${}_L\mathbf{Q}$). In the exact nonlocal conditions (4.6) and (4.9), the first 18 and 7 spatio–temporal amplitudes $u_n(r,t)$ and $\tilde{u}_n^s(z,t)$ of the outgoing waves $U(g,t)$ and $U^s(g,t)$ are taken to be nonzero (see representations (4.3) and (4.7)).

Figure 4.14B illustrates one of many characteristic results of the tests of the algorithms, based on a standard finite-difference discretization schema for the modified problems (3.75), (3.83), and (4.1), (4.6). The antenna (see Fig. 4.14A: $a = 0.0235$; $b = 0.01$; $h_1 = 0.05$; $h_2 = 0$) is excited with a pulsed TEM-wave in a coaxial waveguide with

$$H_\phi^i = U^i(g,t) = \frac{1}{\rho}\exp\left[-\left(\frac{z - t + \tilde{T} + L_2}{\tilde{T}/4}\right)^2\right]; \quad \tilde{T} = 0.05. \tag{4.73}$$

The two problems (4.1), (4.9) (condition (4.9) is formulated in terms of secondary field $U^s(g,t)$, $L_2 = 0.021$, $h = 0.0005$) are considered at the same time. Their computational domain is limited by the arcs $r = L_1 = 0.061$ and $r = P_1 = 0.354$. On the arc $r = L_1$ we have formulated the condition (4.6); the number of spherical harmonics taken into account is equal to ten. During the time of observation $t \leq T$, the excitation does not reach the arc $r = P_1$. Using the solutions U_L and U_P of these problems obtained in each time layer $t_m = ml$, $m = 1, 2, \ldots, M$, $M = 1400$, the averaged densities $W_L(m)$ and $W_P(m)$ of electromagnetic power ($W = [\varepsilon_0|\vec{E}|^2 + \mu_0|\vec{H}|^2]/2$) over the rather large domain $\mathbf{T} \subset \mathbf{Q}_L$ (covered with oblique shadow in Fig. 4.14A) are calculated. The functions $\tilde{W}_P(m) = W_P(m)/\max_m W_P(m)$ are shown in Figure 4.14B by a fine line. Bold lines describe the relative calculation errors $R^W(m) = |W_P(m) - W_L(m)|/W_P(m)$. The errors remain low (less than 10^{-3}) until the moment when the exact value of $W_P(m)$ decreases by 100 to 1000 times. That is as long as the radiation field is present in the computational domain.

The results presented in Figure 4.15 allow us to make judgments about the absolute computational error, and about the choice of time–space grid that will provide the required accuracy of the calculations. Here,

$$D_{\max}(t_m) = \max_{j,k:\{\rho_j,z_k\}\in\mathbf{Q}_L} |U_L(j,k,m) - U_P(j,k,m)|, \tag{4.74}$$

$$D(t_m) = \frac{1}{N}\sum_{j,k:\{\rho_j,z_k\}\in\mathbf{Q}_L} |U_L(j,k,m) - U_P(j,k,m)|, \tag{4.75}$$

$$\tilde{D}_{\max}(t_m) = \max_{j,k:\{\rho_j,z_k\}\in\mathbf{Q}_L} |U_L(2j,2k,2m;h) - U_L(j,k,m;2h)|, \tag{4.76}$$

$$\tilde{D}(t_m) = \frac{1}{N}\sum_{j,k:\{\rho_j,z_k\}\in\mathbf{Q}_L} |U_L(2j,2k,2m;h) - U_L(j,k,m;2h)|, \tag{4.77}$$

$$U_{L,\text{norm}}(\tilde{g},t_m;h) = U_L(\tilde{g},m;h)\big/U_{L,\max}(\tilde{g};h);$$

$$U_{L,\max}(\tilde{g};h) = \max_{m:t_m\leq 0.1} U_L(\tilde{g},m;h), \tag{4.78}$$

$$D_{\text{norm}}(t_m) = \left|U_{L,\text{norm}}(\tilde{g},t_{2m};h) - U_{L,\text{norm}}(\tilde{g},t_m;2h)\right|. \tag{4.79}$$

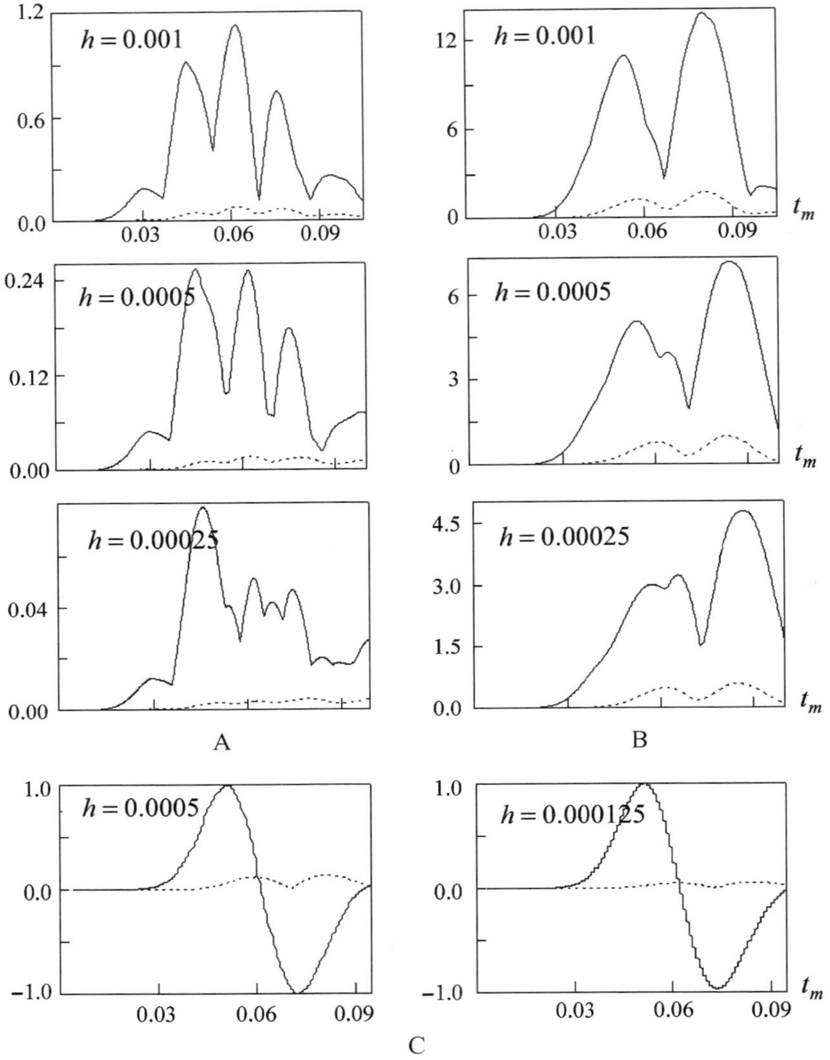

FIGURE 4.15. Analysis of computational errors in the numerical solution of the problem of axially symmetric pulse antenna. A: the behavior of absolute errors, caused by closing of the analysis domain with virtual boundaries; B: behavior of the errors of the finite difference schema; C: determination of the computational schema parameters according to the required accuracy of the solution.

The antenna (see Fig. 4.14A: $a = 0.012$; $b = 0.004$; $h_1 = 0.008$; $h_2 = 0$) is excited by the pulsed TEM-wave (4.73). The first set of figures (see Fig. 4.15A) shows (for different values of $h = h_\rho = h_z = 2l$) the behavior of absolute maximal ($D_{\max}(t_m)$, solid lines) and averaged ($D(t_m)$, dashed lines) errors caused by closing the analysis domain in the problem (4.1) with boundaries $\mathbf{L}_1$ and $\mathbf{L}_2$ with exact conditions (4.6) and (4.9) on them. In (4.74), (4.75) $f(j, k, m)$ are the grid approximate functions $f\left(\rho_j, z_k, t_m\right)$; $U_P(j, k, m)$ can serve as a template for solutions to the problem (4.1) with zero conditions on the boundaries $r = P_1 = 0.125$ in the domain of radiation of the antenna and $z = -P_2 = -0.104$ in the coaxial waveguide (the excitation does not reach the boundaries within the time intervals considered); $U_L(j, k, m)$ is the solution to the problem (4.1), (4.6), (4.9) with $L_1 = 0.025$ and $L_2 = 0.004$; N is the general number of nodes of the grid covering the analysis domain $\mathbf{Q}_L$.

The results presented in the second set of figures (see Fig. 4.15B) specify the behavior of the errors of the finite difference schema itself. Solid lines correspond to the functions $\tilde{D}_{\max}(t_m)$, and dashed lines to $\tilde{D}(t_m)$. In (4.76), (4.77) $U_L(j, k, m; h)$ is the approximate solution to the problem (4.1), (4.6), (4.9) in the grid with space step $h = h_\rho = h_z = 2l$. Together with the results partly presented in the third set of figures (see Fig. 4.15C; only one parameter is changed here: $L_1 = 0.018$) they allow us to establish the parameter values of the computation, which will give the required accuracy of the computations. The solid curves in Figure 4.15C describe the functions $U_{L,\mathrm{norm}}(\tilde{g}, t_m; h)$, and the dashed ones the functions $D_{\mathrm{norm}}(t_m)$. The function $U_L(\tilde{g}, m; h)$ in (4.78), (4.79) is the approximate solution to the problem (4.1), (4.6), (4.9) on the grid with space step h in the test point $\tilde{g} \in \mathbf{Q}_L$ at time t_m (in the example considered $\tilde{g} = \{\tilde{\rho}, \tilde{z}\} = \{0.08, 0\}$).

The tests we have carried out allow us to make the following conclusions. The errors caused by closing the analysis domain using the exact absorbing conditions and appearing as a result of the limitation of the mode structure of waves, outgoing through the boundaries $\mathbf{L}_1$, $\mathbf{L}_2$ and because of the approximation of conditions (4.6) at the grid nodes $\{\rho_j, z_k, t_m\}$, are 10 to 100 times less than the error caused by the discretization of the problem and decrease with decreasing step $h = 2l$ considerably faster than the latter ones. The total errors of the schema implementing algorithms (4.1), (4.6), (4.9) are in strict conjunction with the step value h and can be adjusted according to the requirements of the desired accuracy of the numerical solutions.

5
Transform Operators in Space of Signal's Evolutionary Basis: A Time-Domain Analogue of the Generalized Scattering-Matrix Method

5.1. Introduction

The approach presented in this chapter is based on the description of the scattering properties of the discontinuities in free space and regular wave-guiding structures in terms of transform operators. These transform operators are qualitatively the same for all scatterers that are characterized by the discrete spatial spectrum of the signal's evolutionary basis. By the notion of scatterers with discrete spectrum we imply such electromagnetic objects that have solutions of the corresponding boundary value problems in the form of a series of spherical or cylindrical harmonics, or of plane wave or waveguide modes. In FD this approach is known as the generalized scattering-matrix technique and is widely used nowadays in the context of automatic simulation and design.

All the algorithmization schemes of the initial boundary value problems presented in this chapter are replicas of the approaches that have been implemented logically and successfully for problems in the frequency domain. They are only modified to be suitable for time-domain problems. Thus, for example, in Section 5.2 we have introduced the fundamental notions needed to describe transient processes in terms of the transform operators of signal evolution bases, and we have as well sketched the general algorithmization scheme for problems of increased complexity, that is, we have suggested a direct analogue to the method of generalized scattering matrices.

The main idea of Section 5.3 is the technique of developing a series of efficient rigorous approaches to the computation of electromagnetic characteristics of elementary inhomogeneities (space–time analogues of the mode-matching or Fourier method, the residue calculation method, and the analytical regularization method). In the statement of the initial boundary value problems we narrow down the class of waveguide configurations considered and focus on circular and coaxial cylinders and furthermore we consider the simplest case when they are excited by symmetric electromagnetic TE_{0p}-waves.

Section 5.4 presents the algorithms for computing the transient characteristics of complex resonance inhomogeneities (space–time analogues to the algorithms of the method of generalized scattering matrices) in circular and coaxial waveguides.

Section 5.5 is devoted to the extension of the obtained results to the initial boundary value problems of electromagnetic grating theory. The transfer to the Fourier images of the functions describing physically realizable fields and sources enables one to reduce these problems to those of scattering transient plane waves by the inhomogeneities of the Floquet channel, that is, a plane-parallel waveguide with specific boundary conditions.

5.2. Evolutionary Signal Basis and Transform Operators

The utilization of algorithms from the generalized scattering-matrix technique in FD considerably increases the range of problems for which rigorous analytic–numerical approaches can be applied efficiently. It is well known that the area of application of rigorous methods is rather restricted and has been useful mainly for idealized scatterers of rather simple geometry. The application of the generalized scattering-matrix technique allows us to study complicated structures connected by intervals of regular domains (section of regular waveguide, open space, etc.) with the same efficiency as separate units, making up the complicated junction [3,5,12,137]. Taking into account the interaction between units leads to the solution of the canonic Fredholm system of algebraic equations with exponentially decreasing entries (along both lines and rows [9]). It is presumed that a similar role in the TD problems may be played by the method of transform operators. This section dwells on its description. Several principal issues of the method have already been published in [21,30,117,118]. Below we present for the first time a schema that provides the solution of 3-D electromagnetic problems.

5.2.1. The Field of Given Sources in Hollow Waveguide of Arbitrary Cross-Section

Consider first an applied problem (see, e.g., Section 4.4): find the field $\{\vec{E}(g,t), \vec{H}(g,t)\}$ of given current and momentary ($\vec{E}(g,0)$ and $\vec{H}(g,0)$) sources in a hollow ($\varepsilon(g) \equiv 1$, $\sigma(g) \equiv 0$) regular waveguide with perfectly conducting walls $\mathbf{S} = \mathbf{S}_z \times (-\infty < z < \infty)$:

$$\left\{ \begin{array}{ll} \eta_0 \mathrm{rot}\vec{H} = \dfrac{\partial \vec{E}}{\partial t} + \vec{J}, \quad \mathrm{rot}\vec{E} = -\eta_0 \dfrac{\partial \vec{H}}{\partial t}; & g \in \mathbf{Q}, \quad t > 0 \\ E_{tg}(g,t)\big|_{g\in\mathbf{S}} = 0, \quad \left.\dfrac{\partial H_{tg}(g,t)}{\partial \vec{n}}\right|_{g\in\mathbf{S}} = 0; & t \geq 0 \end{array} \right. . \tag{5.1}$$

Here, $\mathbf{Q} = \{g = \{x, y, z\} : \{x, y\} \in \mathrm{int}\,\mathbf{S}_z, -\infty < z < \infty\}$ is the inside part of the waveguide; $\vec{J} = \eta_0 \vec{j}$, $\vec{j} \equiv \vec{j}(g,t)$ is the vector of density of external currents (see

Section 1.2.1); $\mathbf{S}_z$ is the boundary contour of the waveguide cross-section; int $\mathbf{S}_z$ is the simply connected domain in the plane $z = \text{const}$, bounded by the contour $\mathbf{S}_z$. In (5.1) it is supposed that the supports of the source functions are bounded in $\mathbf{Q}$.

From (5.1), using equations (1.2), (1.3), we arrive at (see also Section 4.4.3)

$$\left\{\begin{array}{l} \left[-\dfrac{\partial^2}{\partial t^2}+\Delta\right]\left\{\begin{array}{c}\vec{E}\\ \vec{H}\end{array}\right\}=\left\{\begin{array}{c}\partial\vec{J}/\partial t+\operatorname{grad}\rho_2\\ -\operatorname{rot}\vec{j}\end{array}\right\}=\left\{\begin{array}{c}\vec{F}^E\\ \vec{F}^H\end{array}\right\};\\ g=\{x,y,z\}\in\mathbf{Q},\quad t>0\\ \left\{\begin{array}{c}\vec{E}(g,0)\\ \vec{H}(g,0)\end{array}\right\}=\left\{\begin{array}{c}\vec{\varphi}^E\\ \vec{\varphi}^H\end{array}\right\},\quad \left\{\begin{array}{l}\partial\vec{E}(g,t)/\partial t\Big|_{t=0}=\eta_0\operatorname{rot}\vec{H}(g,0)\\ \partial\vec{H}(g,t)/\partial t\Big|_{t=0}=-\eta_0{}^{-1}\operatorname{rot}\vec{E}(g,0)\end{array}\right\}=\left\{\begin{array}{c}\vec{\psi}^E\\ \vec{\psi}^H\end{array}\right\};\ .\\ g\in\bar{\mathbf{Q}}\\ \left.\left\{\begin{array}{l}E_{tg}(g,t)\\ \partial H_{tg}(g,t)/\partial\vec{n}\end{array}\right\}\right|_{g\in\mathbf{S}}=0;\quad t\geq 0 \end{array}\right. \tag{5.2}$$

We have to find the values of the longitudinal field $\{\vec{E},\vec{H}\}$ components E_z and H_z in all points g of the domain $\mathbf{Q}$ and for all observation times $t > 0$. For this, let us utilize the following scalar initial boundary value problems, which follows from (5.2).

$$\left\{\begin{array}{l} \left[-\dfrac{\partial^2}{\partial t^2}+\Delta\right]\left\{\begin{array}{c}E_z\\ H_z\end{array}\right\}=\left\{\begin{array}{c}F_z^E\\ F_z^H\end{array}\right\};\quad g=\{x,y,z\}\in\mathbf{Q},\quad t>0\\ \left\{\begin{array}{c}E_z(g,0)\\ H_z(g,0)\end{array}\right\}=\left\{\begin{array}{c}\varphi_z^E\\ \varphi_z^H\end{array}\right\},\quad \left\{\begin{array}{l}\partial E_z(g,t)/\partial t\big|_{t=0}\\ \partial H_z(g,t)/\partial t\big|_{t=0}\end{array}\right\}=\left\{\begin{array}{c}\psi_z^E\\ \psi_z^H\end{array}\right\};\quad g\in\bar{\mathbf{Q}}\ .\\ \left.\left\{\begin{array}{l}E_z(g,t)\\ \partial H_z(g,t)/\partial\vec{n}\end{array}\right\}\right|_{g\in\mathbf{S}}=0;\quad t\geq 0 \end{array}\right. \tag{5.3}$$

In problems (5.3), let us separate the transverse variables (x and y) and present the solutions in the form

$$\left\{\begin{array}{c}E_z(g,t)\\ H_z(g,t)\end{array}\right\}=\sum_{n=\left\{\begin{smallmatrix}1\\0\end{smallmatrix}\right\}}^{\infty}\left\{\begin{array}{c}v_{n,z}^E(z,t)\\ v_{n,z}^H(z,t)\end{array}\right\}\left\{\begin{array}{c}\mu_n^E(x,y)\\ \mu_n^H(x,y)\end{array}\right\}, \tag{5.4}$$

where $\left\{\mu_n^E(x,y)\right\}_{n=1}^{\infty}$ and $\left\{\mu_n^H(x,y)\right\}_{n=0}^{\infty}$ (we have preserved the numbering of modes introduced in Section 2.4.1) are complete orthonormal systems of solutions to Sturm–Liouville problems for the equation $\left(\partial^2/\partial x^2+\partial^2/\partial y^2+\lambda^2\right)\mu = 0$ in the domain int $\mathbf{S}_z$ with Dirichlet ($\mu^E(x,y)|_{\{x,y\}\in\mathbf{S}_z}=0$) and Neumann ($\partial\mu^H(x,y)/\partial\vec{n}|_{\{x,y\}\in\mathbf{S}_z}=0$) conditions on its boundary $\mathbf{S}_z$. These solutions have corresponding eigenvalues λ_n^E and λ_n^H. Finding the scalar functions $v_{n,z}^E(z,t)$ and $v_{n,z}^H(z,t)$ is reduced to finding the solution of the following Cauchy problems for

1-D Klein–Gordon equations.

$$\begin{cases} \left[-\dfrac{\partial^2}{\partial t^2} + \dfrac{\partial^2}{\partial z^2} - \begin{Bmatrix} (\lambda_n^E)^2 \\ (\lambda_n^H)^2 \end{Bmatrix} \right] \begin{Bmatrix} v_{n,z}^E(z,t) \\ v_{n,z}^H(z,t) \end{Bmatrix} = \begin{Bmatrix} F_{n,z}^E \\ F_{n,z}^H \end{Bmatrix}; \quad t > 0 \\ \begin{Bmatrix} v_{n,z}^E(z,0) \\ v_{n,z}^H(z,0) \end{Bmatrix} = \begin{Bmatrix} \varphi_{n,z}^E \\ \varphi_{n,z}^H \end{Bmatrix}, \quad \dfrac{\partial}{\partial t} \begin{Bmatrix} v_{n,z}^E(z,t) \\ v_{n,z}^H(z,t) \end{Bmatrix} \Bigg|_{t=0} = \begin{Bmatrix} \psi_{n,z}^E \\ \psi_{n,z}^H \end{Bmatrix}; \quad , \quad -\infty < z < \infty. \\ n = \begin{Bmatrix} 1, 2, 3, \ldots \\ 0, 1, 2, \ldots \end{Bmatrix} \end{cases} \tag{5.5}$$

Here, $F_{n,z}^E$, $\varphi_{n,z}^E$, $\psi_{n,z}^E$ and $F_{n,z}^H$, $\varphi_{n,z}^H$, $\psi_{n,z}^H$ are the Fourier coefficients of functions F_z^E, φ_z^E, ψ_z^E and F_z^H, φ_z^H, ψ_z^H, respectively, in the bases $\{\mu_n^E(x,y)\}$ and $\{\mu_n^H(x,y)\}$.

After continuation of the functions $v_{n,z}^E(z,t)$, $v_{n,z}^H(z,t)$ and $F_{n,z}^E(z,t)$, $F_{n,z}^H(z,t)$ by zero over the semi-axis $t < 0$, we formulate (5.5) as the generalized Cauchy problems [34]:

$$\begin{aligned} B\begin{pmatrix} \lambda_n^E \\ \lambda_n^H \end{pmatrix} \begin{bmatrix} v_{n,z}^E(z,t) \\ v_{n,z}^H(z,t) \end{bmatrix} &\equiv \left[-\frac{\partial^2}{\partial t^2} + \frac{\partial^2}{\partial z^2} - \begin{Bmatrix} (\lambda_n^E)^2 \\ (\lambda_n^H)^2 \end{Bmatrix} \right] \begin{Bmatrix} v_{n,z}^E(z,t) \\ v_{n,z}^H(z,t) \end{Bmatrix} = \begin{Bmatrix} F_{n,z}^E \\ F_{n,z}^H \end{Bmatrix} \\ &- \delta^{(1)}(t) \begin{Bmatrix} \varphi_{n,z}^E \\ \varphi_{n,z}^H \end{Bmatrix} - \delta(t) \begin{Bmatrix} \psi_{n,z}^E \\ \psi_{n,z}^H \end{Bmatrix} = \begin{Bmatrix} f_{n,z}^E \\ f_{n,z}^H \end{Bmatrix}; \quad n = \begin{Bmatrix} 1, 2, 3, \ldots \\ 0, 1, 2, \ldots \end{Bmatrix}, \\ &-\infty < z < \infty, \quad -\infty < t < \infty. \end{aligned} \tag{5.6}$$

The convolution of the fundamental solution $G(z,t,\lambda) = -(1/2)\chi(t - |z|)J_0(\lambda\sqrt{t^2 - z^2})$ of the operator $B(\lambda)$ (see Section 1.2.5) with the right-hand side of equations (5.6) allows us to present their solutions in the form

$$\begin{aligned} \begin{Bmatrix} v_{n,z}^E(z,t) \\ v_{n,z}^H(z,t) \end{Bmatrix} &= \begin{Bmatrix} G(z,t,\lambda_n^E) \\ G(z,t,\lambda_n^H) \end{Bmatrix} * \begin{Bmatrix} f_{n,z}^E \\ f_{n,z}^H \end{Bmatrix} \\ &= -\frac{1}{2} \Bigg[\int_{-\infty}^{\infty} \int_{-\infty}^{\infty} \chi\left[(t-\tau) - |z - \omega|\right] J_0 \left(\begin{Bmatrix} \lambda_n^E \\ \lambda_n^H \end{Bmatrix} \sqrt{(t-\tau)^2 - (z-\omega)^2} \right) \\ &\times \left(\begin{Bmatrix} F_{n,z}^E(\omega,\tau) \\ F_{n,z}^H(\omega,\tau) \end{Bmatrix} - \delta^{(1)}(\tau) \begin{Bmatrix} \varphi_{n,z}^E(\omega) \\ \varphi_{n,z}^H(\omega) \end{Bmatrix} - \delta(\tau) \begin{Bmatrix} \psi_{n,z}^E(\omega) \\ \psi_{n,z}^H(\omega) \end{Bmatrix} \right) d\omega d\tau \Bigg]; \\ &-\infty < z < \infty, \quad t \geq 0, \quad n = \begin{Bmatrix} 1, 2, 3, \ldots \\ 0, 1, 2, \ldots \end{Bmatrix}. \end{aligned} \tag{5.7}$$

Let us now go outside the bounded domain **Q**, where all sources are concentrated, and move into the domain $\mathbf{G} \subset \mathbf{Q}$, where the waves induced by these sources can propagate freely. Here the following representations (see Sections 1.2.3

and 2.4.1)

$$\vec{E} = \left(\frac{\partial^2 U^E}{\partial x \partial z} - \frac{\partial^2 U^H}{\partial y \partial t}\right)\vec{x} + \left(\frac{\partial^2 U^E}{\partial y \partial z} + \frac{\partial^2 U^H}{\partial x \partial t}\right)\vec{y} + \left(\frac{\partial^2 U^E}{\partial z^2} - \frac{\partial^2 U^E}{\partial t^2}\right)\vec{z},$$

$$\eta_0 \vec{H} = \left(\frac{\partial^2 U^E}{\partial y \partial t} + \frac{\partial^2 U^H}{\partial x \partial z}\right)\vec{x} + \left(-\frac{\partial^2 U^E}{\partial x \partial t} + \frac{\partial^2 U^H}{\partial y \partial z}\right)\vec{y} + \left(\frac{\partial^2 U^H}{\partial z^2} - \frac{\partial^2 U^H}{\partial t^2}\right)\vec{z} \tag{5.8}$$

are valid. In (5.8),

$$U^{E,H}(g,t) = \sum_{n=1,0}^{\infty} u_n^{E,H}(z,t)\, \mu_n^{E,H}(x,y) \tag{5.9}$$

the Borgnis scalar functions have the properties that $[\Delta - \partial^2/\partial t^2][\partial U^{E,H}(g,t)/\partial t] = 0$. The relations (5.4), (5.7) to (5.9) define the field $\{\vec{E}, \vec{H}\}$ in all points g of the domain **G** at any time $t > 0$. In fact, inasmuch as at time $t = 0$ the domain **G** has not been excited, then $[\Delta - \partial^2/\partial t^2]U^{E,H} = 0;\ g \in \mathbf{G},\ t > 0$. From this statement, together with (5.8), (5.9), we get the following relations.

$$E_z = \frac{\partial^2 U^E}{\partial z^2} - \frac{\partial^2 U^E}{\partial t^2} = -\left(\frac{\partial^2 U^E}{\partial x^2} + \frac{\partial^2 U^E}{\partial y^2}\right) = \sum_{n=1}^{\infty} \left(\lambda_n^E\right)^2 u_n^E \mu_n^E,$$

$$\eta_0 H_z = \frac{\partial^2 U^H}{\partial z^2} - \frac{\partial^2 U^H}{\partial t^2} = -\left(\frac{\partial^2 U^H}{\partial x^2} + \frac{\partial^2 U^H}{\partial y^2}\right) = \sum_{n=0}^{\infty} \left(\lambda_n^H\right)^2 u_n^H \mu_n^H.$$

Taking (5.4) into account, we derive $u_n^E(z,t) = \left(\lambda_n^E\right)^{-2} v_{n,z}^E(z,t)$, $u_n^H(z,t) = \eta_0 \left(\lambda_n^H\right)^{-2} v_{n,z}^H(z,t)$. The functions $U^{E,H}(g,t)$ are defined. Together with these functions, the transverse components of the field $\{\vec{E}, \vec{H}\}$ are defined by relations (5.8).

5.2.2. Key Statements

Consider now the model problem of definition of the field $\{\vec{E}(g,t), \vec{H}(g,t)\}$ of momentary ($\vec{E}(g,0)$ and $\vec{H}(g,0)$) and current sources in an open waveguide resonator **Q**, $\mathbf{S} = \bar{\mathbf{Q}} \backslash \mathbf{Q}$, that is, the conjunction of two semi-infinite (for $z_j > 0$, $j = 1, 2$) regular waveguides **A** and **B** connected by a compact inhomogeneity $\mathbf{Q}_L \subset \mathbf{Q}$ (see Fig. 5.1):

$$\begin{cases} \eta_0 \,\mathrm{rot}\, \vec{H} = \varepsilon \dfrac{\partial \vec{E}}{\partial t} + \sigma \vec{E} + \vec{J}, \quad \mathrm{rot} \vec{E} = -\eta_0 \dfrac{\partial \vec{H}}{\partial t}; \\ g = \{x, y, z\} \in \mathbf{Q}, \quad t > 0 \\ E_{tg}(g,t)\big|_{g \in \mathbf{S}} = 0, \quad \left.\dfrac{\partial H_{tg}(g,t)}{\partial \vec{n}}\right|_{g \in \mathbf{S}} = 0; \quad t \geq 0 \end{cases} . \tag{5.10}$$

It is assumed that the supports of the functions $\varepsilon(g) - 1$ and $\sigma(g)$ (which are finite in **Q**) are located in $\mathbf{Q}_L$ and are field sources far away in waveguide **A**. These sources produce the wave $\{\vec{E}_1^i(g,t), \vec{H}_1^i(g,t)\}$ that is coming onto boundary $z_1 = 0$

FIGURE 5.1. The geometry of model problem.

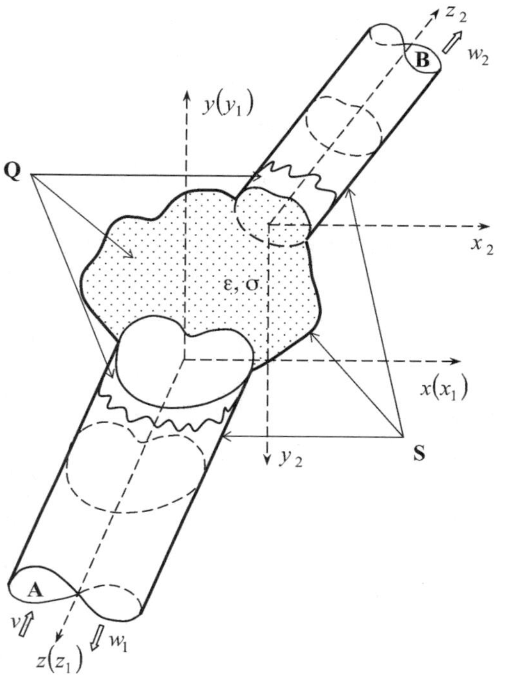

of the inhomogeneity $\mathbf{Q}_L$. Denote by $\left\{\vec{E}_j^s(g,t), \vec{H}_j^s(g,t)\right\}$ the field appearing in waveguides **A** and **B** as a result of scattering of the incident wave by the resonator $\mathbf{Q}_L$. Here, for waveguide **A**,

$$\{\vec{E}(g,t), \vec{H}(g,t)\} = \left\{\vec{E}_1^i(g,t), \vec{H}_1^i(g,t)\right\} + \left\{\vec{E}_1^s(g,t), \vec{H}_1^s(g,t)\right\}$$

and for waveguide **B**,

$$\{\vec{E}(g,t), \vec{H}(g,t)\} = \left\{\vec{E}_2^s(g,t), \vec{H}_2^s(g,t)\right\}.$$

It has been shown in Section 5.2.1 that the fields in all domains are uniquely defined by their longitudinal (directed along waveguide generatrix) components that can be presented in the form

$$\begin{Bmatrix} E_{z_1}^i(g,t) \\ H_{z_1}^i(g,t) \end{Bmatrix} = \sum_{n=\left\{\begin{smallmatrix}1\\0\end{smallmatrix}\right\}}^{\infty} \begin{Bmatrix} v_{n1}^E(z_1,t) \\ v_{n1}^H(z_1,t) \end{Bmatrix} \begin{Bmatrix} \mu_{n1}^E(x_1,y_1) \\ \mu_{n1}^H(x_1,y_1) \end{Bmatrix}; \quad z_1 \geq 0, \quad t \geq 0, \tag{5.11}$$

$$\begin{Bmatrix} E_{z_j}^s(g,t) \\ H_{z_j}^s(g,t) \end{Bmatrix} = \sum_{n=\left\{\begin{smallmatrix}1\\0\end{smallmatrix}\right\}}^{\infty} \begin{Bmatrix} w_{nj}^E(z_j,t) \\ w_{nj}^H(z_j,t) \end{Bmatrix} \begin{Bmatrix} \mu_{nj}^E(x_j,y_j) \\ \mu_{nj}^H(x_j,y_j) \end{Bmatrix};$$

$$z_j \geq 0, \quad t \geq 0, \quad j = 1, 2. \tag{5.12}$$

Here, $\{\mu_{nj}^E(x_j, y_j)\}_{n=1}^{\infty}$ and $\{\mu_{nj}^H(x_j, y_j)\}_{n=0}^{\infty}$ are complete orthonormal systems of the solutions to the Sturm–Liouville problems for the equations $(\partial^2/\partial x_j^2 + \partial^2/\partial y_j^2 + \lambda_j^2)\mu_j = 0$ in domains $\operatorname{int}\mathbf{S}_{z_j}$ with Dirichlet $(\mu_j^E(x_j, y_j) = 0)$ and Neumann $(\partial\mu_j^H(x_j, y_j)/\partial\vec{n}_j = 0)$ conditions at the boundaries $\mathbf{S}_{z_j}$. The eigenvalues λ_{nj}^E and λ_{nj}^H correspond to these solutions. $\mathbf{S}_{z_j}$ are the boundary contours of transverse sections of waveguides $\mathbf{A}$ ($j = 1$) and $\mathbf{B}$ ($j = 2$). $\operatorname{int}\mathbf{S}_{z_j}$ are simply connected domains in the planes $z_j = \text{const} \geq 0$, bounded by contours $\mathbf{S}_{z_j}$. The scalar functions $w_{nj}^E(z_j, t)$ and $w_{nj}^H(z_j, t)$ in domains $z_j > 0$ are defined as the solution of the following Klein–Gordon problems.

$$\left\{\begin{array}{l} \left[-\dfrac{\partial^2}{\partial t^2} + \dfrac{\partial^2}{\partial z_j^2} - \left\{\begin{array}{c}\left(\lambda_{nj}^E\right)^2 \\ \left(\lambda_{nj}^H\right)^2\end{array}\right\}\right]\left\{\begin{array}{c} w_{nj}^E(z_j, t) \\ w_{nj}^H(z_j, t)\end{array}\right\} = 0; \\ t > 0, \quad z_j > 0 \\ \left\{\begin{array}{c} w_{nj}^E(z_j, 0) \\ w_{nj}^H(z_j, 0)\end{array}\right\} = 0, \quad \dfrac{\partial}{\partial t}\left\{\begin{array}{c} w_{nj}^E(z_j, t) \\ w_{nj}^H(z_j, t)\end{array}\right\}\Bigg|_{t=0} = 0; \\ z_j \geq 0, \quad n = \left\{\begin{array}{c} 1, 2, 3, \ldots \\ 0, 1, 2, \ldots\end{array}\right\} \end{array}\right. ; \quad j = 1, 2. \tag{5.13}$$

Let us construct the sets $v_1(z_1, t) = \{v_{n1}(z_1, t)\}_{n=-\infty}^{\infty}$, $w_j(z_j, t) = \{w_{nj}(z_j, t)\}_{n=-\infty}^{\infty}$ and $\{\lambda_{nj}\}_{n=-\infty}^{\infty}$, where $v_{n1}(z_1, t) = v_{n1}^E(z_1, t)$, $w_{nj}(z_j, t) = w_{nj}^E(z_j, t)$, $\lambda_{nj} = \lambda_{nj}^E$ for $n = 1, 2, 3, \ldots$ and $v_{n1}(z_1, t) = v_{-n1}^H(z_1, t)$, $w_{nj}(z_j, t) = w_{-nj}^H(z_j, t)$, $\lambda_{nj} = \lambda_{-nj}^H$ for $n = 0, -1, -2, \ldots$. We call the sets $v_1(z_1, t)$ and $w_j(z_j, t)$ the evolutionary basis of signals $\{\vec{E}_1^i(g, t), \vec{H}_1^i(g, t)\}$ and $\{\vec{E}_j^s(g, t), \vec{H}_j^s(g, t)\}$ as they describe completely the evolution of transient waves over any finite interval of the guiding structure.

Now, we introduce by means of the relations

$$\begin{aligned} w'_{nj}(0, t) &\equiv \frac{\partial}{\partial z_j} w_{nj}(z_j, t)\bigg|_{z_j=0} \\ &= \int_0^t \sum_{m=-\infty}^{\infty} \left[R_{nm}^{\mathbf{AA}}(t-\tau)\delta_j^1 + T_{nm}^{\mathbf{BA}}(t-\tau)\delta_j^2\right] v_{m1}(0, \tau)d\tau; \\ &\quad t \geq 0, \quad n = 0, \pm 1, \pm 2, \ldots, \quad j = 1, 2, \end{aligned} \tag{5.14}$$

the boundary transform operators $R^{\mathbf{AA}}$ and $T^{\mathbf{BA}}$ (at the boundaries of the inhomogeneity $z_j = 0$) of the evolutionary basis $v_1(z_1, t)$ of the transient wave arriving from the waveguide $\mathbf{A}$:

$$w'_j(0, t) = \left\{w'_{nj}(0, t)\right\} = \left[R^{\mathbf{AA}}\delta_j^1 + T^{\mathbf{BA}}\delta_j^2\right][v_1(0, \tau)]; \quad t \geq 0, \quad j = 1, 2. \tag{5.15}$$

The elements $R_{nm}^{\mathbf{AA}}(t-\tau)$ and $T_{nm}^{\mathbf{BA}}(t-\tau)$ of these operators express how the space–time distribution of the energy of the incident wave $\{\vec{E}_1^i(g, t), \vec{H}_1^i(g, t)\}$

gives rise to the corresponding distribution in the scattered fields that have passed through the bounded waveguide resonator. There are two types of distributions: the distribution over the regions (the upper index: from the region whose identifier is on the right, into the region with the identifier on the left) and the distribution over the modes (the lower indices: on the right there is the mode number of the incident wave; on the left there is the mode number of the secondary field).

It is evident that $R^{\mathbf{AA}}$ and $T^{\mathbf{BA}}$, acting on the evolutionary signal bases, are characteristic of the waveguide junction itself and properly sum up the results of multiple elementary disturbances, that can be used to form any primary signal $\{\vec{E}_1^i(g,t), \vec{H}_1^i(g,t)\}$. Thus, if $v_{n1}(0,t) = \delta_n^p \delta(t-\eta)$, where p is an integer and $\eta > 0$, then $w'_{n1}(0,t) = R_{np}^{\mathbf{AA}}(t-\eta)$ and $w'_{n2}(0,t) = T_{np}^{\mathbf{BA}}(t-\eta)$. The use of such an abstract, physically unrealizable signal can be excused for methodological reasons, namely, it is used for elementary excitation of the structure that enables us to extract, in the generated field, the "pure" $R_{np}^{\mathbf{AA}}(t-\tau)$ and $T_{np}^{\mathbf{BA}}(t-\tau)$ components of the transform operators.

The problem of extracting the elements of the matrix functions $R^{\mathbf{AA}}$ and $T^{\mathbf{BA}}$ can be solved in an alternative way, by selecting a sequence of values p in the evolutionary basis $v_1(z_1,t) = \{\delta_n^p v_{n1}(z_1,t)\}_{n=-\infty}^{\infty}$ of the excitation wave, where $v_{n1}(z_1,t)$ are sufficiently random functions. The amplitudes $w'_{nj}(0,t)$ of the secondary field, that correspond to the fixed value p, are related to the sought values by the relation

$$w'_{nj}(0,t) = \int_0^t \left[R_{np}^{\mathbf{AA}}(t-\tau)\,\delta_j^1 + T_{np}^{\mathbf{BA}}(t-\tau)\,\delta_j^2\right] v_{p1}(0,\tau)\,d\tau; \quad j = 1,2,$$

which follows from (5.14). Inversion by the operator method yields

$$R_{np}^{\mathbf{AA}}(t)\,\delta_j^1 + T_{np}^{\mathbf{BA}}(t)\,\delta_j^2 = L^{-1}\left[\frac{L\left[w'_{nj}(0,t)\right](s)}{L\left[v_{p1}(0,t)\right](s)}\right](t); \quad j = 1,2,$$

where $L[\ldots]$ and $L^{-1}[\ldots]$ are the direct and inverse Laplace transforms (2.15).

There are numerous possibilities to realize the respective representations with a minimum error and an acceptable computational burden. The determining here is the choice of functions $v_{p1}(0,t)$: in an ideal case (for simple—canonical—inhomogeneities) it should be such that one could present the required integral transforms in analytical form.

The operators $R^{\mathbf{AA}}$ and $T^{\mathbf{BA}}$ are boundary operators that determine all the peculiarities of the transient processes immediately at the boundary of the inhomogeneity in a regular directing structure. The secondary field, leaving this boundary, propagates freely in the semi-infinite regular channels and also becomes deformed (see [138] and Chapter 6). The spatio–temporal amplitudes of the modes $\{w_{nj}(z_j,t)\}$ (the elements of the evolutionary basis of the signals $\{\vec{E}_j^s(g,t); \vec{H}_j^s(g,t)\}$) vary individually as the time and distance change, for different values of the indices n and j. These variations can be followed over any finite interval of waveguides **A** and **B** by using the diagonal transport operators $Z_{0\to z_1}^{\mathbf{A}}$ and $Z_{0\to z_2}^{\mathbf{B}}$, that are subject to the rule

$$w_j(z_j,t)=\{w_{nj}(z_j,t)\}=\left[Z^{\mathbf{A}}_{0\to z_1}\delta^1_j+Z^{\mathbf{B}}_{0\to z_2}\delta^2_j\right][w'_j(0,\tau)];\quad j=1,2. \tag{5.16}$$

The structure of these operators is detailed by the formula

$$w_{nj}(z_j,t)=-\int_0 J_0\left[\lambda_{nj}\left((t-\tau)^2-z_j^2\right)^{1/2}\right]\chi\left[(t-\tau)-z_j\right]w'_{nj}(0,\tau)\,d\tau;$$
$$t\ge 0,\quad z_j\ge 0,\quad n=0,\pm1,\pm2,\ldots,\quad j=1,2, \tag{5.17}$$

that reflects a common feature of the solutions to homogeneous problems such as (5.13). These are solutions that satisfy the zero initial data and are free of components propagating toward decreasing z_j. The technique of the derivation of the relations (5.17) is described in detail in Chapter 2 (see Section 2.2.1).

5.2.3. Operator Method for Problems of Cascades of Elementary Discontinuities

The operators introduced above completely describe the scattering properties of an inhomogeneity that is excited from channel **A**. Now we determine the transform operators $R^{\mathbf{BB}}$ and $T^{\mathbf{AB}}$, similar to $R^{\mathbf{AA}}$ and $T^{\mathbf{BA}}$, of the evolutionary basis $v_2(z_2,t)=\{v_{n2}(z_2,t)\}_{n=-\infty}^{\infty}$ of the wave $\{\vec{E}^i_2(g,t),\vec{H}^i_2(g,t)\}$, arriving through waveguide **B** at the boundary $z_2=0$:

$$w'_{nj}(0,t)=\int_0^t\sum_{m=-\infty}^{\infty}\left[T^{\mathbf{AB}}_{nm}(t-\tau)\delta^1_j+R^{\mathbf{BB}}_{nm}(t-\tau)\delta^2_j\right]v_{m2}(0,\tau)\,d\tau;$$
$$t\ge 0,\quad n=0,\pm1,\pm2,\ldots,\quad j=1,2. \tag{5.18}$$

Assuming that the sets $\left\{R^{\mathbf{AA}},T^{\mathbf{BA}},R^{\mathbf{BB}},T^{\mathbf{AB}}\right\}$ of transform operators are known for separate simple inhomogeneities, we shall develop an algorithm that solves the problem of determining the scattering properties of a complex junction composed of such inhomogeneities. In the simulated case, sketched in Figure 5.2, the

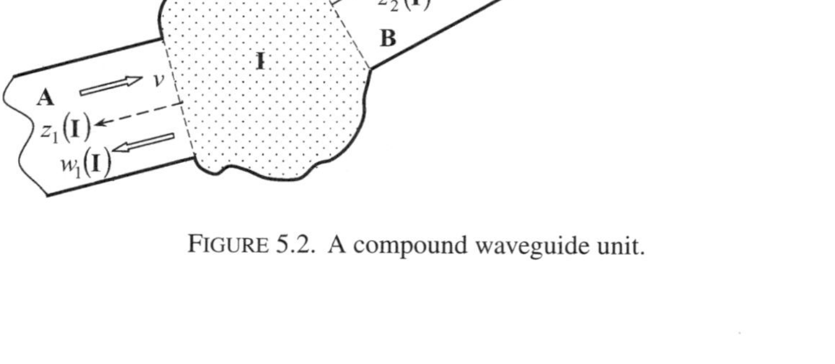

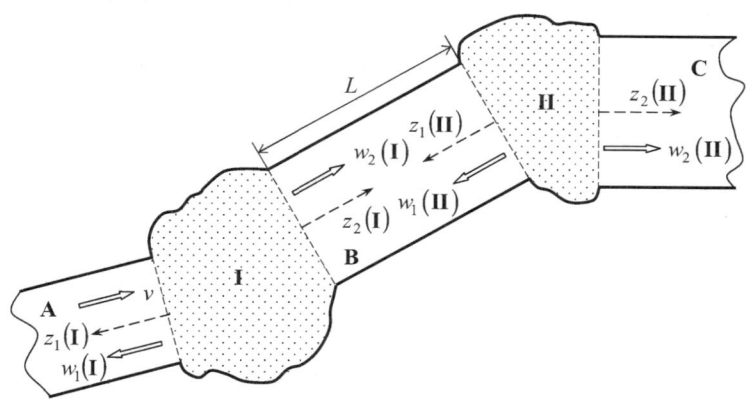

FIGURE 5.2. A compound waveguide unit.

configuration contains two consecutive inhomogeneities **I** and **II**, connected by section **B** of a regular waveguide of finite length L, and excited by a wave of type (5.11), arriving at the boundary $z_1(\mathbf{I}) = 0$ from waveguide **A**. Retaining the notations introduced above (obvious changes are due to the presence of two different inhomogeneities **I** and **II**), we present (symbolically) the solution of the corresponding initial boundary value problem in the regular regions **A**, **B**, and **C** in the form

$$U(\mathbf{A}) = \sum_{n=-\infty}^{\infty} \left[v_{n1}\left(z_1(\mathbf{I}), t\right) + w_{n1}\left(z_1(\mathbf{I}), t\right)\right] \mu_{n1}\left(x_1(\mathbf{I}), y_1(\mathbf{I})\right),$$

$$U(\mathbf{B}) = \sum_{n=-\infty}^{\infty} \left[w_{n2}\left(z_2(\mathbf{I}), t\right) + w_{n1}\left(z_1(\mathbf{II}), t\right)\right] \mu_{n2}\left(x_2(\mathbf{I}), y_2(\mathbf{I})\right),$$

$$U(\mathbf{C}) = \sum_{n=-\infty}^{\infty} w_{n2}\left(z_2(\mathbf{II}), t\right) \mu_{n2}\left(x_2(\mathbf{II}), y_2(\mathbf{II})\right).$$

Here, the first groups of terms correspond to waves moving from left to right, and the second group corresponds to waves moving in the opposite direction (see Fig. 5.2).

Adhering to the definitions (5.14) to (5.18) and using notations of the kind

$$w_j'(\mathbf{I}) \equiv \frac{\partial}{\partial z_j(\mathbf{I})} w_j\left(z_j(\mathbf{I}), t\right)\Big|_{z_j(\mathbf{I})=0}, \quad w_j(\mathbf{I}) = \left\{w_{nj}\left(z_j(\mathbf{I}), t\right)\right\}\Big|_{z_j(\mathbf{I})=0},$$

we obtain the following set of operator equations.

$$\begin{cases} w_1'(\mathbf{I}) = R^{\mathbf{AA}}(\mathbf{I})\left[v_1(\mathbf{I})\right] + T^{\mathbf{AB}}(\mathbf{I})\, Z^{\mathbf{B}}_{z_1(\mathbf{II})=0\to L}\left[w_1'(\mathbf{II})\right] \\ w_2'(\mathbf{I}) = T^{\mathbf{BA}}(\mathbf{I})\left[v_1(\mathbf{I})\right] + R^{\mathbf{BB}}(\mathbf{I})\, Z^{\mathbf{B}}_{z_1(\mathbf{II})=0\to L}\left[w_1'(\mathbf{II})\right] \\ w_1'(\mathbf{II}) = R^{\mathbf{BB}}(\mathbf{II})\, Z^{\mathbf{B}}_{z_2(\mathbf{I})=0\to L}\left[w_2'(\mathbf{I})\right] \\ w_2'(\mathbf{II}) = T^{\mathbf{CB}}(\mathbf{II})\, Z^{\mathbf{B}}_{z_2(\mathbf{I})=0\to L}\left[w_2'(\mathbf{I})\right] \end{cases}. \qquad (5.19)$$

This representation clearly demonstrates all the stages of forming the total response of a complex structure to excitation by a signal $\left\{\vec{E}_1^i(g,t), \vec{H}_1^i(g,t)\right\}$ having the evolutionary basis $v_1\left(z_1(\mathbf{I}), t\right) = \left\{v_{n1}\left(z_1(\mathbf{I}), t\right)\right\}_{n=-\infty}^{\infty}$ (or, simply, to the signal $v_1(\mathbf{I})$). Thus, for instance, the first equation in the set can be interpreted as follows. The signal $w_1(\mathbf{I})$ (the secondary field in the region **A**) is a sum of signals, the first one of which is a result of reflection of the primary signal $v_1(\mathbf{I})$ from the inhomogeneity **I**, and the second one is determined by the signal $w_1(\mathbf{II})$ that has been deformed while propagating through the regular region **B** and inhomogeneity **I**.

By the elimination method, the set (5.19) is reduced to an operator equation of the second kind with respect to an unknown vector function $w_2'(\mathbf{I})$

$$w_2'(\mathbf{I}) = T^{\mathbf{BA}}(\mathbf{I})\left[v_1(\mathbf{I})\right] + R^{\mathbf{BB}}(\mathbf{I})\, Z^{\mathbf{B}}_{z_1(\mathbf{II})=0\to L} R^{\mathbf{BB}}(\mathbf{II})\, Z^{\mathbf{B}}_{z_2(\mathbf{I})=0\to L}\left[w_2'(\mathbf{I})\right] \qquad (5.20)$$

and to a countable set of formulas, determining all the components of the field generated by the inhomogeneity. The operator on the right-hand side of (5.20) influences the sought vector functions $w_2'(\mathbf{I})$, whose argument τ is strictly less than the observation time t in the argument of the same function on the left-hand side (the delay effect due to the finite velocity of propagation of the disturbance). Consequently, the numerical solution of the final equation can be obtained within the framework of a standard scheme of stepwise motion through the time layers, and the initial complex problem has been adequately reformulated to allow a direct inversion by using conventional computational tools. The complex junction is now classified as an elementary basic unit, after having computed the elements of its boundary operators by the formulas (5.14), (5.18).

Let us recall once more the formulas (5.14) to (5.18) which lead to questions about why the boundary transform operators are determined in this particular fashion. Actually, these operators act in a way that is different from their analogues in the frequency domain. In principle, it is possible also to follow tradition in the time domain, by associating, through the boundary operators, the pair "field $\to$ field", and not "field $\to$ derivative of the field along the propagation direction", as was done in (5.14) and (5.18). But let us consider now the structure of transport operators $Z^{\mathbf{A}}_{0\to z_1}$ and $Z^{\mathbf{B}}_{0\to z_2}$ (formulas (5.16), (5.17)). The optimal scheme, in terms of computational efficiency, of incorporating them into the algorithm that describes the complex junction (see equations (5.19), (5.20)), was the factor that predetermined the not completely physical choice preferred in (5.14), (5.18). The transform "field $\to$ field" is implemented by multiplying the operators ZR or ZT directly, without involving the intermediate operator of differentiation in the line of wave propagation.

5.3. Canonical Problems of the Time Domain

In Section 5.2.3, we have considered an example of the conjunction of two-port waveguide resonators. It is clear that a similar approach can be used in more general cases: the operator description of electromagnetic properties of multiport junctions allows carrying out an efficient analysis of rather complicated cascade junctions of a diverse range of configurations, as has been done in the FD computer-aided system of electromagnetic simulation [139] focused on analysis and design of antenna feeder structure. Moreover, as the evolutionary bases are qualitatively the same for all structures with discrete special spectra (see previous chapters), then the operator method can be used also for the study of compact and non-compact objects separated in open space, multilayered periodic structures, and the like. The major part of the problem in FD is solved within the framework of operator approaches [14] that allow a considerable extension of application of rigorous semi-analytic approaches.

The essential problem of operator methods (both in FD and in TD) is the calculation of operators $R^{\mathbf{AA}}$, $T^{\mathbf{BA}}$, and so on, for a rather wide variety of elementary discontinuities. In this section we consider problems that are suitable for algorithmic supplementation of the analysis of complicated axially symmetric resonant junctions in circular and coaxial waveguides. The rigorous semi-analytic method suggested herein allows us to determine the electromagnetic characteristics of several canonical problems: thin irises, coaxial bifurcations, dielectric windows, and the like. For calculation of the characteristics of elementary resonator cells with rather arbitrary geometric and constitutive parameters it is necessary to apply conventional numerical methods such as, for example, the finite difference method with explicit absorbing conditions on virtual boundaries **L**. Analysis of complicated junctions compounded of such cells has to be based on numerical routines of operator methods: the computational domain decreases considerably when the regular domains, connecting elementary discontinuities, are excluded.

5.3.1. Formulation of the Problems

The study of TE_0 symmetric ($\partial/\partial\phi \equiv 0$) electromagnetic waves in cylindrical structures shown in Figures 2.2, 5.3, and 5.4, is reduced (see Sections 1.2 and 2.3) to the solution of the following initial boundary value problems such as (2.29),

$$\left\{\begin{array}{l}\left[-\varepsilon(z)\dfrac{\partial^2}{\partial t^2}-\sigma(z)\dfrac{\partial}{\partial t}+\dfrac{\partial^2}{\partial z^2}+\dfrac{\partial}{\partial\rho}\left(\dfrac{1}{\rho}\dfrac{\partial}{\partial\rho}\rho\right)\right]=0;\\ g=\{\rho,z\}\in\mathbf{Q},\quad t>t\\ U(\rho,z,0)=U^i(\rho,z,0),\quad \dfrac{\partial}{\partial t}U(\rho,z,t)\Big|_{t=0}=\dfrac{\partial}{\partial t}U^i(\rho,z,t)\Big|_{t=0};\\ g\in\bar{\mathbf{Q}}\\ U(z,\rho,t)|_{g\in\mathbf{S}}=0;\quad t\geq 0\end{array}\right. . \tag{5.21}$$

Here, $U(\rho, z, t) = E_\phi(\rho, z, t)$, $E_z = E_\rho = H_\phi = 0$, and the nonzero components of the magnetic field strength vector are defined by the ratios

$$\frac{\partial}{\partial t}H_\rho = \frac{1}{\eta_0}\frac{\partial}{\partial z}E_\phi, \qquad \frac{\partial}{\partial t}H_z = -\frac{1}{\eta_0\rho}\frac{\partial}{\partial\rho}\left(\rho E_\phi\right).$$

The domain of analysis **Q** is the part of semi-plane $\phi = \text{const}$, bounded by the contour **S**. The surface of "perfect conductors" is defined by $\mathbf{S} \times [0 \leq \phi \leq 2\pi]$. The excitation wave $U^i(z, \rho, t) = E^i_\phi(\rho, z, t)$ is assumed to be nonzero only in the regular semi-infinite waveguide (in the region **A** for the structures presented in Fig. 5.3, and in the region **B** for the structures in Fig. 5.4), from which this wave at time $t > 0$ runs over the inhomogeneity. In this waveguide, $U(\rho, z, t) = U^i(\rho, z, t) + U^s(\rho, z, t)$. In other waveguides, $U(\rho, z, t) = U^s(\rho, z, t)$. The geometry and the variable coefficients in the equations of problems (5.21) are such that their general solution in all partial regular regions can be represented as follows.

$$U^i(\rho, z, t) = \sum_n v_n(z, t)\,\mu_n(\rho), \qquad U^s(\rho, z, t) = \sum_n u_n(z, t)\,\mu_n(\rho); \quad t \geq 0, \tag{5.22}$$

where $\{\mu_n(\rho)\}$ is an orthonormalized basis of transverse functions in the corresponding plane region int **S** (circular or ring). It is determined by a set of nontrivial solutions to the homogeneous (spectral) problems (2.31) or (2.32).

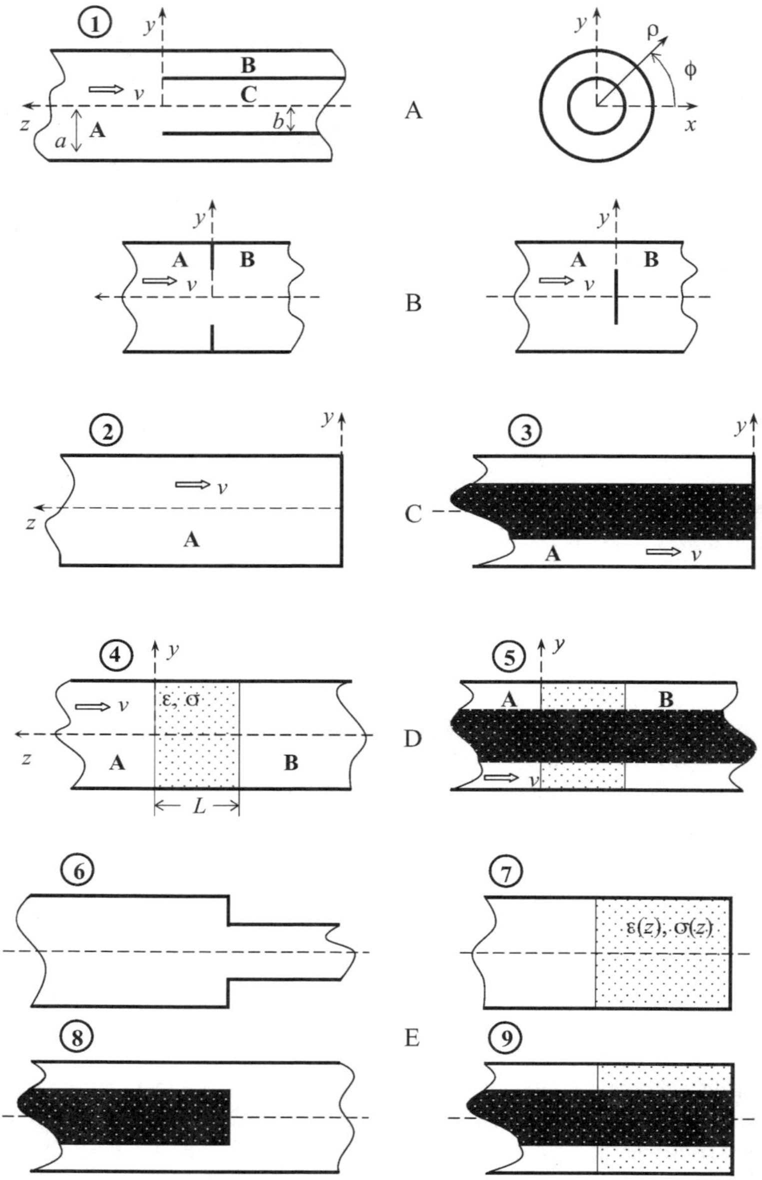

FIGURE 5.3. Simple discontinuities in circular and coaxial waveguides: (A) coaxial bifurcation of circular waveguide; (B) thin metal irises; (C) transverse shorting; (D) dielectric windows; (E) another step-type junction.

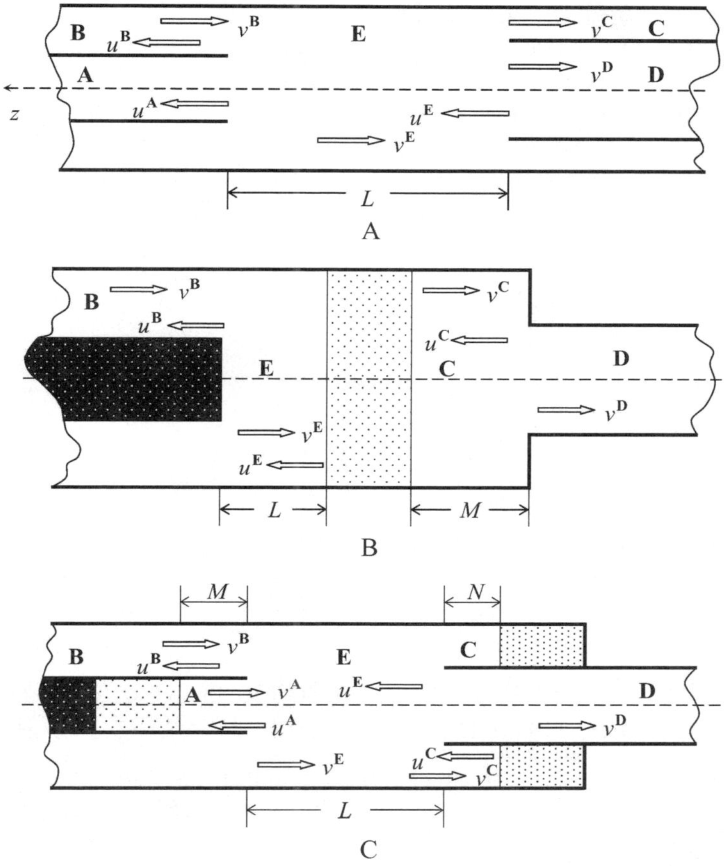

FIGURE 5.4. (A) Break in coaxial waveguide and (B, C) other resonant discontinuities in circular waveguides.

The spatio–temporal amplitudes $u_n(z, t)$ satisfy the equations

$$B(\lambda_n)[u_n] \equiv \left[-\frac{\partial^2}{\partial t^2} + \frac{\partial^2}{\partial z^2} - \lambda_n^2\right] u_n = \left[(\varepsilon(z) - 1)\frac{\partial^2}{\partial t^2} + \sigma(z)\frac{\partial}{\partial t}\right] u_n;$$
$$t > 0, \quad n = 1, 2, \ldots. \tag{5.23}$$

One can distinguish three types of regions int **S** (see Figs. 2.2 and 5.3) in the considered problems: two circular ($\rho < a$ and $\rho < b$, $a > b$) and one ring ($b < \rho < a$) region. The sets $\{\mu_n\}, \{\lambda_n\}$ of solutions to the corresponding spectral problems are well known for all these cases (see Section 2.3.1).

In the following considerations, we choose a wave of type (5.22) with $v_n(z, t) = \delta_n^p v_n(z, t)$, $n = 1, 2, \ldots$, $v_n(0, 0) = 0$, and $p \in \{1, 2, \ldots\}$ a fixed integer as the excitation wave $U^i(\rho, z, t)$. In the case when $v_n(0,t) = \delta_n^p \delta(t-\eta)$, $\eta > 0$, the amplitudes $u_n'(0,t)$ of the secondary fields explicitly determine the elements of the boundary operators $R^{\mathbf{AA}}$, $T^{\mathbf{BA}}$, and so on (see Section 5.2). The formulas for computing these amplitudes contain the function $v_p'(0,t) = \partial v_p(z,t)/\partial z\big|_{z=0}$. Hence, in order to verify that all operations which determine $R^{\mathbf{AA}}$, $T^{\mathbf{BA}}$, and so on, can be performed accurately we should demonstrate that, by choosing $v_p(0,t) = \delta(t-\eta)$, we can also uniquely define $v_p'(0,t)$. The required result

$$v_p'(0,t) = \left(\lambda_p^2 + \frac{\partial}{\partial t^2}\right)[J_0[\lambda_p(t-\eta)]\,\chi(t-\eta)] \tag{5.24}$$

we obtain from the Volterra equation of the first kind

$$\delta(t-\eta) = \int_0 J_0[\lambda_p(t-\tau)]\,\chi\,(t-\tau)\,v_p'(0,\tau)\,d\tau \tag{5.25}$$

while inverting it by the operational calculus method. (5.25) is a relation of the same type as (5.17), where the direction of propagation of the signal is accounted for.

5.3.2. *Thin Diaphragms: Exact Solution of the Initial Boundary Value Problems by the Mode-Matching Method*

Let an axis-symmetric ring or circular diaphragm be situated in plane $z = 0$ of a cross-section of a circular waveguide having the radius a (see Fig. 5.3B). According to (5.17) and (5.22), the complete field formed in region **A** ($z > 0$) and passed into the region **B** ($z < 0$) can be written as

$$U\begin{pmatrix}\mathbf{A}\\ \mathbf{B}\end{pmatrix}$$
$$= \mp \sum_n \mu_n(\rho) \int_0 J_0[\lambda_n((t-\tau)^2 - z^2)^{1/2}]\,\chi\,[(t-\tau)\mp z]\begin{pmatrix}\left(u_n^{\mathbf{A}}\right)'\\ \left(u_n^{\mathbf{B}}\right)'\end{pmatrix}(0,\tau)\,d\tau$$
$$+ \begin{pmatrix}\mu_p(\rho)\,v_p(z,t)\\ 0\end{pmatrix}.$$

Hereafter, the dash above the identifier stands for differentiating with respect to z. Thus, for instance, $(u_n^{\mathbf{A}})'(0,\tau) = \partial u_n^{\mathbf{A}}(z,\tau)/\partial z|_{z=0}$ and so on.

From the boundary condition $E_\phi(g,t) = 0$ at the metallic parts of the diaphragm and the continuity condition of the component $E_\phi(g,t)$ of the electric field strength vector at the aperture in the plane $z = 0$, it follows that

$$\left(u_n^A\right)'(0,t) + \left(u_n^B\right)'(0,t) = \delta_n^p v_p'(0,t). \tag{5.26}$$

The substitution of (5.26) into the continuity condition for the tangential component of the magnetic field strength vector $(\partial H_\rho/\partial t)$ yields the equality

$$\sum_n \left(u_n^{\mathbf{A}}\right)'(0,t)\,\mu_n(\rho) = 0, \tag{5.27}$$

which is true for values ρ, which correspond to the aperture that connects regions **A** and **B**. The range of variation of the variable ρ in (5.27) can be extended up to the whole range $0 < \rho < a$, by differentiating the boundary condition $U(\mathbf{A}) = 0$ at the metallic parts of the diaphragm with respect to t. As a result we obtain

$$\sum_n \left(u_n^{\mathbf{A}}\right)'(0,t)\mu_n(\rho) = \begin{cases} \displaystyle\sum_n \left\{\lambda_n \int_0 J_1[\lambda_n(t-\tau)]\,\chi(t-\tau)\left(u_n^{\mathbf{A}}\right)'(0,\tau)\,d\tau \right. \\ \displaystyle\left. +\,\delta_n^p \frac{\partial v_p(0,t)}{\partial t}\right\}\mu_n(\rho); \quad \rho \in \text{metal} \\ 0; \quad \rho \in \text{aperture} \end{cases} \tag{5.28}$$

In terms of the Fourier coefficients of the matching functions an equivalent expression for (5.28) is:

$$\left(u_n^{\mathbf{A}}\right)'(0,t) = 2\pi \sum_m F_{nmj} \left\{\lambda_m \int_0 J_1[\lambda_m(t-\tau)]\,\chi(t-\tau)\left(u_m^{\mathbf{A}}\right)'(0,\tau)\,d\tau \right.$$
$$\left. +\,\delta_m^p \frac{\partial v_p(0,t)}{\partial t}\right\}, \quad n = 1,2,\ldots, \quad j = 1,2. \tag{5.29}$$

Here,

$$F_{nmj} = \int_{\delta_j^1 b}^{\delta_j^1 a + \delta_j^2 b} \mu_n(\rho)\,\mu_m(\rho)\,\rho\,d\rho,$$

and for $j = 1$, formula (5.29) describes the properties of a circular aperture, and for $j = 2$ those of a ring (see Fig. 5.3B).

The elements of the unknown vector function $\{(u_m^{\mathbf{A}})'(0,\tau)\}$ on the right-hand side of equations (5.29) determine the value of this function at the time t by their values at moments $\tau < t$ (i.e., τ is strictly less than t). The elimination of the momentary effect allows us to consider the right-hand side as known and interpret the result (5.29) as an exact solution to the problem. Sampling of the variable t turns (5.29) into a simple recurrence formula. The computational error generated by this can easily be evaluated and minimized.

In our formulation and algorithmization of initial boundary value problems we use the idea of a generalized solution, whose exact definition is based on the concept of a generalized derivative and, generally, on the concept of a generalized function (see [34] and the Chapter 1). It is commonly recognized that such an approach substantially simplifies the analysis of PDEs, because it separates the issues of existence and uniqueness from the study of the analytical properties of solutions. It also simplifies the transforms required for constructing the solutions and it gives us a greater number of practically interesting and important problems which belong to the class that can be treated correctly using these concepts. Partially, we have already used the corresponding possibilities, but the requirements of the numerical implementation of the developed algorithms draw our attention to some peculiar features of the representations we have obtained. Therefore, it is useful to recall the formulas of this section and use them to exemplify some typical situations.

If for the excitation wave $U^i(\rho, z, t) = \sum v_n(z,t)\,\mu_n(\rho)$ we assume $v_n(0,t) = \delta_n^p \delta(t-\eta)$, $\eta > 0$, then $\left(u_n^{\mathbf{A}}\right)'(0,t) = R_{np}^{\mathbf{AA}}(t-\eta)$ and $\left(u_n^{\mathbf{B}}\right)'(0,t) = -T_{np}^{\mathbf{BA}}(t-\eta)$. However, the final formulas (5.26), (5.29) will include singular (not regular) generalized functions (5.24) and $\delta^{(1)}(t-\eta)$. The direct discretization of the problem then becomes impossible. To obtain a numerical solution one should apply the approximating features of the regular (normal) functions $f_\varepsilon(x)$ such that $f_\varepsilon(x) \to \delta(x)$ as $\varepsilon \to 0$ (convergence in space of the generalized functions). The choice of functions $f_\varepsilon(x)$ is wide enough [34], and it allows us to minimize the errors caused by the approximation. The computation of the elements of the matrix functions $R^{\mathbf{AA}}$ and $T^{\mathbf{BA}}$ can certainly be performed as suggested in Section 5.2.2, but, by such an approach, an approximate substitution of the singular generalized functions by regular ones is inevitable.

Let us consider one additional way to settle such a standard, frequently arising problem. The most natural and efficient method implies using only fundamental properties of the generalized functions, and, particularly, the properties of the convolution $(\partial^\alpha f) * g = \partial^\alpha (f * g) = f * (\partial^\alpha g)$. The essence of this is the elimination of singular generalized functions from the computation procedures, while retaining their usual symbolic content at all stages where transform operators are used. Let us illustrate this statement with simple example.

By the definition of $R^{\mathbf{AA}}$, the operator equation (5.29) can be written as

$$R_{np}^{\mathbf{AA}}(t-\eta) = 2\pi \sum_m F_{nmj} \left\{ \lambda_m \int_0 J_1\left[\lambda_m (t-\tau)\right] \chi(t-\tau)\, R_{mp}^{\mathbf{AA}}(\tau-\eta)\, d\tau + \delta_m^p \delta^{(1)}(t-\eta) \right\}. \tag{5.30}$$

After the substitution $S_{np}^{\mathbf{AA}}(t-\eta) = R_{np}^{\mathbf{AA}}(t-\eta) - 2\pi F_{npj}\delta^{(1)}(t-\eta)$, (5.30) reads

$$S_{np}^{\mathbf{AA}}(t-\eta) = 2\pi \sum_m \lambda_m F_{nmj} \int_0 J_1\left[\lambda_m (t-\tau)\right] \chi(t-\tau)\, S_{mp}^{\mathbf{AA}}(\tau-\eta)\, d\tau + P_{npj}(t-\eta), \tag{5.31}$$

where $P_{npj}(t-\eta) = 4\pi^2 \sum_m \lambda_m F_{nmj} F_{mpj} \alpha_m (t-\eta)$, and

$$\alpha_m (t-\eta) = \int_0 J_1 [\lambda_m (t-\tau)] \chi (t-\tau) \delta^{(1)} (\tau-\eta) \, d\tau$$

$$= \int_0 \frac{\partial}{\partial t} \{J_1 [\lambda_m (t-\tau)] \chi (t-\tau)\} \delta (\tau-\eta) \, d\tau$$

$$= \frac{\partial}{\partial t} \{J_1 [\lambda_m (t-\eta)] \chi (t-\eta)\}$$

$$= \chi (t-\eta) \lambda_m [J_0[\lambda_m (t-\eta)] - J_1[\lambda_m (t-\eta)][\lambda_m (t-\eta)]^{-1}]$$

(here it is taken into account that the point $\tau = 0$ belongs to the zero set of the function $\delta (\tau - \eta)$, and that the product $J_1 [\lambda_m (t-\eta)] \delta (t-\eta)$ yields a function that is identically equal to zero). Equation (5.31) does not contain singular generalized functions and allows us, using the standard procedure, to calculate the elements $S^{\mathbf{AA}}_{nm} (t-\tau)$, that is, the regular part of the reflection operator $\{R^{\mathbf{AA}}_{nm} (t-\tau)\}$. This is sufficient for all further use of $R^{\mathbf{AA}}$ in all the algorithms of the transform operator method. Thus, for instance, formula (5.14), that is used to determine the amplitudes $(u^{\mathbf{A}}_n)' (0, t)$ of the reflected wave when a primary wave characterized by an arbitrary set of amplitudes $v_n (z, t)$, is realized in the following way.

$$\left(u^{\mathbf{A}}_n\right)' (0, t) = \int_0^t \sum_m R^{\mathbf{AA}}_{nm} (t-\tau) \, v_m (0, \tau) \, d\tau$$

$$= \sum_m \left[\int_0^t S^{\mathbf{AA}}_{nm} (t-\tau) \, v_m (0, \tau) \, d\tau + 2\pi F_{nmj} \alpha_m (t) \right];$$

$$\alpha_m (t) = \int_0^t \delta^{(1)} (t-\tau) \, v_m (0, \tau) \, d\tau$$

$$= -\delta (t-\tau) \, v_m (0, \tau)|_0^t + \int_0^t \delta (t-\tau) \frac{\partial v_m (0, \tau)}{\partial \tau} d\tau = \frac{1}{2} \frac{\partial v_m (0, t)}{\partial t}$$

(the product $\delta (t-\tau) \, v_m (0, \tau)$ at $\tau = 0$ and $\tau = t$ yields a function that is identically equal to zero).

The algorithm (5.26), (5.29) has been developed by matching the fields at the boundary of two equal semi-infinite regular waveguides, which has simplified the solution considerably. The resulting computation scheme of the method is somehow complicated if employed for analyzing other canonical stepwise inhomogeneities in circular and coaxial waveguides (the semi-infinite extension, semi-infinite bifurcation (splitting), etc.; see Fig. 5.3E). However, even in such cases this scheme is much simpler to implement and provides prompt results more easily than other known numerical–analytical methods and approaches.

5.3.3. Residue Calculation Method and Analytical Regularization Method for Canonical Problems in the Time Domain

Let us first consider the problem concerning a coaxial semi-infinite bifurcation of a circular waveguide (see Fig. 5.3A), which is a classical problem for the frequency domain. The excitation wave $U^i(\rho, z, t) = v_p(z, t)\,\mu_p(\rho)$, propagating in region **A** ($z > 0$), induces in the waveguides **A**, **B**, and **C** the field

$$U(\mathbf{A}) = -\sum_n \mu_n(\rho) \int_0 J_0\left[\lambda_n((t-\tau)^2 - z^2)^{1/2}\right] \chi[(t-\tau) - z] u'_n(0,\tau)\, d\tau + U^i(g,t),$$

$$U(\mathbf{B}, \mathbf{C}) = \sum_n \mu_{nj}(\rho) \int_0 J_0\left[\lambda_{nj}((t-\tau)^2 - z^2)^{1/2}\right] \chi[(t-\tau) + z]\, u'_{nj}(0,\tau)\, d\tau; \quad j = 1, 2. \tag{5.32}$$

The formulas (5.32) are obtained from the common representations $U(\mathbf{A}) = \sum \mu_n(\rho)\, u_n(z,t) + U^i(g,t)$ and $U(\mathbf{B}, \mathbf{C}) = \sum \mu_{nj}(\rho)\, u_{nj}(z,t)$; $j = 1, 2$, while substituting (5.17) into them and taking into account the change of direction, along which we take the derivatives of the amplitudes $u_{nj}(z,t)$. The functions $\mu_n(\rho)$, $\mu_{nj}(\rho)$ and their associated values λ_n, λ_{nj} are determined by the formulas from Section 2.3.1. The value $j = 1$ corresponds to the region **B** ($z < 0$, coaxial waveguide); all values $j = 2$ refer to the region **C** ($z < 0$, a circular waveguide of radius b). In the case when $v_p(0,t) = \delta(t-\eta)$, the amplitudes of the secondary fields explicitly determine the elements of the boundary operators $R^{\mathbf{AA}}$, $T^{\mathbf{BA}}$, and $T^{\mathbf{CA}}$: $u'_n(0,t) = R^{\mathbf{AA}}_{np}(t-\eta)$, $u'_{n1}(0,t) = -T^{\mathbf{BA}}_{np}(t-\eta)$ and $u'_{n2}(0,t) = -T^{\mathbf{CA}}_{np}(t-\eta)$.

Imposing the continuity conditions on the functions U and $\partial U/\partial z$ in the plane $z = 0$, which provide the uniqueness of the extension of the solution to (5.21) from one partial domain (**A**) into the two others (**B** and **C**), results in functional equations, one of the forms of which can be expressed in terms of the Fourier coefficients as

$$\sum_n F_{nmj} \int_0 \{J_0[\lambda_n(t-\tau)] + J_0[\lambda_{mj}(t-\tau)]\}\, \chi(t-\tau)\, u'_n(0,\tau)\, d\tau$$
$$= F_{pmj}\left\{ v_p(0,t) - \int_0 J_0[\lambda_{mj}(t-\tau)]\, \chi(t-\tau)\, v'_p(0,\tau)\, d\tau \right\};$$
$$m = 1, 2, \ldots, \quad j = 1, 2, \tag{5.33}$$

$$u'_{mj}(0,t) = \sum_n F_{nmj}\left[u'_n(0,t) + \delta^p_n v'_p(0,t)\right]; \quad m = 1, 2, \ldots, \quad j = 1, 2. \tag{5.34}$$

Here,

$$F_{nmj} = \int\limits_{\delta_j^1 b}^{\delta_j^1 a+\delta_j^2 b} \mu_n(\rho)\mu_{mj}(\rho)\rho\, d\rho; \quad j = 1, 2. \tag{5.35}$$

The expression (5.33) is in itself a pair operator equation with respect to a set of unknown functions $\{u_n'(0,t)\}$; (5.34) are a countable set of formulas for the determination of the amplitudes of transient fields in the first (**B**, $j = 1$) and second (**C**, $j = 2$) waveguide channels in the region $z < 0$. Differentiating (5.33) with respect to t, we obtain

$$\sum_n F_{nmj} u_n'(0,t) = f_{mj}(t); \quad m = 1, 2, \ldots, \quad j = 1, 2. \tag{5.36}$$

It is obvious that the elements of the unknown vector function $\{u_n'(0,\tau)\}$ that are contained in the right-hand side

$$f_{mj}(t) = \frac{1}{2}\left\{\sum_n F_{nmj}\int\limits_0^t \{\lambda_n J_1[\lambda_n(t-\tau)] + \lambda_{mj} J_1[\lambda_{mj}(t-\tau)]\}\chi(t-\tau)u_n'(0,\tau)\,d\tau\right.$$
$$\left. + F_{pmj}\frac{\partial}{\partial t}\left[v_p(0,t) - \int\limits_0^t J_0[\lambda_{mj}(t-\tau)]\,\chi(t-\tau)\,v_p'(0,\tau)\,d\tau\right]\right\}$$

of equation (5.36) determine the value of this function at time t by their values at times $\tau < t$ (i.e., τ is strictly less than t). This means that if we fix t (e.g., at the next step by moving over time layers), we can consider (5.36) as a pair infinite system of linear algebraic equations of the first kind with respect to a set of unknowns $\{u_n'(0,t)\}$ with a known right-hand side $\{f_{mj}(t)\}$. The elements F_{nmj} do not depend on the time parameters, and hence the solution to the problem for various values of t is reduced to a single inversion of a corresponding operator.

Let's introduce new notations as follows (see the representations for transverse functions and relevant eigenvalues in Section 2.3.1 and formula (5.35)).

$$\xi_n = \lambda_n a, \quad \gamma_n = \lambda_{n1} a, \quad \theta = b/a, \quad R_n = 2u_n'(0,t)\,J_1(\xi_n\theta)/J_0(\xi_n),$$
$$\omega_{m2} = -f_{m2}\,\theta/\xi_m, \quad K_q(\lambda, \rho/a) = G_q(\lambda/a, \rho),$$
$$\omega_{m1} = -f_{m1}\left[K_0^2(\gamma_m, 1) - \theta^2 K_0^2(\gamma_m, \theta)\right]^{1/2}/[\theta\gamma_m K_0(\gamma_m,\theta)].$$

Then, equation (5.36) takes the form

$$\left\{\begin{array}{l} \sum\limits_n R_n\left(\xi_n^2 - \gamma_m^2\right)^{-1} = \omega_{m1} \\ \sum\limits_n R_n\left(\xi_n^2 - \xi_m^2/\theta^2\right)^{-1} = \omega_{m2} \end{array}\right. ; \quad m = 1, 2, \ldots. \tag{5.37}$$

For inversion of the system (5.37) we apply the residue calculation method [5,12] that is based on the Mittag–Leffler theorem about the expansion of a meromorphic function in the series of the principal parts [67]. Assume that there are meromorphic functions $Q_{rj}(w)$ of the complex variable w such that: $Q_{rj}(w)$ have simple poles at the points $w = \xi_n^2, n = 1, 2 \ldots; Q_{r1}(\gamma_m^2) = -\delta_r^m, Q_{r1}(\xi_m^2/\theta^2) = 0, Q_{r2}(\gamma_m^2) = 0, Q_{r2}(\xi_m^2/\theta^2) = -\delta_r^m, r, m = 1, 2 \ldots; Q_{rj}(w)$ decrease on arbitrary circles $\mathbf{C}_{|w|}$ in the w-plane as their radii $|w|$ grow. Also assume that the elements ω_{rj} decrease as the index r grows, sufficiently fast to make the series $\sum_r Q_{rj}(w)\omega_{rj}$ converge uniformly in the region of analyticity of the functions $Q_{rj}(w)$. Making these assumptions, we have

$$R_n = \sum_{r,j} \operatorname{Res} Q_{ri}\left(\xi_n^2\right)\omega_{rj}; \quad n = 1, 2 \ldots. \tag{5.38}$$

Indeed, because

$$\lim_{|w|\to\infty} \frac{1}{2\pi i} \oint_{\mathbf{C}_{|w|}} \left[\sum_{r,i} Q_{ri}(w)\,\omega_{ri}\right] \frac{dw}{\left(w - a^2\lambda_{mj}^2\right)}$$

$$= \sum_n \left[\sum_{r,i} \operatorname{Res} Q_{ri}\left(\xi_n^2\right)\omega_{ri}\right] \frac{1}{\left(\xi_n^2 - a^2\lambda_{mj}^2\right)} - \omega_{mj} = 0$$

for all $m = 1, 2, \ldots$ and $j = 1, 2$, the set $\{R_n\}$ in (5.38) gives the solution to (5.37).

The functions $Q_{mj}(w)$, possessing all the said properties, can be written as

$$Q_{m1}(w) = -\frac{2\gamma_m J_1(w^{1/2}\theta) K_1(w^{1/2}, 1) J_1(\gamma_m)}{J_1(w^{1/2}) J_1(\gamma_m\theta) K_1'(\gamma_m, 1)(w - \gamma_m^2)}; \quad m = 1, 2, \ldots,$$

$$Q_{m2}(w) = -\frac{2\xi_m J_1(w^{1/2}\theta) K_1(w^{1/2}, 1) J_1(\xi_m/\theta)}{\theta^2 J_1(w^{1/2}) K_1(\xi_m/\theta, 1) J_0(\xi_m)(w - \xi_m^2/\theta^2)}; \quad m = 1, 2, \ldots. \tag{5.39}$$

Here, $K_1'(w, 1) = \partial K_1(w, 1)/\partial w$. A direct check, using the properties of cylindrical functions [120], reveals that the functions $Q_{mj}(w)$ have simple zeros and poles at the required points and only there. The peculiarity at the point $w = 0$ (bifurcation of the function $w^{1/2}$ and the simple pole of the function $N_1(w)$) can be eliminated. The required normalization is ensured by the factor introduced in (5.39), which does not depend on w. For large $|w|$, n, and m we have $\omega_{mj} = O\left(m^{-2}\right)$ and

$$Q_{mj}(w) = O(m^2 w^{-1/2}(m^2 - w)^{-1}), \qquad \operatorname{Res} Q_{mj}(\xi_n^2) = O(m^2(m^2 - n^2)^{-1}).$$

Thus, the problem is solved. It is obvious that the study of the scattering properties of an inhomogeneity that is excited from the opposite side (from region **B** or (and) **C**) does not require any fundamental changes in the algorithm. Actually, the same operator should be inverted as in (5.37), but now with right-hand sides that are different from ω_{mj}. Having determined the complete set of transform operators for this structure, we can, according to a scheme similar to (5.19), (5.20), solve more complicated problems of analyzing the electromagnetic characteristics of an open waveguide resonator (see, e.g., Fig. 5.4A: a discontinuity of the interior conductor in a coaxial waveguide).

In fact, this and other similar schemes of the transform operator method implement, in the final analytic form, the procedure of extracting and inverting the contribution of separate elementary scatterers to the general electromagnetic field that is generated by a complex unit. This procedure simplifies the formal presentation of the solution and, what is most important, regularizes the numerical algorithm. We obtain all the necessary data through a direct sequential moving over time layers (i.e., without inversion of operators).

Although building solutions to the considered canonical problems, the regularization has been performed by differentiating with respect to t of the integral convolution operators,

$$\int_0 J_0\left[\lambda(t-\tau)\right]\chi(t-\tau)\,u(\tau)\,d\tau \overset{\partial/\partial t}{\rightarrow} u(t)-\lambda\int_0 J_1\left[\lambda(t-\tau)\right]\chi(t-\tau)\,u(\tau)\,d\tau,$$

and by inversion of a part of a functional equation (see the transfer (5.28) $\rightarrow$ (5.29)) or a matrix ((5.37) $\rightarrow$ (5.38)) equation. As a result, the current states $u(t)$ were strictly separated from the previous ones ($u(\tau)$ with $\tau < t$), and the parts of the operators associated with them were explicitly inverted. The final operator equations of the second kind allow us to follow easily the whole history of the transient processes, when seeking one time layer after another, using only direct computational operations.

Note that the operator $F = \{F_{nmj}\}$ of problem (5.36), as well as any other convolution operator that occurs under re-expansion of a function determined in one basis with respect to another basis, can be easily inverted by using the standard techniques of the theory of Fourier series. Applying the techniques of the theory of meromorphic functions in this case, we just point to methods that in an essential way widen the possibilities of treating problems for even more complex structures cases efficiently (see, e.g., [4,5,9,140,141]).

5.3.4. Inhomogeneities Preserving Mode Structure of the Field

Dielectric inserts with the coordinate boundaries ($z =$ const) and $\varepsilon = \varepsilon(z)$, $\sigma = \sigma(z)$, and transverse metal walls (end caps) in a regular waveguide (see Fig. 5.3C,D) do not influence the mode composition of a transient wave: in any cross-section of the structure the sets of eigenfunctions $\{\mu_n(\rho)\}$ (or $\{\mu_{n1}(\rho)\}$) and eigenvalues $\{\lambda_n\}$ (or $\{\lambda_{n1}\}$), necessary for representation of the incident and scattered field, are the same. Formally this results in the fact that the excitation field $U^i(\rho, z, t) = v(z, t)\,\mu(\rho)$ induces a complete field $U = (\rho, z, t) = u(z, t)\,\mu(\rho)$ of the same kind as the incident one. Thus hereafter the inferior indices p (for circular waveguide) and $p1$ (for coaxial waveguide) of the values λ, $\mu(\rho)$, $v(z, t)$, $u(z, t)$, that do not matter in the considerations, are omitted.

On the assumption that the source $f(z, t)$ that generates the wave $v(z, t)$, at any $t \geq 0$ is situated in the region $z > 0$ (i.e., in region **A**), we reduce the analysis of the scattering properties of a dielectric inhomogeneity (see (5.23) and Fig. 5.3D) to the solution of the following generalized Cauchy problem.

$$B(\lambda)[u] = f(z, t) + \left[(\varepsilon(z) - 1)\frac{\partial^2}{\partial t^2} + \sigma(z)\frac{\partial}{\partial t} \right] u(z, t);$$
$$-\infty < t < \infty, \qquad -\infty < z < \infty. \tag{5.40}$$

According to the definitions (5.14), (5.15), if $v(0, t) = \delta(t - \eta)$, we obtain

$$R_{np}^{\mathbf{AA}}(t - \eta) = \delta_n^p \left[u'(0, t) - v'(0, t) \right], \quad T_{np}^{\mathbf{BA}}(t - \eta) = -\delta_n^p u'(-L, t).$$

Through inversion, using a respective fundamental solution $G(z, t, \lambda)$, of the operator $B(\lambda)$ in (5.40), we have an equivalent formulation of the problem:

$$u(z, t) = v(z, t) - \frac{1}{2}\int_L^0\!\!\int_0 J_0\left[\lambda\left((t - \tau)^2 - (z - \omega)^2\right)^{1/2}\right]\chi\left((t - \tau) - |z - \omega|\right)$$
$$\times \left[(\varepsilon(\omega) - 1)\frac{\partial^2}{\partial \tau^2} + \sigma(\omega)\frac{\partial}{\partial \tau} \right][u(\omega, \tau)]\, d\tau\, d\omega; \quad t \geq 0, \quad |z| < \infty. \tag{5.41}$$

The first component in (5.41) determines the field generated by the source $f(z, t)$ in a regular waveguide, and the second component represents the correction to it due to the scattering by the inhomogeneity localized in the region $-L < z < 0$. The features of the convolution allow a substitution of problem (5.41) by the following equivalent one.

$$u(z, t) = v(z, t) - \frac{1}{2}\int_{-L}^0\!\!\int_0 [(\varepsilon(\omega) - 1)F_\varepsilon(z - \omega, t - \tau)$$
$$+ \sigma(\omega)F_\sigma(z - \omega, t - \tau)]\frac{\partial u(\omega, \tau)}{\partial \tau}\, d\tau\, d\omega;$$

$$F_\varepsilon(z, t)$$
$$= J_0\left[\lambda(t^2 - z^2)^{1/2}\right]\delta[t - |z|] - \lambda t(t^2 - z^2)^{-1/2} J_1\left[\lambda(t^2 - z^2)^{1/2}\right]\chi[t - |z|],$$
$$F_\sigma(z, t) = J_0\left[\lambda(t^2 - z^2)^{1/2}\right]\chi[t - |z|]. \tag{5.42}$$

If $z \notin [-L; 0]$, the formulas (5.42) determine the field outside the inhomogeneity with respect to the values $u(\omega, \tau)$ for every $\omega \in [-L; 0]$ and time τ, that are less than the observation time t. These values are determined by the expressions (5.42) if $z \in [-L;0]$ as solutions of a two-dimensional integrodifferential equation of the second kind. The delay effect due to the finite velocity of propagation of the excitation in the systems described by the problems (5.21), enables one to build relatively simple, direct computational schemes to invert equation (5.42) (see, e.g., [142]), providing also an evaluation of the accuracy of the obtained results.

Another simple computation scheme is based on a standard direct discretization of problem (5.40), using the method of finite differences. However, the problem of limiting the computation region (see Chapter 2) is solved by using the following exact conditions at the boundaries in the planes $z = 0$ and $z = -L$ (see Fig. 5.3D):

$$\begin{cases} u(0,t) - v(0,t) = -\int\limits_0 J_0[\lambda(t-\tau)]\,\chi(t-\tau)\,[u'(0,\tau) - v'(0,\tau)]d\tau \\ u(-L,t) = \int\limits_0 J_0[\lambda(t-\tau)]\,\chi(t-\tau)\,u'(-L,\tau)\,d\tau \end{cases}. \tag{5.43}$$

Conditions (5.43) are obtained from (2.34), by taking into account the propagation directions of the complete field components $U(\rho, z, t) = u(z,t)\,\mu(\rho)$ in regions **A** and **B**. The idea to use (2.34) in such a way was first suggested by the authors of [18]. Note that a sequence of simple operations allows us to transfer from (5.43), first to the nonlocal (in terms of t) conditions

$$\begin{cases} \left(u^{\mathbf{A}}\right)'(0,t) + \dfrac{\partial u^{\mathbf{A}}(0,t)}{\partial t} = \lambda\int\limits_0 J_1[\lambda(t-\tau)]\,\chi(t-\tau)\left(u^{\mathbf{A}}\right)'(0,t)\,d\tau \\ -u'(-L,t) + \dfrac{\partial u(-L,t)}{\partial t} = -\lambda\int\limits_0 J_1[\lambda(t-\tau)]\,\chi(t-\tau)\,u'(-L,\tau)\,d\tau \end{cases}, \tag{5.44}$$

$$\begin{cases} \left(u^{\mathbf{A}}\right)'(0,t) + \dfrac{\partial u^{\mathbf{A}}(0,t)}{\partial t} = -\lambda\int\limits_0 \dfrac{J_1[\lambda(t-\tau)]}{t-\tau}\chi(t-\tau)\,u^{\mathbf{A}}(0,t)\,d\tau \\ -u'(-L,t) + \dfrac{\partial u(-L,t)}{\partial t} = -\lambda\int\limits_0 \dfrac{J_1[\lambda(t-\tau)]}{t-\tau}\chi(t-\tau)\,u(-L,t)\,d\tau \end{cases}, \tag{5.45}$$

and then (see Chapter 2) also to the local conditions of such a type. Here $u^{\mathbf{A}}(0,t) = u(0,t) - v(0,t)$.

A transverse electric wall (end cap) of any waveguide channel with transverse eigennumbers $\{\alpha_n\}$ is completely described by the diagonal reflection operator $R^{\mathbf{AA}}$ with

$$R^{\mathbf{AA}}_{np}(t-\eta) = \delta^p_n\left(\alpha_p^2 + \frac{\partial}{\partial t^2}\right)[J_0[\alpha_p(t-\eta)]\,\chi(t-\eta)]. \tag{5.46}$$

Expression (5.46) is obtained according to the same procedure as was used for $v'_p(0, t)$ in the case when $v_p(0, t) = \delta(t - \eta)$ (see formulas (5.24), (5.25)): the condition that the tangential component of the complete electric field $U(\mathbf{A}) = [v_p(z, t) + u_{p\alpha}(z, t)]\mu_{p\alpha}(\rho)$ should be zero in the plane region $z = 0$ reduces the problem of determining $R_{np}^{\mathrm{AA}}(t - \eta)$ to the inversion of the Volterra integral equation (5.25). If $\alpha_p = \lambda_p$ and $u_{p\alpha}(z, t) = u_p(z, t)$, the result of (5.46) refers to a circular waveguide of the radius a (see Fig. 5.3C), and if $\alpha_p = \lambda_{p1}$ and $u_{p\alpha}(z, t) = u_{p1}(z, t)$, it refers to a coaxial waveguide. Formally, the same results can be obtained from equation (5.30) by closing the whole cross-section of the waveguide with a metal diaphragm.

5.4. Algorithms for Calculating Transient Characteristics of Resonant Inhomogeneities

The list of elementary inhomogeneities for which the boundary operators of transforming the evolutionary basis of transient waves are calculated by the methods considered in Section 5.3, can be substantially extended, namely, to the structures whose geometry is given in Figure 5.3E. They can all also be identified as elementary, inasmuch as none of them has features that require the implementation of nonstandard dispersion laws, which lead to more complicated (and physically interesting) abnormal or resonance scattering modes [6,7,9]. Indeed, although dielectric inhomogeneities may form a connecting volume between semi-infinite regular channels, still they do not enrich the mode content of the field; their boundary transform operators are diagonal.

The dynamics of the components of frequency spectra (the behavior of complex eigenfrequencies by the variation of parameters) of open waveguide resonators presented in Figure 5.4 is quite complicated and diversified. As a result, the spectrum of physical effects taking place by monochromatic wave scattering by such a structure is also very diversified [7,143,144]. In the time domain the effects of super-high-Q free oscillations, the effects of linear interaction of free oscillations in zones of concentration of spectrum points, and so on, can as well give rise to nonstandard wave-scattering modes never recorded before. This is validated, in particular, by the results of analysis undertaken in Chapter 6. The detailed study of such effects, peculiar features of spatio–temporal transformations of electromagnetic fields, should be of great interest and importance for the fundamental and applied aspects of many sciences. Here there is no reasonable alternative to the use of computational experiments. However, a close study of the nature of the processes as they emerge from the computational experiments should be a valuable basis for developing more detailed mathematical models for those processes.

Below, we give examples of developing algorithms for analyzing transient field characteristics of the field in complex resonance inhomogeneities of circular and coaxial waveguides. The role of elementary structures is here played by those considered in Section 5.3, and we assemble them into complex units using the scheme discussed in Section 5.2.

Let the inhomogeneity presented in Figure 5.4A, be exposed to the wave $U^i(\rho, z, t) = \sum_n v_n^{\mathbf{B}}(z, t)\, \mu_n^{\mathbf{B}}(\rho)$ coming from the coaxial waveguide **B**. The complete field $U(\rho, z, t)$ that is induced through the scattering of the excitation wave $U^i(\rho, z, t)$, is in each partial region represented as follows.

$$\begin{aligned}
U(\mathbf{A}) &= \sum_n u_n^{\mathbf{A}}(z, t)\, \mu_n^{\mathbf{A}}(\rho),\\
U(\mathbf{B}) &= \sum_n \left(v_n^{\mathbf{B}}(z, t) + u_n^{\mathbf{B}}(z, t)\right) \mu_n^{\mathbf{B}}(\rho),\\
U(\mathbf{C}) &= \sum_n v_n^{\mathbf{C}}(z, t)\, \mu_n^{\mathbf{C}}(\rho),\\
U(\mathbf{D}) &= \sum_n v_n^{\mathbf{D}}(z, t)\, \mu_n^{\mathbf{D}}(\rho),\\
U(\mathbf{E}) &= \sum_n \left(v_n^{\mathbf{E}}(z, t) + u_n^{\mathbf{E}}(z, t)\right) \mu_n^{\mathbf{E}}(\rho).
\end{aligned} \tag{5.47}$$

This agrees with (5.22) and gives the general solution to (5.21). The elements of the evolutionary basis of waves propagating toward increasing z are denoted $u_n(z, t)$, and those propagating toward decreasing z are denoted $v_n(z, t)$. Retaining the notations used before for known values $v^{\mathbf{B}} = \{v_n^{\mathbf{B}}(0, t)\}$, we describe the vector functions to be determined as

$$\left(u^{\mathbf{A}}\right)' = \left\{ \frac{\partial u_n^{\mathbf{A}}(z, t)}{\partial z}\bigg|_{z=0} \right\}, \quad \left(v^{\mathbf{C}}\right)' = \left\{ -\frac{\partial v_n^{\mathbf{C}}(z, t)}{\partial z}\bigg|_{z=-L} \right\}$$

and so on, taking into account the direction of propagation of the waves and the location of the virtual planes at which they appear. In these notations, the use of boundary and transport operators (see Section 5.2) reduces the problem to the solution of a self-congruent complete set of operator equations

$$\begin{cases}
(u^{\mathbf{B}})' = R^{\mathbf{BB}}(1)\,[v^{\mathbf{B}}] + T^{\mathbf{BE}}(1)\, Z^{\mathbf{E}}(L)\,[(u^{\mathbf{E}})']\\
(u^{\mathbf{A}})' = T^{\mathbf{AB}}(1)\,[v^{\mathbf{B}}] + T^{\mathbf{AE}}(1)\, Z^{\mathbf{E}}(L)\,[(u^{\mathbf{E}})']\\
(v^{\mathbf{E}})' = T^{\mathbf{EB}}(1)\,[v^{\mathbf{B}}] + R^{\mathbf{EE}}(1, 0)\, Z^{\mathbf{E}}(L)\,[(u^{\mathbf{E}})']\\
(u^{\mathbf{E}})' = R^{\mathbf{EE}}(1, -L)\, Z^{\mathbf{E}}(L)\,[(v^{\mathbf{E}})']\\
(v^{\mathbf{C}})' = T^{\mathbf{CE}}(1) Z^{\mathbf{E}}(L)\,[(v^{\mathbf{E}})']\\
(v^{\mathbf{D}})' = T^{\mathbf{DE}}(1) Z^{\mathbf{E}}(L)\,[(v^{\mathbf{E}})']
\end{cases}. \tag{5.48}$$

From (5.48), we derive an operator equation of the second kind

$$(v^{\mathbf{E}})' = T^{\mathbf{EB}}(1)\,[v^{\mathbf{B}}] + R^{\mathbf{EE}}(1, 0)\, Z^{\mathbf{E}}(L)\, R^{\mathbf{EE}}(1, -L)\, Z^{\mathbf{E}}(L)\,[(v^{\mathbf{E}})'], \tag{5.49}$$

which is the basis of the algorithm of numerical solution to the problem, and the countable set of formulas that determine the field $U(\rho, z, t,)$ in all partial regions. In (5.48) and (5.49) the index m in $R^{\mathbf{BB}}(m)$, $T^{\mathbf{BE}}(m)$, and so on determines the number of the elementary inhomogeneities, whose boundary operators are used (see Fig. 5.3). To distinguish between, generally, different operators $R^{\mathbf{EE}}(1)$, that are applied at the virtual boundaries $z = 0$ and $z = -L$, we have introduced additional identifiers, via $R^{\mathbf{EE}}(1, 0)$ and $R^{\mathbf{EE}}(1, -L)$. In this section we also use simplified (compared to those in the definitions (5.16), (5.17) and in the system (5.19)) identifications for transport operators: $Z^{\mathbf{B}}(L)$ instead of $Z^{\mathbf{B}}_{z_1(\mathbf{II})=0\to L}$ and so on.

For the structures presented in Figures 5.4B and C (which are still excited by the wave $U^i(\rho, z, t) = \sum_n v_n^{\mathbf{B}}(z, t)\,\mu_n^{\mathbf{B}}(\rho)$ from the coaxial waveguide **B**), omitting details, we can in a straightforward way write down the final formulas:

$$\begin{cases} (v^{\mathbf{E}})' = T^{\mathbf{EB}}(8)\,[v^{\mathbf{B}}] + R^{\mathbf{EE}}(8)\,Z^{\mathbf{E}}(L)\,[R^{\mathbf{EE}}(4)\,Z^{\mathbf{E}}(L)\,[(v^{\mathbf{E}})'] \\ \qquad\qquad + T^{\mathbf{EC}}(4)\,Z^{\mathbf{E}}(M)\,[(u^{\mathbf{C}})']] \\ (u^{\mathbf{C}})' = R^{\mathbf{CC}}(6)\,Z^{\mathbf{E}}(M)\,[T^{\mathbf{CE}}(4)\,Z^{\mathbf{E}}(L)\,[(v^{\mathbf{E}})'] + R^{\mathbf{CC}}(4)\,Z^{\mathbf{E}}(M)\,[(u^{\mathbf{C}})']] \end{cases},$$

and

$$\begin{cases} (v^{\mathbf{C}})' = R^{\mathbf{CC}}(1)\,Z^{\mathbf{C}}(N) R^{\mathbf{CC}}(9)\,Z^{\mathbf{C}}(N)\,[(v^{\mathbf{C}})'] + T^{\mathbf{CE}}(1) Z^{\mathbf{E}}(L)\,[(v^{\mathbf{E}})'] \\ (v^{\mathbf{E}})' = T^{\mathbf{EB}}(1)\,[v^{\mathbf{B}}] + R^{\mathbf{EE}}(1,0)\,Z^{\mathbf{E}}(L) \\ \qquad \times [T^{\mathbf{EC}}(1)\,Z^{\mathbf{C}}(N)\,R^{\mathbf{CC}}(9)\,Z^{\mathbf{C}}(N)\,[(v^{\mathbf{C}})'] + R^{\mathbf{EE}}(1,-L)\,Z^{\mathbf{E}}(L)\,[(v^{\mathbf{E}})']] \\ \qquad + T^{\mathbf{EA}}(1)\,Z^{\mathbf{A}}(M)\,R^{\mathbf{AA}}(7)\,Z^{\mathbf{A}}(M)\,[(u^{\mathbf{A}})'] \\ (u^{\mathbf{A}})' = T^{\mathbf{AB}}(1)\,[v^{\mathbf{B}}] + T^{\mathbf{AB}}(1)\,Z^{\mathbf{E}}(L) \\ \qquad \times [T^{\mathbf{EC}}(1)\,Z^{\mathbf{C}}(N)\,R^{\mathbf{CC}}(9)\,Z^{\mathbf{C}}(N)\,[(v^{\mathbf{C}})'] + R^{\mathbf{EE}}(1,-L) Z^{\mathbf{E}}(L)\,[(v^{\mathbf{E}})']] \\ \qquad + R^{\mathbf{AA}}(1)\,Z^{\mathbf{A}}(M)\,R^{\mathbf{AA}}(7)\,Z^{\mathbf{A}}(M)\,[(u^{\mathbf{A}})'] \end{cases}$$

These systems seem to be more complicated than (5.49) and the first example (5.20). However, this complication does not touch upon the fundamental characteristics of the algorithms for the numerical solution to the corresponding initial boundary value problems. The computations utilize the direct explicit scheme implying moving stepwise along the time axis t.

It is obvious that the final formulas that are equivalent to those written above can equally well be presented in another, simpler form if one wants to gradually increase the complexity of the analyzed units: the elementary inhomogeneities are used to assemble a unit of medium complexity, which is then considered to be an elementary one and so on.

5.5. Signals in the Floquet Channel: The Transform Operators and Some Canonical Initial Boundary Value Problems

The Floquet channel does not actually differ from the conventional closed waveguide channels. That is why most of the results obtained by analyzing problems for waveguide inhomogeneities (see Sections 5.2, 5.3) are also almost entirely true for grating problems.

5.5.1. The Basic Definitions

The model initial boundary problems

$$\begin{cases} \left[-\varepsilon(g)\dfrac{\partial^2}{\partial t^2} - \sigma(g)\dfrac{\partial}{\partial t} + \dfrac{\partial^2}{\partial z^2} + \dfrac{\partial^2}{\partial y^2}\right] U(g,t) = F(g,t); \\ t > 0, \quad g = \{y,z\} \in \mathbf{Q} \\ U(g,0) = \varphi(g), \quad \dfrac{\partial}{\partial t} U(g,t)\Big|_{t=0} = \psi(g); \quad g \in \bar{\mathbf{Q}} \\ E_{tg}(g,t)\Big|_{g\in\mathbf{S}} = 0, \quad U\left\{\dfrac{\partial U}{\partial y}\right\}(2\pi, z, t) = e^{i2\pi\Phi} U\left\{\dfrac{\partial U}{\partial y}\right\}(0, z, t); \qquad t \geq 0 \end{cases} \tag{5.50}$$

for structures periodic along the y-axis and homogeneous along the x-axis (see Fig. 5.5A) will be considered in space–time coordinates $\{x, y, z, t\}$ (which all have the dimension of length), where the length of the grating period is 2π. It is assumed, as before (see the statements of problems (2.54)), that the functions $F(g,t)$, $\varphi(g)$, $\psi(g)$, $\varepsilon(g) - 1$, and $\sigma(g)$ are finite in $\bar{\mathbf{Q}}$ and that they satisfy the conditions for the unique solvability of problem (5.50) in the Sobolev space $\mathbf{W}_2^1(\mathbf{Q}^T)$. In (5.50) the following notations are used: $U(g,t) = E_x(g,t)$ in the case of the E-polarized field and $U(g,t) = H_x(g,t)$ in the H-case ($\varepsilon(g)$ and $\sigma(g)$ are in this case piecewise constant functions); $\mathbf{S}$ is the boundary contour of the grooves of the grating, constructed from perfect metal, with the period $0 \leq y \leq 2\pi$; $\mathbf{\Phi}$ is the real parameter of the Floquet channel $\mathbf{R} = \{g \in \mathbf{R}^2 : 0 < y < 2\pi\}$; $\mathbf{Q} = \mathbf{R}\backslash\overline{\text{int}\,\mathbf{S}}$, and $\overline{\text{int}\,\mathbf{S}}$ is the closure of regions filled with metal. The complex-valued functions $F(g,t)$, $\varphi(g)$, $\psi(g)$, and $U(g,t)$ (in the same way as in the problems of Section 2.5) are the Fourier transforms $f(g,t,\mathbf{\Phi})$ of the real functions $f_{true}(g,t)$ describing the true sources and scattered fields.

For a Floquet channel $\mathbf{R}$ that is regular over the whole length ($\sigma = \varepsilon - 1 \equiv 0$ and $\overline{\text{int}\,\mathbf{S}} = \varnothing$), the solution of (5.50) can be written as

$$U(g,t) = \sum_{n=-\infty}^{\infty} v_n(z,t)\,\mu_n(y); \quad g = \{y,z\} \in \bar{\mathbf{R}}, \qquad t \geq 0, \tag{5.51}$$

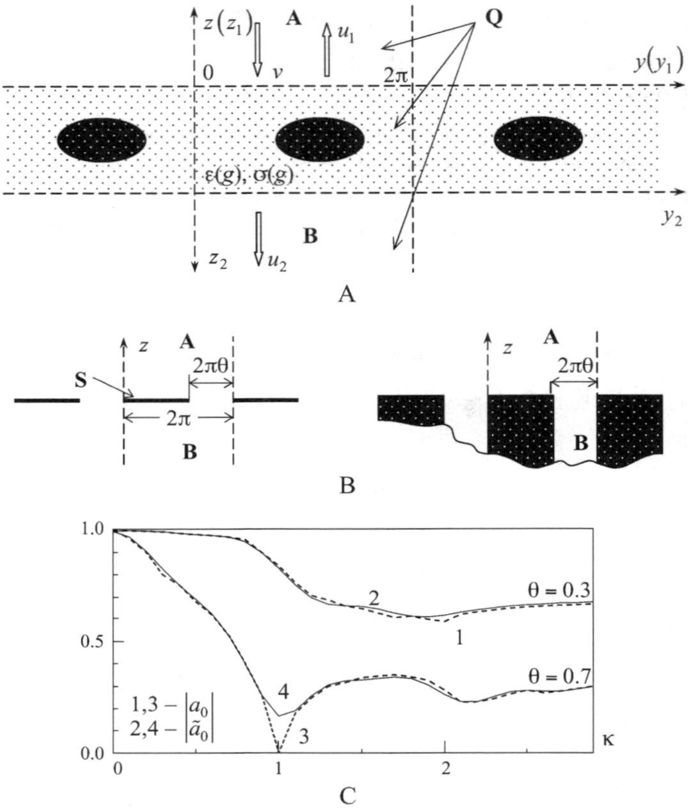

FIGURE 5.5. (A, B) The section of 1-D periodic gratings by the plane$x$ = const and (C) results of comparison of frequency responses of stripe gratings with transparency factors $\theta = 0.7$ and $\theta = 0.3$, obtained by means of methods in FD (dashed lines) and TD (solid lines).

where the elements of the sequence of functions $v\,(z,t) = \{v_n\,(z,t)\}$ are the solutions to the generalized Cauchy problems

$$B\,(\lambda_n)\,[v_n] \equiv \left[-\frac{\partial^2}{\partial t^2} + \frac{\partial^2}{\partial z^2} - \lambda_n^2\right] v_n\,(z,t)$$
$$= a_n\,(z,t) - \delta^{(1)}\,(t)\,b_n\,(z) - \delta\,(t)\,c_n\,(z)\,; \quad t, z \in R^1,$$

and $\{\mu_n(y) = (2\pi)^{-1/2}\exp(i\Phi_n y)\}$, $\{\lambda_n = \Phi_n = n + \Phi\}$, $n = 0, \pm 1, \ldots$ are the sets of eigenfunctions and eigenvalues of the homogeneous boundary problem

$$\begin{cases} \left[\dfrac{d^2}{dy^2} + \lambda^2\right] \mu\,(y) = 0; \quad 0 < y < 2\pi \\ \mu \left\{\dfrac{d\mu}{\partial y}\right\} (2\pi) = e^{i2\pi\Phi} \mu \left\{\dfrac{d\mu}{\partial y}\right\} (0) \end{cases}.$$

Here, $a_n(z,t)$, $b_n(z)$, and $c_n(z)$ are the Fourier coefficients of the functions $F(g,t)$, $\varphi(g)$, and $\psi(g)$ in the orthonormal basis $\{\mu_n(y)\}$ of the space $\mathbf{L}_2[(0;2\pi)]$.

We introduce the transform operators $R^{\mathbf{AA}}$, $T^{\mathbf{BA}}$ and $Z^{\mathbf{A}}_{0\to z_1}$, $Z^{\mathbf{B}}_{0\to z_2}$, describing the scattering properties of a grating that is excited from region **A** (see Fig. 5.5A), according to the scheme developed in Section 5.2.2:

$$\begin{aligned} u'_{nj}(0,t) &\equiv \frac{\partial}{\partial z_j} u_{nj}(z_j,t)\big|_{z_j=0} \\ &= \int_0^t \sum_m \left[R^{\mathbf{AA}}_{nm}(t-\tau)\,\delta^1_j + T^{\mathbf{BA}}_{nm}(t-\tau)\,\delta^2_j\right] v_m(0,\tau)\,d\tau; \\ u_{nj}(z_j,t) &= -\int_0 J_0\left[\Phi_n\left((t-\tau)^2 - z_j^2\right)^{1/2}\right]\chi\left[(t-\tau)-z_j\right] u'_{nj}(0,\tau)\,d\tau; \\ z_j &\geq 0, \quad j=1,2, \quad n,m=0,\pm1,\ldots \end{aligned}$$

or

$$\begin{aligned} u'_j(0,t) &\equiv \{u'_{nj}(0,t)\} = \left[R^{\mathbf{AA}}\delta^1_j + T^{\mathbf{BA}}\delta^2_j\right][v(0,\tau)]; \\ u_j(z_j,t) &\equiv \{u_{nj}(z_j,t)\} = \left[Z^{\mathbf{A}}_{0\to z_1}\delta^1_j + Z^{\mathbf{B}}_{0\to z_2}\delta^2_j\right][u'_j(0,\tau)]; \\ z_j &\geq 0, \quad j=1,2. \end{aligned}$$

Here, $v_n(z,t)$ and $u_{nj}(z,t)$, $n=0,\pm1,\ldots$, $j=1,2$ are elements of the evolutionary bases used for expanding the excitation wave

$$U^i(g,t) = \sum_{n=-\infty}^{\infty} v_n(z,t)\,\mu_n(y); \quad z=z_1\geq 0,$$

and waves of the secondary fields

$$U^s_j(g,t) = \sum_{n=-\infty}^{\infty} u_{nj}(z_j,t)\,\mu_n(y); \quad j=1,2, \quad z_j\geq 0,$$

occurring in the zones of reflection (**A**) and transition (**B**) and propagating toward growing values of z_j.

5.5.2. Stripe Grating and a Grating of Thick Half-Planes: Solution of Initial Boundary Value Problems by Using the Mode-Matching Method

It is evident that only minor changes need to be undertaken in the major results of Sections 5.2 and 5.3 in order to extend them so as to apply to problems of transient processes in periodic structures. Thus, for instance, the formulas

$$R_{np}^{\mathbf{AA}}(t-\eta) = \frac{1}{2\pi i}\sum_{m=-\infty}^{\infty}\frac{e^{i(m-n)2\pi(1-\theta)}-1}{m-n}$$

$$\times\left\{\Phi_m\int_0 J_1\left[\Phi_m(t-\tau)\right]\chi(t-\tau)R_{mp}^{\mathrm{AA}}(\tau-\eta)\,d\tau+\delta_m^p\delta^{(1)}(t-\eta)\right\};$$

$$n=0,\pm1,\ldots,$$

$$T_{np}^{\mathrm{BA}}(t-\eta)=R_{np}^{\mathrm{AA}}(t-\eta)-\delta_n^p v_p'(0,t);$$

$$n=0,\pm1,\ldots,\quad v_p(0,t)=\delta(t-\eta) \tag{5.52}$$

give the exact solution of the problem of nonmonochromatic quasi-periodic excitation of a metal stripe grating (see Fig. 5.5B) in the case of an E-polarized field. It is easy to follow their relations to formulas (5.29) and (5.30), which describe the scattering properties of thin axis-symmetric metal diaphragms in a circular waveguide.

Figure 5.5C illustrates the results of a test of algorithm (5.52). Accurate (rigorous) solution to the problem of diffraction of plane E-polarized waves by the thin stripe gratings with various values of the coefficient of transparency θ (FD) has been compared with the solution, obtained by transformation of TD results into FD. The corresponding numerical experiment is described in detail in Section 4.6.1. Accurate values of the complex amplitudes $a_n(\kappa)$ of the scattered field in the domain of reflection of the grating is obtained by means of one of the most efficient and reliable methods for the analysis of such structures in FD, viz. the method of the Riemann–Hilbert problem of the theory of analytic functions [2,3]. In Figure 5.5C the corresponding frequency responses are depicted with dashed lines; parameter $\mathbf{\Phi}=0$. The results $\tilde{a}_n(\kappa)$ obtained from the solution of the problem in TD are shown by solid lines. The step l of time sampling in algorithm (5.52) was equal to $8.76\cdot10^{-4}$, and the upper limit T for the integration over t in representation of the kind (4.68) was chosen to be equal to 12.0.

The problem for a grating of thick half-planes (see Fig. 5.5B) is somewhat more complex. However, the conventional matching technique also provides in this case a satisfactory result without essential changes in the approach discussed above. Thus, let the grating be excited by an E-polarized wave $U^i(g,t)$ with $v_n(0,t)=\delta_n^p\delta(t-\eta)$. In region **A** the total field is written as

$$U(\mathbf{A})=-\sum_{n=-\infty}^{\infty}\mu_n(y)\int_0 J_0[\Phi_n((t-\tau)^2-z^2)^{1/2}]$$

$$\times\chi[(t-\tau)-z]\,R_{np}^{\mathbf{AA}}(\tau-\eta)d\tau+U^i(y,z,t).$$

In region **B** (plane-parallel waveguide), the following expression is obtained (see Section 2.2.1).

$$U(\mathbf{B}) = -\sum_{m=1}^{\infty} \mu_{m1}(y) \int_0 J_0[\lambda_{m1}((t-\tau)^2 - z^2)^{1/2}]$$
$$\chi[(t-\tau)+z]\, T_{mp}^{\mathbf{BA}}(\tau-\eta)d\tau,$$

where $\lambda_{m1} = m/2\theta$ and $\mu_{m1}(y) = (\pi\theta)^{-1/2} \sin[m(y - 2\pi(1-\theta))/2\theta]$, $m = 1, 2, \ldots$. By matching the time derivatives of the tangential components of the complete field strength vectors ($\partial U/\partial_z$ and $\partial U/\partial t$) on the interval $2\pi(1-\theta) \le y \le 2\pi$ in the plane $z = 0$ and satisfying the boundary condition $(\partial U/\partial t) = 0$ at the step $0 \le y \le 2\pi(1-\theta)$, we obtain

$$\sum_{n=-\infty}^{\infty} \left[R_{np}^{\mathbf{AA}}(t-\eta) + \delta_n^p v_p'(0,t)\right] \mu_n(y) = -\sum_{m=1}^{\infty} T_{mp}^{\mathbf{BA}}(t-\eta)\mu_{m1}(y);$$
$$2(\pi-\theta) \le y \le 2\pi, \tag{5.53}$$

$$\sum_{n=-\infty}^{\infty} \left[R_{np}^{\mathbf{AA}}(t-\eta) - \delta_n^p \delta^{(1)}(t-\eta)\right] \mu_n(y) = \sum_{m=1}^{\infty} T_{mp}^{\mathbf{BA}}(t-\eta)\mu_{m1}(y)$$
$$+ \sum_{n=-\infty}^{\infty} \Phi_n \mu_n(y) \int_0 J_1[\Phi_n(t-\tau)]\chi(t-\tau) R_{np}^{\mathbf{AA}}(\tau-\eta)\, d\tau$$
$$- \sum_{m=1}^{\infty} \lambda_{m1}\mu_{m1}(y) \int_0 J_1[\lambda_{m1}(t-\tau)]\chi(t-\tau) T_{mp}^{\mathbf{BA}}(\tau-\eta)\, d\tau;$$
$$2\pi(1-\theta) \le y \le 2\pi, \tag{5.54}$$

$$\sum_{n=-\infty}^{\infty} \left[R_{np}^{\mathbf{AA}}(t-\eta) - \delta_n^p \delta^{(1)}(t-\eta)\right] \mu_n(y)$$
$$= \sum_{n=-\infty}^{\infty} \Phi_n \mu_n(y) \int_0 J_1[\Phi_n(t-\tau)]\chi(t-\tau) R_{np}^{\mathbf{AA}}(\tau-\eta)\, d\tau;$$
$$0 \le y \le 2\pi(1-\theta). \tag{5.55}$$

In terms of the Fourier coefficients of the joint functions, the equalities (5.53) to (5.55) can be presented in several equivalent forms: simple or more complicated, or forms that immediately give the correct algorithm of determining the electromagnetic characteristics of the structure from the point of view of computations, and furthermore there are those requiring additional analytic transformations in order to serve as bases for reliable and efficient computation schemes. One example of such a form is the set of operator equations of the second kind

$$\begin{cases} R^{\mathbf{AA}}_{np}(t-\eta) = \delta^p_n \delta^{(1)}(t-\eta) + \Phi_n \int\limits_0 J_1\left[\Phi_n(t-\tau)\right]\chi(t-\tau)\, R^{\mathbf{AA}}_{np}(\tau-\eta)\, d\tau \\ \quad + \sum\limits_{m=1}^{\infty} F_{nm} \left\{ T^{\mathbf{BA}}_{mp}(t-\eta) - \lambda_{m1} \int\limits_0 J_1\left[\lambda_{m1}(t-\tau)\right]\chi(t-\tau)\, T^{\mathbf{BA}}_{mp}(\tau-\eta)\, d\tau \right\}; \\ \quad n = 0, \pm 1, \ldots \\ T^{\mathbf{BA}}_{mp}(t-\eta) = -\sum\limits_{n=-\infty}^{\infty} \left[R^{\mathbf{AA}}_{np}(t-\eta) + \delta^p_n v'_p(0,t) \right] F^*_{nm}; \quad m = 1, 2, \ldots \end{cases}$$

with respect to the unknown vectorfunctions $\left\{R^{\mathbf{AA}}_{np}(t-\eta)\right\}_n$ and $\left\{T^{\mathbf{BA}}_{mp}(t-\eta)\right\}_n$. The solution of this set requires the inversion of infinite matrices, whose elements do not depend on the time parameters and are explicitly expressed by the integrals

$$F_{nm} = \int\limits_{2\pi(1-\theta)}^{2\pi} \mu_{m1}(y)\, \mu^*_n(y)\, dy.$$

6
Open Periodic Resonators and Waveguides: Novel Results in Electromagnetic Theory of Gratings

6.1. Introduction

Transient resonance wave scattering by gratings is rather complicated phenomenon; it depends on numerous parameters and has not been studied thoroughly yet. The numerical simulation that takes into account diverse properties of the spectrum (various eigenmodes: natural oscillations and eigenwaves) may serve as the most reliable tool for such investigation. The components of the grating's spectrum may be found through analytical continuation of the solutions to diffraction problems into the domain of complex values of the real physical parameters (such as frequency, propagation constants, and geometrical dimensions). When the parameters vary within the domain of real values (diffraction theory in the frequency domain) properties of complex spectrum usually stay out of the focus of intrest. However, it turns out that spectral properties play the principal role in the formation of the response of an electromagnetic structure to every external excitation. The approach, based on the study of the singularities (spectrum) of resolvent operators of boundary value problems in the frequency domain, allows us to consider and treat various mathematical, electromagnetic, and applied problems of resonant wave scattering using the same methodological background. In this chapter we apply this approach to the analysis of regularities in transient resonance wave scatterings and for the description and treatment of irregular effects and anomalous phenomena associated with the ability of a grating to support regimes with complicated nonclassical dispersion laws.

For the first time clear perspectives provided by such approaches have been indicated in [145–147]. Several examples demonstrating on analytical level the connection between simple resonances of monochromatic wave diffraction and the excitation in the structures of oscillations close to the natural ones are given in these papers. Actually, the progress in theoretical studies of spectral electromagnetic problems, of the physics of resonant scattering processes, was substantially stimulated by implementation of the latest achievements of the theory of nonselfadjoint operators and meromorphic operator functions [68–70,148–150]. One of the first and successful steps in this direction was taken in [151,152]. The analytical results of the spectral theory of grating (of the open periodic resonators and waveguides) obtained on this base are presented in Section 6.2. They can be considered as a tool allowing us to study the physical features of resonant wave scattering by one-dimensional periodic structures on a qualitatively new level. The most typical examples of such analysis in frequency domain are collected in Section 6.3.

Periodical structures, as already mentioned, can be identified as scattering inhomogeneities in an imaginary plane-parallel Floquet channel that is basically identical to the classic closed waveguide channels. The readers, interested in, can easily apply the essential results and conclusions, obtained from the spectral grating theory, to the problems of analyzing different types of waveguide discontinuities (open waveguide resonators). In this context the results of such original works as [5,7,143,144,153–157] can be very useful.

Section 6.4 is devoted to the discussion of the results of numerical experiments focused on the study of the characteristic features and regularities in formation of resonant responses of periodic structures excited by pulses of various duration. Section 6.5 deals with the analysis of pulse distortions (distortions of pulsed spatial harmonics of the field scattered by the grating) on their regular propagation trajectories in the Floquet channels (in the areas of reflection and transition by the structure).

The specific character of the physical processes considered below determines the set of mathematical models, methods, and algorithms suitable for their analysis. The choice of methods is mainly conditioned by requirements to accuracy, efficiency, possibility of focusing of results on specific details, and mechanisms that influence the structure of the secondary fields. Numerical solutions to the diffraction and spectral problems in the frequency domain have been obtained by using the analytical regularization methods [3,5,9,10,140] that implement the classic concept of functional analysis and integral equation theory concerning the explicit inversion of the singular (main) part of an ill posed operator equation of the first kind and enable one to obtain reliable, trustworthy data by variation of parameters in the resonance domains. For solving the problems in the time domain, the methods and algorithms that are described in Chapters 2 to 5 of this book, and in [20,21,30,117,118] have been applied.

6.2. Essential Qualitative Results in the Spectral Theory of Gratings

6.2.1. Formulation of Boundary Value Problems and Principal Definitions

Let an E- or H-polarized time-harmonic plane wave

$$\tilde{U}_p^i(g, \kappa) = \exp[i(\Phi_p y - \Gamma_p(z - 2\pi\delta))]; \quad g = \{y, z\}, \quad z > |2\pi\delta|,$$
$$p = 0, \pm 1, \pm 2, \ldots \tag{6.1}$$

be incident on the gratings depicted in Figures 1.1A and B (the structures are homogeneous along the x-axis). In the classic formulation of the diffraction problem, when all functions are assumed to be sufficiently smooth or constant, the total field is determined by the solution $\tilde{U}(g,\kappa) \in \mathbf{C}^2(\mathbf{Q}) \cap \mathbf{C}^1(\bar{\mathbf{Q}})$ of the two-dimensional Helmholtz equation

$$\left[\frac{\partial^2}{\partial y^2} + \frac{\partial^2}{\partial y^2} + \kappa^2 \tilde{\varepsilon}(g)\right] \tilde{U}(g,\kappa) = 0; \quad g \in \mathbf{Q}, \quad \kappa > 0. \tag{6.2}$$

This solution must satisfy the boundary conditions

$$\tilde{U}\left\{\frac{\partial \tilde{U}}{\partial y}\right\}(2\pi, z, \kappa) = e^{i2\pi\Phi}\tilde{U}\left\{\frac{\partial \tilde{U}}{\partial y}\right\}(0, z, \kappa), \tag{6.3}$$

$$E_{tg}(g,\kappa)\big|_{g\in\mathbf{S}} = 0 \tag{6.4}$$

and the radiation condition

$$\tilde{U}(g,\kappa) = \begin{Bmatrix} \tilde{U}_p^i \\ 0 \end{Bmatrix} + \sum_{n=-\infty}^{\infty} \begin{Bmatrix} R_{np} \\ T_{np} \end{Bmatrix} e^{i(\Phi_n y \pm \Gamma_n (z \mp 2\pi\delta))}; \quad \begin{Bmatrix} z \geq 2\pi\delta \\ z \leq -2\pi\delta \end{Bmatrix} \tag{6.5}$$

that complies with the requirement of the absence of waves arriving from $z = \pm\infty$ in the scattered field.

In the case of piecewise smooth or piecewise constant functions $\tilde{\varepsilon}(g) = \varepsilon(g) + i\sigma(g)/\kappa$, the problem (6.2) to (6.5) has to be complemented with the continuity condition for tangential components of the vectors of electrical field density along the contours $\mathbf{S}_\varepsilon$, where $\tilde{\varepsilon}(g)$ is discontinuous. If 2π-periodic (along y) contours $\mathbf{S}_\varepsilon$ and $\mathbf{S}$ (determining the boundaries of the perfectly conducting inclusions) are not sufficiently smooth, then the additional equation, describing the condition of the integrability of the energy density within any bounded domain $\mathbf{V} \subset \mathbf{Q}$ has to be added to the principal equations of the problems:

$$\int_{\mathbf{V}} (|\tilde{U}|^2 + |\text{grad}\,\tilde{U}|^2)\, dv < \infty. \tag{6.6}$$

Reducing the formal requirements, we consider the generalized solutions to this problem: $\tilde{U}(g,\kappa) \in \mathbf{W}_2^1(\mathbf{Q}_b)$; $\mathbf{Q}_b = \{g \in \mathbf{Q} : |z| < b\}$, $b < \infty$.

Here, κ is a dimensionless frequency parameter, characterizing the ratio between the actual spacing l of the grating and the wave length λ of the incident wave; all the processes are considered in terms of the time–spatial coordinates $\{x, y, z, t\}$, which all have the dimension of length. The period of the grating is 2π, and the time dependence is determined by the factor $\exp(-i\kappa t)$; $\tilde{\varepsilon}(y,z) : \tilde{\varepsilon}(y+2\pi, z) = \tilde{\varepsilon}(y,z), \tilde{\varepsilon}(y, |z| > 2\pi\delta) \equiv 1, \text{Re}\,\tilde{\varepsilon} \geq 1, \text{Im}\,\tilde{\varepsilon}\,\text{Re}\,\kappa \geq 0$ is a complex-valued function characterizing the complex relative dielectric permittivity of the grating material. $\mathbf{Q} = \{g \in \mathbf{R}^2 \backslash (\mathbf{S}_\varepsilon \cup \overline{\text{int}\,\mathbf{S}}) : 0 < y < 2\pi\}$ is a domain in the Floquet channel $\mathbf{R} = \{g \in \mathbf{R}^2 : 0 < y < 2\pi\}$, where $\mathbf{R}^2$ is the plane of variables $g = \{y, z\}$. $\Phi_n = n + \Phi$, $\Gamma_n = (\kappa^2 - \Phi_n^2)^{1/2}$, $\text{Re}\Gamma_n \geq 0$, and $\text{Im}\,\Gamma_n \geq 0$; δ and Φ are real parameters.

For an E-polarized field, $\tilde{U}(g, \kappa) = E_x(g, \kappa)$ is the only nonzero component of the electric field strength vector and (see Section 1.2.2)

$$H_y = \frac{1}{i\kappa\eta_0}\frac{\partial E_x}{\partial z}, \quad H_z = -\frac{1}{i\kappa\eta_0}\frac{\partial E_x}{\partial y}, \quad H_x = 0.$$

For H-polarization, $\tilde{\varepsilon}(g)$ is a piecewise constant function, $\tilde{U}(g, \kappa) = H_x(g, \kappa)$, $E_x = H_y = H_z = 0$, and

$$E_y = -\frac{\eta_0}{i\kappa\tilde{\varepsilon}}\frac{\partial H_x}{\partial z}, \quad E_z = \frac{\eta_0}{i\kappa\tilde{\varepsilon}}\frac{\partial H_x}{\partial y}.$$

The Green's function $\tilde{G}(y, z, y_0, z_0, \kappa, \Phi)$ of the grating in the field of quasi-periodic point sources

$$\tilde{f}(y, z) = \sum_{n=-\infty}^{\infty} \delta(y - y_0 - 2\pi n, z - z_0)\, e^{i\Phi 2\pi n}; \quad 0 < y_0 < 2\pi, \quad |z| < 2\pi\delta \tag{6.7}$$

is determined by the problem obtained by replacing the zero in the right-hand part of (6.2) by the function $\delta(y - y_0, z - z_0)$ and assuming $\tilde{U}_p^i(g, \kappa) \equiv 0$. In the general case, quasi-periodic along y source functions

$$\tilde{f}(g) : \tilde{f}(y + 2\pi, z) = \exp(i2\pi\Phi)\,\tilde{f}(y, z); \quad \tilde{f}(g) \in \mathbf{L}_2(\mathbf{Q})$$

with compact in $\mathbf{Q}$ supports from $\mathbf{Q}_\delta = \{g \in \mathbf{Q} : |z| \le 2\pi\delta\}$ generate the field $\tilde{U}(g, \kappa)$ that is defined by the problem

$$\left[\frac{\partial^2}{\partial y^2} + \frac{\partial^2}{\partial y^2} + \kappa^2\tilde{\varepsilon}(g)\right]\tilde{U}(g, \kappa) = \tilde{f}(g); \quad g \in \mathbf{Q}, \quad \kappa > 0, \tag{6.8}$$

$$\tilde{U}\left\{\frac{\partial \tilde{U}}{\partial y}\right\}(2\pi, z, \kappa) = e^{i2\pi\Phi}\tilde{U}\left\{\frac{\partial \tilde{U}}{\partial y}\right\}(0, z, \kappa), \quad E_{tg}(g, \kappa)\big|_{g\in\mathbf{S}} = 0, \tag{6.9}$$

$$\tilde{U}(g, \kappa) = \sum_{n=-\infty}^{\infty}\left\{\begin{matrix} R_n \\ T_n \end{matrix}\right\} e^{i(\Phi_n y \pm \Gamma_n(z \mp 2\pi\delta))}; \quad \left\{\begin{matrix} z \ge 2\pi\delta \\ z \le -2\pi\delta \end{matrix}\right\}. \tag{6.10}$$

We call the domain $\mathbf{Q}_\delta$, enveloping all discontinuities of the Floquet channel $\mathbf{R}$ and sources $\tilde{f}(g)$, the domain of connection of the area of reflection ($\{g \in \mathbf{R} : z \ge 2\pi\delta\}$) and transition ($\{g \in \mathbf{R} : z \le -2\pi\delta\}$) of the periodic structure.

The amplitudes R_{np} and T_{np} in (6.5) form the so called generalized scattering matrices of the grating: the reflection matrix $R = \{R_{np}\}_{n,p=-\infty}^{\infty}$ and the transmission matrix $T = \{T_{np}\}_{n,p=-\infty}^{\infty}$. They are connected (see [3,6,9,11]) by energy balance relations,

$$\sum_{n=-\infty}^{\infty}\left[|R_{np}|^2 + |T_{np}|^2\right]\left\{\begin{matrix} \operatorname{Re}\Gamma_n \\ \operatorname{Im}\Gamma_n \end{matrix}\right\} = \left\{\begin{matrix} \operatorname{Re}\Gamma_p + 2\operatorname{Im} R_{pp}\operatorname{Im}\Gamma_p \\ \operatorname{Im}\Gamma_p - 2\operatorname{Im} R_{pp}\operatorname{Re}\Gamma_p \end{matrix}\right\} - \frac{\kappa^2}{2\pi}\left\{\begin{matrix} W_1 \\ W_2 \end{matrix}\right\};$$
$$p = 0, \pm 1, \pm 2, \ldots, \tag{6.11}$$

and by reciprocity relations

$$\frac{R_{np}(\Phi)}{\Gamma_p(\Phi)} = \frac{R_{-p,-n}(\Phi)}{\Gamma_{-n}(\Phi)}; \quad n, p = 0, \pm 1, \pm 2, \ldots. \tag{6.12}$$

Here,

$$W_1 = \int_{\mathbf{W}} \operatorname{Im}\tilde{\varepsilon}|\vec{E}|^2 dw, \quad W_2 = \int_{W} [\eta_0^2 |\vec{H}|^2 - \operatorname{Re}\tilde{\varepsilon}|\vec{E}|^2]\, dw$$

for E-polarized fields, and

$$W_1 = \eta_0^{-2} \int_{\mathbf{W}} \operatorname{Im}\tilde{\varepsilon}|\vec{E}|^2 dw, \quad W_2 = \int_{\mathbf{W}} [\eta_0^{-2} \operatorname{Re}\tilde{\varepsilon}|\vec{E}|^2 - |\vec{H}|^2]\, dw$$

for H-polarized fields; $\mathbf{W} = \mathbf{Q}_\delta \times [|x| < 0.5]$. A detailed discussion of relations of the kind (6.11), (6.12) and nontrivial physical consequences, which follow from them, are presented in [6,7].

In studying the diffraction properties of periodic gratings the problem (6.1) to (6.5) is conventionally considered with the condition that in (6.1) $p = 0$ and $\operatorname{Im}\Gamma_0 = 0$. Considering this case separately, farther on we introduce the following notations for amplitudes of spatial harmonics of scattered field (see (6.5)): $a_n = R_{n0}$, $b_n = T_{n0}$ (E-polarization) and $A_n = R_{n0}$, $B_n = T_{n0}$ (H-polarization). According to (6.11), the values $W_m^a = |a_m|^2 \operatorname{Re}\Gamma_m/\Gamma_0$, $W_m^b = |b_m|^2 \operatorname{Re}\Gamma_m/\Gamma_0$ and so on, determine energy capacity of harmonics, that is, the relative part of the energy that is directed by the structure into the relevant spatial radiation channel. The channel, corresponding to the mth harmonic, we call *open* if $\operatorname{Im}\Gamma_m = 0$. The regime when only one channel for a spatial harmonic is open ($m = 0$) we call the *single-mode regime*. For such a regime the basic (or principal) spatial harmonic in the reflection area is called the *specular one*, and its amplitude (a_0 or A_0) is called the *reflection coefficient* (clearly, b_0 or B_0 is called the *transmission coefficient*).

6.2.2. Grating as an Open Periodic Resonator

Let Φ have a fixed value. Then the following Statements 6.1 to 6.6 [9,158–161], that qualitatively characterize the grating as an open periodic resonator, are true.

Statement 6.1 (definition and qualitative characteristics of the frequency spectrum). *The analytic continuation of the solutions $\tilde{U}(g, \kappa)$ of (6.2) to (6.5) and (6.8) to (6.10) into the domain of complex (nonphysical) values of κ produces a meromorphic function $\tilde{U}(\kappa)$ with a natural domain of definition* $\mathbf{K}$ *that is an infinite sheet Riemann surface consisting of the planes $\kappa \in \mathbf{C}$ cut along the directions of* $(\operatorname{Re}\kappa)^2 - (\operatorname{Im}\kappa)^2 - \Phi_n^2 = 0$, $n = 0, \pm 1, \pm 2, \ldots$, $\operatorname{Im}\kappa \leq 0$. *The first sheet of the surface* $\mathbf{K}$ *is unambiguously determined by the radiation conditions (6.5) and (6.10), that is, by the choice of the values* $\operatorname{Re}\kappa \operatorname{Re}\Gamma_n \geq 0$, $\operatorname{Im}\Gamma_n \geq 0$, $n = 0, \pm 1, \pm 2, \ldots$ *at the axis* $\operatorname{Im}\kappa = 0$. *This sheet is depicted in Figure 6.1A. The next sheets (each of them with its own set of compliance pairs* $\{\kappa \to \Gamma_n(\kappa)\}$*), have, unlike the first one, opposite signs (root branches) of* $\Gamma_n(\kappa)$ *for a finite number*

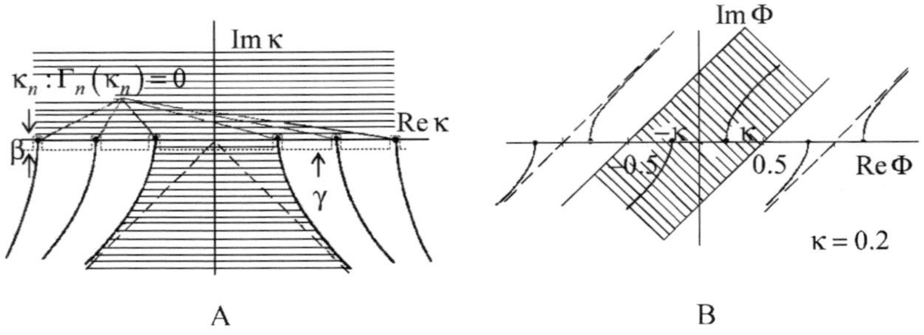

FIGURE 6.1. The definition of Riemann surfaces (A) **K** and (B) **F**.

of values of the index n. The cuts (solid lines) originate from the real algebraic branch points $\kappa_n : \Gamma_n(\kappa) = 0;\ n = 0, \pm 1, \pm 2, \ldots$.

The set of poles $\{\bar{\kappa}_n\} \in \Omega_\kappa$ *of the functions* $\tilde{U}(\kappa)$ *that are meromorphic in the local variables on the surface* **K** *(just a countable set without finite accumulation points) determines the complete frequency spectrum of the grating as an open periodic resonator: if* $\kappa = \bar{\kappa}_n$, *the homogeneous* $(\tilde{U}_p^i(g, \kappa) \equiv 0$ *or* $\tilde{f}(g) \equiv 0 \tilde{f}(g) \equiv 0)$ *problems (6.2) to (6.5) and (6.8) to (6.10) are solvable in a nontrivial way in* **K**, *and the corresponding solutions* $\tilde{U}_n(g, \bar{\kappa}_n)$ *have the meaning of free states (oscillations) of the field in the structure.*

Statement 6.2 (local expansion theorem and analytic description of the Wood anomalies). *Every diffraction characteristic* $W(\kappa)$, $(\tilde{U}(g, \kappa)$, $R_{np}(\kappa)$, *and so on) of the grating in the field of quasi-periodic point sources (in the field of a plane wave), depending linearly on the fundamental solution* $\tilde{G}(g, g_0, \kappa)$*;* $g_0 = \{y_0, z_0\}$ *of the problem (6.8) to (6.10), in the vicinity of any point* $\underline{\kappa} \in$ **K** *can be presented by the Laurent series*

$$W(\kappa) = \sum_{m=-M}^{\infty} w_m \eta^m; \quad \eta = \begin{cases} \kappa - \underline{\kappa}; & \underline{\kappa} \notin \{\kappa_n\} \\ (\kappa - \underline{\kappa})^{1/2}; & \underline{\kappa} \in \{\kappa_n\} \end{cases} \tag{6.13}$$

in the local on **K** *variables* η*; M is the order of the pole in the point* $\underline{\kappa}$.

Statement 6.3 (a rough localization of the spectrum Ω_κ). *The spectrum* Ω_κ *is not empty in the bounded part of the surface* **K**. *For volume gratings* $(\delta \neq 0)$ *with an ideal dielectric loading* $(\mathrm{Im}\,\tilde{\varepsilon} \equiv 0)$ *at* $\mathrm{Im}\,\kappa = 0$, $\mathrm{Re}\,\kappa \neq 0$, *there are no non-trivial solutions of the homogeneous problem (6.8) to (6.10) such, that* $\sum_n (|R_n|^2 + |T_n|^2)\mathrm{Re}\,\Gamma_n \neq 0$, *and at* $\mathrm{Im}\,\kappa \neq 0$, $\mathrm{Re}\,\kappa \neq 0$, *such, that* $\mathrm{Im}\,\kappa\,\mathrm{Re}\,\kappa \sum_n (|R_n|^2 + |T_n|^2)\mathrm{Re}\,\Gamma_n > 0$.

The spectrum of the plane gratings $(\delta = 0)$ *at the first sheet of the surface* **K** *can contain only the points* κ_n, *situated on the real axis, where* $\Gamma_n(\kappa_n) = 0$. *For any parameter values, the domains with* $0 < \arg\kappa < \pi$ *and* $(5/4)\pi < \arg\kappa < (7/4)\pi$ *of the first sheet of* **K** *(crosshatched region in Fig. 6.1A) do not contain the points of the spectrum* Ω_κ. *In the case of nonideal* $(\tilde{\varepsilon} = \mathrm{const}, \mathrm{Im}\,\tilde{\varepsilon} \neq 0)$ *volume* $(\delta \neq 0)$ *structures for* $|\mathrm{Re}\,\kappa| \to \infty$ *on the first sheet of* **K**, *all the eigenfrequencies* $\bar{\kappa}_n$ *move away from the real axis, and the corresponding spacing increases faster than* $|\mathrm{Re}\,\kappa|$.

Note that the radiation conditions (6.5), (6.10) containing the requirement of nonexistence of waves arriving from infinity in the scattered field, have been used for solving the boundary value problems (6.2) to (6.4) and (6.8), (6.9) since the time of Lord Rayleigh. However, the corresponding principle has been shown to be valid, until recently, only for an absorbing medium [3]. The drawbacks of this result have been displayed in the evident, coming out on the analytical level, ambiguity in the solution to the diffraction problems for several canonic geometries [3,9]. The results of the spectral theory clarified this problem.

First, the differences in the spectra of open compact resonators and those with a finite number of energy-consuming (i.e., carrying the field energy) harmonics that serve for connection with the free space (gratings and open waveguide resonators) are clearly defined. The spectra of the open compact resonators (see Section 1.3.1) are free from real $\bar{k}$ starting with the first—physical—sheet of the surface **K**. For periodic and waveguide resonators such real $\bar{\kappa}$ are quite common in the spectrum Ω_κ.

Second, for resonators formed by inhomogeneities in the regular waveguide channels (the Floquet channels and closed waveguides), the radiation principles have been analyzed, as well as the principle of limiting absorption and the limiting amplitude principle, that are usually employed to extract the unique, physical solution of elliptic boundary value problems in nonclassic (unbounded) domains of spaces of different dimensions. The foundations of these principles and the analysis of their applicability is a nontrivial problem from a mathematical point of view. Solution to these problems was obtained only for a limited number of specific cases with typical features that are mainly conditioned by the type of the operator, boundary geometry, and the space dimension [162–167]. Such results are rather rare in problems whose boundaries extend to infinity. A quite comprehensive analysis here was undertaken only for the cases in which one can obtain an explicit solution to the boundary value problems [163,167]. Below we present the results of an analysis of the known methods of deriving the unique solution to the problems (6.2) to (6.4) and (6.8), (6.9) that describe diffraction properties of gratings in the field of E- and H-polarized waves [9,161,168]. The properties of the resolvent operators of these problems are the same (see Statement 6.1), and therefore we formulate the corresponding result for only one of them (E-polarization).

Statement 6.4 (radiation principle). *Let* $\operatorname{Im}\tilde{\varepsilon}(g) > 0$ *on an arbitrary set in* $\mathbf{Q}_\delta$ *with nonzero measure. Then according to the radiation condition (6.10), there is a unique solution to (6.8), (6.9) for every* $\kappa > 0$. *If* $\operatorname{Im}\tilde{\varepsilon}(g) \equiv 0$, *then the solution to (6.8) to (6.10) exists and is unique for every* $\kappa > 0$, *with the possible exception of not more than the countable set* $\{\bar{\kappa}_n\} \in \Omega_\kappa$ *without finite accumulation points. The solution to (6.8) to (6.10) in the point* $\bar{\kappa}_m \in \{\bar{\kappa}_n\}$ *can be obtained only if*

$$\int_{\mathbf{Q}} \operatorname{Res}\tilde{G}(g, g_0, \bar{\kappa}_m)\, \tilde{f}(g_0)\, dg_0 = 0. \tag{6.14}$$

The condition $\mathrm{Im}\,\tilde{\varepsilon} > 0$ at $\kappa > 0$ indicates the presence of absorption of the field energy in the lossy dielectric of which the grating is made. The real poles of the function $\tilde{G}(g, g_0, \kappa)$ (the function $\tilde{U}(g, \kappa)$) and of the resolvent operator function of (6.8) to (6.10) are simple [9]. Hence, the requirement (6.14) can be reformulated in terms of the orthogonality of the source function $\tilde{f}(g)$ with respect to the relevant eigenfunctions (free oscillations of the field) $\tilde{U}_m(g, \bar{\kappa}_m)$ of the homogeneous problem (6.8) to (6.10).

Let $0 < \arg\kappa < \pi/2$ and assume that the solution to (6.8), (6.9) for real positive values of the frequency parameter is given by the limit

$$\tilde{U}(g, \mathrm{Re}\,\kappa) = \lim_{\mathrm{Im}\,\kappa \to 0} \tilde{V}(g, \kappa), \tag{6.15}$$

understood as convergence in $\mathbf{W}_2^1(\mathbf{Q}_b)$. Here $\tilde{V}(g, \kappa)$ is the solution to the equation

$$\tilde{P}\left[\tilde{V}\right] \equiv \left[\frac{\partial^2}{\partial y^2} + \frac{\partial^2}{\partial z^2} + \kappa^2\right] \tilde{V}(g, \kappa) = \tilde{f}(g) + \kappa^2 \left[1 - \tilde{\varepsilon}(g)\right] \tilde{V}(g, \kappa). \tag{6.16}$$

The operator $\tilde{P}[\tilde{V}]$ is defined (for $0 < \arg\kappa < \pi/2$) for functions from $\mathbf{W}_2^1(\mathbf{Q})$, that satisfy the condition $\tilde{V}(g, \kappa)|_{g \in \mathbf{S}} = 0$. This solution exists, is unique and coincides with the solution to (6.8), (6.9) that has been singled out by means of the radiation condition (6.10) at the corresponding κ [9].

Statement 6.5 (limiting absorption principle). *If* $\mathrm{Re}\,\kappa \notin \{\bar{\kappa}_n\} \in \Omega_\kappa$, *then the limit (6.15) exists and singles out a unique solution to (6.8), (6.9). This solution satisfies the radiation condition (6.10). The limiting absorption condition is similar to the radiation condition, as both of them single out the same solution to (6.8), (6.9) on the same set of the real values of* $\kappa > 0$.

As for the classical approach [166], developed within the frame $\mathbf{L}_2$-theory (only the spectrum of self-adjoint problems is taken into consideration), the restrictions on the regions where the radiation and limit absorption conditions can be applied can be formulated in terms of eigenvalues and eigenfunctions of (6.16). This is ensured by the fact that the nontrivial solutions to the homogeneous problem (6.8) to (6.10) for the real points of the spectrum Ω_κ that do not coincide with the branch points κ_n, subject to the uniqueness theorem from [158] (see also Statement 6.3), are elements of the space $\mathbf{L}_2(\mathbf{Q})$.

In the limiting amplitude principle, the solution to (6.8), (6.9) is chosen to be the limit

$$\tilde{U}(g, \kappa) = \lim_{t \to \infty} U(g, t)\, e^{i\kappa t}; \quad \mathrm{Im}\,\kappa = 0, \tag{6.17}$$

where $U(g,t)$ is the solution of the initial boundary value problem

$$\begin{cases} \left[-\varepsilon\frac{\partial^2}{\partial t^2}-\sigma\frac{\partial}{\partial t}+\frac{\partial^2}{\partial z^2}+\frac{\partial^2}{\partial y^2}\right]U=F(g,t); \quad t>0, \quad g\in\mathbf{Q} \\ U\left\{\frac{\partial U}{\partial y}\right\}(2\pi,z,t)=e^{i2\pi\Phi}U\left\{\frac{\partial U}{\partial y}\right\}(0,z,t), \quad E_{tg}(g,t)\big|_{g\in\mathbf{S}}=0; \quad t\geq 0 \\ U(g,0)=\varphi(g), \quad \frac{\partial}{\partial t}U(g,t)\Big|_{t=0}=\psi(g); \quad g\in\bar{\mathbf{Q}} \end{cases} \tag{6.18}$$

(see also the similar problem (2.54)) for$F(g,t)=\tilde{f}(g)e^{-i\kappa t}$, $\varepsilon(g)=\operatorname{Re}\tilde{\varepsilon}(g)$, $\sigma(g)=\kappa\operatorname{Im}\tilde{\varepsilon}(g)$, and $\varphi(g)=\psi(g)\equiv 0$.

Statement 6.6 (limiting amplitude principle). *Assume that κ is not a branch point of the surface* **K** *(see Statement 6.1) and that the Green function $G(g,g_0,\kappa)$ has no real poles on the first (physical) sheet* **K** *(this second requirement is always satisfied if $\operatorname{Im}\tilde{\varepsilon}(g)\not\equiv 0$). Then the limit (6.17) exists and gives a unique solution to (6.8), (6.9). This solution satisfies the radiation condition (6.10).*

If the second requirement is not satisfied, but the point $\kappa\notin\{\bar{\kappa}_n:\operatorname{Im}\bar{\kappa}_n=0\}$ $\subset\Omega_\kappa$ and the source function $\tilde{f}(g)$ are such that the integral $\int_{\mathbf{Q}}\operatorname{Res}\tilde{G}(g,g_0,\bar{\kappa})$ $\tilde{f}(g_0)\,dg_0$ vanishes for all real $\bar{\kappa}\in\{\bar{\kappa}_n\}$, then (6.17) is also valid. In other cases the limiting amplitude principle cannot be applied to obtain the physical solution to (6.8), (6.9).

In Section 6.2.4, we return to the limiting amplitude principle (or rather, to cases contradicting this principle). The asymptotic (at large t) representations $U(g,t)$ are used to analyze anomalous spatio–temporal field transformations: transient processes described by (6.18) where $F(g,t)=\tilde{f}(g)e^{-i\kappa t}$ do not always provide the harmonic mode at $t\to\infty$ if there are real eigenfrequencies $\bar{\kappa}_n$ in the spectrum Ω_κ of open periodic resonators.

Let us return to Statements 6.1 to 6.4. There are many papers dealing with the analysis of threshold effects in classic electromagnetics (the Wood's anomalies). However, only the analytic continuation of the solutions $\tilde{U}(g,\kappa)$ of the elliptic boundary value problems (6.2) to (6.5) into the domain of complex values of the frequency parameter allow us to describe, in the most natural way, the behavior of the complex amplitudes $R_{np}(\kappa)$ and $T_{np}(\kappa)$ of the partial components of the diffraction field near the points κ_n where there appear new homogeneous (propagating) waves (see Statement 6.2 and [159]). In general, the results of the spectral theory, partially cited in Statements 6.1 to 6.4, make many of the previously obtained solutions free from ambiguity. It enables us to study problems of the dynamic theory of open periodic resonators in a framework of formally correct analytic and computational procedures that suggests an optimal way to the goal. Thus, due to the fact that $\tilde{U}(\kappa)$ (as well as $\tilde{G}(\kappa)$, $R_{np}(\kappa)$, and $T_{np}(\kappa)$) are meromorphic functions, one should inevitably, while studying any kind of irregularity in the behavior of the relevant function in the real frequency domain, pay attention first of all to the specific points of the analytic continuation of the function into the complex frequency domain. The analysis of local abnormal effects and phenomena is based on fundamental results of function theory. In this way there appears a possibility to get to the sources of origin of such phenomena, and the possibility of considerably reducing the use of traditional time-consuming diffraction theory approaches. In this way, it becomes possible to describe analytically the mechanisms of forming the resonance response of structures in the form of local theorems about representation and expansion. Here are two simple examples [9] to confirm this statement.

Let a reflective grating (see Fig. 1.1B) be excited by an E-polarized wave $\tilde{U}_0^i(g,\kappa)$ (see (6.1)) in a regime of the parameters where $|\Phi| \leq 0.5$ (the fundamental interval of variation of Φ; see Section 1.2.4) and in a scattered field there is only one propagating harmonic, the principal one ($\operatorname{Re}\Gamma_0 > 0$ and $\operatorname{Im}\Gamma_n > 0$ for $n = 0, \pm 1, \pm 2, \ldots$; it is the single-wave mode). For a reflective structure in such a regime $|a_0(\kappa)| = 1$, that is, $a_0^*(\kappa)\, a_0(\kappa) = 1$ (hereafter $*$ as an upper index denotes complex conjugation). We continue $a_0(\kappa)$ into the domain of complex values of κ on the first sheet of the surface $\mathbf{K}$. From $\ln\left[a_0^*(\kappa)\, a_0(\kappa)\right] = 0$ for $\operatorname{Im}\kappa = 0$, using the symmetry principle [67], we obtain $\ln a_0(\kappa^*) = -\ln a_0^*(\kappa)$ or $a_0(\kappa^*)\, a_0^*(\kappa) = 1$. If the function $a_0(\kappa)$ has a simple pole in the point $\kappa = \bar{\kappa}$, then $a_0(\bar{\kappa}^*) = 0$. If only the first two terms in the expansion of $a_0(\kappa)$ into a Laurent series are non-zero, we derive, in view of the above, the following representation that is true in a certain vicinity $\mathbf{D}$ of the point $\bar{\kappa}$.

$$a_0(\kappa) = \frac{r_{-1}}{\kappa - \bar{\kappa}} + r_0 = e^{i \arg r_0} \frac{\kappa - \bar{\kappa}^*}{\kappa - \bar{\kappa}} = e^{i(\arg r_0 - 2\arg(\kappa - \bar{\kappa}))}; \quad \operatorname{Im}\kappa = 0. \tag{6.19}$$

From (6.19) follows that, if $\operatorname{Im}\bar{\kappa} \ll 1$, by varying the real frequency parameter κ in a small interval containing the point $\kappa = \operatorname{Re}\bar{\kappa}$, the phase of the reflection coefficient $a_0(\kappa)$ changes by approximately 2π. This dynamic phase effect can be used in the design of tunable dispersive open resonators with selective mirror gratings and also other radio-engineering devices in the millimeter and submillimeter wave range. From this effect, one can also obtain reliable information about the eigenfrequency $\bar{\kappa}$ of the structure.

Multipole representations such as (6.19) can be applied (see [9]) in almost the whole single-wave range $\{\kappa : \operatorname{Re}\Gamma_0 > 0 \text{ and } \operatorname{Im}\Gamma_n > 0; n \neq 0\}$ both for the reflecting and semi-transparent structures that are symmetric with respect to the y-axis when the analytical continuation $a_0(\kappa)$ (or $b_0(\kappa)$) from a respective interval of real κ into the first sheet $\mathbf{K}$ reveals a finite number of eigenfrequencies of free field oscillations. Analytic estimation of the error is impossible in solutions to the diffraction problems obtained using this method, but according to computational experiments, this error usually does not exceed 5% over the whole single-wave range. The most essential discrepancy from the rigorous results is encountered far from the real parts of the complex eigenfrequencies. The behavior of the

FIGURE 6.2. Reconstruction of granting's frequency response by points $\bar{\kappa} \in \Omega_\kappa$ ($\delta = 0.25$, $\tilde{\varepsilon} = 2.07$, $\theta = 0.503$, $\Phi = 0$).

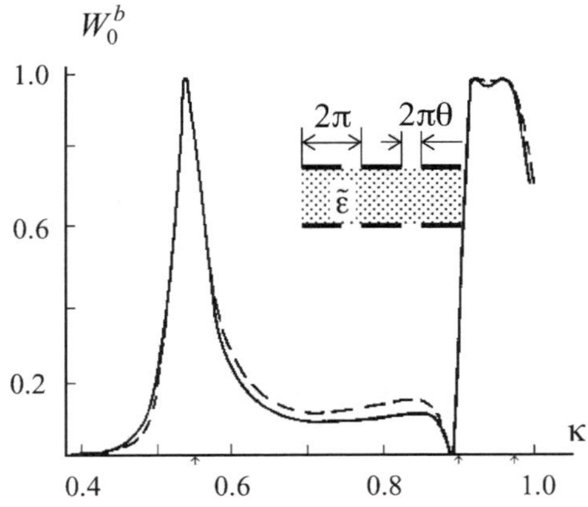

diffraction characteristics in the domain of the resonance parameters is described very accurately.

In Figure 6.2, the solid line shows the dependence $W_0^b(\kappa)$ for a double-layered band grating with a dielectric loading of the coupling zone $\mathbf{Q}_\delta$, which was obtained from the rigorous solution of (6.2) to (6.5). The dashed line shows the same characteristic reconstructed by using the pole representation [169]. The location of the real values of the eigenfrequencies $\bar{\kappa}_1 = 0.542 - i0.0248$, $\bar{\kappa}_2 = 0.907 - i0.0114$, $\bar{\kappa}_3 = 0.977 - i0.0764$ is marked with arrows on the κ-axis. Passing a few steps forward, we note that a semi-transparent structure does not reflect an incident wave at frequencies close to $\operatorname{Re}\bar{\kappa}_1$ and $\operatorname{Re}\bar{\kappa}_3$. Near $\operatorname{Re}\bar{\kappa}_2$ there is a coupled resonance that immediately changes the total reflection mode into the mode of complete transition of the energy of the wave incident on the grating.

Statements 6.3 and 6.4 constitute essentially the uniqueness and existence theorems for the solutions to (6.2) to (6.5) and (6.8) to (6.10) and they mainly serve to guarantee a correct numerical implementation. All the data presented in this and the next sections are obtained in computational experiments using the algorithms of the analytic regularization method [3,5,10,140]. This method, which is relatively computationally inexpensive, allows us to achieve the required numerical accuracy in quite wide ranges of variation of the main parameter values, even in the quasi-resonance situations that are crucial for many direct methods.

The basis of the analytic regularization method is the operations of extracting and analytic inversion of the singular part of operators of ill posed matrix equations of the method of partial domains (mode-matching method) or boundary integral equations of potential theory. As a result, the elliptic boundary value problems are reduced to an operator equation of the second kind with a finitely meromorphic operator function $A(\kappa)$. For spectral problems this operator equation is homogeneous ($(I + A(\kappa))c = 0$; $\kappa \in \mathbf{K}$) and for excitation problems it is inhomogeneous ($(I + A(\kappa))c = d$; $\operatorname{Im}\kappa = 0$, $\operatorname{Re}\kappa > 0$). Here I is the identity operator, and the infinite matrix function $A(\kappa) : l_2 \to l_2$, where

$$l_2 = \left\{ a = \{a_n\}_{n=-\infty}^{\infty} : \sum_n |a_n^2| < \infty \right\},$$

defines a kernel operator or Koch matrix. It allows us to build the algorithm for searching for the components of the spectrum Ω_κ (of eigenfrequencies $\bar{\kappa}_n$) on the basis of an approximate solution to the characteristic equation

$$h(\kappa) = \det(I + A(\kappa)) = 0; \quad \kappa \in \mathbf{K}, \tag{6.20}$$

and to use finite-dimensional analogues (up to the order N) of corresponding infinite-dimensional Fredholm operator equations of the second kind to find the unknown sequence $c = \{c_n\} \in l_2$, associated directly with the complex amplitudes

$$\{R_{np}\}_n, \{T_{np}\}_n, \{R_n\}_n, \{T_n\}_n \in \tilde{l}_2 = \left\{ a = \{a_n\}_{n=-\infty}^{\infty} : \sum_n |a_n|^2 (1 + |n|) < \infty \right\}$$

of the diffraction fields and the free oscillation fields. The error according to the norm of the space of infinite sequences $\tilde{l}_2$ depends on the properties of the operator $A(\kappa)$ in every specific problem, and in the cases considered below, this error was, after a rough analytic estimation, found to be from an order of $N^{-1/2}$ up to $\exp(-\text{const} \cdot N)$. Also note that $\tilde{l}_2$ is, in a certain sense, a class of correctness for problems (6.2) to (6.5) and (6.8) to (6.10): if the sequences of amplitudes $\{R_{np}\}_n, \{T_{np}\}_n, \{R_n\}_n, \{T_n\}_n \in \tilde{l}_2$, then the energy condition (6.6) is satisfied. This is a determining factor in extracting the physical solution in the case when the contours $\mathbf{S}_\varepsilon$ and $\mathbf{S}$ are not smooth.

6.2.3. Grating as an Open Periodic Waveguide

Now change the roles of the parameters κ and Φ ($\kappa > 0$ is fixed; Φ becomes a spectral parameter) and consider the grating as an open periodic waveguide. The following statements are true [9,170,171].

Statement 6.7 (definition and qualitative characteristics of eigenwaves). *The analytic continuation of the solutions $\tilde{U}(g, \Phi)$ to problems (6.2) to (6.5) and (6.8) to (6.10) into the domain of complex values of Φ results in a meromorphic functions $\tilde{U}(\Phi)$ whose natural domain of definition* $\mathbf{F}$ *is an infinite sheet Riemann surface consisting of planes $\Phi \in \mathbf{C}$ cut along the directions $\kappa^2 - (n + \operatorname{Re}\Phi)^2 + (\operatorname{Im}\Phi)^2 = 0$; $n = 0, \pm 1, \pm 2, \ldots$, $\operatorname{Im}\Phi \operatorname{Re}(n + \Phi) \geq 0$. The first sheet of the surface is determined by the values $\operatorname{Im}\Gamma_n(\Phi) \geq 0$, $\operatorname{Re}\Gamma_n(\Phi) \geq 0$; $n = 0, \pm 1, \pm 2, \ldots$ on the axis $\operatorname{Im}\Phi = 0$. This is illustrated in Figure 6.1B. The distinction of the next sheets is that for the finite number of values of the index n, the signs (root branches) $\Gamma_n(\Phi)$ change to the opposite ones. The cuts originate in the real algebraic branch points $\Phi_n : \Gamma_n(\Phi_n) = 0$.*

The system of poles $\{\overline{\Phi}_n\} \in \Omega_\Phi$ of the functions $\tilde{U}(\Phi)$, meromorphic in $\mathbf{F}$ *(no more than a countable set of poles, without finite accumulation points), determines* *the complete spectrum (the eigenpropagation constants) of the grating as an open periodic waveguide: if* $\Phi = \overline{\Phi}_n$, *the homogeneous problems (6.2) to (6.5) and (6.8) to (6.10) in* **F** *have nontrivial solutions; the corresponding solutions* $\tilde{U}_n\left(g, \overline{\Phi}_n\right)$ *are eigenwaves of a periodic directing structure.*

Statement 6.8 (rough localization of the spectrum Ω_Φ). *All the information on the eigenwaves associated with the eigenvalues of* $\overline{\Phi}$, *located in an arbitrary point of the surface* **F**, *can be obtained by solving the homogeneous problems (6.2) to (6.5) or (6.8) to (6.10) in the spaces that are separated at the surfaces (sheets) of* **F** *by an arbitrary string (half-string* $\text{Im}\,\Phi \geq 0$ *or* $\text{Im}\,\Phi \leq 0$ *if the structure possesses the symmetry planes* $y = \text{const}$*) with the range of* $\text{Re}\,\Phi$ *being equal to unity. One of the possible versions of the selected strip is shown in Figure 6.1B.*

For $\text{Im}\,\tilde{\varepsilon} \not\equiv 0$, *the real axis of the first sheet of the surface* **F** *does not contain the eigenvalues of* $\overline{\Phi}$. *The eigenpropagation coefficients for the plane gratings* ($\delta = 0$) *can here coincide only with the branch points* Φ_n. *If an ideal volume grating* ($\text{Im}\,\tilde{\varepsilon} \equiv 0, \delta \neq 0$) *supports an eigenwave having* $\text{Im}\,\overline{\Phi} = 0$ *(the first sheet of* **F**), *then the amplitudes* R_n, T_n *of the partial field components of this wave become zero for numbers n such that* $\text{Re}\,\Gamma_n > 0$.

Statement 6.9 (energy of eigenwaves). *All eigenwaves whose eigenvalues are located on the real axis of the first sheet and do not coincide with the branch points are surface waves (i.e., waves with an exponentially decaying field in the direction away from the grating). These waves, which propagate along the grating, can only be slow waves. In the case of E- (H-) polarization, the energy of their electric field in the domain* $\mathbf{Q}_\delta$, *connecting the radiation zones, always exceeds (respectively, is inferior to) the magnetic field energy. The value of*

$$\text{Re}\,P(\tilde{U}_n(g, \overline{\Phi}_n), y, \vec{y}) = \text{Re} \int_{-\infty}^{\infty} ([\vec{E}(y, z) \times \vec{H}^*(y, z)] \cdot \vec{y}) dz,$$

determining the magnitude and the direction of the energy transfer by such waves, is constant for any complete cross-section of the guiding structure.

For $\text{Im}\tilde{\varepsilon} \not\equiv 0$, *there are no surface waves* $\tilde{U}_n\left(g, \overline{\Phi}_n\right)$, *that transfer energy into the positive (for* $\text{Im}\,\overline{\Phi}_n < 0$*) and negative (for* $\text{Im}\,\overline{\Phi}_n > 0$*) directions, respectively, of the y-axis. For* $\text{Im}\,\tilde{\varepsilon} \equiv 0$ *there are no complex* ($\text{Im}\,\overline{\Phi}_n \neq 0$) *surface waves that transfer the energy along the structure. The surface waves* $\tilde{U}_n\left(g, \overline{\Phi}_n\right)$ *and* $\tilde{U}_n\left(g, -\overline{\Phi}_n\right)$ *with the real* $\overline{\Phi}_n$ *(real waves) transfer energy into the opposite direction of the y-axis.*

If $\tilde{U} = \sum_j \tilde{U}_j(g, \overline{\Phi}_j)$, *where* $\tilde{U}_j(g, \overline{\Phi}_j)$ *are surface waves with differing real* $\overline{\Phi}_j$, *then* $\text{Re}\,P(\tilde{U}, \vec{y}) = \sum\limits_j \text{Re}\,P(\tilde{U}_j(g, \overline{\Phi}_j), \vec{y})$.

Statement 6.10. *For* $\text{Im}\,\tilde{\varepsilon}\,(g) \equiv 0$ *the points of spectrum* Ω_Φ *are grouped in quadruples* $\{\overline{\Phi}, -\overline{\Phi}, \overline{\Phi}^*, -\overline{\Phi}^*\}$. *This means that together with the point* $\overline{\Phi}$, *located, for example, on the first sheet of the surface* **F**, *the points* $-\overline{\Phi}, \overline{\Phi}^*, -\overline{\Phi}^*$ *of the certain (the same or another) sheet of* **F** *also belong to the set* Ω_Φ.

The special feature of the approach, whose results have been partially formulated in Statements 6.7 to 6.10, that distinguishes it from the previously used methods, is that the corresponding boundary value problems are considered in the natural domain of variation of Φ. The complete statement (on the Riemann surface) of the spectral problems with properly extended radiation conditions eliminates subjective elements that inevitably appear in attempts to limit the variations of Φ to simple, physical regions, and allows us to analyze eigenwaves of all types: surface waves, leaky waves, with the field exponentially increasing as one moves away from the grating; and piston-type waves, whose field does not change in directions normal to the plane of the structure. The classical approach, concerned with formulating and solving self-adjoint problems, also has similar possibilities. Still, to realize these possibilities, one should overcome some additional difficulties, for example, continuation through continuous spectrum [147].

The results stated above not only characterize the physical discrete spectrum Ω_Φ qualitatively, they also substantially simplify the numerical analysis. They reduce the search area for eigenpropagation constants and suggest characteristic features of their dynamics with variation of the parameters, thereby providing the necessary basis for constructing rigorous and efficient computational procedures [9]. Thus, for example, from Statement 6.10 follows that the propagation constants $\overline{\Phi}$ of the eigenwaves move over the surface **F** in quadruples. If the imaginary part of $\overline{\Phi}$ is zero, then, in order for the respective eigenwave to be able to turn from a real into a complex one by variation of parameters, this wave should collide with another real eigenwave: only two waves from the point $\overline{\Phi} = \overline{\Phi}_1$, $\operatorname{Im} \overline{\Phi}_1 = 0$, where their propagation constants coincide, can simultaneously enter the domain of complex values of the spectral parameter $\overline{\Phi}$. Or, rather, only the projections of the propagation constants on the first sheet of the surface **F** coincide; the points $\overline{\Phi}_1$ themselves that correspond to both waves, can in the general case belong to different sheets of **F**. The collided real waves turn into complex ones in different half-planes of the sheets of the surface **F**: the propagation constants have increments to the imaginary part that are equal in the absolute value, but with opposite signs. Such qualitative analysis of the spectral characteristics of open periodic waveguides can be done by using other results of this section. However, the main tool for analyzing wave processes in such structures, described by complex boundary value problems, is numerical experiments with models that enable us to obtain all necessary information quickly and with the required accuracy.

Let us consider, as an example, the following result (we explain the exact meaning of the terms used in the next section, which is almost completely based on such numerical results). Figure 6.3B shows a part of the spectral surface $\overline{\Phi}(\kappa, \delta)$, whose points determine the conditions of existence for real surface eigenwaves in a lamellar grating composed of metal rectangular bars with dielectric loading of the channels coupling the reflection and transition zones (see Fig. 6.3A). Two sections of the surface have one common point in the plane $\Phi = 0.5$. This is the point of intersection of the spectral lines $\overline{\Phi}(\kappa, \delta) = 0.5$ of eigenwaves of H_{013}- and H_{021}-modes. These two waves belong to the same class of symmetry with respect

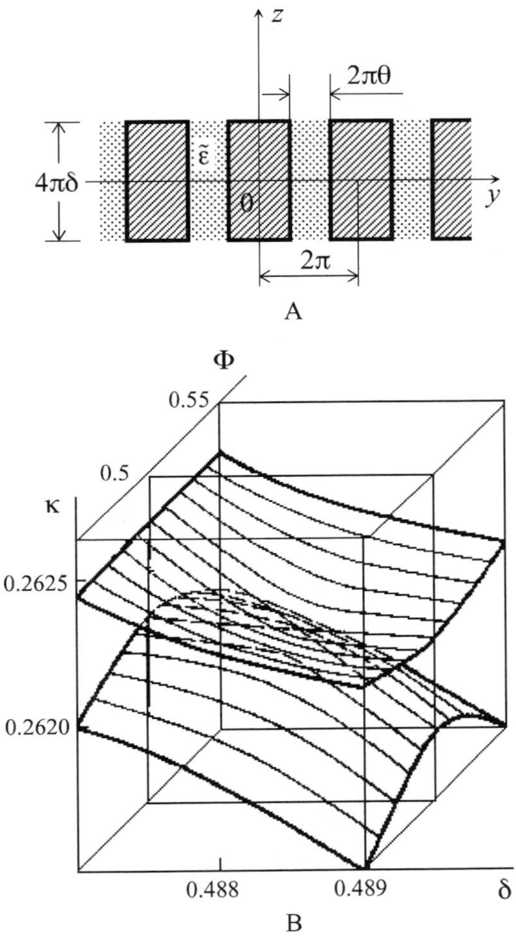

FIGURE 6.3. Fragment of spectral surface $\bar{\Phi}$ (κ, δ) (B), points of which define the conditions of existence of regular real eigenwaves of the lamellar metal grating (A) with dielectric filling of coupling channels.

to the plane $z = 0$ (they have the same parity of the third index in the identifier of the mode H_{0mn}), but to different classes of symmetry with respect to the planes $y =$ const, that halve the grating slots (the parities of the second indices do not coincide). They have a propagation mode that is typical for decelerating periodical structures without attenuation and coupling (all channels of radiation into the zones $|z| > 2\pi\delta$ are closed; i.e., $\mathrm{Re}\,\Gamma_n = 0$, $\mathrm{Im}\,\Gamma_n > 0$ for every $n = 0, \pm 1, \pm 2, \ldots$).

Taking into account general behavior patterns of the elements of the spectral sets [9], one can easily imagine the extension of a part of a spectral surface beyond the boundaries of the sections presented in Figure 6.3B. This will be a cellular structure that completely determines the propagation conditions for slow waves in a grating with fixed values of $\tilde{\varepsilon}$ and θ. The cross-sections of this structure by the planes $\delta =$ const reveal the same distribution pattern of spectral lines $\overline{\Phi}(\kappa)$ as in an infinite passive periodic medium [172].

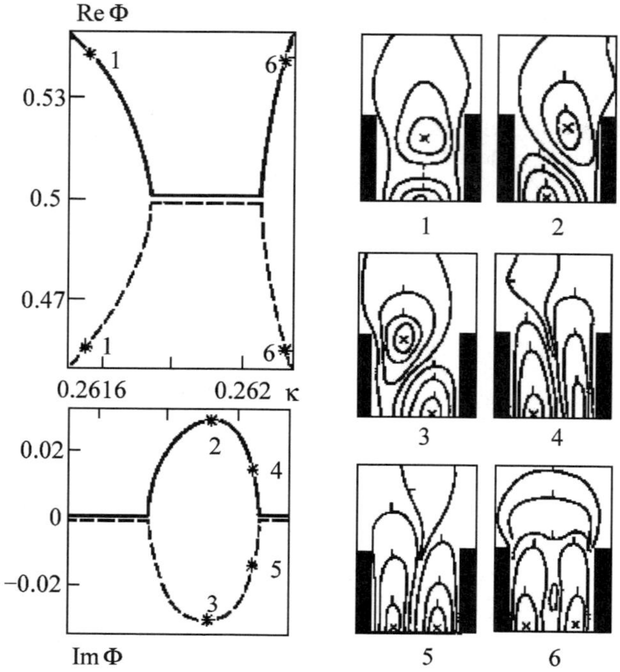

FIGURE 6.4. Slow eigenwaves of lamellar grating with closed for energy radiation channels: $\tilde{\varepsilon} = 25, \theta = 0.8$.

Consider one such cross-section having $\delta = 0.489$ (Fig. 6.4). Up to $\kappa \approx 0.2617$ the grating sustains two slow H_{013}-waves with propagation constants $\overline{\Phi}$ and $1 - \overline{\Phi}$, respectively (the fields of the eigenwaves are identified by the pattern of lines of equal level $|\tilde{U}(g, \overline{\Phi})|$, $z > 0$, $0 \leq y \leq 2\pi$, that were recorded at the points on the spectral curves marked with crosses). If $\overline{\Phi} = 0.5$, the real waves collide and turn into complex waves. After colliding, the value of Re $\overline{\Phi}$ remains constant for both waves and is equal to 0.5, and the values of Im$\overline{\Phi}$ have opposite signs. The fields of the eigenwaves also transform in opposite directions, and they get the typical shape for the field of an H_{021}-wave (see the field patterns 2 to 5 in Fig. 6.4). Near the point $\kappa = 0.262$ the waves collide again, turning now from complex into real ones. The latter can be classified as slow H_{021}-waves (see field patterns 6 in Fig. 6.4).

All features mentioned here concerning the behavior of the components of the grating's spectrum Ω_Φ are related to general regularities revealed by the qualitative analysis of spectral characteristics of open periodic waveguides [9]. However, the result of coupling between eigenwaves is somewhat unexpected: on a short interval of variation of the frequency parameter κ, both waves change the mode H_{013} to H_{021}. The point is that usually (see Section 6.3.1) the interaction leads either to local changes of the wave mode (pure mode → hybrid mode → recovering of the preceding pure mode), or to the exchange of the wave modes (two different pure modes → hybrid modes → exchange of the pure modes).

The components of the spectrum Ω_Φ cannot be made to be zero anywhere in the finite part of **F** by varying the main material and geometric parameters of the grating [9,150]. Hence, slow eigenwaves should, without exception, manage to pass the nontransmission zones of an open periodic waveguide. Turning for a while from real into the complex waves, they implement the only possible variant of a smooth transition into the domain of parameter values, where they can propagate again in their inherent mode.

Let $\tilde{G}(g, g_0, \Phi)$ be the Green's function of a grating in the field of quasiperiodical point sources $\tilde{f}(y, z)$ (see formula (6.7)). Using the superposition principle and taking into account the representation

$$\delta(y - y_0, z - z_0) = \int_{-0.5}^{0.5} \tilde{f}(y, z)\, d\Phi, \qquad (6.21)$$

we obtain

$$\underset{\circ}{\tilde{G}}(g, g_0) = \int_{-0.5}^{0.5} \tilde{G}(g, g_0, \Phi)\, d\Phi. \qquad (6.22)$$

Here, $\underset{\circ}{\tilde{G}}(g, g_0)$ is the Green's function of the grating in the field of a single point source (6.21), that is, the fundamental solution of the operator of the problem of excitations of an infinite periodic structure by waves from compact lumped sources or plane wave beams. These are the key problems in the analysis of model synthesis of various quasi-optical devices (beamformers, absorbing and rescattering coatings, open dispersive resonators with a substantially rarefied spectrum), where the gratings serve as resonant elements, which efficiently select signals with respect to frequency, polarization, and space [9,173]. A correct analysis of these problems is only possible when the radiation conditions have been taken into account, which should, first, provide a physically valid solution and, second, allow us to use the standard methods of potential theory to transfer from differential to integral representations.

The formal approach to the problem using the transformation (6.22) does not yield quite accurate results [12,174], as it is necessary to have some information on the singularities of the integrand (the function $\tilde{G}(g, g_0, \Phi)$) along the pass integration, and about acceptable ways of going around these singularities, and so on. Such information can be obtained only through the spectral theory that considers the gratings as open periodic waveguides (see previous results of this Section and [9,170,171]). Here the following statement applies (to simplify the case, we formulate it for reflective gratings having a geometry shown in Fig. 1.1B).

Statement 6.11. *Assume that the frequency $\kappa > 0$ is not critical for a one-dimensional periodic waveguide; that is, the section $|\Phi| \leq 0.5$ does not contain any eigenvalues of $\overline{\Phi}_n$ corresponding to eigenwaves $\tilde{U}_n(g, \overline{\Phi}_n)$ that do not transfer energy along the structure. We also assume that among the elements of the* *set* $\Omega_\Phi \cap [-0.5; 0.5]$ *there are no branch points of the surface* **F**. *Then, if* $r \to \infty$ $(|\alpha| < \pi/2)$ *and* $|y| \to \infty$ $(|\alpha| = \pi/2)$,

$$\underset{\circ}{\tilde{G}}(g, g_0) = \left(\frac{2\pi\kappa}{r}\right)^{1/2} e^{i(r\kappa - \pi/4)} \left\{ \sum_{n:|\kappa \sin\alpha + n| \le 0.5} \eta_n R_n(g_0, -\kappa\sin\alpha - n) \right\} \cos\alpha$$

$$+ O\left(r^{-1}\right) + \begin{cases} \pi i \sum\limits_{\overline{\Phi}_m \in M^+} \eta_m \text{Res}\tilde{G}\left(g, g_0, \overline{\Phi}_m\right); & y > 0 \\ -\pi i \sum\limits_{\bar{\Phi}_m \in M^-} \eta_m \text{Res}\tilde{G}\left(g, g_0, \overline{\Phi}_m\right); & y < 0 \end{cases}. \qquad (6.23)$$

Here, η_m *is a number (1 or 0.5) that is used for differentiating the contribution of poles and points of the stationary phase depending on their location (inside or at the end of the integration interval);* $R_n(g_0, \Phi)$ *are the amplitudes of the Green's function* $\tilde{G}(g, g_0, \Phi)$ *harmonics in expressions like (6.10);* M^+ *and* M^- *are finite sets of real eigennumbers* $|\overline{\Phi}_m| \le 0.5$ *corresponding to eigenwaves, which transfer the energy along the grating in the positive and negative directions of the* y*-axis, respectively.*

By exciting the grating by a compact (in the plane $\mathbf{R}^2$) source, by scattering of plane waves, plane wave bunches, and eigenwaves of a periodic open waveguide in a system consisting of grating–compact discontinuity, the radiation conditions such as (6.23) single out the unique solution of the corresponding diffraction problems. This solution satisfies the physically valid requirement due to which the scattered field should be free from waves arriving from infinity at the zone where the sources and efficient scatterers are concentrated.

The first set of terms on the right-hand side of (6.23) represents a diverging cylindrical inhomogeneous wave with the amplitude (field pattern)

$$D(\alpha) = \cos\alpha \sum_{n:|\kappa\sin\alpha + n| \le 0.5} \eta_n R_n(g_0, -\kappa\sin\alpha - n). \qquad (6.24)$$

Due to (6.24), the field pattern in the far zone is completely determined by the amplitudes of the propagating harmonics of the quasi-periodic Green's function $\tilde{G}(g, g_0, \Phi)$. A similar result has been obtained in [174,175]. It is substantially complemented by the fields of real eigenwaves of the structure, that are present in (6.23) and decrease exponentially for $z > 2\pi\delta$, but are of the order $O(1)$ for bounded z and $|y| \to \infty$.

6.2.4. Natural Resonances and Transient Processes in Open Periodic Structures: Examples for Analysis

In order to study the influence of natural resonances (free oscillation of the field) on the processes of wideband signal scattering, one should find an analytical relation between the spatio–temporal representations ($U(g, t)$) of the fields generated by the grating and the spatio–frequency representations ($\tilde{U}(g, \kappa)$). One of the possible ways to achieve this is based on the Laplace integral transform (see Section 1.3.1), that match the solutions $U(g, t)$ of the original initial boundary value problems (6.18) by using (1.36), with the solutions $\tilde{U}(g,\kappa)$ of the elliptic problem (6.8) to (6.10) where

$$\tilde{f}(g) = \tilde{F}(g,\kappa) + i\kappa\tilde{\varepsilon}(g)\varphi(g) - \varepsilon(g)\psi(g); \quad \tilde{F}(g,\kappa) = L[F(g,t)](-i\kappa),$$

and $\tilde{\varepsilon} = \varepsilon + i\sigma/\kappa$. In (6.18), as before (see (6.2) to (6.5)), Φ is a real parameter of the Floquet channel $\mathbf{R}$. It is also assumed that the supports of the functions $F(g,t)$, $\varphi(g)$, and $\psi(g)$ belong to $\overline{\mathbf{Q}}_\delta$, and that the functions themselves satisfy the conditions of Statement 1.1 (the problem (6.18) has a unique solution with respect to $U(g,t)$ in the Sobolev space $\mathbf{W}_2^1(\mathbf{Q}^T)$).

In Chapter 1, we have already mentioned that the spectra of compact open resonators that are analogues of discrete spectral sets $\{\bar{\kappa}_n\} = \Omega_\kappa$ do not contain real eigenelements in the physical domain of variation of the frequency parameter. In the spectrum Ω_κ of ideal gratings (i.e., free of absorbing inserts) the existence of such elements $\bar{\kappa}$ is possible. One of the reasons, as shown in Section 6.3.1, is the possible influence of a so-called compensation mechanism in the formation of the radiation fields of free oscillations. The contribution of various partial components of the free oscillation field in the region $\mathbf{Q}_\delta$ into some energy-consuming (propagating) spatial harmonic of the radiation field (see the representation (6.10)) can be summed up in the antiphase, in fact, closing the formally open channel of energy radiation (the number of such channels is always finite) from the near zone of the structure.

The presence of real eigenfrequencies $\bar{\kappa}_n$ and frequencies $\bar{\kappa}_n$ with $|\mathrm{Im}\,\bar{\kappa}_n| \ll 1$, the possibility that they converge in the metric of the complex space $\mathbf{K}$ by varying the geometric and material parameters, and actually the structure of $\mathbf{K}$ itself, predetermine many features in the space-time transformations of electromagnetic fields. We illustrate these statements with a variety of examples applicable to a wide enough range of periodic gratings (arbitrary geometry of the boundaries $\mathbf{S}$ and $\mathbf{S}_\varepsilon$, piecewisesmooth functions $\varepsilon(g)$ and $\sigma(g)$) that can serve as useful models in computational experiments and in the analysis of their results of these experiments.

Assuming that the elements of the subset $\{\bar{\kappa}_n\}$ do not accumulate (for $|\mathrm{Re}\,\kappa| \to \infty$) near the $\mathrm{Im}\kappa = 0$ axis of the first (physical) sheet of the surface $\mathbf{K}$, and moving the integration contour in the representation (1.36) for $U(g,t)$ up to the position γ in Figure 6.1A, we obtain

$$\begin{aligned} U(g,t) &= \frac{1}{2\pi}\int\limits_{i\alpha-\infty}^{i\alpha+\infty}\left[\int\limits_{\mathbf{Q}}\tilde{G}(g,g_0,\kappa)\tilde{f}(g_0,\kappa)\,dg_0\right]\mathrm{e}^{-i\kappa t}d\kappa \\ &= \frac{1}{i}\left\{\sum_n\int\limits_{\mathbf{Q}}\operatorname*{Res}_{\kappa\in\{\bar{\kappa}_n:-\beta<\mathrm{Im}\,\bar{\kappa}_n\le 0\}}\left[\tilde{G}(g,g_0,\kappa)\tilde{f}(g_0,\kappa)\mathrm{e}^{-i\kappa t}\right]dg_0\right. \\ &\quad\left.+\sum_m\int\limits_{\mathbf{Q}}\operatorname*{Res}_{\kappa\in\{\bar{\kappa}_m:-\beta<\mathrm{Im}\,\bar{\kappa}_m\le 0;\bar{\kappa}_m\notin\{\bar{\kappa}_n\}\}}\left[\tilde{G}(g,g_0,\kappa)\tilde{f}(g_0,\kappa)\mathrm{e}^{-i\kappa t}\right]dg_0\right\} \\ &\quad+\frac{1}{2\pi}\int\limits_{\gamma}\tilde{U}(g,\kappa)e^{-i\kappa t}d\kappa. \end{aligned} \tag{6.25}$$

Here, $\tilde{\kappa}_m$ are the poles of the function $\tilde{f}(g,\kappa)$ that do not coincide with the elements of the spectral set Ω_κ. All the features involved in the summation in (6.25) are situated on the first sheet of the surface $\mathbf{F}$ above the contour γ. Due to the above assumption, their total number is finite. In the case when $F(g,t) = f(g)\exp(-i\tilde{\kappa}t)$; $\operatorname{Im}\tilde{\kappa} = 0$ and $\psi(g) = \varphi(g) \equiv 0$, the representation (6.25) can be rewritten as

$$U(g,t) = \sum_n \int_{\mathbf{Q}} \operatorname*{Res}_{\kappa\in\{\bar{\kappa}_n:\operatorname{Im}\bar{\kappa}_n=0\}} \left[\tilde{G}(g,g_0,\kappa)\frac{f(g_0)}{\kappa-\tilde{\kappa}}\mathrm{e}^{-i\kappa t}\right]dg_0$$

$$+\sum_n \int_{\mathbf{Q}} \operatorname*{Res}_{\kappa\in\{\bar{\kappa}_n:-\beta<\operatorname{Im}\bar{\kappa}_n<0\}} \left[\tilde{G}(g,g_0,\kappa)\frac{f(g_0)}{\kappa-\tilde{\kappa}}\mathrm{e}^{-i\kappa t}\right]dg_0$$

$$+\eta\mathrm{e}^{-i\tilde{\kappa}t}\left[\int_{\mathbf{Q}} \tilde{G}(g,g_0,\tilde{\kappa})f(g_0)dg_0\right] + P_\beta(g,t); \quad g\in\mathbf{Q}_a,$$

where $\mathbf{Q}_a = \{g\in\mathbf{Q} : |z| < a\}$, and η is zero if $\tilde{\kappa}\in\{\bar{\kappa}_n\}$ and otherwise unity. If $t\to\infty$, $2\pi\delta \le a \ll t$, and $\{\kappa_n\}\cap\{\bar{\kappa}_n\} = \emptyset$, the contribution of function $P_\beta(g,t)$ according to the norm of the space $\mathbf{W}_2^1(\mathbf{Q}_a)$ is estimated as follows [9]: for any finite a and β, there is T such that if $t > T$ then $\|P_\beta(g,t)\| = O(t^{-1})$, if $\tilde{\kappa}\notin\{\kappa_n\}$, and $\|P_\beta(g,t)\| = O(1)$, if $\tilde{\kappa}\in\{\kappa_n\}$.

Consider more closely the solution $U(g,t)$ in the near zone of the structure ($g\in\mathbf{Q}_a$) and at large t ($t\to\infty$), as it is determined by the obtained expression. Assume that an open periodic resonator does not have real eigenfrequencies on the first sheet of the surface $\mathbf{K}$ ($\{\bar{\kappa}_n : \operatorname{Im}\bar{\kappa}_n = 0\} = \emptyset$) and that the excitation frequency $\tilde{\kappa}$ does not coincide with any of the branch points ($\tilde{\kappa}\notin\{\kappa_n\}$). Then,

$$U(g,t) = \tilde{U}(g,\tilde{\kappa})\mathrm{e}^{-i\tilde{\kappa}t} + O\left(t^{-1}\right), \tag{6.26}$$

where $\tilde{U}(g,\kappa)$ is the solution to the elliptic problem (6.8) to (6.10) with the function $f(g)$ on the right-hand side of equation (6.8). From (6.26) it follows that for sufficiently large t, the field scattered by a periodic structure turns into a harmonic mode at any finite distance from this structure; this is a realization of the limiting amplitude principle (see Statement 6.6). Obviously, the violation of any of the assumptions made in the derivation of (6.26) drastically changes the nature of the process.

The most interesting case is when the real spectrum is not empty ($\{\bar{\kappa}_n : \operatorname{Im}\bar{\kappa}_n = 0\}$). Because the elements of this subset of the spectrum Ω_κ are associated only with simple poles of the function $\tilde{G}(g,g_0,\kappa)$ [9], it follows that

$$U(g,t) = \tilde{U}(g,\tilde{\kappa})\mathrm{e}^{-i\tilde{\kappa}t} + \sum_{n:\operatorname{Im}\bar{\kappa}_n=0} \frac{A_n(f)}{\bar{\kappa}_n-\tilde{\kappa}}\tilde{U}_n(g,\bar{\kappa}_n)\mathrm{e}^{-i\bar{\kappa}_n t}$$
$$+\begin{cases} O(t^{-1}); & \tilde{\kappa}\notin\{\kappa_n\} \\ O(1); & \tilde{\kappa}\in\{\kappa_n\} \end{cases} \tag{6.27}$$

in the case when the excitation frequency $\tilde{\kappa}$ is not an eigenfrequency, and

$$U(g,t) = \sum_{n\neq m} \frac{A_n(f)}{\bar{\kappa}_n - \tilde{\kappa}} \tilde{U}_n(g,\bar{\kappa}_n)\mathrm{e}^{-i\bar{\kappa}_n t} - iA_m(f)\,t\,\tilde{U}_m(g,\bar{\kappa}_m)\mathrm{e}^{-i\tilde{\kappa}t}$$

$$+\mathrm{e}^{-i\tilde{\kappa}t}\left\{\int_Q \frac{\partial}{\partial\kappa}\left[\tilde{G}(g,g_0,\kappa)(\kappa-\bar{\kappa}_m)\right]\Big|_{\kappa=\bar{\kappa}_m} f(g_0)\,dg_0\right\} + O\left(t^{-1}\right);$$

$$g \in \mathbf{Q}_a \tag{6.28}$$

for $\tilde{\kappa} = \bar{\kappa}_m \in \{\bar{\kappa}_n : \operatorname{Im}\bar{\kappa}_n = 0\}$. Here $A_n(f)$ are the amplitude coefficients whose values depend on the function $f(g)$, and on the geometric and material parameters of the grating; $\tilde{U}_n(g,\bar{\kappa}_n)$ are the eigenfunctions (free field oscillation) corresponding to the eigenvalues (eigenfrequencies) $\bar{\kappa}_n$, and

$$A_n(f)\,\tilde{U}_n(g,\bar{\kappa}_n) = \int_Q \operatorname{Res}\tilde{G}(g,g_0,\bar{\kappa}_n)\,f(g_0)\,dg_0.$$

If the excitation frequency $\tilde{\kappa}$ is close to one of the eigenfrequencies $\bar{\kappa}_m \in \{\bar{\kappa}_n : \operatorname{Im}\bar{\kappa}_n = 0\}$, but does not coincide with it, the main term in the expansion (6.27) is determined by the value

$$R(g,t) = \frac{A_m(f)}{\bar{\kappa}_m - \tilde{\kappa}}\tilde{U}_m(g,\bar{\kappa}_m)\left(\mathrm{e}^{-i\bar{\kappa}_m t} - \mathrm{e}^{-i\tilde{\kappa}t}\right).$$

It dominates in the field formed by the grating, but not for all times: when the excitation frequency comes close to one of the eigenfrequencies of the structure, this results in the occurrence of beats having $\min|R(g,t)| = 0$ at $(\bar{\kappa}_m - \tilde{\kappa})t = 2\pi j;\ j = 0, \pm 1, \ldots,$ and $\max|R(g,t)| = 2|A_m\tilde{U}_m(g,\bar{\kappa}_m)(\bar{\kappa}_m - \tilde{\kappa})^{-1}|$ at $(\bar{\kappa}_m - \tilde{\kappa})t = (2j+1)\pi$. At the time when the maximum value of $|R(g,t)|$ is reached, the secondary field is almost completely determined by the field of free oscillation in the structure at the frequency $\kappa = \bar{\kappa}_m$. With increasing t, the ratio between the energies of fields determined by the major and the background parts of $U(g,t)$ remains almost constant (in terms of the order of t). The energy of a continuous source is efficiently reradiated into the far zone $|z| \gg 1$.

If $\tilde{\kappa} = \bar{\kappa}_m, t \to \infty$ and $A_m(f) \neq 0$, the term $R(g,t) = -iA_m(f)t\tilde{U}_m(g,\bar{\kappa}_m)\mathrm{e}^{-i\tilde{\kappa}t}$ dominates in the expansion (6.28). The energy of a continuous source is not scattered into the space surrounding the grating, but stays in the immediately adjoining region. This is a realization of the effect of energy accumulation in the near zone of the structure.

Next, we study transient processes on a limited initial interval of variation of the observation time t. The results of the qualitative analysis cannot be applied directly here. However, they still allow some important conclusions.

First, a sufficiently complete analytical description of the pulse deformation is scarcely possible, as it requires computations whose results should be properly considered. Second, the considered processes develop very dynamically and are influenced by many factors, therefore, a strong background of computational experiments is necessary, and their results should be somehow predicted beforehand (you should know what you expect to see, to in fact see it). Third, the asymptotic representations refine the process by focusing on only part of its key components, but these components are present also in the transition period: if monochromatic components and possible free oscillations play such an important role in forming $U(g,t)$ for large t, then they are equally important for the formation of transient states of the field. Hence, a reliable and efficient analysis of such states is impossible without the supporting basis of a state-of-the-art theory of resonant scattering of time-harmonic waves and the spectral theory of open periodic, compact and waveguide resonators (see in the references the papers and books by V.P. Shestopalov, L.N. Litvinenko, V.G. Sologub, S.A. Masalov, A.A. Kirilenko, S.L. Prosvirnin, L.A. Rud, A.E. Poyedinchuk, P.N. Melezhik et al.). The corresponding results give us a reasonable chance to carry out a well-motivated computational experiment in the time domain and save us from an inefficient search through arbitrary possible situations.

By developing the theory of transient processes, specific parameters and characteristics of media and structures where the electromagnetic waves are propagating and scattered are taken into account. In this chapter, we have focused our attention on signals formed by classic one-dimensional periodic structures. The effects and phenomena that have been identified and studied certainly do not represent the full spectrum of possible common and also more unusual spatio–temporal and spatio–frequency transformations of electromagnetic fields in gratings. Rather, they were only first steps made in the study of the physics of time-domain processes. These steps are based on general and specific data on the distribution and dynamics of the elements of spectral sets, on the configuration and the mechanisms of forming free oscillation fields, and on the effect that these oscillations have on the diffraction (monochromatic) field. Some results from the frequency domain that should be involved in such an analysis, are given in the next section.

We return to the representation (6.25) in order to clarify some aspects that simplify the organization of computational experiments and the interpretation of their results. The last term in (6.25) characterizes in the whole the contribution of the singularities of the function $\tilde{G}(g, g_0, \kappa, \Phi)$; they cannot be accounted for by the deformation of the contour γ. They cannot be estimated and discarded for finite times t. Preliminary conclusions can still be made on the basis of the first two terms. Thus, in a numerical experiment, when the frequencies of the exciting signals never coincide exactly with the real eigenfrequencies, and the observation time interval $[0; T]$ is not too long, the contributions of possible free oscillations $\tilde{U}_n(g, \bar{\kappa}_n)$ with $\operatorname{Im}\bar{\kappa}_n = 0$ and $|\operatorname{Im}\bar{\kappa}_n| \ll 1$ in the formation of a complete field of $U(g,t)$, will be comparable. Any such free oscillation can be increased to the level of a major one in the field $U(g,t)$, by adjusting the function $F(g,t)$ at suitable values of the geometric and material structure parameters.

All such free oscillations may contribute toward forming beats similar to those described earlier in this section. The characteristics of such beats enable us to define more exactly the parameters of a free oscillation (field configuration, values of the imaginary and real parts of the eigenfrequency) and thus to come to the next higher level of the computational experiment. Essentially, every component of the field $U(g, t)$ in (6.25), can be made more pronounced by a reasonable choice of the source $F(g, t)$, so as to develop a more pronounced contrast to the inevitable background and thus simplify the continued analysis and investigations. When momentary sources of wideband signals $\varphi(g)$ and $\psi(g)$ are used, such a selection becomes more complicated. Examples that confirm these observations are considered in Sections 6.4 and 6.5.

6.3. Dynamic Patterns of Spectral Points in the Frequency Domain

A numerical experiment is a big help in problems of analysis if it is based on sufficiently efficient algorithms with sufficiently simple model geometries described by a small number of parameters. In this section we use waveguide gratings as such a model, particularly a grating composed of metal rectangular bars with a dielectric ($\tilde{\varepsilon} = \text{const}$) loading of waveguide channels of the region $\mathbf{Q}_\delta$, coupling the reflection and transition zones of the structure (see Fig. 6.3A). The waveguide gratings allow us to study model situations that explain, quite comprehensively, the effects of the creation and existence of free field oscillations (hereafter, sometimes, just oscillations). They also enable us to follow the dynamics of the variations of the eigenfrequencies (EFs), classify oscillations, and to proceed, on the basis of the lemma about a continuous dependence of the poles of the resolving operator in problems such as (6.2) to (6.5) and (6.8) to (6.10) [9], to the analysis of regularities in the behavior of spectral sets of structures, in whose geometry it is impossible to separate regular channels coupling the energy radiation zones.

Below, we present the essential results of [6,9] concerning oscillations with E-polarization of the field (TE-oscillations). We also consider specific features of TM-oscillations (H-polarization) that influence the scattering pattern of plane monochromatic waves.

6.3.1. Specific Properties of the Dynamics of Elements of the Spectral Sets: Effects of Existence of Super-High-Q Oscillations and Surface Waves in Periodic Structures with Open Energy Radiation Channels

In waveguide gratings, the mode of oscillation is described by the identifier H_{0mn}, where H_{0m} stands for a wave of the $\mathbf{Q}_\delta$ channel (a section of a plane-parallel waveguide) that dominates in the eigenmode field, and n is the number of field variations in the region $|z| \leq 2\pi\delta$ of the structure. The eigenfields with similar indices m we classify as belonging to one family; n determines the sequence number of the oscillation in this family. If the structure and value of the parameter Φ are such that the problem has the planes $z = 0$ and (or) $y = \text{const} + 2\pi$ as symmetry planes, then the indices m and n determine the same class of the oscillation symmetry

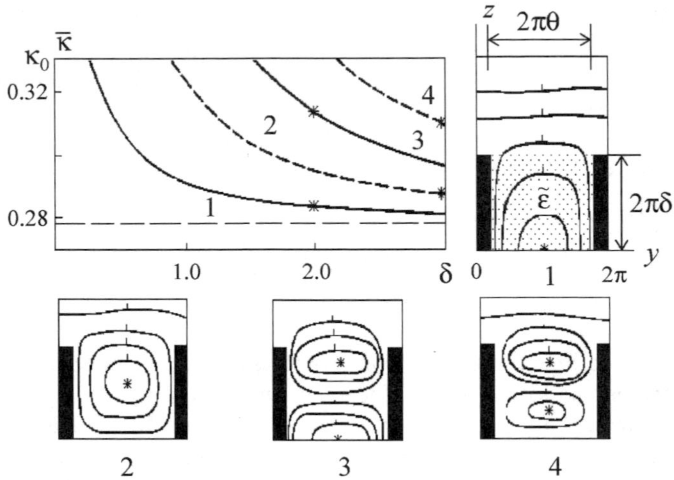

FIGURE 6.5. The first family of eigenoscillations in grating with closed for energy radiation channels (real-valued eigenfrequencies).

(whether the configuration of the oscillation field is symmetric or antisymmetric with respect to the relevant planes).

Figures 6.5 ($\Phi = 0.33$; $\theta = 0.8$; $\tilde{\varepsilon} = 5$), 6.6, and 6.7 ($\Phi = 0.1$; $\theta = 0.8$; $\tilde{\varepsilon} = 3.89$) show, in the coordinates $\underline{\kappa}$, δ, and partially on the surface $\mathbf{K}$, the standard behavior of curves of EFs $\bar{\kappa}(\delta)$ of lower-mode oscillations in a grating composed of rectangular metal bars. The 2-D fragments here show the lines of equal values of the field strength of corresponding oscillations ($|E_x(g)| = |U(g, \bar{\kappa})| = \text{const}$), recorded at the points of spectral curves that are marked with crosses. The structure is shown in its half-height, the dashes on the lines $|E_x(g)| = \text{const}$ show the direction of decrease of the field amplitude, and the crosses show the positions of local maxima.

By an unlimited increase of δ, all EFs of oscillations of the first and second families tend to the cutoff points of the waves in channel $\mathbf{Q}_\delta$, which dominate in the field of the respective eigenmode. The structure of the oscillation fields is clearly pronounced and is uniquely determined by the fields of H_{01}-(the first family) and H_{02}-waves (second family). The transformation of the field of the first family of oscillations can be followed easily and is associated only with the transfer from pure and almost pure resonances of region $\mathbf{Q}_\delta$ (for large δ at complex EFs and in the case of real EFs the field is concentrated in the slots of the structure) to the resonances covering zones $|z| > 2\pi\delta$. By moving to the region of small δ, the spectrum of EFs becomes rarefied; the Q-factor of oscillations that is determined by the values of $|\text{Im}\,\bar{\kappa}|^{-1}$ decreases substantially.

The diffraction Q-factor of the oscillations of the second family (curves 1 and 2 in Figs. 6.6 and 6.7) is approximately two to three orders higher. That is why EFs split up in the complex plane even where the lines $\text{Re}\,\bar{\kappa}(\delta)$ corresponding to the oscillations of different families intersect for certain values of δ. Before and after

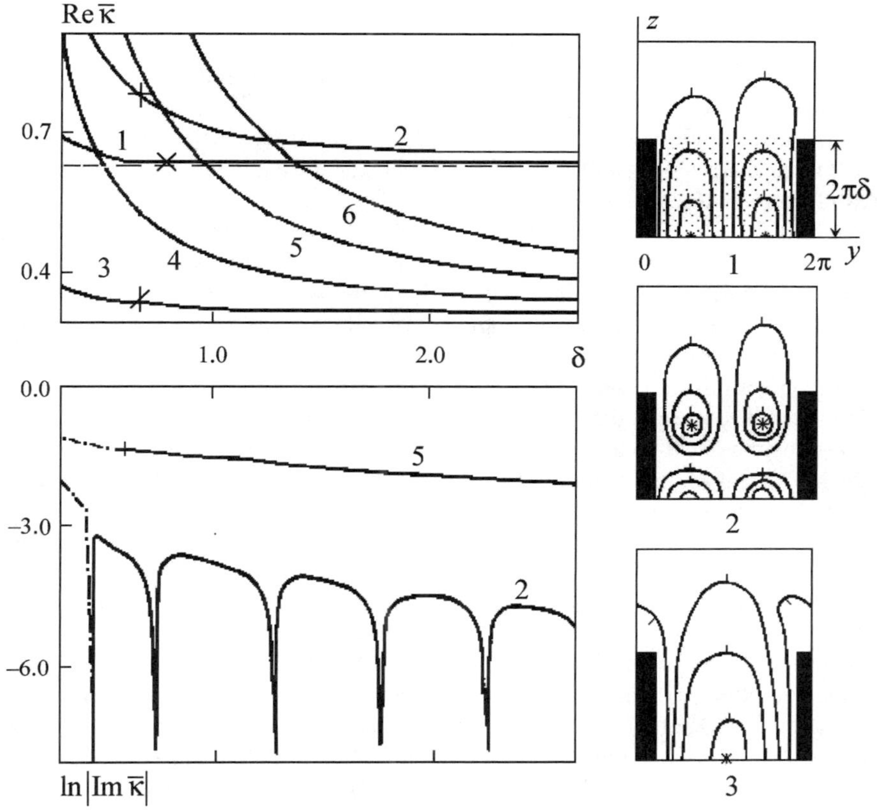

FIGURE 6.6. Eigenfrequencies of symmetric eigenoscillations: First (lines 3 to 6) and second (lines 1 and 2) families.

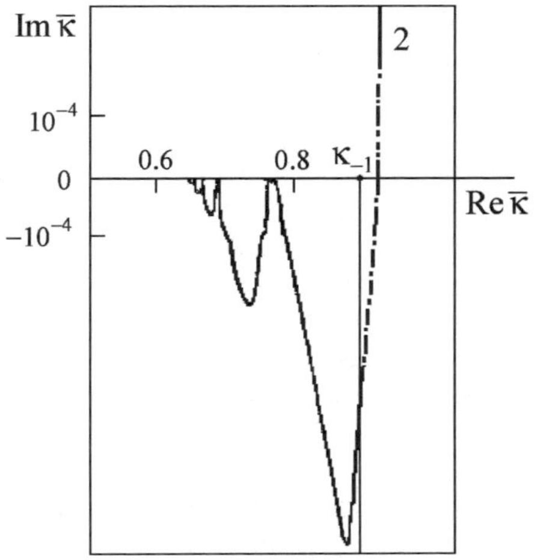

FIGURE 6.7. The passage of EF of the eigenmode belonging to the second family into the second sheet of surface **K**.

the intersection of the lines $\operatorname{Re}\bar{\kappa}(\delta)$, all the characteristic features of the related oscillations remain. However, near the coincidence of the real values of EFs of oscillations of the first and second families (and only for oscillations belonging to similar classes of symmetry) the field geometry of the latter (hybrid mode) is essentially distorted, and their Q-factor increases rapidly. It can be said that in the respective points, it is almost infinite. According to Statement 6.3 and a physically clear understanding of the case at this moment, the amplitudes R_0 and T_0 of the partial components of the oscillation field that propagate in the region $|z| > 2\pi\delta$ should become zero (the energy should not be radiated into the far zone). This fact is proved by the numerical results.

In the considered case, the Q-factors of the oscillations of the first and second families (the corresponding EFs are denoted by $\bar{\kappa}_1$ and $\bar{\kappa}_2$, respectively) are determined by the radiation losses due to the radiation of both H_{01}- and H_{02}-waves. For distant $\bar{\kappa}_1$ and $\bar{\kappa}_2$, the radiation losses dominate, which are integrally determined by the resonating waveguide wave (H_{01} for the first and H_{02} for the second family). When $\operatorname{Re}\bar{\kappa}_1$ and $\operatorname{Re}\bar{\kappa}_2$ lie close to each other, the relative excitation level of the H_{01}- and H_{02}-waves in the oscillation fields becomes more even. As a result, the contribution of these waves to the radiation losses becomes more even too. This may lead to reduced total losses (i.e., an increase of the Q-factor), when the contributions are summed up having phases, differing by π. Exactly such a mechanism, implying a compensation of contributions, results in an increased Q-factor of the oscillation of the second family under conditions that are close to those for existence of the oscillation of the first family. The fact that at this stage there are no significant changes in the Q-factor of the oscillation of the first family is to be attributed to the incommensurability of the absolute values of its average radiation losses with the compensating contribution due to the intensified influence of H_{02}-waves. Note that the super-high-quality oscillations occur in the region $\kappa > \kappa_0$, that is, where the structure is generally open: the field energy can be carried off by the propagating harmonics (components of (6.10)), toward larger $|z|$, without attenuation. A more detailed analysis of the presence of such oscillations in this kind of structure with open energy radiation channels is given below.

By decreasing δ, the next peak of the Q-factor of oscillation 2 of the second family occurs now on the second sheet of the surface **K** (see Fig. 6.7). The respective spectral curve is not just touching the real axis κ, but crosses it and goes into the domain of κ with $\operatorname{Im}\kappa > 0$ (to distinguish sections of the trajectory that belong to different sheets of **K**, we mark this curve with dashes and dots). When $\bar{\kappa}$ transfers to the upper half-space, the oscillation field structure remains practically unchanged. The transfer of the point $\bar{\kappa} \in \Omega_\kappa$ into the upper half-spaces of the upper sheets of the surface **K** does not contradict Statement 6.3 or the results of the general theory described in [9]. We return later to the peculiarities of such dynamics and discuss their relation to the effect of complete transformation of wave packets.

In regular situations, the oscillation mode described by the identifier H_{0mn} is a sufficiently stable characteristic with respect to smooth variations of the parameters (by transferring to the real axis of the EF of the H_{02n}-mode oscillation in the previous example, the chain "pure mode → hybrid mode → recovering of the previous pure mode" falls on small intervals of variation of the parameter δ). In anomalous situations occurring due to the conversion of EFs of several free oscillations of one symmetry class in the metric of the corresponding complex space, a H_{0mn}-mode can be changed fundamentally. The accompanying effects are considered below.

Let $\mathrm{Im}\,\tilde{\varepsilon} = 0$, and consider the interval $\kappa_0 < \kappa < \min \kappa_{\pm 1}$ (there is then only one open radiation channel for a field (6.10) harmonic with $n = 0$ that propagates without attenuation). Furthermore, let the parameters of the spectral problem be such that this interval can contain three critical points of the H_{0m}-waves ($m = 1, 2, 3$). Then, in the region $\mathrm{Re}\,\kappa > 0$ of the first sheet of the surface **K**, between the first and second cuts, we find the EFs of oscillation of three families (oscillations of H_{01}-, H_{02}-, and H_{03}-modes). If we neglect minor details in the trajectories of complex EFs $\bar{\kappa}(\delta)$, then their dynamics appears to be of a standard type, similar to that shown in Figures 6.5 and 6.6, but having one level more, that corresponds to oscillations of H_{03n}-modes (the third oscillation family). However, these details in the closest region near the expected intersection points $\mathrm{Re}\,\bar{\kappa}(\delta)$ of EFs of the second and third oscillation families, are of great importance. Actually, the Q-factors of the oscillation of the second and higher families are of the same order and these result in a qualitatively new effect: intermode relations of oscillations (mode coupling). This occurs when the oscillations on a small section of closely spaced continuous spectral curves exchange all their characteristic features, that is, modes (fields patterns), EFs variation dynamics, and so on (see Fig. 6.8: the EFs of one oscillation are shown by solid lines, and for the other one by dashed lines). The imaginary parts of EFs before the expected intersection of lines $\mathrm{Re}\,\bar{\kappa}(\delta)$ start to come closer.

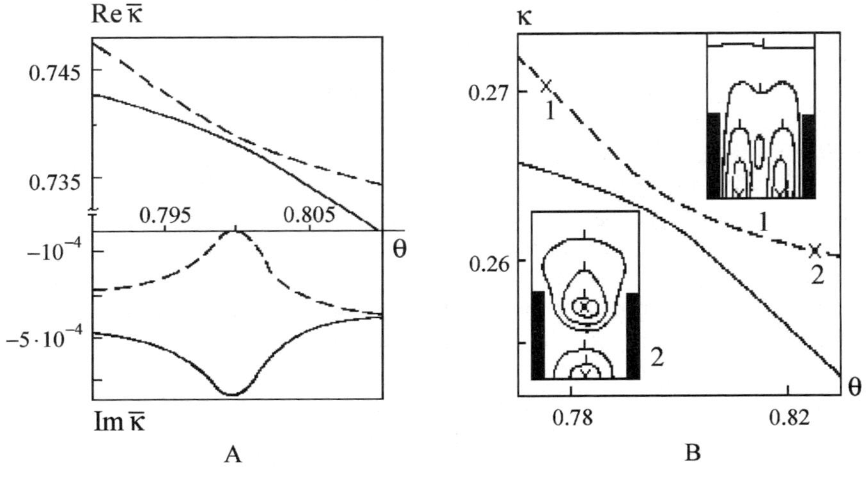

FIGURE 6.8. Mode coupling phenomenon: (A) Modes with complex-valued EFs ($\Phi = 0.1$, $\delta = 0.39$, $\tilde{\varepsilon} = 6.92$); (B) modes with real-valued EFs ($\Phi = 0.33$, $\delta = 0.48$, $\tilde{\varepsilon} = 25$).

Oscillations having a similar Q-factor begin to interact actively while intensifying the trend toward unlimited growth of the Q-factor of one oscillation and decreasing the Q-factor of the other. Between the continuous lines $\mathrm{Re}\,\bar{\kappa}\,(\delta)$, there is a zone of silence of the grating in the form of an open periodic resonator. By switching from a direction typical for one family to that typical for a different one, the structure of the oscillation field is changed dynamically: "pure mode $\rightarrow$ hybrid mode $\rightarrow$ other pure mode."

In order for two oscillations belonging to one symmetry class to interact, their diffraction Q-factor should be made equal. That is why for complex values of $\bar{\kappa}$, only those oscillations are related that are generated by contradirectional upper H_{0m}-waves ($m > 1$). The first family oscillations have a significantly lower diffraction Q-factor and do not take part in the mode coupling. At the same time, field oscillations can occur that belong both to the first (see Fig. 6.5), and to the next families, at real frequencies $\kappa < |\kappa_0|$. Regardless of whether they belong to one family or the other, these oscillations have similar (theoretically, infinite) Q-factors, and in such a case, the H_{01n}-modes also take part in the mode coupling (see Fig. 6.8B).

In the areas of parameter variations, where two, three, or more free space radiation channels are open, the convergence of EFs required for interaction is not always possible. To activate the compensation mechanism for directing one of the EFs to the real axis, at least three oscillations of different families and one symmetry class should take part in the interaction.

An approximate mathematical model of the mode coupling in closed resonators with excited boundaries was given for the first time in [176]. There the results of classical excitation theory of self-adjoint operators are used. For gratings, the analysis of spectral characteristics is based on solutions to non-self-adjoint problems, EFs are complex, and the eigenfunctions in the general case do not have the basis property. Therefore, because the model of [176] cannot be applied, the main research tool here is the computational experiment. The results of [73,144,153–156,177] prove the general character of the analyzed effect. Evidently, the conditions required for the existence of this effect can be created in almost any open resonator: waveguide, periodic, with compact or infinitely expanded boundaries. The effect of mode coupling allows the spectral lines of the oscillations that are comparable regarding the diffraction Q-factor, to separate in the complex space without intersecting, avoiding, in this way, degeneration, that is, the possibility of existence of different oscillations at one frequency $\bar{\kappa}$. In [176], the deviation at the resonator boundaries and in its material parameters from ideal ones are considered to be necessary conditions of the existence of intermode relations. On the other hand, [178] proves the instability of degenerate states. Combination of these two results allows us to consider the effect of intermode relation as a reaction of a nonideal system (in our case, free field oscillations in an open periodic resonator) as it approaches an unstable state. The mode coupling points out the possibility of degeneration of the oscillations in a certain ideal case that cannot be achieved by varying the parameters in the physical domain (real Φ, δ, θ, and the values of $\tilde{\varepsilon}$ such that $\mathrm{Re}\,\tilde{\varepsilon} > 0$ and $\mathrm{Im}\,\tilde{\varepsilon}\,\mathrm{Re}\,\kappa \geq 0$).

The search for degenerate states in open systems is almost always associated with the requirement of analytic continuation of the solutions to model spectral problems into the domain of complex (nonphysical) values of nonspectral parameters. In the simplest cases, the effect of removing degeneration and transfer to the mode coupling can be modeled also with physical parameter values. Thus, for instance, for half-integer Φ, the spectrum of free field oscillations in a lamellar grating is divided into two additional symmetry classes associated with the planes $y = n\pi; n = 0, \pm 1, \ldots$ (see Fig. 6.3A). At least in the case of real eigenfrequencies $\bar{\kappa} < |\kappa_0|$, this allows us to observe the degeneration in the intersection point of spectral curves corresponding to various oscillation modes of different symmetry classes (with respect to the planes $y = \text{const}$), as well as the effect of degeneration removing and transfer to the mode coupling when the symmetry is broken under the quasi-periodicity conditions (6.3), (6.9) (the value of Φ is not a half-integer).

The effect of mode coupling between oscillations is sufficiently stable with respect to changes of the independent parameter along which changes usually follow the spectral characteristics of one-dimensional periodic open resonators. For variations of the parameter $\text{Im}\,\tilde{\varepsilon}$ that characterizes the energy absorption rate in a lossy dielectric in the region $\mathbf{Q}_\delta$, the geometry of the field of the oscillations and the real parts of their eigenfrequencies changes slightly. When $\text{Im}\,\tilde{\varepsilon}$ increases, $|\text{Im}\,\bar{\kappa}|$ increases also, namely, linearly with respect to $\text{Im}\,\tilde{\varepsilon}$ and uniformly with respect to other independent parameters of the structure. This uniformity remains, mainly, in the regions where the Q-factor of the oscillations changes in an anomalous way. The dynamics of changes in the spectral characteristics in the interaction area remains generally the same as in the case of $\text{Im}\,\tilde{\varepsilon} = 0$ [see 9]: the oscillations are coupled, exchange of identifiers of the mode type occurs, and so on.

Numerical experiments, whose results were used to analyze the effect of existence of super-high-Q oscillation, do not settle the question about the limiting computation accuracy for small values of $\text{Im}\,\bar{\kappa}$, that is, the question of whether the dependence $\bar{\kappa}(\eta)$, where η is a varying parameter, comes to the real axis at a certain point η_1, or if $\text{Im}\,\bar{\kappa}$ does not vanish although it reaches very small values. The analysis of the corresponding dispersion equations, taking into account only the propagating waves of the coupling channel $\mathbf{Q}_\delta$, allows us to estimate the probability of occurrence of oscillations at real frequencies and, thus, partially gives the answer to the question [9,179]: if there is one open channel of radiation into free space at the principal harmonics of the field (6.10), and in the slots at least two waves (H_{0m} and H_{0p}, $n \neq p$) are propagating without attenuation, then, for a countable number of values of δ, the required phase conditions for the longitudinal resonances on the H_{0m}- and H_{0p}-waves can be satisfied simultaneously, for a cross-longitudinal resonance, when by a complete passing $z = 2\pi\delta \to z = -2\pi\delta \to z = 2\pi\delta$, half a cycle is realized at the H_{0m}-wave, and the second half at the H_{0p}-wave. At these values of δ the spectral line $\bar{\kappa}\,(\delta)$ of the oscillation of one of the families, associated with H_{0m}- or H_{0p}-waves, comes out to the real axis. The error of the analytic result obtained by this method decreases exponentially with increasing δ; that is, it can be made arbitrarily small.

Several general remarks on the super-high-Q oscillations in open resonators are in order. If we disregard the trivial case when the energy-radiating channels are electromagnetically closed (for gratings at $|\Phi| \leq 0.5$ it is the region $\kappa < |\Phi|$), one can say that the effect of existence of super-high-Q oscillations is realized only in structures whose radiation field always contains only a finite number of energy-consuming harmonics (i.e., waves that carry the energy to infinity). This illustrates the qualitative difference between the spectra of open periodic and waveguide resonators, and, on the other hand, compact open resonators. The latter (see Section 1.3.1) can sustain (on the physical sheet of $\mathbf{K}$) only attenuating ($\mathrm{Im}\,\bar{k} \neq 0$) field oscillations.

The super-high-Q oscillations ($\mathrm{Im}\,\bar{\kappa} = 0$) are, at the same time, surface eigenwaves of the grating that propagate without attenuation toward the direction where the structure is periodical (see Statement 6.7; the eigenpropagation constant is $\bar{\Phi} = \Phi$; the wavelength is determined by the values of $\kappa = \mathrm{Re}\,\bar{\kappa}$). The area where one or several radiation channels are open is for $|\Phi| \leq 0.5$ (the main variation interval of Φ) given by the inequality $\kappa > |\bar{\Phi}|$ and classified in the literature as "forbidden" for real ($\mathrm{Im}\,\bar{\Phi} = 0$) slow waves. If the channel is open only at the principal harmonics ($\mathrm{Re}\,\Gamma_0 > 0$ and $\mathrm{Im}\,\Gamma_n > 0$ for $n = \pm 1, \pm 2, \ldots$), then the amplitudes R_0 and T_0 of plane waves—partial field components of a surface eigen wave in regions $|z| > 2\pi\delta$ (see (6.10))—vanish (Statements 6.3, 6.8). In this case, the region $\kappa > |\bar{\Phi}|$ does not differ from the region $\kappa < |\bar{\Phi}|$ (traditional regions where slow eigenwaves are present) as regards the energy exchange between the near grating and free space, and the existence here of a slow wave is no longer considered an unusual effect. Thus, because super-high-Q oscillations have been revealed in a grating with open energy radiating channels, the traditional idea about the boundaries of the region of existence of surface slow waves for open periodic resonators has been essentially extended. This result can be useful for solving many applied problems. It expands the area of applications of gratings as decelerating and directing systems in various devices in microwave engineering, electronics, and optics.

In Section 1.3.2, we have already noticed that the multiplicity of the root $\bar{\kappa}$ of the scalar equation (6.20) determines the multiplicity of the eigenvalue of $\bar{\kappa}$ of a homogeneous operator equation $(I + A(\kappa))\,c = 0$; $\kappa \in \mathbf{K}$. In the numerical experiments, when problem (6.20) was divided according to symmetry classes, no root with a multiplicity order higher than one was found. If we assume that the poles of the resolvent of the problem (6.8) to (6.10) (problem (6.2) to (6.5)) in $\mathbf{K}$ are simple, then the increased multiplicity results in the degeneration of EF of $\bar{\kappa}$: one such eigenvalue corresponds to several linearly independent free field oscillations in a periodic open resonator of one symmetry class. Usually, this situation does not come about by changing the parameter values in the physical region. Rather, in the metric of the complex space, two EFs $\bar{\kappa}_1$ and $\bar{\kappa}_2$, that correspond to oscillations of different modes of the same symmetry class, can be situated arbitrarily closely. As a result, the oscillations begin to interact which leads to substantial local and global changes in their spectral characteristics. In particular, the regular features of the spectral curves $\bar{\kappa}(\eta)$ are violated, at which the zero lines of a similar level of function $f(\kappa, \eta) = \det(I + A(\kappa, \eta))$ (as a mapping $f : \mathbf{K} \times \mathbf{R}^1 \to \mathbf{C}$) are realized. Here $\mathbf{C}$ is a complex plane, $\mathbf{R}^1$ is the real number domain, and $\eta \in \mathbf{R}^1$ is a nonspectral parameter. The violation of the regular course of the corresponding curve means that the trajectory $\bar{\kappa}(\eta)$ passes near the critical point (for image f) [180].

The use of well-known results of the theory of singularities of smooth mappings [181] showed [10,182] that the oscillation interaction area that is the result of a characteristic nonclassical dispersion law (see, e.g., Fig. 6.8), that is, through a specific behavior of two EFs as functions of the parameter η at $\operatorname{Im}\eta = 0$, contains an isolated nondegenerate (Morse) critical point $\{\kappa_0, \eta_0\}$ of the image $f(\kappa, \eta) : \mathbf{C} \times \mathbf{C} \to \mathbf{C}$ (the boundaries of the physical region of the variation area of values η are extended). At $f(\kappa_0, \eta_0) = 0$ the root $\bar{\kappa}$ has the second order of degeneracy; the oscillations of the open periodic resonator do not interact. If $f(\kappa_0, \eta_0) \neq 0$, we obtain the dispersion law for the behavior of spectral curves that is typical for interaction, mode coupling of oscillations. Thus, the value $f(\kappa_0, \eta_0)$ determines the rate of mutual influence of oscillations when their EFs approach each other in the complex space. Numerical experiments on the function $f(\kappa_0, \eta_0)$ confirm the described regularities [183]. In [183] it was also shown that the local structure of solutions $\bar{\kappa}(\eta)$ to (6.20) near the Morse critical point $\{\kappa_0, \eta_0\}$ can be successfully reconstructed by the values of $f(\kappa_0, \eta_0)$.

6.3.2. Gallery of Anomalous Spatio-Frequency Transformations of Electromagnetic Field

The analytical continuation of the resolvent of the diffraction problems (6.2) to (6.5) into the domain of complex values of κ reveals a certain set of singularities $\{\bar{\kappa}_n\} \in \Omega_\kappa$, which according to Statement 6.1 may be qualified as isolated (for $|\kappa| < \infty$) poles of finite order of relevant operator functions in local variables on the surface $\mathbf{K}$. This fundamental result of spectral theory allows us, in particular, to formulate a sequential and well-proved approach to the study of local anomalous changes in the electromagnetic fields formed by a diffraction grating (see Statement 6.2). This type of changes usually takes place within extremely short intervals of variations of the free parameters and, as a rule, are accompanied by sharp changes, reaching limiting values of the principal diffraction characteristics. In this section we have used the advantages of the spectral theory results techniques for the analysis and treatment of the most pronounced phenomena of the set mentioned above, such as total transition and reflection of electromagnetic waves, appearing in diffraction by semi-transparent structures; the regimes of complete transformation of wave packets by open periodic resonators and so on. We do not dwell here on general (let us say regular) behavior of diffraction characteristics, which has been described rather explicitly in [3,6,11,14].

Let κ be a real positive frequency parameter, and let an arbitrary semitransparent (Fig. 1.1A) or reflecting (Fig. 1.1B) structure be excited by a harmonic plane E- or H-polarized wave $\tilde{U}_0^i(g, \kappa)$ (6.1), $\Phi = \kappa \sin\alpha$, $|\alpha| < \pi/2$. The complex amplitudes a_n, b_n of the spatial harmonics composing the diffraction field (6.5) (the H-case is marked by capital A_n and B_n) are complicated functions of κ and α, as well as of the geometry and the material parameters of the grating.

The first term in (6.5) for the reflection zone $z > 2\pi\delta$ corresponds to a wave incident on the grating. The infinite series for the zones $z > 2\pi\delta$ and $z < -2\pi\delta$ determine the secondary (scattered) field. The terms of these series are usually referred to as partial components of the spatial spectrum of the structure or as spatial (diffraction) harmonics of a scattered field. Every harmonic for which $\operatorname{Im}\Gamma_n = 0$ and $\operatorname{Re}\Gamma_n > 0$, is a homogeneous plane wave, propagating away from the grating at the angle $\alpha_n = -\arcsin\left(n\kappa^{-1} + \sin\alpha\right)$ in the reflection zone and at the angle $\alpha_n = \pi + \arcsin\left(n\kappa^{-1} + \sin\alpha\right)$ in the grating's passing zone (all the angles are measured in the plane $y0z$, anticlockwise from the z -axis, see Fig. 1.1A). The angle α is an angle of incidence of the homogeneous excitation wave $\tilde{U}_0^i(g, \kappa)$ on the grating. It is obvious that the direction of propagation of the energy-consuming (homogeneous) harmonics of the secondary field depends on their number n, and on the values of κ and α. The relative share of energy carried by each of them at $\operatorname{Re}\Gamma_0 > 0$ (excitation by a homogeneous wave) is determined by the modulus of the corresponding complex amplitude (see Section 6.2.1). The angle between the directions of propagation of the primary and the minus nth reflected plane wave $\alpha - \alpha_{-n} = \pm 2\varphi$, $\varphi \geq 0$ is determined from the equation $2\kappa\sin(\alpha \mp \varphi)\cos\varphi = n$.

Particularly, at $\varphi = 0$ the corresponding harmonic propagates toward the incident wave. The creation of such a nonspecular reflecting mode is called autocollimation. Not all the amplitudes a_n, b_n (A_n, B_n) are equally useful for physical analysis: in the far zone, the secondary field is formed only by the propagating harmonics with the numbers n such that $\operatorname{Re}\Gamma_n \geq 0$. The radiation field in the immediate proximity of the grating requires taking into account the contribution of damped harmonics $(n : \operatorname{Im}\Gamma_n > 0)$. Moreover, in some situations (resonance mode) this contribution is the dominating one.

The number of propagating harmonics $N = \sum_n \operatorname{Re}\Gamma_n/|\Gamma_n|$ is determined by the number of channels that are open for radiation into free space, and is the most general characteristic of the diffraction process. If the grating geometry is such that the channels in region $\mathbf{Q}_\delta$ with known wave propagation conditions can be isolated, then it is possible to introduce one more analogous identifier M which denotes the number of electromagnetically open channels at one period of the structure.

The joint qualitative characteristic $\{N, M\}$ taking into account the most common properties of gratings that are described by the consequences of the complex power theorem and the Lorenz lemma (see [3,6,11,173] and Section 6.2.1), enables us, in some cases, to predict rather precisely the feasibility of one or other scattering modes, whose specifics are determined by purely quantitative energy parameters, for example, the ratio W_0^a/W_0^b (the energy conservation law requires $\sum_n (W_n^a + W_n^b) = 1$). In particular, it can easily be shown [6] that in the area of parameter values that correspond to the vector $\{N, M\}$ with $N = M = 1$, a countable number of values $\delta \neq 0$ can always be found such that

$$|a_0| = 0, \quad |b_0| = 1 \quad (|A_0| = 0, \quad |B_0|) = 1, \tag{6.29}$$

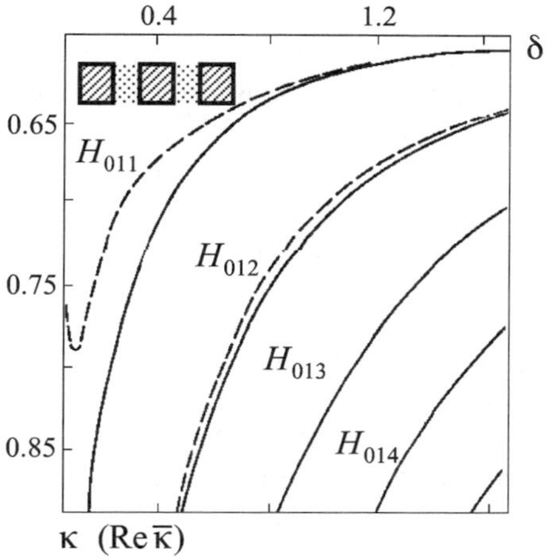

FIGURE 6.9. Overlapping of spectral and diffraction characteristics of the lamellar metal grating ($\theta = 0.8$, $\tilde{\varepsilon} = 1.108$, $\Phi = 0.1$).

independent of the general configuration of the boundaries of a semi-transparent structure.

If (6.29) is fulfilled in some point in the region $\{1,1\}$, it means that the primary wave propagates entirely (without reflection) into the zone $z < -2\pi\delta$. The value $\arg b_0$ ($\arg B_0$) determines the equivalent phase incursion caused by the presence of the grating. The effect of complete transition is usually accompanied by an increased field strength in region $\mathbf{Q}_\delta$, and the more pronounced it is, the narrower the coupling channels and the higher the structure. Such observations suggested to many researchers the idea of a resonant nature of such fully transparent modes. However, this suggestion has been strictly proven only in [3]. The parallel analysis of the numerical solutions of spectral (homogeneous) problems and plane wave diffraction problems allows us to generalize the conclusions made in [3,6] and to give a clear interpretation of other regularities mentioned earlier [9].

Figure 6.9 shows, in coordinates κ, δ (Re $\bar{\kappa}$, δ), the lines $|b_0| = 1$ (solid lines) and spectral lines Re $\bar{\kappa}$ (δ) (dashed lines). In the sections where they coincide, only the level lines $|b_0| = 1$ are shown. The limits of the variation of the frequency parameter κ here are such that $N = M = 1$. In the corresponding region of the first sheet of the surface **K** there is only one family of free oscillations, whose dispersion curves come together as $\delta \to \infty$ in the cut-off point of the H_{01}-wave of the coupling channel, or otherwise to the bottom boundary of region $\{1, 1\}$. The spectral lines of this family related to the oscillations of both symmetry classes, are plotted in Figure 6.9. The higher the Q-factor of free oscillations, the more closely the spectral lines overlap with the level lines $|b_0(\kappa, \delta)| = 1$. The Q-factor of the H_{011}-mode oscillation for small δ does not suffice to change the indicated at $\delta \approx 1$ trends of behavior of the line $|b_0| = 1$. The line at $\delta < 0.08$ is almost parallel to the κ-axis, determining in this way the minimum value of the height of the grating that still allows us to realize the complete transition mode. This mode at $\kappa > 0.8$ can no longer be called a resonant one. In all other cases the occurrence of the complete transition effect in region $\{1,1\}$ is conditioned by the excitation in the structure of oscillations that are close to the eigenoscillations.

The establishment of such a connection and the application of the results obtained while studying the EFs behavior (see the above section), enable us to give a quite well-reasoned explanation to all the characteristic features of the effects of complete resonant wave transition: the increased number of points where these effects take place at one and the same section of the frequency parameter when δ increases; their almost periodic repetition (with respect to δ for a fixed κ); the increased strength of the near field of the structure (determined by the value of $|\mathrm{Im}\,\bar{\kappa}|^{-1}$) in the moment when it becomes fully transparent, and so on.

If a plane H-polarized wave is incident on a grating composed of perfectly conducting cylinders of arbitrary cross-section (see Fig. 1.1A for $\tilde{\varepsilon}(y,z) \equiv 1$) in the case when the relations

$$\theta = \cos\alpha; \quad \kappa < (1 + |\sin\alpha|)^{-1}, \quad \delta > 0 \tag{6.30}$$

are satisfied, one can observe almost complete transition ($|A_0| \ll 1$) of the primary wave through such a semi-transparent structure (the Malyuzhinets effect). Here $2\pi\theta$ is the minimum spacing between the generatrices of the grating, and the limitation on κ is determined by the single-wave mode in the radiation zones, that is, the mode with $N = 1$. In the case (6.30), the smaller κ is, the closer $|B_0|$ is to unity, and it remains constant within wide limits of variation of other parameters. The development of rigorous analytic and numerical methods of studying the problems of wave diffraction on periodic structures allowed us to confirm the regularity of the effect and specify the conditions required for it to be realized on gratings whose geometry differs from the canonical one [6,184]. Particularly, it was established that the dielectric filling of the coupling channels in $\mathbf{Q}_\delta$ leads to a reduction proportional to $\tilde{\varepsilon}^{-1/2}$ ($\mathrm{Im}\,\tilde{\varepsilon} = 0$) of the efficient open part of the period and growth of the observation angle $|\alpha|$ of the Malyuzhinets effect. The corresponding changes are reflected in conditions similar to (6.30), viz. $\theta\tilde{\varepsilon}^{-1/2} = \cos\alpha$ for a grating of bars (Fig. 6.10A: $\kappa = 0.3$; $\theta = 0.6$; $\tilde{\varepsilon} = 4.0$; curve 1 – $\delta = 0.2$; 2 – 0.4) and $\theta\tilde{\varepsilon}_1^{-1/2} + (1-\theta)\,\tilde{\varepsilon}_2^{-1/2} = \cos\alpha$ for a knife grating with a complex period structure (Fig. 6.10B: $\kappa = 0.05$; $\tilde{\varepsilon}_1 = 1.0$; $\tilde{\varepsilon}_2 = 9.0$; $\delta = 0.4$). In compliance with the above results, the condition (6.30) can easily be generalized to the case of a grating with an arbitrary number of waveguide channels of the same period, which are filled with different dielectric materials. The occurrence of higher propagating harmonics outside the grating (going out into the region with $N > 1$) always results in a failure in creating the effect. The presence of higher propagating waves, coupling the radiation zones ($M > 1$), leads to the failure of this phenomenon only in the case of resonances.

Thus, for semi-transparent structures (see Fig. 1.1A) the described effect is not an exception: it is realized for an arbitrary ratio between the slot and the period. It almost does not depend on the grating height and in the long-wave region it does not depend on the frequency. The two latter features essentially distinguish this effect from those of complete resonant wave transitions that were considered earlier. The conditions $\delta > 0.25$ and $\kappa < 0.3$ give a quantitative characterization of the notions of nonzero height in expressions such as (6.30) and of the long-wave

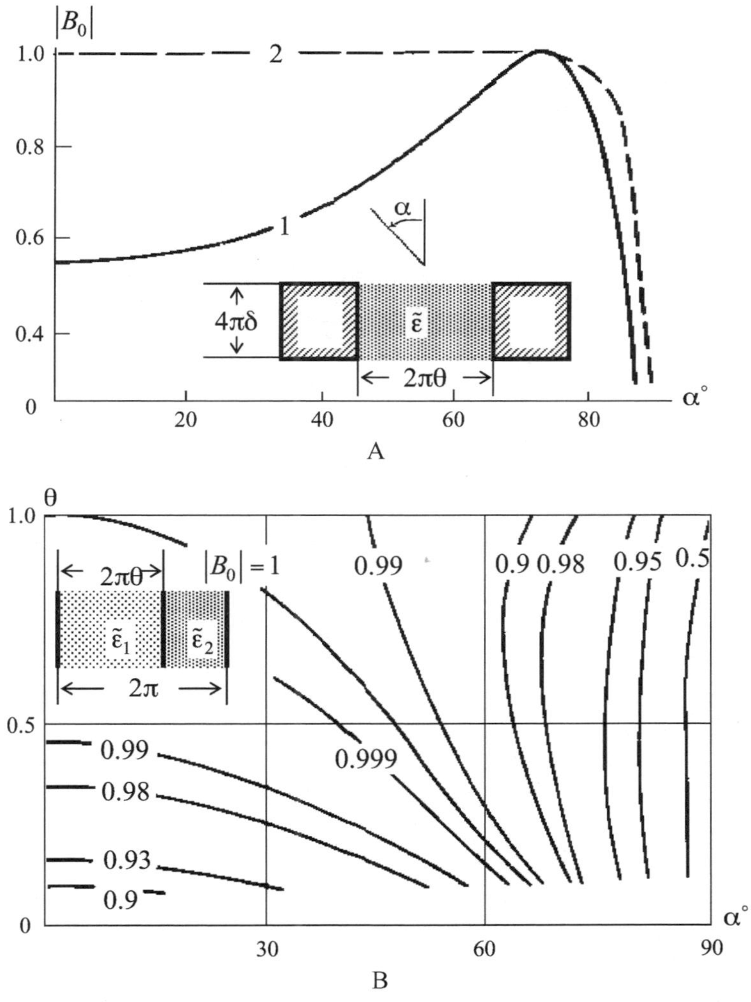

FIGURE 6.10. Malyuzhinetz's phenomena in densely periodic gratings: (A) Changes of $|B_0|$ when α is varying and (B) lines of equal values of $|B_0(\theta, \alpha)|$.

region. If for normal incidence ($\alpha = 0$) the parameters κ and δ are such that the field, as a resonance, goes fully through the grating, then $|B_0|$ stays very close to unity in the interval of the angle α from zero to the value at which the Malyuzhinets effect appears (see Fig. 6.10A, curve 2).

In the previous items, we have described the simplest variants of a grating operating as a selective unit. In fact, the potentialities of periodic structures as a means to achieve efficient spatial, frequency, and polarization selection of signals seem to be inexhaustible. In computational experiments, the effects of complete resonant transition and reflection of the energy of incident E- and H-polarized plane waves at semi-transparent structures (i.e., open for field penetration into the transition

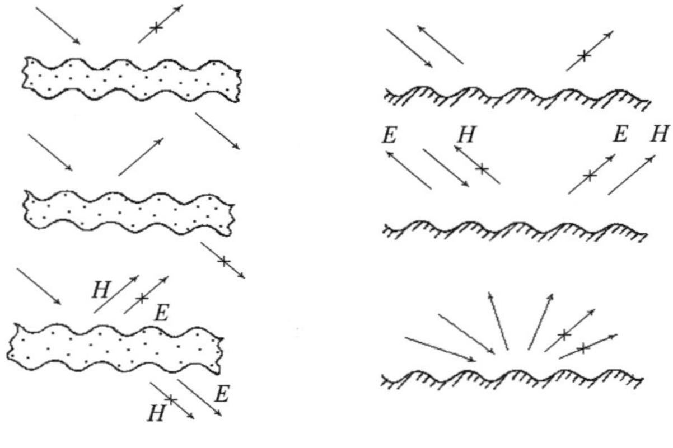

FIGURE 6.11. Analysis of gratings' selective properties in the resonant domain of parameters.

zone $z < -2\pi\delta$) were discovered. These effects are shown schematically in left fragment of Figure 6.11. The effects of complete nonspecular (autocollimation) wave reflection and complete transformation of wave packets by periodic surfaces impenetrable for electromagnetic field (see right part of Fig. 6.11), and so on, have been revealed and studied in depth [3,6,9,184–189]. Note that the corresponding modes can be both narrowband and wideband with respect to each of the variable parameters. One and the same grating can react in different ways in such a mode to two waves having similar length but different polarization.

Figure 6.12 presents computed results for a certain structure of the comb-type: the energy characteristics of the autocollimation (inverse) reflection of the first but

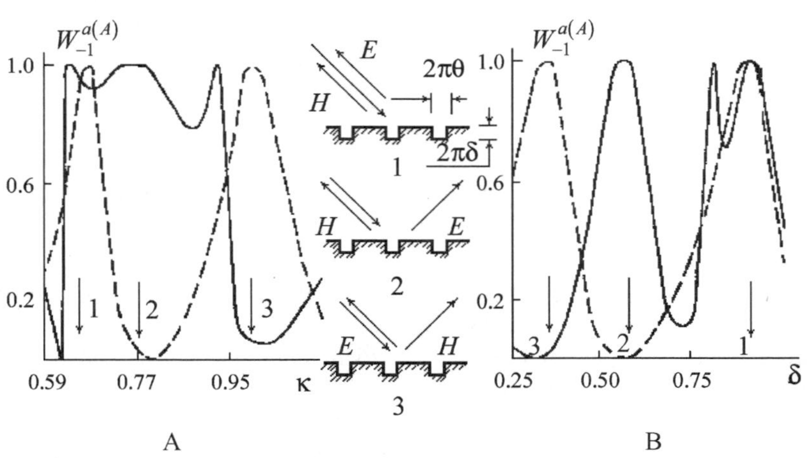

FIGURE 6.12. Spatial and polarization selection of signals in autocollimation regime of reflection of the minus first harmonic: Dashed lines correspond to E-polarized field, solid lines to H-polarization; $\tilde{\varepsilon} = 2$. (A) $\theta = 0.6$, $\delta = 0.69$; (B) $\theta = 0.5$, $\kappa = 0.95$.

one harmonic (only the specular and autocollimating first but one harmonic are energy-carrying in the scattered field (6.5)). In case 1, the grating is not sensitive to the polarization of the signal in the modes of complete nonspecular reflection. In cases 2 and 3, E- and H-polarized components are completely separated by the radiation channels that are coupled with the specular (zero) harmonic and that propagating toward the incident wave.

The methods of classical diffraction theory used in the study of the above-mentioned and other abnormal effects and phenomena, were a tool for gathering statistical data that can be analyzed to give answers to questions such as "What happens?" and "What may happen?" The mechanisms of forming resonant feedback of structures to the external excitation could be treated only hypothetically. The fact is that a study of real κ (the diffraction theory domain) cannot provide comprehensive information on all the specific features of solutions $\tilde{U}(g, \kappa)$ to the problems (6.2) to (6.5) as functions of the complex κ. And these specific features, as we have repeatedly mentioned, are, in fact, the ID of the structures that determine their electromagnetic characteristics. The rest of this section is devoted to the justification of this thesis (the case of frequency domain).

If we extend the values of the parameter domain $\{N, M\}$ to $N = 1$ and $M \geq 2$, qualitatively new effects appear, namely, the effects of complete reflection of the primary wave by semi-transparent structures ($|a_0| = 1$ or $|A_0| = 1$). For a grating having a simple periodic geometry (in such a geometry there is only one waveguide channel coupling the radiation zones), the complete reflection can be observed only near those values of δ and κ, at which the height of channel $\mathbf{Q}_\delta$ is a function of the number of half-waves of one of the higher waves that propagates there [6]. For a periodic dielectric layer with $\tilde{\varepsilon} = \tilde{\varepsilon}(y) = 2 + \sin y$ these are the higher propagating harmonics of a Floquet channel that is inhomogeneous in the cross-direction (see Fig. 6.13A: lines of equal level of the reflection coefficient $|a_0(\kappa, \delta)|$ in region $\{1, 2\}$, the bottom boundary of this region is marked with a dashed line; $\Phi = 0.1$). For waveguide gratings (see Fig. 6.13B illustrating the dependence $|b_0(\theta)| = \left(1 - |a_0(\theta)|^2\right)^{1/2}$ for various coupling modes of radiation zones of gratings with a complex period structure; $\kappa = 0.9$; $\alpha = 5.5°$; $\tilde{\varepsilon}_1 = 4.93$; $\tilde{\varepsilon}_2 = 1.0$; $\delta = 0.55$) these are H_{0m}-waves, $m = 2, 3, \ldots$ (E_{0m}-waves, $m = 1, 2, \ldots$ in the case of H-polarization). The larger δ, the better the coincidence. With increasing κ or δ, the external Q-factor of the modes becomes higher; the number of the complete reflection point increases in one and the same interval of the values of the variable parameter. The key regularities listed above and analyzed in detail in [6] differ from the regularities of the fully transparent resonance modes only in that they are in this case adjacent, not to the principal wave of region $\mathbf{Q}_\delta$, but to the higher waves propagating here. The effect of complete reflection is due to free oscillations of the field of the second and next families. This conclusion is confirmed by the dynamics of the variations of the corresponding EFs.

Figure 6.14 presents, by the level lines $|b_0(\kappa, \delta)| = 1$ (solid lines, a completely transparent structure) and $|b_0(\kappa, \delta)| = 0$ (dashed lines, complete reflection mode),

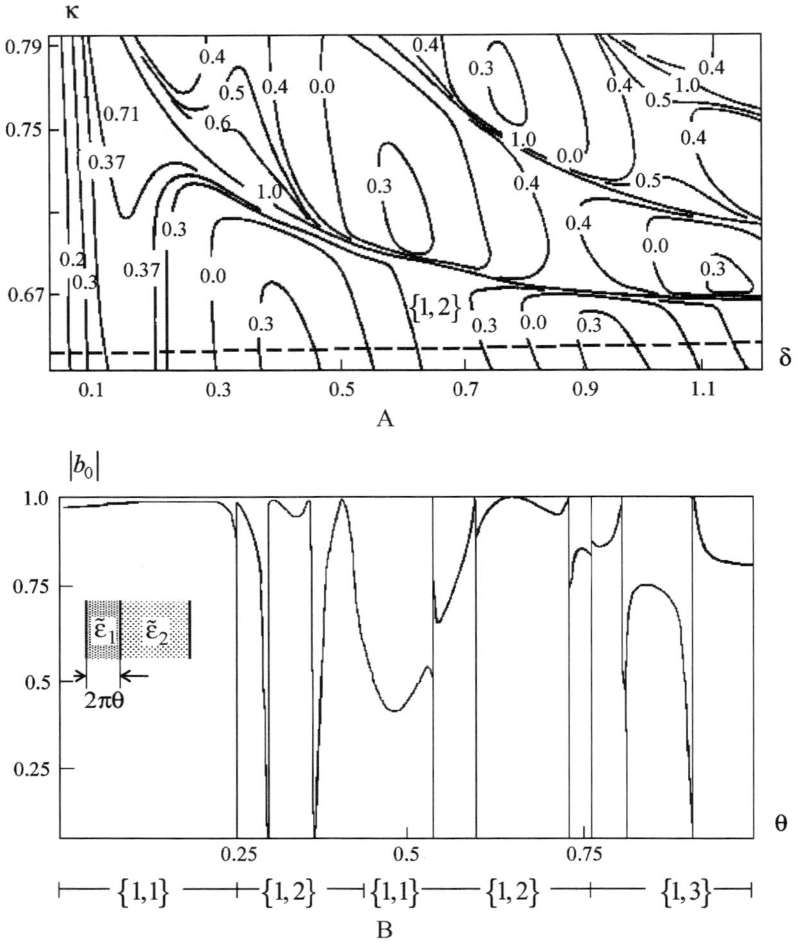

FIGURE 6.13. Total resonant reflection of electromagnetic waves by semi-transparent gratings: (A) Dielectric periodic layer and (B) lamellar grating with compound structure of period.

spectral lines $\mathrm{Re}\,\bar{\kappa}\,(\delta)$ of the first and second families of the symmetric (plus sign) and antisymmetric (minus sign) free field oscillations in a grating of metal bars (see Fig. 6.3A). It turned out that the spectral lines $\mathrm{Re}\,\bar{\kappa}\,(\delta)$ of the second family of oscillations (H_{02m}-oscillations) coincided (with graphical accuracy) with the lines $|b_0\,(\kappa, \delta)| = 0$. This means that the complete reflection mode in region {1,2} is a resonant mode and happens due to the excitation of oscillations that are close to free oscillations whose field is formed by the first higher waves propagating in the region $\mathbf{Q}_\delta$. The spectral lines $\mathrm{Re}\,\bar{\kappa}\,(\delta)$ of the first family of oscillations that converge, for increasing δ, to the cut-off point of the H_{01}-wave, overlap with the lines $|b_0\,(\kappa, \delta)| = 1$.

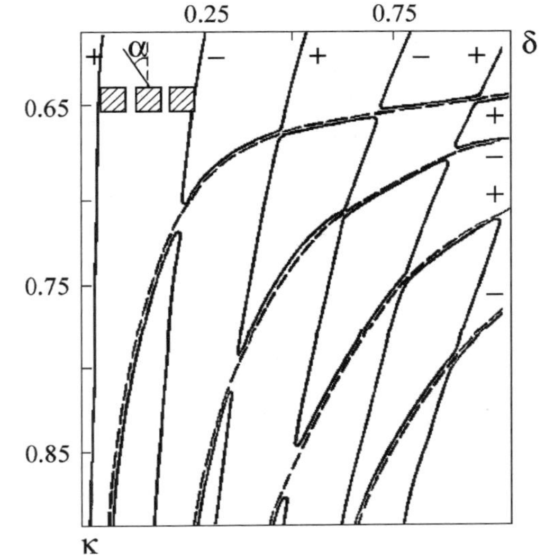

FIGURE 6.14. Description of the nature of the resonant scattering regimes in the region {1,2} ($\theta = 0.8$, $\tilde{\varepsilon} = 3.89$, $\Phi = 0.1$, $\sin\alpha = \Phi/\kappa$).

This situation is similar to that observed in region {1, 1}, but the agreement is not continuous here. The continuity of the agreement is violated in the area of intersection of the lines $\mathrm{Re}\,\bar{\kappa}\,(\delta)$ of the first and second families of oscillations. The modes of complete transparency along the lines $\mathrm{Re}\,\bar{\kappa}\,(\delta)$ of the first family oscillations are separated by the modes of complete reflection of the primary wave and are forced to join the latter, and they jointly form a pair resonance (see also Fig. 6.13). The closed lines $|b_0\,(\kappa, \delta)| = 1$ stratify the plane κ, δ, and determine the islands with the local minima of the transition coefficient.

When the value of M increases, the number of oscillation families increases too, and their EFs are revealed by the analytic continuation of the diffraction problems to the first sheet of the surface **K** from the corresponding section of the real axis κ. The spectral lines $\mathrm{Re}\,\bar{\kappa}$ of the oscillations of all the next following families (for $M > 2$) almost exactly locate the points of realization of the modes of complete reflection of plane waves by a semi-transparent structure. For $M \geq 3$, free oscillations of $H_{0m\,n}$-modes, $m \geq 2$, belonging to various families but having the same symmetry class, interact by converging EFs, exchanging their key characteristic features. It influences the scattering process in the following way: the lines of complete reflection $|b_0\,(\kappa, \delta)| = 0$ follow the changes of $\mathrm{Re}\,\bar{\kappa}\,(\delta)$ (see Figs. 6.15 and 6.8), and transparent windows appear in the structure between two tongues of lines $|b_0\,(\kappa, \delta)| = 0$. In the regions where the lines $|b_0\,(\kappa, \delta)| = 0$ ($|b_0\,(\kappa, \delta)| = 1$) change their direction, wideband modes of almost complete reflection (transition) of the wave incident on the grating are quite possible.

The correspondence between spectral and abnormal diffraction characteristics has been established but the mechanism that creates this correspondence has not been studied yet. Modeling such a mechanism is quite an intricate problem.

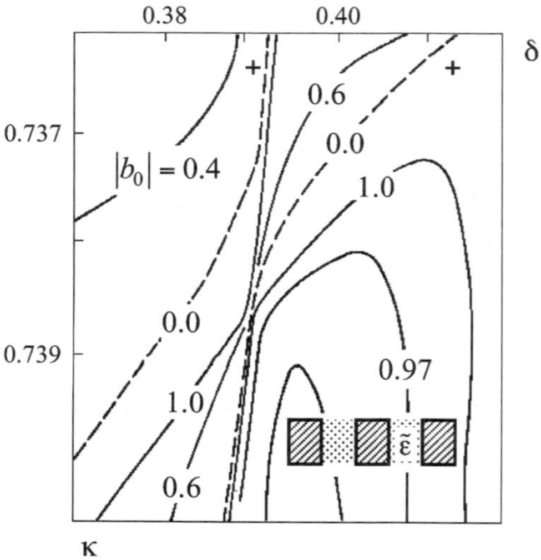

FIGURE 6.15. Lines of equal values of $|b_0(\kappa, \delta)|$, regime $\{1, 3\}$: Spectral curves $\operatorname{Re}\bar{\kappa}(\delta)$ coincide with curves $|b_0(\kappa, \delta)| = 0$. The region of mode coupling of symmetric (sign +) oscillations of second and third families ($\theta = 0.8$, $\tilde{\varepsilon} = 6.92$, $\Phi = 0.1$).

Therefore we focus on one simple situation, allowing us to estimate the major factors that are also relevant for more complicated cases. The validity of the simplifications in the modeling can be justified as all the qualitative conclusions have been verified by results of numerical experiments with rigorous solutions of corresponding boundary value problems.

Let the geometry of the structure and the excitation parameters be such that the coupling between zones $|z| > 2\pi\delta$ is realized by one propagating wave, generated for a complex $\bar{\kappa}$ of the field of the first (lower) family of free oscillations. As was mentioned before, in such a case, for values of κ close to $\operatorname{Re}\bar{\kappa}$, the grating becomes completely transparent. The contributions to the radiation field in the zone $z > 2\pi\delta$ (into the propagating harmonics of the spatial spectrum of the structure), caused by the reflection of an incident wave from the grating aperture in plane $z = 2\pi\delta$ and the radiation of the propagating wave of the coupling channel $\mathbf{Q}_\delta$, coincide in absolute value but differ in sign. Full compensation is possible if the dynamic phase effect applies (see (6.19) in Subsection 6.2.2) in one of the components in the following representation for the reflection coefficient [9].

$$2a_0(\kappa) = \exp\left(i \arg\left(r_0 \mp t_0\right)\right) + \exp\left(i\left[\arg\left(r_0 \pm t_0\right) - 2\arg\left(\kappa - \bar{\kappa}\right)\right]\right); \quad |\kappa - \bar{\kappa}| \ll 1. \tag{6.31}$$

This is a single pole representation similar to (6.19) for semi-transparent structures that are symmetrical with respect to the plane $z = 0$. The upper sign is chosen if the pole $\bar{\kappa}$ corresponds to a symmetrical free field oscillation, and the lower sign for an antisymmetric one). If the point $\bar{\kappa} \in \Omega_\kappa$ is such that the unknown values of r_0 and t_0 in (6.31) can be determined by using the known solution to the problem for the value of $\kappa = \Phi$ (this approach is described in [6,9]), then the condition for the realization of the complete transition effect can be written as

$$\arg(\Phi - \bar{\kappa}) - \arg(\kappa - \bar{\kappa}) = \pi/2. \tag{6.32}$$

From (6.32), it follows that $|a_0(\kappa)|$ vanishes if κ exceeds the values of $\mathrm{Re}\,\bar{\kappa}$ somewhat (see Fig. 6.9).

By a multichannel coupling of the reflection and transition zones, the condition required for the mode of complete reflection of a plane wave by a semi-transparent structure to take place ($M \geq 2$) remains unchanged, but should be specified (for an example of a grating having two channels at the period, see Fig. 6.13B). We note that in the case of an H-polarized field, channels having different geometry can be electromagnetically equivalent for the lower mode waves (the propagation constants of these waves coincide). A possible situation: $M \geq 2$, but the analytic continuation of the solutions to the respective boundary value problem reveals spectral points of only one family of free oscillations (a synchronous resonance of identical coupling channels), and near the EFs of these oscillations, only the modes of complete transparency of the structure are observed. Thus, in gratings with a complex period structure, electromagnetically identical channels should always be taken into account in the index M as a single one [9].

Let us illustrate this consideration with the following result. The point $\theta = 0.375$ in Figure 6.16A is a common boundary of the regions that can be formally described by the vectors $\{1, 3\}$ and $\{1, 2\}$. When we cross the cut-off point of the E_{01}-wave in the channel having the width of $2\pi(1-\theta)$, the complete reflection effect vanishes, and the structure becomes completely transparent. Formally, the scattering process takes place in region $\{1, 2\}$, but actually, in region $\{1, 1\}$, because the coupling channels become electromagnetically equivalent. The violation of this equivalence (Fig. 6.16B) results in the effect of complete resonance reflection, whose diffraction Q-factor increases as $|\tilde{\varepsilon}_2 - \tilde{\varepsilon}_1|$ decreases. In the case presented in Figure 6.16B, the second spectral family is composed of free oscillations of the field in the channel filled with a dielectric. The value $\tilde{\varepsilon}_2 = \tilde{\varepsilon}_1$ is, in fact, the boundary between the regions $\{1, 2\}$ and $\{1, 1\}$. By approaching such a boundary between regions by any of the variable parameters, the diffraction Q-factor of the complete reflection effect increases without bounds. When the boundary is crossed, the effect just vanishes and, as a result, the branches of the resonance curve merge.

In the case of an E-polarized field, it may happen that the channels connecting the radiation zones are identical. An obvious example: a knife grating with a complex period structure (see Figs. 6.13B and 6.16A) if $\theta = 0.5$ and $\tilde{\varepsilon}_1 = \tilde{\varepsilon}_2$. If the equality $\tilde{\varepsilon}_1 = \tilde{\varepsilon}_2$ is not satisfied and θ deviates from $\theta = 0.5$, the coupling channels are excited asynchronously, and the resonance in one of them (with the cut-off point of the propagating wave being larger than that of the other one) causes the effect of complete reflection (see Fig. 6.16C). The transfer to the values $\theta = 0.5$ in the case presented in Figure 6.16C, makes the complete reflection effect vanish. The scattering process is accompanied only by the complete transition resonances ($|b_0| = 1$) associated with a single family of free field oscillations in the structure.

Let us highlight one additional specific feature of such gratings. If they are excited by an H-polarized wave, the transition to the region $\{1,2\}$ can proceed

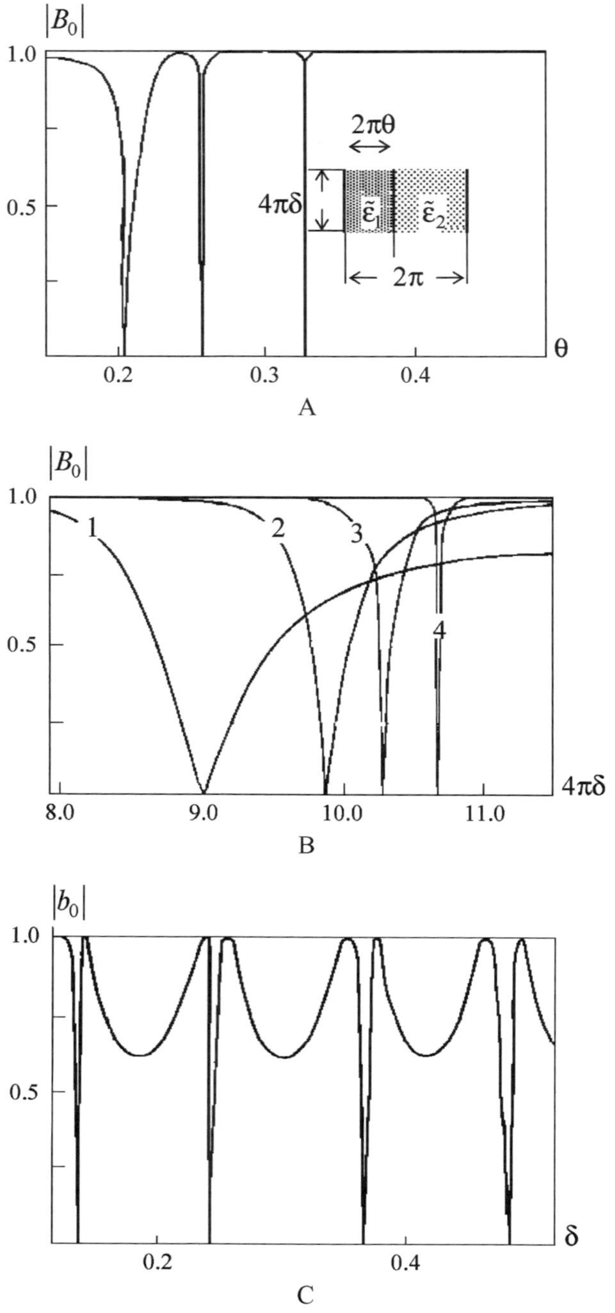

FIGURE 6.16. Peculiarities of resonant scattering by gratings with compound structure of period: (A) H-polarization, $\kappa = 0.8$, $\alpha = 11.3^{\circ}$, $\tilde{\varepsilon}_1 = \tilde{\varepsilon}_2 = 1$, $\delta = 0.8$; (B) H-polarization, $\kappa = 0.5$, $\alpha = 0^{\circ}$, $\theta = 0.5$, $\tilde{\varepsilon}_1 = 1$, $\tilde{\varepsilon}_2 = 2(1), 1.5(2), 1.3(3), 1.1(4)$; (C) E-polarization, $\kappa = 0.8$, $\alpha = 0^{\circ}$, $\tilde{\varepsilon}_1 = \tilde{\varepsilon}_2 = 9$, $\theta = 0.45$.

for almost all small values of κ. Thus, for instance, for a grating as in Figure 6.16, it suffices to take $\tilde{\varepsilon}_1 \neq \tilde{\varepsilon}_2$. This also means that in the long-wave region of the range, such semi-transparent structures are able to completely reflect (transmit) an incident wave even if $\left|\tilde{\varepsilon}_j - 1\right| \ll 1$. However, the closer κ is to zero (i.e., to the cut-off point that is the limit point as $\delta \to \infty$ for the spectral lines $\bar{\kappa}(\delta)$ of the oscillation of the first and second families), the larger is the value of δ, for which the respective mode can be realized.

An efficient physical analysis of various diffraction effects in the multiwave region $\{N, M\}$ with $N \geq 2$ is complicated because of a great number of parameters, the presence of several channels open for energy radiation, none of which has a clearly enough pronounced priority. A study of the diffraction characteristics of semi-transparent structures in the region $\{N, M\}$ with $N > 1$ reveals one feature that is most typical for such situations. Namely, the occurrence of new higher propagating spatial harmonics (for constant M) leads to the reduction in the total number of resonance modes of wave scattering and a substantial drop of the Q-factor of other modes. The corresponding changes are determined by the ratio between N and M and do not depend on the structure of the period of the grating. Spectral methods give the following explanation to this fact. The number M specifies the number of oscillation families, whose EFs are revealed by the analytic continuation of the solutions to the diffraction problems from the observation section of real κ to the first sheet of the surface $\mathbf{K}$. The number N of channels open for energy radiation directly influences the Q-factor of these oscillations: it decreases when new propagating harmonics in the diffraction field (6.5) appear (at the corresponding real κ). When a new component is added to N, one of the oscillation families falls out of play. The Q-factor of this family's oscillations falls down and crosses the limit beyond which it can exercise no essential influence on the scattering process.

Let us discuss briefly one of the simplest cases when $N = 2$. In the case of an E-polarized field at $\theta = 0.7$, $\tilde{\varepsilon} = 1.208$, $\Phi = 0.4$, the analytic continuation from the section of the real axis $1.3 < \kappa < 1.4$ of the solution to the diffraction problem (6.2) to (6.5) for a grating of metal bars reveals eigenfrequencies $\bar{\kappa}$ of two oscillation families, whose fields are formed in the coupling region $\mathbf{Q}_\delta$ by contradirectional H_{01}- and H_{02}-waves. The spectral lines $\mathrm{Re}\,\bar{\kappa}(\delta)$ of the oscillations of these families are shown in Figure 6.17 (the solid and dashed lines correspond

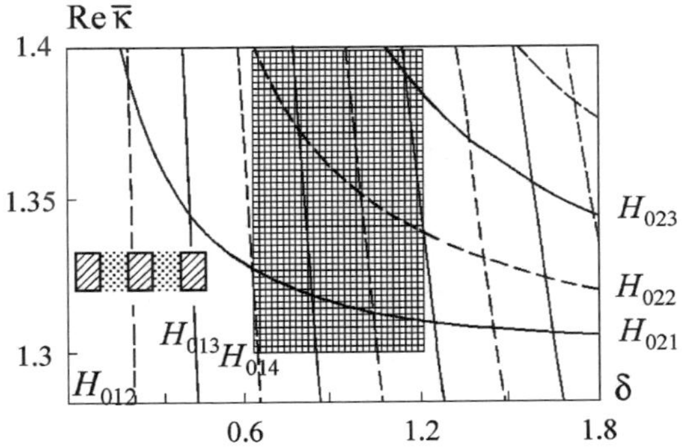

FIGURE 6.17. Real parts of complex-valued EFs of gratings with two opened for energy radiation channels.

to symmetrical and asymmetrical oscillations, respectively). The general view is similar to that considered in Section 6.3.1 by analyzing EFs situated under the region $\kappa_0 < \kappa < \kappa_{-1}$ of the first sheet of surface **K** ($N = 1$). Essential changes are only due to the fact that the value of $|\mathrm{Im}\,\bar{\kappa}\,(\delta)|$ has increased, on average, by an order of magnitude, the gap between the values $|\mathrm{Im}\,\bar{\kappa}\,(\delta)|$ for the oscillations of the first and second families has widened, and near the intersection point of the lines $|\mathrm{Re}\,\bar{\kappa}\,(\delta)|$ there are no clearly pronounced interactions between oscillations belonging to different families and the same symmetry class. The number of waves propagating in the coupling channels is not enough to activate the compensation mechanism that was used to generate super-high-Q oscillations at $N = 1$. Note that such effects are also possible at $N = 2$, but one should then transfer to the region $\{2, M\}$ with $M \geq 3$, while increasing, say, the relative dielectric permittivity $\tilde{\varepsilon}$ of the material.

Let us use a well-tested method and plot, on the grid of lines $\mathrm{Re}\,\overline{\kappa}\,(\delta)$, the lines of constant level of two major energy-scattering characteristics in region $\{2, 2\}$ (see Fig. 6.18: the sections of the plane κ, δ with $W > 0.9$ are crosshatched with vertical lines, and those with $W < 0.1$, with horizontal lines). All the characteristics show very limited reaction to the presence of H_{01m}-oscillations in the structure. Their influence is substantially decreased because of the presence of the second ($n = -1$) propagating spatial harmonic in the radiation zones. The role of the lower family

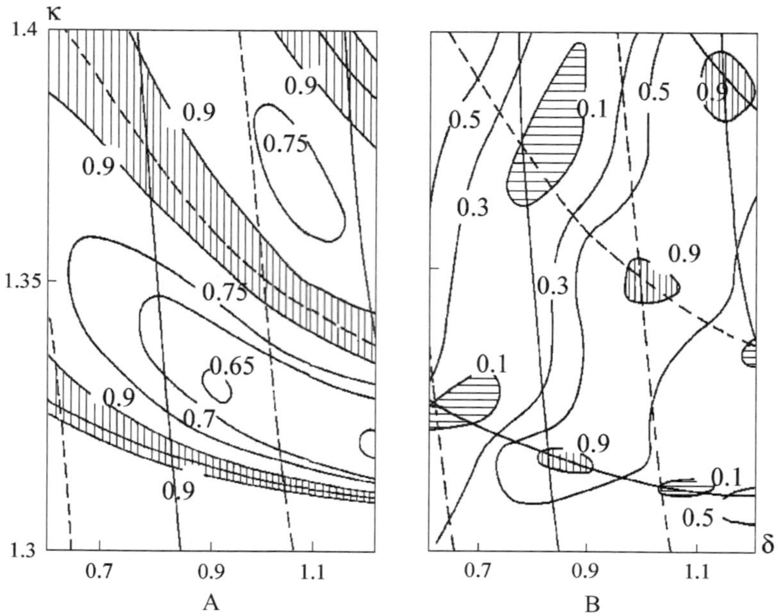

FIGURE 6.18. Definition of connection between spectral and diffraction characteristics of grating within operating regime $\{2, 2\}$: (A) lines of equal values of $W_0^b + W_{-1}^b = \text{const}$; (B) $W_0^b = \text{const}$.

is taken over by the H_{02m}-oscillations. As for $N = 1$, the lines $\mathrm{Re}\,\bar{\kappa}\,(\delta)$ of this family's oscillations determine the areas of maximum energy concentration of the scattered field in harmonics going out under the grating (see Fig. 6.18A). The contribution of W_0^b and W_{-1}^b to the total energy in these areas is not equal. Rather, it is determined by the generalized spectral characteristics including the characteristics of the oscillations both of the first and second families: the energy W_0^b (W_{-1}^b) is the maximum (minimum) in the area where the values $\mathrm{Re}\,\bar{\kappa}$ of the oscillations belonging to the same symmetry class coincide and the minimum (maximum) in the area where lines $\mathrm{Re}\,\bar{\kappa}\,(\delta)$ of the oscillations of different symmetry classes intersect (see Fig. 6.18B).

In certain points of the regions with extreme levels of W_0^b (W_{-1}^b), the corresponding function reaches the values $W_0^b \approx 1$ ($W_{-1}^b \approx 0$) or $W_0^b \approx 0$ ($W_{-1}^b \approx 1$). Thus, by combining the resonance conditions for the H_{01}- and H_{02}-waves, the contributions to all the propagating harmonics can be compensated, except for one harmonic ($n = 0$ or $n = -1$) that carries all the energy into the transition zone of the grating. For the parameters $\{2, 2\}$, the energy characteristics of scattering in the reflection zone of the structure vary within substantially narrower limits. The function $W_{-1}^a\,(\kappa, \delta)$ is more dynamic. The characteristic W_0^a plays the role of a constant background on the level $W_0^a \approx 0.2$.

In the course of studying the diffraction characteristics of periodic structures and discontinuities in multimode waveguides, it was revealed that one wave can be completely or almost completely transformed into another. In the electromagnetic theory of reflecting gratings, the most interesting effects of such kind are the nonspecular wave reflection effects in the autocollimation modes (see the example in Fig. 6.12 and [6,9,187,188]) and with a large telescoping coefficient [190]. In the waveguide structures, these are the effects of intense transformation of lower waves into higher ones that have been found in simple and complex corner discontinuities, coaxial junctions of circular waveguides with various diameters [7], and other types of waveguide resonators. In the search for such modes, efficient algorithms for the analysis of diffraction problems have been applied. These algorithms were developed on the basis of the semi-inversion method [2,3,5], and involved the use of parametric optimization methods. The mechanisms for realization of such modes are studied within the framework of the spectral theory [9,153], and the main cause is seen to be the existence of complex eigenfrequencies of corresponding open resonators on the first physical sheet of the surface **K**. In the case considered below, the field scattered by the grating is uniquely determined by the features of the analytic continuation of the resolvents of the diffraction problems (6.2) to (6.5), lying on the upper, nonphysical sheets of the surface **K** [191]. Because they usually have complex EFs, open periodic and waveguide resonators have specific sets of real eigenfrequencies that are directly associated with the effects of complete transformation of waves or wave packets.

Assume that for certain parameter values, one EF $\bar{\kappa}_m$ of a periodic open resonator comes to the real axis in the region $\mathrm{Re}\,\kappa > 0$ of one of the higher-order sheets of the surface **K**. Consider the partial components of a free oscillation

$$\tilde{U}_m(g, \bar{\kappa}_m) = \sum_{n=-\infty}^{\infty} \left\{ \begin{matrix} R_n \\ T_n \end{matrix} \right\} e^{i[\Phi_n y \pm \Gamma_n (z \mp 2\pi\delta)]}; \quad \left\{ \begin{matrix} z > 2\pi\delta \\ z < -2\pi\delta \end{matrix} \right\}, \tag{6.33}$$

corresponding to the spectral point $\bar{\kappa}_m$ (of the nontrivial solution of the homogeneous problem (6.2) to (6.5) at $\kappa = \bar{\kappa}_m$). According to Statement 6.1, part of these partial components with the numbers n such that $\bar{\kappa}_m < |\Phi_n|$, are inhomogeneous plane waves that attenuate exponentially (if $\mathrm{Im}\,\Gamma_n(\bar{\kappa}_m) > 0$) or increase (if $\mathrm{Im}\,\Gamma_n(\bar{\kappa}_m) < 0$) as $|z|$ increases. The set of partial components (6.33) that is complementary to the first set, with n such that $\bar{\kappa}_m > |\Phi_n|$, combines the homogeneous plane waves arriving at the grating (if $\mathrm{Re}\,\Gamma_n(\bar{\kappa}_m) < 0$) or going out of it into free space (if $\mathrm{Re}\,\Gamma_n(\bar{\kappa}_m) > 0$).

From (6.33), we form two wave packets $\tilde{U}_{m,1}$ and $\tilde{U}_{m,2}$, which are nonintersecting at the set $n = 0, \pm 1, \ldots$:

$$\tilde{U}_{m,j}(g, \bar{\kappa}_m) = \sum_{n \in N_j} \left\{ \begin{matrix} R_n \\ T_n \end{matrix} \right\} e^{i[\Phi_n y \pm \Gamma_n (z \mp 2\pi\delta)]}; \quad \left\{ \begin{matrix} z > 2\pi\delta \\ z < -2\pi\delta \end{matrix} \right\}, \quad j = 1, 2. \tag{6.34}$$

Here, $N_1 + N_2 = \{n\}_{-\infty}^{\infty}$; that is, $\tilde{U}_{m,1} + \tilde{U}_{m,2} = \tilde{U}_m$; $N_1 = \{n : \mathrm{Im}\,\Gamma_n(\bar{\kappa}_m) < 0$ or $\mathrm{Re}\,\Gamma_n(\bar{\kappa}_m) < 0\}$, $N_2 = \{n : \mathrm{Im}\,\Gamma_n(\bar{\kappa}_m) > 0$ or $\mathrm{Re}\,\Gamma_n(\bar{\kappa}_m) > 0\}$. Consider the existence of a free field oscillation in a grating at EF $\bar{\kappa}_m$ in terms of the common problem of plane wave diffraction at one-dimensional periodical structures (see (6.2) to (6.5)). In view of (6.34), we obtain: if the grating is excited by a plane wave packet $\tilde{U}_{m,1}(g, \kappa)$ at the frequency $\kappa = \bar{\kappa}_m$ (here κ is the projection of point $\bar{\kappa}_m$ onto the first sheet of the surface **K**), the secondary (scattered) field coincides with the plane wave packet $\tilde{U}_{m,2}(g, \kappa)$. Thus, the point $\bar{\kappa}_m \in \Omega_\kappa$ determines the frequency at which the grating transforms one plane wave packet into another. These packets are composed of different harmonics of the spatial spectrum of the structure.

Hence, the solution to the problem of complete transformation can be reduced to the search for real eigenfrequencies lying on the nonphysical sheets of the Riemann surface **K**. The characteristics of the packets $\tilde{U}_{m,1}$ and $\tilde{U}_{m,2}$ determine the sheet of the surface and the section of the real axis (between two adjacent branch points) that are the search area. The efficiency of the spectral approach to the problem of synthesis of a structure that completely transforms wave packets containing several harmonics propagating in different directions, is conditioned by its ability to provide comprehensive data on the scattering process. Examples of such data are the complete diffraction field (free oscillation field); the amplitudes of the wave packet components (amplitudes of the partial field components in the radiation zones); the working frequency (the projection of a real EF $\bar{\kappa}_m$ to the first sheet of the surface **K**); and the structure parameters at which the mode is realized.

As an example we suggest a result shown in Figure 6.19, but first we return to Figure 6.7. With decreasing δ, the spectral line 2 of an H_{023}-oscillation goes under the cut whose origin is the branch point κ_{-1}, and goes out into the upper half-space of the second sheet of **K**. A similar behavior is inherent in the spectral curve of

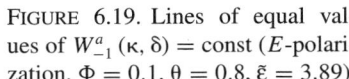

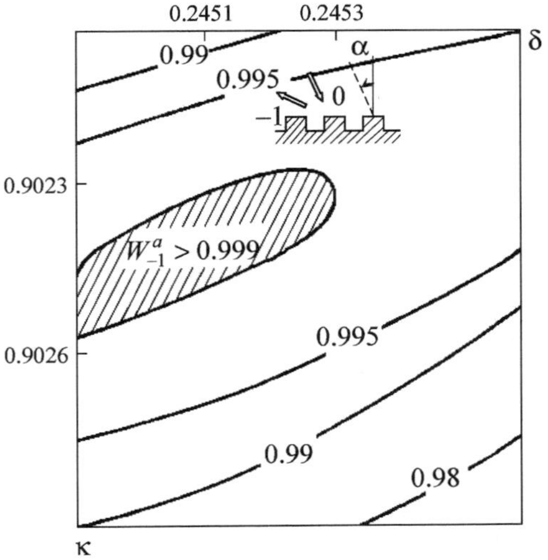

FIGURE 6.19. Lines of equal values of $W^a_{-1}(\kappa, \delta) = \text{const}$ (E-polarization, $\Phi = 0.1, \theta = 0.8, \tilde{\varepsilon} = 3.89$).

the H_{022}-oscillation. It crosses the real axis of this sheet at $\delta = \bar{\delta} \approx 0.245$ in the point $\kappa = \bar{\kappa} \approx 0.9024$. Here $\text{Re}\,\Gamma_0 > 0$, $\text{Im}\,\Gamma_0 = 0$, $\text{Re}\,\Gamma_{-1} < 0$, $\text{Im}\,\Gamma_{-1} = 0$ and for the rest n is $\text{Im}\,\Gamma_n > 0$, $\text{Re}\,\Gamma_n = 0$. The H_{022}-oscillation is not affected by the substitution of the symmetry plane of a semi-transparent grating by a perfectly conducting one. Hence, this oscillation is an eigenoscillation also for the reflective grating obtained as a result of such a substitution. This means that, on such a reflective structure, with the values of δ and κ being close to $\bar{\delta}$ and $\bar{\kappa}$, (but now in the physical domain of variation of κ), the effect of complete transformation of the first but one incident harmonic into the zero outgoing one and v.v. should take place. The feasibility of the inverse transformation is guaranteed by the reciprocity principle (see Section 6.2.1 and [3,6,11]). This is the working mode of the reflective grating. This fact is verified by the data in Figure 6.19, obtained by numerical solution of the diffraction problem (6.2) to (6.5).

In the simplest case considered, each of the packets $\tilde{U}_{m,j}$ contains only one homogeneous plane wave. The results of application of spectral analysis of the transformation properties of the gratings, in the case when the packet $\tilde{U}_{m,1}$ contains only homogeneous waves (one to three waves), and the packet $\tilde{U}_{m,2}$ contains an infinite set of inhomogeneous waves and one or two homogeneous waves, can be found in [9], together with the complete theoretical basis of this method. The theory and the results obtained by using this method for studying the transformation properties of open waveguide resonators are described in [191].

The ability of gratings to vary the transparency within maximal limits, and thus to controllably redistribute the energy between plane waves propagating in different directions, provides great possibilities for using them as polarization and frequency filters, location angle filters, screens, selective mirrors of dispersive open resonators, plane pattern-forming devices, and so on [9,21,173]. One of the major practical applications of the selective properties of gratings is in vacuum and

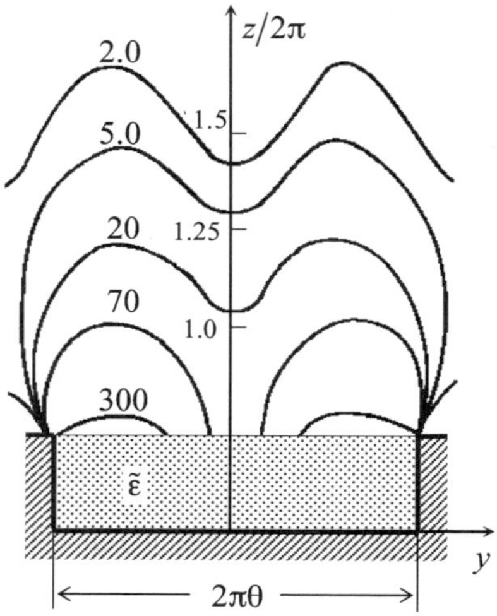

FIGURE 6.20. Anomalous rise of near field density when the value of excitation frequency κ is close to EF of one of the eigenmodes from the second family of oscillations (H-polarization, $\kappa = 0.1$, $\theta = 0.8$, $\delta = 0.7147$, $\alpha = 3°$, $\tilde{\varepsilon} = 51.2$).

solid-state electronics, in particular, diffraction electronics [192], whose devices and tools utilize the effect of transforming inhomogeneous waves (a field of the density-modulated electron beam) into homogeneous waves. The modeling of processes occurring in such devices and tools is based on the solution of the key problem about the diffraction of H-polarized homogeneous and inhomogeneous plane waves by reflective gratings of different types. The design criteria can certainly be expected to be very diverse. However, an appropriate combination of the methods of classical diffraction theory with those of spectral theory can be expected to be successful.

The high-Q resonances of the coupling channels at the higher propagating waves cause an abnormal increase in the near field strength of waveguide gratings [193]. The solution of one of the model problems that enabled us to digitize the corresponding effect is shown in Figure 6.20. The figure shows the lines of equal levels of $|E_y(g)|$ of the complete diffraction field normalized by the amplitude E_y^i of an incident field (H-polarization). The extraordinary increase of the field strength is due to the presence of a free oscillation at EF $\bar{\kappa} = 0.0978 - i4.81 \cdot 10^{-5}$. When the effect occurs, all the field components increase, which enhances their interaction with an electron beam. In such a way one can optimize the performance of some devices: diffraction radiation sources, resonance back-wave tubes, and others.

The second example is associated with establishing the existence of surface real waves in gratings with open channels of energy radiation into free space (see Section 6.3.1). We recall that for open periodic waveguides, the dimensionless frequency parameter κ determines the ratio between the actual period length l of the structure and the working wavelength λ. The occurrence of slow waves in the region "banned" for them (i.e. moving toward large κ) means that gratings whose fabrication has already become a routine practice, for example, for millimeter wavelength ranges, can be successfully used in the submillimeter wavelength ranges as well. This result is of great significance for spreading the possibilities of vacuum electronics into shorter wavelength ranges, as the production of operational periodic structures (in the traditional sense) tends to be more problematic.

The third example deals with the synthesis of reflective gratings for devices of relativistic diffraction electronics [9,189]. The problem is solved within the model problem describing the radiation properties of periodic structures being excited by a plane inhomogeneous H-polarized wave $\tilde{U}_p^i(g,\kappa) = H_x^i(g,\kappa)$ (see (6.1)) with the $\mathrm{Im}\,\Gamma_p > 0$ and $\mathrm{Re}\,\Gamma_p = 0$. Due to the excitation, the secondary field (radiation field) $H_x = \sum_{n=-\infty}^{\infty} R_{np} \exp\left[i\left(\Phi_n y + \Gamma_n(z - 2\pi\delta)\right)\right]$ occurs in the reflection zone of the gratings ($z > 2\pi\delta$). Some of the waves have numbers n such that $\mathrm{Im}\Gamma_n = 0$. These are energy-consuming spatial harmonics for which the energy is radiated into free space. The relation

$$\sum_n |R_{np}|^2 \frac{\mathrm{Re}\,\Gamma_n}{|\Gamma_p|} = 2\,\mathrm{Im}\,R_{pp} \tag{6.35}$$

for such an interaction between the field $\tilde{U}_p^i(g,\kappa)$ with the grating ($\mathrm{Im}\,\tilde{\varepsilon}(g) \equiv 0$) is an analogue of the energy conservation law (see Section 6.2.1 and [6,9,192]). The value on the righ-thand side of (6.35) characterizes the total radiated energy, and every component in the left-hand side determines the part of energy carried off by a corresponding harmonic (homogeneous plane wave). If we write $\beta = \kappa/(p - \kappa\sin\alpha)$, the link between the problem about inhomogeneous plane wave scattering and that of diffraction radiation (in terms of the concept of a specified current) as a result of excitation of the gratings by a density-modulated monochromatic electron beam becomes evident (see, e.g., [192]). In this case, β characterizes the relative electron velocity, λ—the wavelength of the beam modulation; and α (see Figs. 1.1A and B) determines the angle at which the part of energy lost by the beam is moving off the grating at the zero spatial harmonic. In order to determine the value $\bar{W}_0^R = |R_{0p}|^2 \mathrm{Re}\,\Gamma_0/|\Gamma_p|$ that characterizes the efficiency of scattering (diffraction radiation) in the direction α, $|R_{0p}|$ should be found.

Nonresonant scattering of inhomogeneous plane waves by gratings of different design is considered in considerable detail in [192,194]. In [9,189] attention is concentrated on the occurrence of resonances as β varies ($\beta > 0.9$: the beam velocity is close to the velocity of light). This is interesting, first of all, from the point of view of studying the possibility of developing efficient relativistic devices of diffraction electronics. For such values of β, the length of the excitation wave is comparable to the characteristic grating dimensions, and the key role is now played by various resonant effects associated with the excitation of electromagnetic field oscillation that are close to eigenoscillations. The solutions of model problems demonstrated from an engineering aspect most favorably trend toward smaller depth of grating profiling that is required for optimal radiation characteristics, for an increasing parameter β (beam velocity).

Requirements concerning the stability of the electron beam velocity have been formulated that allow us to avoid drops in the diffraction radiation intensity. Easy-to-make structures have been synthesized (combs and echelettes, see Figs. 6.21 and 6.22) that, for the given intervals of variation of β, provide a constantly high level of energy extraction and its concentration in a direction close to the normal.

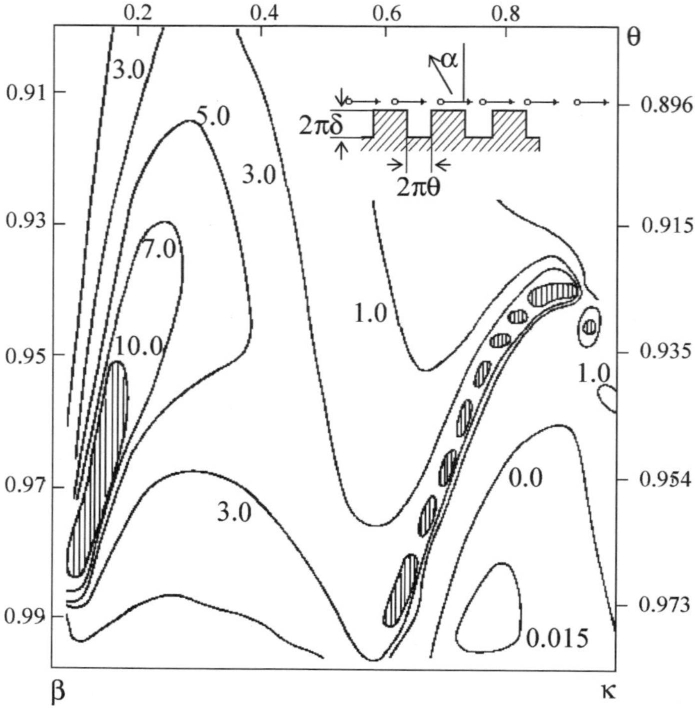

FIGURE 6.21. Lamellar grating: Intensity radiated in almost normal to grating direction. Lines of equal values of $\bar{W}_0^R\,(\beta, \theta)$, $\kappa = (\sin\alpha + 1/\beta)^{-1}$. The regions where $\bar{W}_0^R > 10$ are shaded ($\delta = 0.135$, $\alpha = 1^\circ$, $p = 1$).

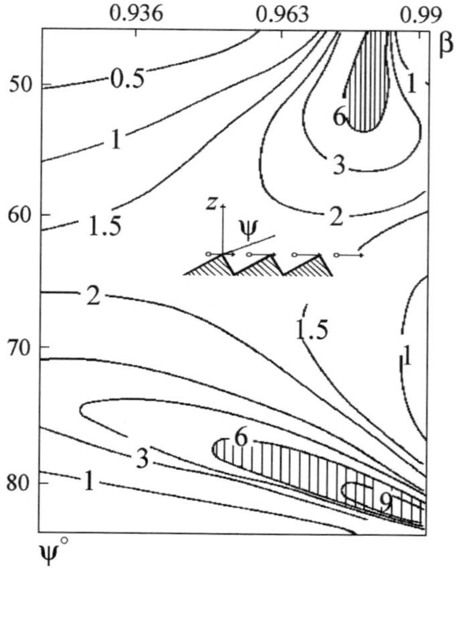

FIGURE 6.22. Zones of high radiation intensity for echelette gratings: Lines of equal values of $\bar{W}_0^R\,(\psi, \beta)$ ($\alpha = 1^\circ$, $p = 1$).

It was proven that at the echelettes, even in those cases when a periodic structure is transformed into an almost plane flake surface (see Fig. 6.22: $\psi \approx 75^\circ \div 85^\circ$), radiation characteristics can be reached that are close to those for a comb grating. This result is important, because, for shorter excitation waves, the production of comb-type gratings with an optimized profile becomes problematic, whereas the technology of producing echelettes is a routine practice, both in the optical and microwave wavelength ranges.

6.4. Gratings in Transient Wave Fields: Establishing of Regularities

6.4.1. Scattering of Narrowband Signals: Dynamical Images of Spectral Points in the Time Domain

Let the solution to the diffraction problem (6.1) to (6.5) (for any diffraction characteristic $W(\kappa)$) be known in a certain not too wide band of frequencies κ. We try to discover how the features of the processes of resonance wave scattering that were observed in the frequency domain (see Section 6.3.2), show up in the time domain. Such a trace of the individual anomalous or resonance scattering mode can easily be followed by imposing on the characteristic $W(\kappa)$ (on the excitation wave $\tilde{U}_p^i(g,\kappa)$ in the plane $z = 2\pi\delta$) a quite narrowband window $\tilde{v}_0(\kappa)$. The analogue of $\tilde{v}_0(\kappa)\, W(\kappa)$ in the time domain, that is, the characteristic $V(t)$, we obtain after the integral Fourier transform

$$V(t) = \int_{-\infty}^{+\infty} W(\kappa)\, \tilde{v}_0(\kappa)\, \mathrm{e}^{-i\kappa t} d\kappa.$$

Relations such as

$$\tilde{v}_0(\kappa) = \frac{1}{2\pi} \int_{-\infty}^{+\infty} V_0(t)\, \mathrm{e}^{i\kappa t} dt \leftrightarrow V_0(t) = \int_{-\infty}^{+\infty} \tilde{v}_0(\kappa)\, \mathrm{e}^{-i\kappa t} d\kappa \tag{6.36}$$

determine the correspondence between harmonic and nonharmonic signals, particularly, between the signals that excite the grating. Thus, for example, in the case of E-polarization for $p = 0$ and $\operatorname{Im}\Gamma_0 = 0$,

$$a_n(\kappa)\, \tilde{v}_0(\kappa) \leftrightarrow w_n(2\pi\delta, t) = w_n(t), \quad b_n(\kappa)\, \tilde{v}_0(\kappa) \leftrightarrow u_n(-2\pi\delta, t) = u_n(t),$$

where $w_n(z,t)$ and $u_n(z,t)$ are the spatio–temporal amplitudes of the electromagnetic field

$$U(g,t) = \left\{ \begin{matrix} U_0^i(g,t) \\ 0 \end{matrix} \right\} + \sum_{n=-\infty}^{\infty} \left\{ \begin{matrix} w_n(z,t) \\ u_n(z,t) \end{matrix} \right\} e^{i\Phi_n y} = \left\{ \begin{matrix} U_0^i(g,t) \\ 0 \end{matrix} \right\} + U^s(g,t);$$
$$\left\{ \begin{matrix} z \geq 2\pi\delta \\ z \leq -2\pi\delta \end{matrix} \right\}, \tag{6.37}$$

that is generated by the signal $U_0^i(g,t) = v_0(z,t)\exp(i\Phi_0 y)$; $z \geq 2\pi\delta$ such that $v_0(z,t) = V_0(t)$.

Choosing

$$\tilde{v}_0(\kappa) = \mathrm{e}^{-\tilde{\alpha}^2(\tilde{\kappa}-\kappa)^2}\mathrm{e}^{i\tilde{T}\kappa}; \quad \operatorname{Im}\tilde{\alpha} = \operatorname{Im}\tilde{T} = 0, \tag{6.38}$$

where $\tilde{\kappa}$ is the central frequency, we obtain

$$V_0(t) = \mathrm{e}^{-i\tilde{\kappa}(t-\tilde{T})}\frac{\pi^{1/2}}{\tilde{\alpha}}\mathrm{e}^{-(t-\tilde{T})^2/4\tilde{\alpha}^2}, \tag{6.39}$$

where $\tilde{\kappa}$ now determines the high-frequency content of the signal, and $\tilde{T}$ is the delay time. The requirement

$$\tilde{T} > 0; \quad |V_0(0)|/\left|V_0(\tilde{T})\right| = \text{const} \ll 1 \tag{6.40}$$

correlates the beginning of the contact between the signal (6.39) and the structure to the time $t = 0$. Note that the term "signal" here (unlike in [138]) is used irrespective of the analytic properties of the complex envelopes as a function of frequency or time. The values of the constant in (6.40) and of the parameter $\tilde{\alpha}$, defining the spectral content of the signal $V_0(t)$, are determined by the conditions of the computational experiments, especially by the effective width of the band encompassing all the peculiarities of the characteristic function $W(\kappa)$ in the frequency domain that are of interest for our study.

First, let us consider Figure 6.23, presenting the absolute values of the coefficients of transition ($|b_0(\kappa)|$) and reflection ($|a_0(\kappa)|$) of a grating of metal bars with dielectric loading of waveguide channels of the height $4\pi\delta$ (see Fig. 6.3A: $\theta = 0.8$, $\tilde{\varepsilon} = 3.89$, $\delta = 0.75$) that were obtained in the numerical solution of (6.1) to (6.5) for $\Phi = 0.1$.

The response of the grating placed in the field of a plane E-polarized wave of unit amplitude, up to the critical point of the H_{02}-wave in channel $\mathbf{Q}_\delta$ ($\kappa \approx 0.634$), is determined mainly by resonances associated with the excitation of

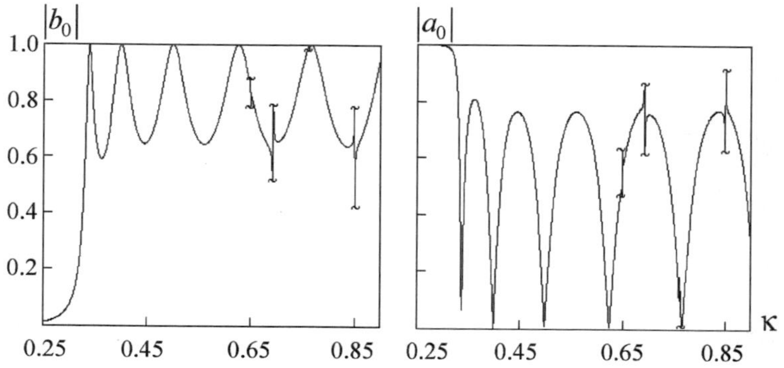

FIGURE 6.23. The changes of transparency of grating in the domain of frequency parameter κ variation providing regimes {1,1} and {1,2}; E-polarization.

quasi-eigen H_{01n}-oscillations. These resonances allow the incident wave to pass into the domain $z < -2\pi\delta$ without reflection. Above this point, in the frequency interval corresponding to the regime$\{1, 2\}$ (the first and second indices here indicate the number of propagating harmonics in regions $|z| > 2\pi\delta$, and in the region $|z| < 2\pi\delta$, respectively), the two different modes operate: pure and cross [7,195] ($\kappa \approx 0.76$) resonances at H_{01}- and H_{02}-waves that completely open and close the Floquet channel **R**. Not all the features of the functions $|b_0(\kappa)|$ and $|a_0(\kappa)|$ can be presented on the scale used in Figure 6.23. In the zones indicated by wavy lines, the curves are interrupted, and the missing details are given in Figures 6.24 to 6.28. The envelopes of the excitation signals in the frequency and time domains are marked with dashed lines.

The narrowband pulse $V_0(t)$ having the carrier frequency $\tilde{\kappa}$ from the frequency region $\{1,1\}$, passes through the grating almost without being distorted (see Fig. 6.24; we consider the pulse $u_0(t) \leftrightarrow b_0(\kappa)\,\tilde{v}_0(\kappa)$ that is the principal component of the transient field in the plane $z = -2\pi\delta$). The variation of $\arg b_0(\kappa)$ provides a physically correct time delay $\Delta t = 4\pi\delta\sqrt{\bar{\varepsilon}}$ in the observation of the excitation in the transition region of the structure. The high-frequency content as it is shown by the behavior of the function $\overline{\arg}\, u_0(t)$ (hereinafter, $\overline{\arg}\, V(t) = \arg\{V(t)\exp[i\tilde{\kappa}(t - \tilde{T})]\}$), remains through the whole duration of the pulse. If the amplitude center of $\tilde{v}_0(\kappa)$ does not coincide with the point that realizes the complete transition of a plane monochromatic wave, noticeable changes can take place in the pulse energy only: the absolute values of its spectral components will decrease.

The spectral amplitudes of the pulse $w_0(t) \leftrightarrow a_0(\kappa)\,\tilde{v}_0(\kappa)$ generated by the signal $V_0(t)$ in the reflection zone of the periodic structure are almost an order lower (see Fig. 6.24). The complete transition mode divides its amplitude center into two parts without considerably changing the high-frequency content. The pulse in the waist, that is, in the middle, shrinks to zero. This is confirmed by the jump of magnitude π of the value of $\overline{\arg}\, u_0(t)$. The qualitative analytic description of this effect yields the following approximate representation,

$$a_0(\kappa) \approx K \sin\left[(\kappa - \tilde{\kappa})\,\tilde{T}_\beta\right] e^{i\kappa\tilde{T}_\gamma}, \tag{6.41}$$

that allows for all the peculiar features in the behavior of $a_0(\kappa)$ in the spectral domain of the pulse $V_0(t)$: the absolute value of the local peaks (real number K), the zero points of $|a_0(\kappa)|$ and the positions of the jumps of $\arg a_0(\kappa)$ by 180° (by means of the real number $\tilde{T}_\beta$; in this case $\tilde{T}_\beta \approx 32$), the linear change of the phase $\arg a_0(\kappa)$ near the carrier frequency $\kappa = \tilde{\kappa}$ (the real number $T_\gamma \approx 30$). By using (6.41), we obtain

$$w_0(t) \approx \frac{K\pi^{1/2}}{2i\tilde{\alpha}} e^{-i\tilde{\kappa}(t-\tilde{T}-\tilde{T}_\gamma)} \left[e^{-(t-\tilde{T}-\tilde{T}_\gamma-\tilde{T}_\beta)^2/4\tilde{\alpha}^2} - e^{-(t-\tilde{T}-\tilde{T}_\gamma+\tilde{T}_\beta)^2/4\tilde{\alpha}^2} \right]. \tag{6.42}$$

The formula (6.42), in turn, gives a reasonably exact description of the pulse in the reflection region of the grating: the zero value of $|w_0(t)|$ at $t = \tilde{T} + \tilde{T}_\gamma$, the

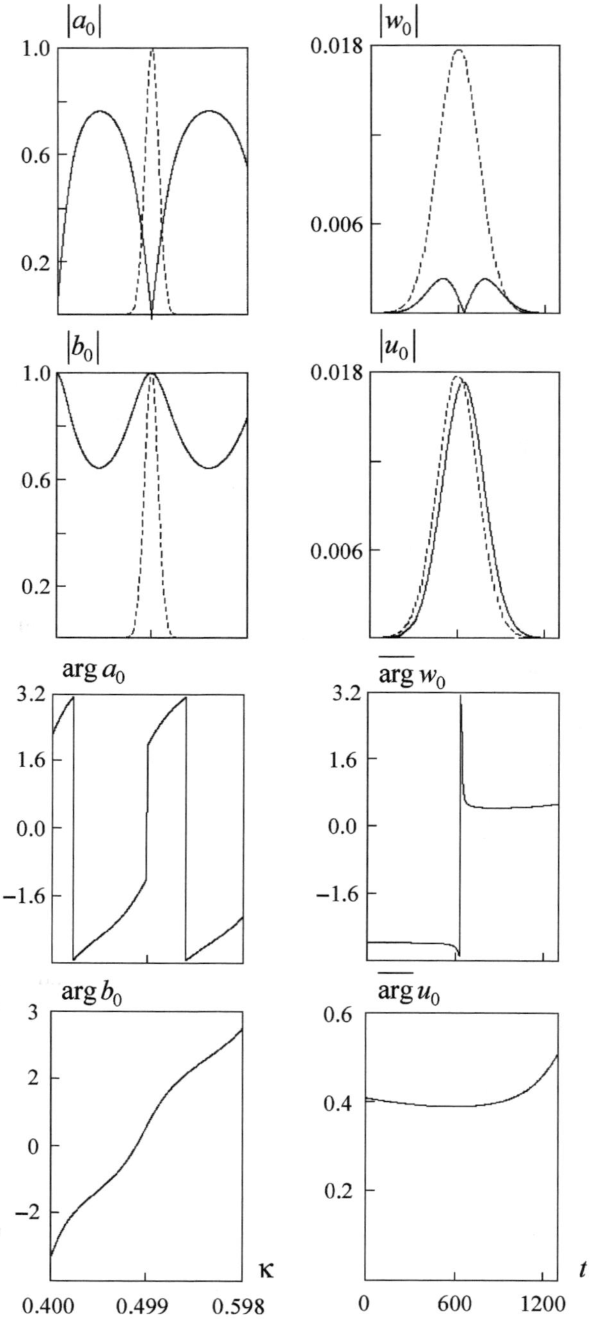

FIGURE 6.24. Reflection and transmission of narrowband signal with spectral content from the domain $\{1,1\}$: $\tilde{\alpha} = 100$; $\tilde{\kappa} = 0.499$.

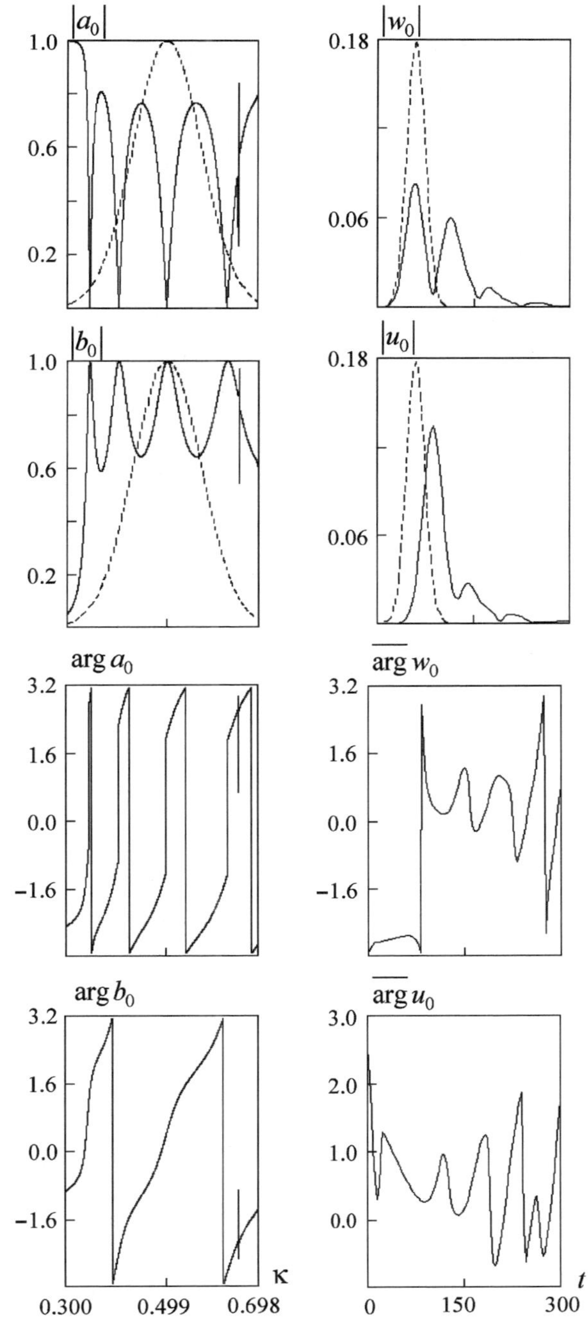

FIGURE 6.25. Transient processes in grating excited with narrowband pulse of shorter duration: $\tilde{\alpha} = 10$; $\tilde{\kappa} = 0.499$.

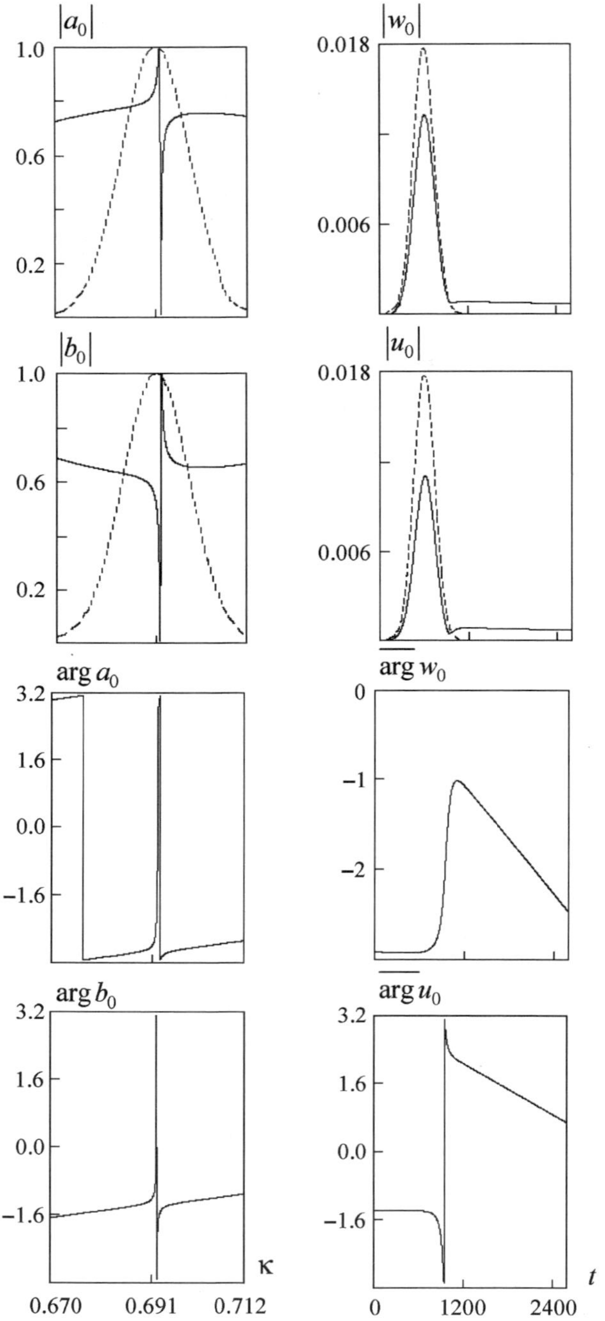

FIGURE 6.26. Reflection and transmission of narrowband pulse with spectrum from the region {1,2}: $\tilde{\alpha} = 100$; value of frequency parameter $\tilde{\kappa} = 0.691$ is close to the real part of H_{02}-mode EF.

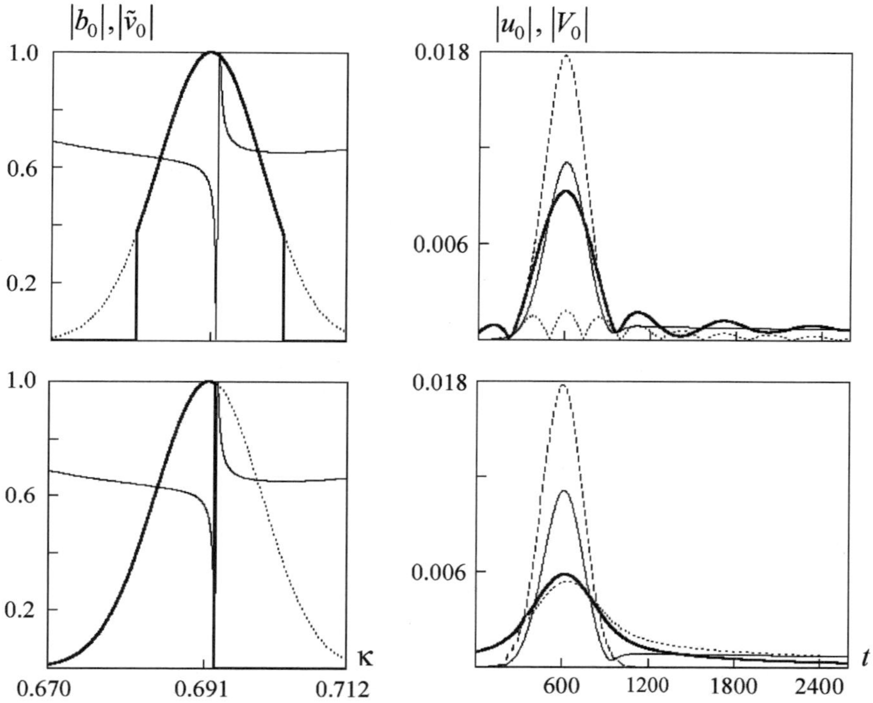

FIGURE 6.27. The grating's response to the changing of spectral content of pulse: $\tilde{\kappa} = 0.691$, $\tilde{\alpha} = 100$.

location of the bifurcated amplitude center, the position of $\overline{\arg}\, w_0(t)$ jump by 180°, and so on.

By reducing the value of the parameter $\tilde{\alpha}$, the duration of the excitation pulse $V_0(t)$ is substantially shortened (see Fig. 6.25), and the frequency band taken by this pulse is extended. If $\tilde{\alpha} = 10$, this band already contains several points, where the complete resonance transition modes of plane monochromatic waves are realized, and partially cover the region $\{1,2\}$, where the resonances of H_{02}-waves extend the range of possible values of the diffraction characteristics $a_0(\kappa)$ and $b_0(\kappa)$ to the maximum (the modes of complete transparency of the structure can be substituted by the modes of complete reflection of plane waves). According to (6.42), the pulse $V_0(t)$ is predictably deformed by reflection and transition: the amplitude center of the reflected pulse is bifurcated, as in the previous case, but the duration of the signal is increased significantly. The reason is that for $\tilde{\alpha} = 10$, the values $\tilde{T}$, $\tilde{T}_\beta$, and $\tilde{T}_\gamma$ become close ($\tilde{T} \approx 60$), and only the front part, before the waist, of the reflected pulse can be situated on the interval of times t corresponding to the duration of the efficient excitation pulse $V_0(t)$. According to (6.42), the high-frequency content of the pulse $w_0(t)$ should not change much. This is true for its front part (see the picture for $\overline{\arg}\, w_0(t)$), but the gradually attenuating tail is

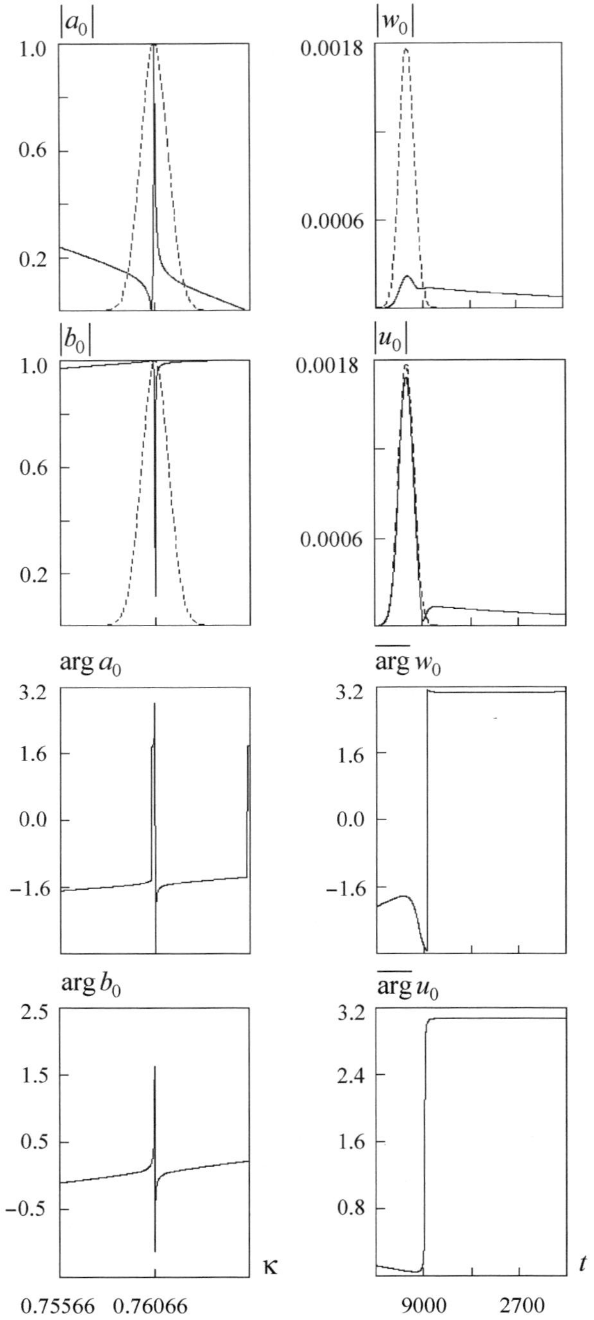

FIGURE 6.28. Reflection and transmission of narrowband pulse with spectrum covering the region of intercrossed resonance of H_{01}- and H_{02}-modes: $\tilde{\kappa} = 0.76066$, $\tilde{\alpha} = 1000$.

influenced by the resonance at the H_{02}-wave, that occurs in the spectral area of the excitation pulse and is not taken into account in the formula (6.41).

The transfer of the carrier $\tilde{\kappa}$ to the frequency region $\{1, 2\}$, that is, the resonant one, changes the pattern considerably. The resonance of an H_{02}-wave of the coupling channel $\mathbf{Q}_\delta$ (see Fig. 6.26) results in the occurrence, in the reflection and transition areas, of pulses $w_0(t)$ and $u_0(t)$, whose main parts differ only slightly from the excitation pulse $V_0(t)$: the energy spectrum is weakened, but the high-frequency content is retained. The main parts are followed by a tail, weakly attenuating in time, whose high-frequency content slightly differs from that of $V_0(t)$ and remains practically constant. These effects are described on the basis of the single pole representations (see [9] and formulas (6.19), (6.31)), that are valid in the vicinity of an isolated complex singular point of the resolvent of the elliptic boundary value problem. Let us apply one of them and represent the spectral components of the pulse $w_0(t)$ in the form

$$a_0(\kappa)\,\tilde{v}_0(\kappa) = \frac{1}{2}\tilde{v}_0(\kappa)\left[e^{i\arg C} + e^{i\arg D}\right] + i\,\mathrm{Im}(\bar{\kappa})\,e^{i\arg D}\left[\frac{\tilde{v}_0(\kappa)}{\kappa-\bar{\kappa}}\right]. \tag{6.43}$$

Here, C and D are complex numbers, and $\bar{\kappa} = \mathrm{Re}(\bar{\kappa}) + i\,\mathrm{Im}(\bar{\kappa})$ is the eigenfrequency of a free field oscillation at H_{02}-waves in the channel $\mathbf{Q}_\delta$. From (6.43) obtain

$$\begin{aligned} w_0(t) &= \frac{1}{2}(e^{i\arg C} + e^{i\arg D})V_0(t) + i\,\mathrm{Im}(\bar{\kappa})\,e^{i\arg D}\int_{-\infty}^{+\infty}\frac{\tilde{v}_0(\kappa)}{\kappa-\bar{\kappa}}e^{-i\kappa t}\,d\kappa \\ &= w_{0,1}(t) + w_{0,2}(t) \end{aligned}$$

As $(\kappa-\bar{\kappa})^{-1} \leftrightarrow -2\pi i\exp[-i\bar{\kappa}t]\,\chi(t)$, where χ is the Heaviside function, then

$$w_{0,2}(t) = \mathrm{Im}(\bar{\kappa})\,e^{i\arg D}\int_{-\infty}^{t} V_0(\tau)\,e^{-i\bar{\kappa}(t-\tau)}\,d\tau.$$

For $t > 2\tilde{T}$, the part $w_{0,2}(t) = \mathrm{const}(\tilde{\kappa}, \tilde{\alpha}, \bar{\kappa})\exp(-i\bar{\kappa}t)$ completely describes the tail of the pulse $w_0(t)$: the level of the envelope, small attenuation in time ($\mathrm{Im}\,\bar{\kappa} < 0$, $|\mathrm{Im}\,\bar{\kappa}| \ll 1$), change of the high-frequency content, and so on. By the changes of the high-frequency content one can judge about the deviation of the carrier of pulse $V_0(t)$ from the real part of the complex EF $\bar{\kappa}$. In the case considered (see the figure showing $\overline{\arg}\,w_0(t)$) it is $\tilde{\kappa} - \mathrm{Re}\,\bar{\kappa} \approx -10^{-3}$. Up to a certain time t within the interval $0 < t < 2\tilde{T}$, the part $w_{0,2}(t)$ is the background, and the main part of the pulse $w_0(t)$ is determined by the component $w_{0,1}(t)$, which differs from the excitation pulse $V_0(t)$ only by a constant complex factor. By smoothing the levels of $w_{0,1}(t)$ and $w_{0,2}(t)$, their contribution to $w_0(t)$ is partially compensated, and, for clear reasons, at this moment the phase envelope of the pulse changes rapidly.

The response of the grating to the excitation by a narrowband signal is predetermined by the dynamics of the elements of its spectral set Ω_κ. Any change of the envelope of the signal $\tilde{v}_0(\kappa)$ leads to easily predictable changes in the pulses $w_0(t)$

and $u_0(t)$. The analysis is carried out according to the same scheme, but here we should take into account the changes in the excitation pulse $V_0(t) \leftrightarrow \tilde{v}_0(\kappa)$. Thus, already before the numerical data are obtained, a qualitative analysis of the pulses $u_0(t)$ can be performed, whose amplitude envelopes are shown in Figure 6.27, in the right row of figures. In the left row $|b_0(\kappa)|$ are shown (with thin solid lines) and $|\tilde{v}_0(\kappa)|$, modules of complementary windows to fit the standard bell-shape (dotted and bold solid lines). The same lines in the right row of pictures show the pulses $u_0(t)$ generated by the corresponding pulses. The initial pulses (dashed lines) and pulses $u_0(t)$ generated by them (thin solid lines) under the standard shape (6.38) of the window $\tilde{v}_0(\kappa)$ are shown in those figures also.

The deformation of a narrowband pulse, whose energy spectrum is concentrated in the cross resonance region at H_{01}- and H_{02}-waves, is generally the same as in the cases discussed above (compare Figs. 6.26 and 6.28). The analytic description of such a deformation is given also by a single pole representation (with the pole at the point $\bar{\kappa}$, that is, an eigenfrequency of H_{02n}-oscillation), the second pole (the complex eigenfrequency of an H_{01n}-oscillation) is situated far from the real axis (see Section 6.3.1) and it cannot be taken into account if the window $\tilde{v}_0(\kappa)$ is very narrow. The behavior of the tails of the pulses $w_0(t)$ and $u_0(t)$ (see Fig. 6.28) allows us to state that in this case the amplitude center of the pulse $V_0(t)$ (the carrier frequency $\tilde{\kappa}$) coincides with $\mathrm{Re}\,\bar{\kappa}$, and the eigenfrequency $\bar{\kappa}$ is in the immediate proximity of the real κ-axis.

It can hardly be expected that for concrete values of θ, Φ, $\tilde{\varepsilon}$, δ, the spectral region of a narrowband pulse $V_0(t)$ covers the purely real ($\mathrm{Im}\,\bar{\kappa} = 0$) eigenfrequency of an H_{02n}-oscillation. But if it does happen, then the principal part of the pulses $w_0(t)$ and $u_0(t)$ resembles (with a decrease in the maximum) the pulse $V_0(t)$. The amplitude envelope of the tail will be parallel to the t-axis and its high-frequency content will be determined by the real EF $\bar{\kappa}$. Here the effect of capturing part of the energy of a pulse in the near field of the structure takes place. The analytic description of this effect can be obtained by using the technique applied in Section 6.2.4 and substituting a constantly operating current source by a momentary one. Note that the relative levels of the principal and tail parts of the pulses $w_0(t)$ and $u_0(t)$ either in this case or in those illustrated in Figures 6.26 to 6.28, can change within quite a wide range. The values of the complex constants C and D (see (6.43)) that are linearly connected to the expansion coefficients of the diffraction characteristics $a_0(\kappa)$ and $b_0(\kappa)$ in a Laurent series in a small neighborhood of the points $\bar{\kappa}$ of the spectrum Ω_κ have a decisive significance here.

6.4.2. Wideband Signal Scattering

A drastic extension of the spectral region of the signals requires an adequate revision of the methods and techniques used in the numerical analysis: the application of the Fourier transform becomes quite a problem due to the presence of high and super-high frequencies in the spectrum of the signal. The individual influence of natural resonances that was taken into account within the framework of the fully justified local representations for $\tilde{U}(g, \kappa)$ in small neighborhoods of points

$\bar{\kappa} \in \Omega_\kappa$, now gives way to the collective effect of the main part of elements composing the set Ω_κ. To separate the contribution of one or a certain single group of elements of Ω_κ and thus to understand the physics of the process for a relatively short time interval (the analysis of the transient states of the field), is quite a big challenge, requiring significant efforts to carry out comprehensive problem-oriented computation experiments and to develop a special method of processing and analyzing the results. The results presented herein and in the next section are actually only intended to illustrate that we are ready to deal with the following problem: to find a reliable background for an efficient physical analysis and satisfactory algorithms for visualization of transient states of fields in the near zone of gratings that are irradiated by wideband pulses. In Figures 6.29 and 6.30 this is

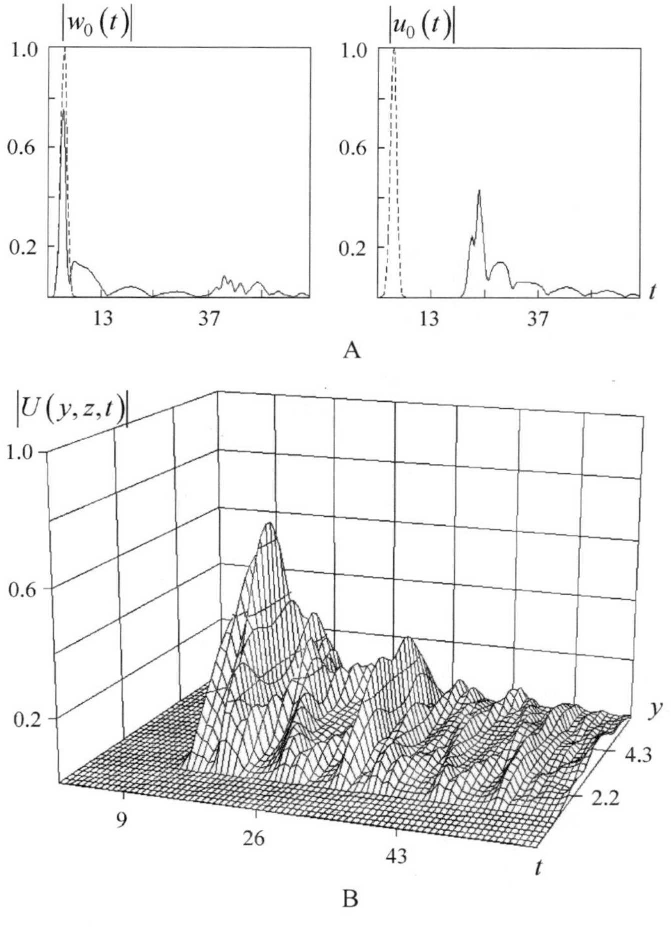

FIGURE 6.29. Grating in the field of wideband pulse: (A) Absolute values of complex-valued envelopes of pulses $U^i(g, t)$, $z = 2\pi\delta$ (dashed lines) and $w_0(t)$, $u_0(t)$ (solid lines); (B) pattern of the $U(y, z, t)$ in the plane of symmetry of structure $z = 0$.

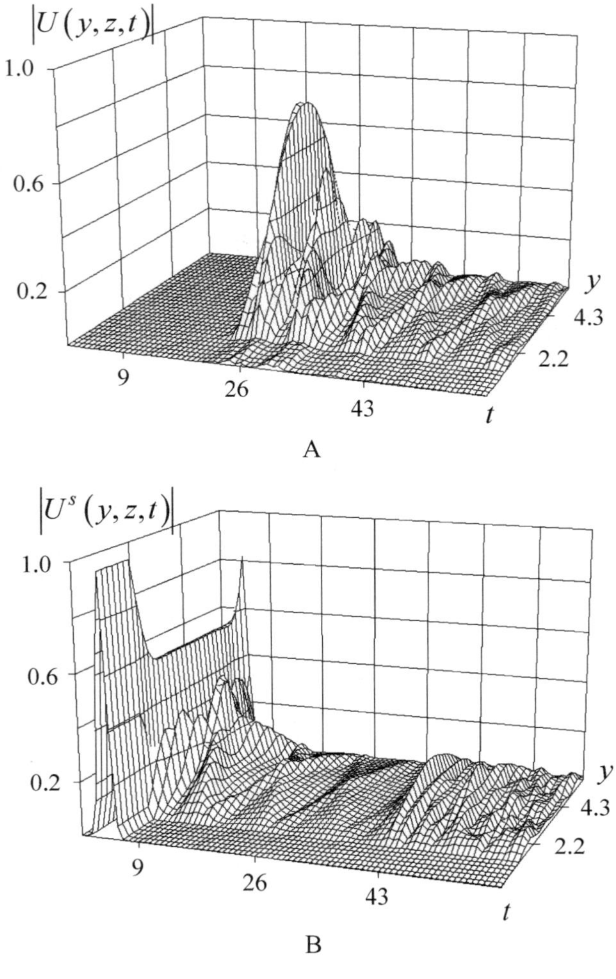

FIGURE 6.30. Grating in the field of wideband pulse: (A) Pattern of the field $U(y, z, t)$ in the plane $z = -2\pi\delta - \gamma$, $0 < \gamma \ll 1$; (B) reflected field $U^s(y, z, t) = U(y, z, t) - U^i(y, z, t)$ in the plane $z = 2\pi\delta$.

the pulse (E-polarization)

$$U^i(g, t) = f(t + z)\,\mathrm{e}^{i\Phi y} - \frac{\Phi}{2}\mathrm{e}^{i\Phi y} \times \int_{z-t}^{z+t} f(z_0)(t - z + z_0)\frac{J_1\left(\Phi\sqrt{t^2 - (z - z_0)^2}\right)}{\sqrt{t^2 - (z - z_0)^2}}dz_0,$$

where $f(t) = \exp[-(t - \tilde{T})^2/\tilde{\alpha}^2]\exp[-i\tilde{\kappa}(t - \tilde{T})]$; $\Phi = 0.1$; $\tilde{T} = 7.8$; $\tilde{\kappa} = 0.76066$; $\tilde{\alpha} = 1$, and J_1 is the Bessel function. Its amplitude envelope in the plane

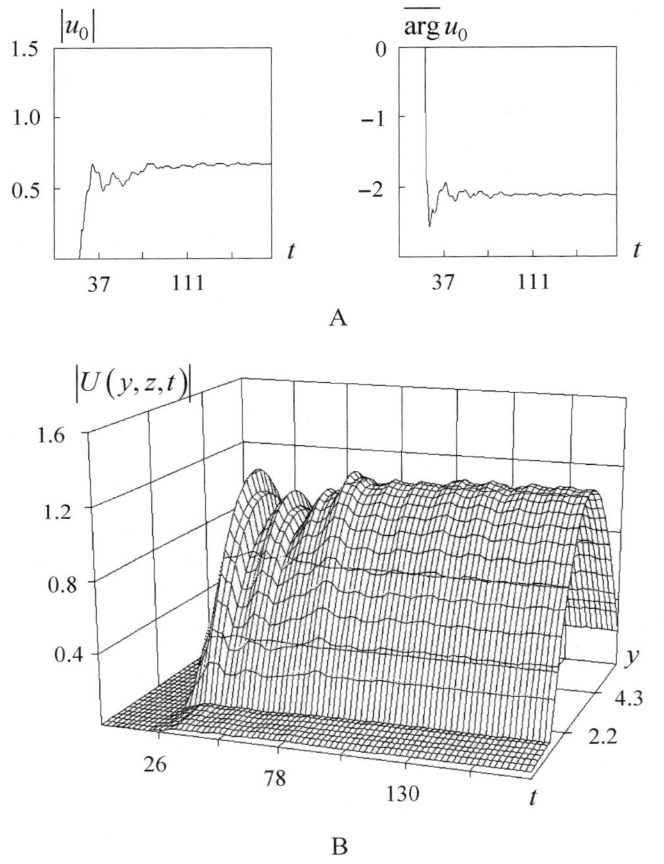

FIGURE 6.31. Grating in the field of continuously operating source: (A) Amplitude of the principal partial component in the transmission zone of structure; (B) field $U(y, z, t)$ in the plane $z = -2\pi\delta - \gamma$, $0 < \gamma \ll 1$.

$z = 2\pi\delta$ is plotted in Figure 6.29A with a dashed line. The value of the height δ, as well as the values of other parameters of the periodic structure remains unchanged (see Section 6.4.1: the lamellar grating with slots filled with a dielectric). The fields $U(g, t)$ (see formula (6.37)) are described by the solutions to the initial boundary value problem (6.18) with $\varepsilon = \mathrm{Re}\,\tilde{\varepsilon}$, where in this case $F(g, t) \equiv 0$ and $U(g, 0) = U^i(g, 0)$, and for Figure 6.31 we have $\varphi(g) = \psi(g) \equiv 0$. The pulses $w_0(t)$ and $u_0(t)$ are, as before, the amplitudes of the fundamental partial components of the field reflected by the grating and transient field in planes $z = 2\pi\delta$ and $z = -2\pi\delta$, tat is, the zero-order coefficients of Fourier expansion of the corresponding functions over the basis $\{\exp(i\Phi_n y)\}_n$, that is complete in the interval $0 \leq y \leq 2\pi$.

The central frequency $\tilde{\kappa}$ of the pulse $U^i(g, t)$ coincides with the real part of the eigenfrequency $\bar{\kappa}$ of the H_{02n}-oscillation. The imaginary part of $\bar{\kappa}$ is small (the cross-resonance zone at H_{02}- and H_{01}-waves). These circumstances do not significantly influence the space–time configuration of the field $U(g, t)$ for such an excitation and for the times considered. Results similar to those presented in Figures 6.29 and 6.30 were also obtained for values $\tilde{\kappa} = 0$; $\tilde{\kappa} = 0.499$ (the resonance in the grating slot at H_{01}-wave), and $\tilde{\kappa} = 0, 69182$ (pure resonance for an H_{02}-wave). The spectral region of the pulse is too wide, which eliminates the distinctions due to the relatively small shift of its amplitude center. These distinctions can most likely be exposed for large observation times t and for the field patterns, calculated all over the region $\mathbf{Q}_\delta$.

The result illustrated by Figure 6.31, is described analytically in Section 6.2.4. A continuous source (starting from $t = 0$)

$$F(g, t) = \begin{cases} 10\,(i\tilde{\kappa})^2\, e^{-i\tilde{\kappa}t} e^{i\Phi y}; & |z - 7.8| \leq 0.1 \\ 0; & |z - 7.8| > 0.1 \end{cases};$$

$$\tilde{\kappa} = 0.76066, \qquad \Phi = 0.1$$

generates a field beginning to sustain, in a while, a harmonic mode having the frequency $\kappa = \tilde{\kappa}$. Such a mode can be established quite soon due to the following three factors. The first one: $\tilde{\kappa}$ coincides (or almost coincides, see the last point in Section 6.4.1) with the real part of eigenfrequency $\bar{\kappa}$ of an H_{02n}-oscillation. The second one: the contribution to the common field of this oscillation does not match with the value of the imaginary part of its eigenfrequency inasmuch as the functions $F(g, t)$ and $\tilde{U}(g, \bar{\kappa})$, roughly speaking, belong to different symmetry classes with regard to the plane traversing the middle of the slots of the structure (the coefficient $A(f)\tilde{U}(g, \bar{\kappa})$ in expressions such as (6.27) is small). The third one: there are no competing harmonic components, inasmuch as the grating does not sustain free field oscillations with purely real eigenfrequencies (the first wave, connecting the reflection and transition zones, occurs only if $\kappa > 0.317$, i.e., in the region $\kappa > \Phi$, where the energy radiation channel into free space is already open), and the complex eigenfrequencies of other oscillations are much farther away from the real axis in the complex κ plane.

6.4.3. Visualization of Transient Fields: Reflecting Gratings

The analysis of transient fields is substantially simplified by their visualization in the region of space containing efficient scatterers, at different observation times t. An evolution of the field patterns is presented as a sequence of its most characteristic configurations, registering all essential features of the spatio-temporal field transformations, which allow us to associate the changes occurring with concrete reasons causing the changes.

The conclusions below are the result of an analysis of graphical data obtained in computational experiments on visualization of fields $U(g, t) = E_x(g, t)$ in the near zone of reflective gratings. The functions $U(g, t)$, describing the transient processes of the E-polarized field of classic one-dimensional periodic structures, are the solution to the following initial boundary value problems.

$$\begin{cases} \left[-\varepsilon(g)\dfrac{\partial^2}{\partial t^2}+\dfrac{\partial^2}{\partial z^2}+\dfrac{\partial^2}{\partial y^2}\right]U(g,t)=F(g,t,\tilde{\kappa}); \quad t>0, \quad g\in \mathbf{Q} \\ U(g,t)|_{t=0}=U^i(g,t)\big|_{t=0}=\varphi(g,\tilde{\kappa}), \\ \left.\dfrac{\partial U(g,t)}{\partial t}\right|_{t=0}=\left.\dfrac{\partial U^i(g,t)}{\partial t}\right|_{t=0}=\psi(g,\tilde{\kappa}); \quad g\in\bar{\mathbf{Q}} \\ U\left\{\dfrac{\partial U}{\partial y}\right\}(2\pi,z,t)=e^{i2\pi\Phi}U\left\{\dfrac{\partial U}{\partial y}\right\}(0,z,t), \quad U(g,t)|_{g\in\mathbf{S}}=0; \quad t\geq 0 \end{cases} \tag{6.44}$$

Here, ε is a relative permittivity of the dielectric filling the grooves; $\tilde{\kappa}$ is the amplitude center of the incident signal in frequency domain; $U^i(g,t)$ is the incident wave; $F(g,t,\tilde{\kappa})$ is the current source; and Φ is the parameter of the Floquet channel $\mathbf{R}=\{g=\{y,z\}:0<y<2\pi\}$. Furthermore $\Phi=0$, and thus $\operatorname{Im}U(g,t)=0$ for real $F(g,t,\tilde{\kappa})$, $\varphi(g,\tilde{\kappa})$, and $\psi(g,\tilde{\kappa})$. The analysis domain $\mathbf{Q}$ (the part of the strip $\mathbf{R}$) is bounded from below by the perfectly reflecting grating surface $\mathbf{S}$. The computation is carried out in the domain $\mathbf{Q}_L=\{g\in\mathbf{Q}:z<L\}$. In the reflection zone $z\geq 0$ (all the discontinuities of the Floquet channel $\mathbf{R}$ are located in the domain $z<0$) field $U(g,t)$ can be presented in the form (see Section 2.5.2),

$$U(g,t)=\sum_{n=-\infty}^{\infty} w_n(z,t)\,\mu_n(y); \quad \mu_n(y)=(2\pi)^{-1/2}e^{i\Phi_n y}, \quad \Phi_n=n+\Phi.$$

The problems (6.44) have been solved by means of a finite difference technique with the exact boundary conditions (2.61) on the boundaries $z=L>0$ in the reflection zones. The patterns of the field $U(g,t)$; $g\in\mathbf{Q}_L$ for various observation times t are presented in continuous-tone figures. The period of the grating has been chosen equal to 2π, the depth of grooves is $2\pi\delta$, $2\pi\theta$ is the width of lamellar grating grooves, and ϑ is the angle of the echelette right facet (counted clockwise from the axis z).

Figures 6.32 to 6.34 show reflective gratings of different types in the field of plane wave pulses $U^i(g,t)$, nonresonance scattering. For all structures shown in Figure 6.32 the value of δ is equal to 0.5, and the grooves are empty ($\varepsilon=1$). The profiles $\mathbf{S}$ of the gratings on the interval $0\leq y\leq 2\pi$ are described with functions:

$$z=-2\pi\delta\chi(\pi(1+\theta)-y)\chi(y-\pi(1-\theta))$$

(comb type gratings, $\theta=0.9$);

$$z=-\sqrt{\pi^2-(y-\pi)^2}$$

(the gratings with half cylindrical grooves);

$$z=\pi\delta(\cos y-1)$$

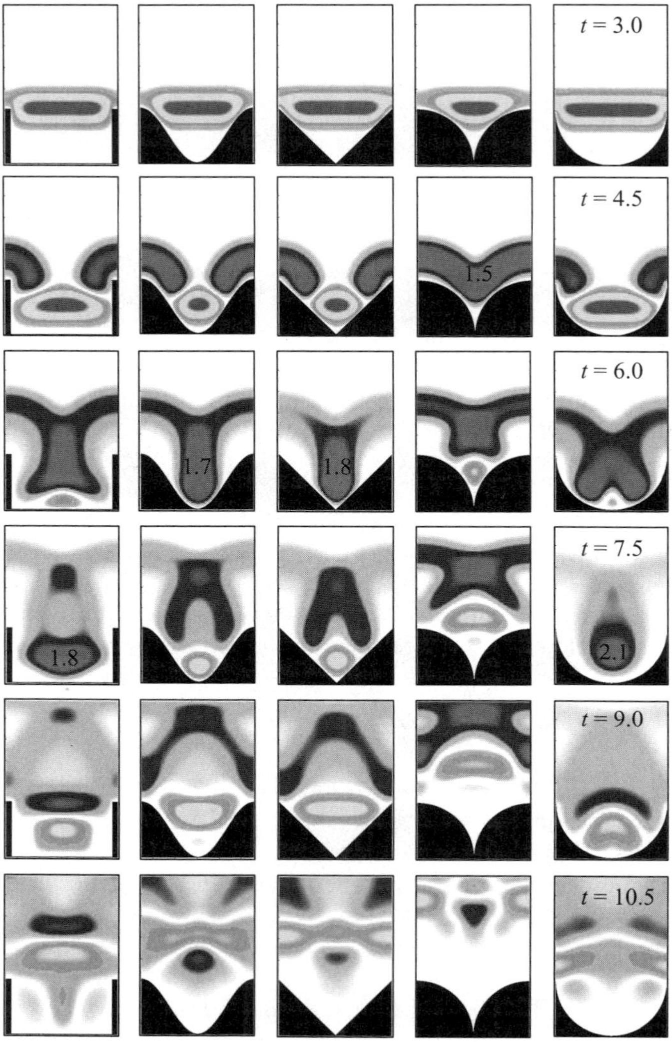

FIGURE 6.32. Reflective gratings in the field of plane pulsed wave (6.45): $\tilde{\alpha} = 1$, $\tilde{T} = 3$, $\tilde{\kappa} = 0.9$.

(sinusoidal gratings);

$$z = -ytg(\vartheta)\chi\left[(2\pi - y)\operatorname{ctg}(\vartheta) - y\operatorname{tg}(\vartheta)\right] - (2\pi - y)\operatorname{ctg}(\vartheta)\chi\left[y\operatorname{tg}(\vartheta) - (2\pi - y)\operatorname{ctg}(\vartheta)\right]$$

(echelette gratings with angles between facets $\vartheta = \pi/4$);

$$z = -\pi + \sqrt{\pi^2 - y^2}\chi(\pi - y) + \sqrt{\pi^2 - (y - 2\pi)^2}\chi(y - \pi)$$

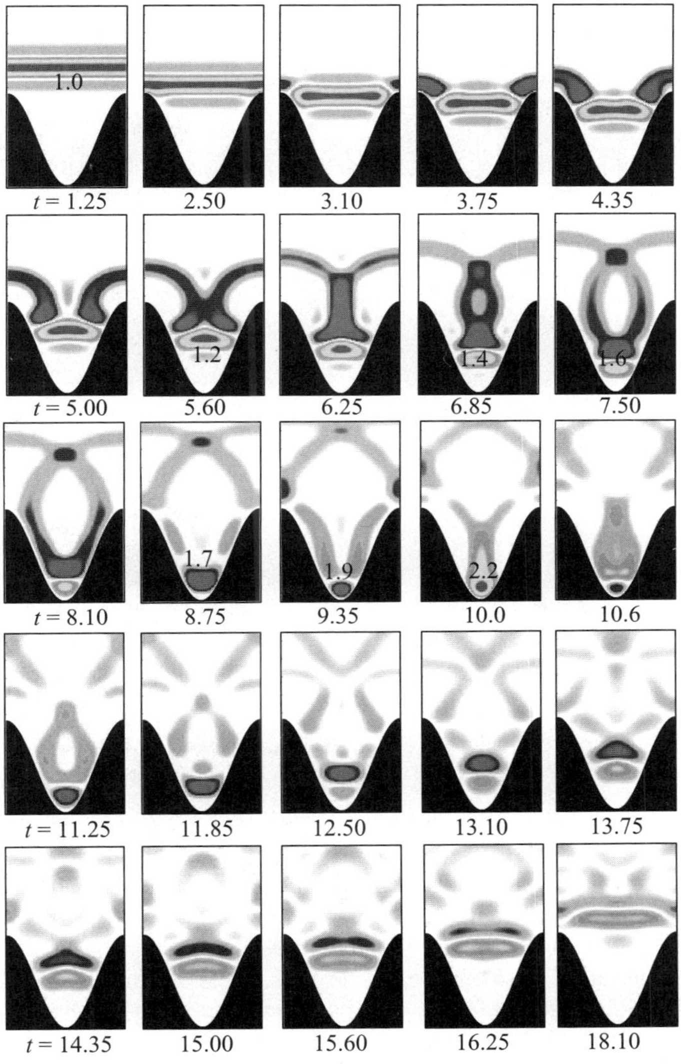

FIGURE 6.33. Sinusoidal ($z = \pi\,[\cos(y) - 1]$) grating in the field of plane pulsed wave (6.45): $\tilde{\alpha} = 1$, $\tilde{T} = 3$, $\tilde{\kappa} = 2$.

(gratings with half-cylindrical grids). The incident wave is the plane pulse

$$U^i(z,t) = \exp[-(t + z - \tilde{T})^2/\tilde{\alpha}^2]\cos[\tilde{\kappa}(t + z - \tilde{T})] \tag{6.45}$$

with $\tilde{\alpha} = 1$, $\tilde{T} = 3$, and $\tilde{\kappa} = 0.9$. The numbers in the figures indicate absolute values of the extreme values of field density.

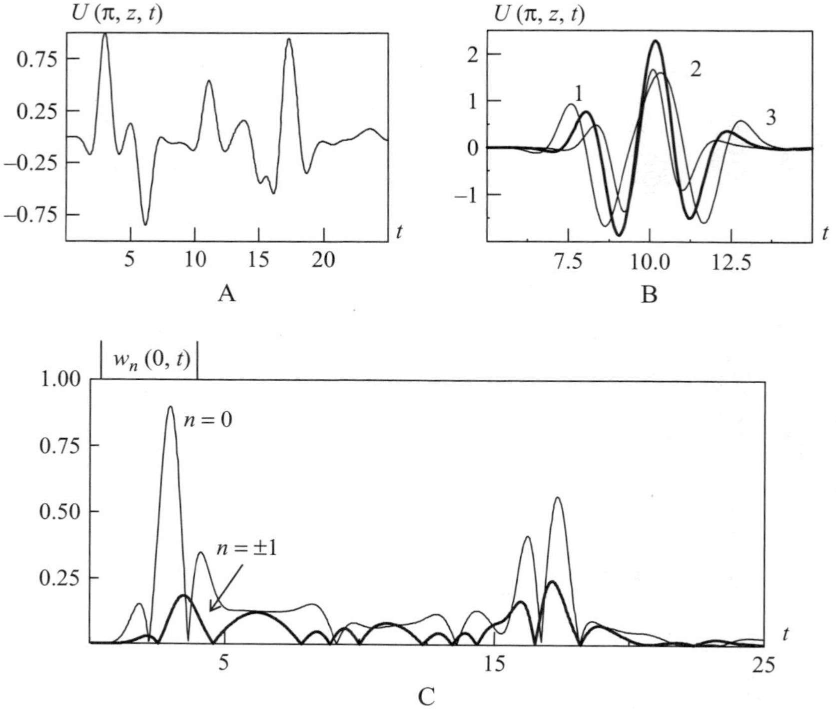

FIGURE 6.34. (Complementation to Fig. 6.33). (A, B) Time variations of field's density and (C) amplitudes of spatial harmonics of sinusoidal grating. (A) $z = 0$; (B) $z = -4.8\,(1), -5.3\,(2), -5.8\,(3)$.

Figure 6.33 illustrates a similar transient process with more details. Sinusoidal grating $z = \pi\,[\cos(y) - 1]$ ($\delta = 1$) is excited with the wave (6.45) with $\tilde{\alpha} = 1$, $\tilde{T} = 3$, and $\tilde{\kappa} = 2$. The numbers in the figures reflect the dynamics of the increase of the field density. Figure 6.34 completes the general picture with several linear graphs, describing the field density variation within the various intervals of the analysis domain $\mathbf{Q}_L$.

The local scattering centers generate partial components that, when added, lead to (at a certain point of space $g = \{y, z\} = g_{foc}$ and at a certain time $t = t_{foc}$) a maximum space-time focusing of the field $U(g, t)$:

$$\left|U\left(g_{foc}, t_{foc}\right)\right| = \max_{g,t} |U(g, t)|.$$

The value of the ratio $|U(g_{foc}, t_{foc})|/\max_g |U^i(g, 0)|$ depends on the number of scattering centers of the structure, and on their relative positions. In the cases considered the ratio varies from 1.1 to 2.2. If $t > t_{foc}$, regions with a relatively high level of $|U(g, t)|$ can exist long enough, and move quite large distances in space.

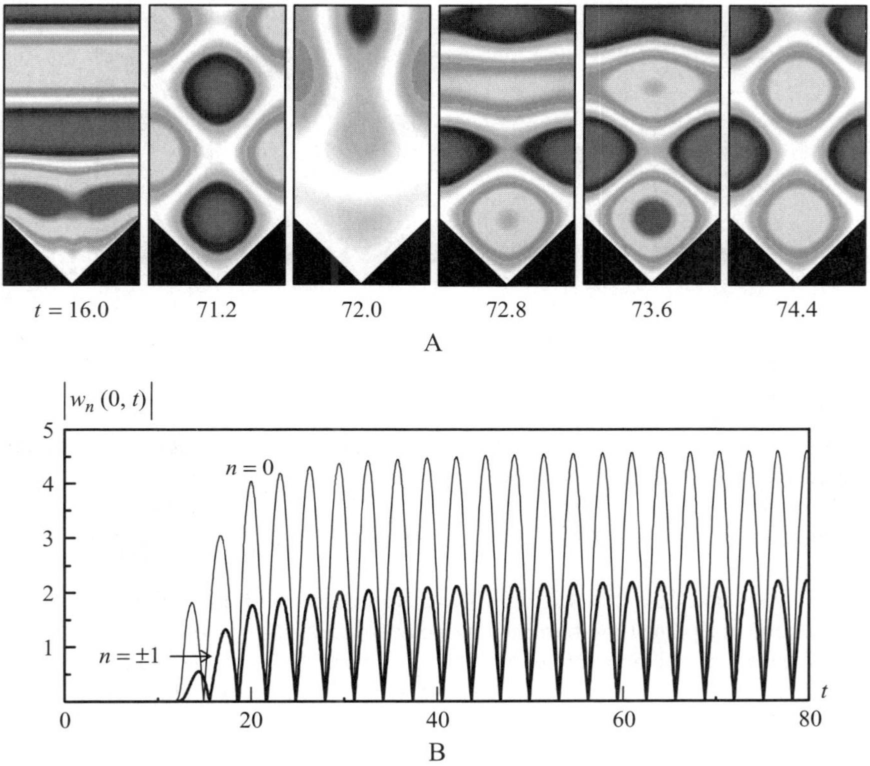

FIGURE 6.35. Geometrical resonance in the symmetric ($\vartheta = \pi/314$) echelette grating with 90° apex angle: (A) Pattern of near field in various instants of time t; (B) time–spatial amplitudes of principal harmonics.

Figures 6.35 and 6.36 demonstrate the implementation of the effects of the frequency domain in the time domain. Figure 6.35 presents the symmetric echelette excited with current source

$$F(g, t, \tilde{\kappa}) = 10\chi(12.5 - z)\chi(z - 12)\cos(\tilde{\kappa}t); \quad \tilde{\kappa} = 1, \quad \varphi = \psi \equiv 0.$$

The parameters for the computation experiment have been taken from Table 3 of the book [3], where several sections are devoted to the description of so-called geometric resonances of echelettes and gratings of the jalousie type: obvious solutions to the corresponding diffraction problems in the frequency domain with very simple (easy to guess) structures of scattered field for E- or H-polarization.

In the case under consideration, a so-called geometric resonance effect occurs: a superposition of four plane E-polarized time-harmonic waves $\tilde{U}_1 = \exp(-i\kappa z)$, $\tilde{U}_2 = \exp(i\kappa z)$, $\tilde{U}_3 = -\exp(-i\kappa y)$, and $\tilde{U}_4 = -\exp(i\kappa y)$ propagating in free space, produces a field whose zero-level line at $\kappa = 1$ can be matched with the

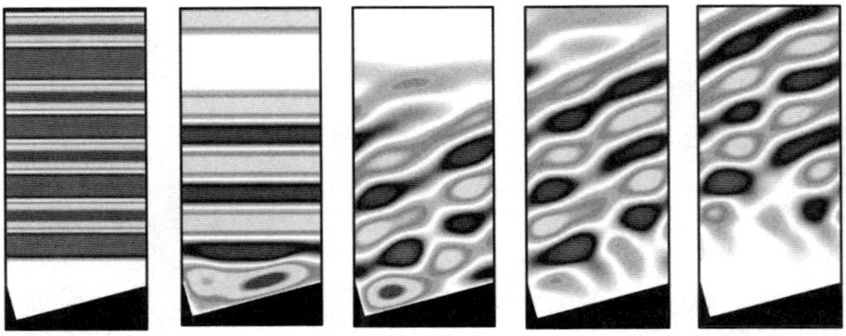

FIGURE 6.36. Nonspecula reflection of electromagnetic waves by nonsymmetric ($\vartheta = 75°$) echelette.

surface of a symmetric echelette with rectangular toothing. Hence, the complete diffraction field having a respective structure excited by the wave $\tilde{U}^i = \tilde{U}_1$, coincides with $\tilde{U} = \sum \tilde{U}_j$, whereas the waves $\tilde{U}_2$, $\tilde{U}_3$, and $\tilde{U}_4$ are the principal (zero), the first but one, and the first spatial harmonics in the field $\tilde{U}^s = \tilde{U} - \tilde{U}^i$ scattered by the grating.

In the second case (see Fig. 6.36), we are dealing with the effect of a strong conversion of a plane time-harmonic wave $\tilde{U}^i = \exp(-i\kappa z)$, normally incident on an asymmetric echelette, into one of the higher harmonics of the spatial spectrum of the structure. This effect is displayed through the distorted field symmetry and belongs to the effects of a complete (or almost complete) nonmirror reflection of waves by one-dimensional periodic gratings.

In Figure 6.36, the echelette is excited with a rectangular pulse of current

$$F(g, t, \tilde{\kappa}) = 10\chi(10.5 - z)\chi(z - 10)\chi(t_1 - t)\cos(\tilde{\kappa}t);$$
$$\tilde{\kappa} = 2.3, \quad \varphi = \psi \equiv 0.$$

The choice of the value $t_1 = 10.93$ allows us to get rid of the tail of the pulsed wave $U^i(g, t)$ that is generated by the source $F(g, t, \tilde{\kappa})$ (see Fig. 6.51 in Section 6.5.2), and to clearly describe in such a way the reflected wave for times $t > 20$, when the scattered field $U^s(g, t)$ is the dominant part of the total field $U(g, t) = U^i(g, t) + U^s(g, t)$. Parameters $\tilde{\kappa}$ and ϑ repeat the values of frequency and geometrical parameters, for which the effect of strong transformation of the plane monochromatic wave is realized that is normally exciting the nonsymmetric echelette into the spatial harmonic of the minus one order (see Fig. 105 in the book [3]).

The effects in the frequency domain that served as a basis of our analysis, are well studied (see, e.g., [3,6]). The choice of the amplitude centers of wideband signals in the computational experiments whose results are shown in Figures 6.35 and 6.36 guarantees that the main qualitative characteristics of the corresponding effects remain unchanged also in the processes in the time domain.

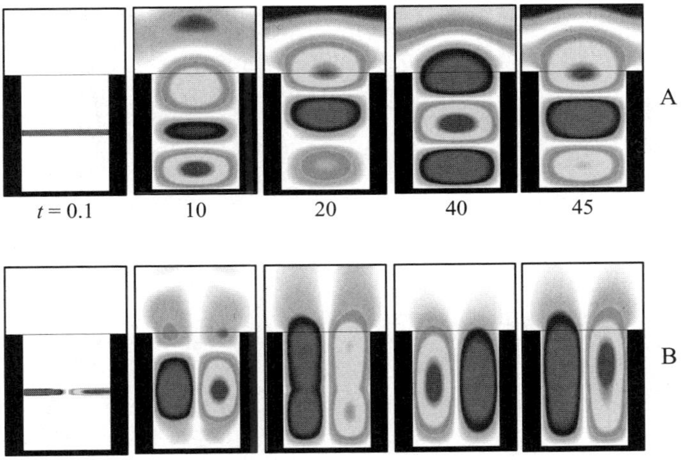

FIGURE 6.37. The excitations of the eigen (A) H_{013}- and (B) H_{021}-modes in the same configuration of grating with the same central frequency $\tilde{\kappa} = 0.74$ of incident signal.

The possible free field oscillations in a comb grating we visualize by Figures 6.37 and 6.38. Central frequency $\tilde{\kappa}$, configuration and location of current sources

$$F(g, t, \tilde{\kappa}) = 10\chi(-3 - z)\chi(z + 3.28)\chi(1.73\pi - y)\chi(y - 0.27\pi)\sin(y/2)\cos(\tilde{\kappa}t)$$

(see Fig. 6.37A) and

$$F(g, t, \tilde{\kappa}) = 10\chi(-3 - z)\chi(z + 3.28)\chi(1.73\pi - y)\chi(y - 0.27\pi)\sin(y)\cos(\tilde{\kappa}t)$$

(see Fig. 6.37B), and the structures parameters ($\theta = 0.73$, $\varepsilon = 3.89$, $\delta = 1$), are chosen so as to provide the most favorable conditions for the excitation in the grating of the oscillations that are close to natural ones (see Fig. 2.11 in the book [9]). There are no momentary sources. Points $g = \{y, z\}$ in the graphs of functions $U(g, t)$ in Figure 6.38 (and Figures 6.39 to 6.41; see below) coincide (or almost coincide) with local extreme points of the field density of the oscillations considered.

The central frequency $\tilde{\kappa} = 0.74$ of the current sources $F(g, t, \tilde{\kappa})$ is the same, but the geometry of the fields $U(g, t)$ is different and resembles the fields $\tilde{U}(g, \bar{\kappa})$ of free oscillations of different (H_{013}- and H_{021}-)modes from different symmetry classes. The deviation Φ from zero eliminates the difference in the symmetry classes; the oscillations whose EFs are very close, interact, and this can induce super-high-Q oscillations (see Section 6.3.1). In our case, the infinite Q-factor of an H_{021}-oscillation is conditioned by the difference in the symmetry classes for $\tilde{U}(g, \bar{\kappa})$ and the wave at which the energy can be reradiated into the reflection zone of the structure. The eigenfrequency $\bar{\kappa}$ is real, but $\tilde{\kappa} \neq \bar{\kappa}$.

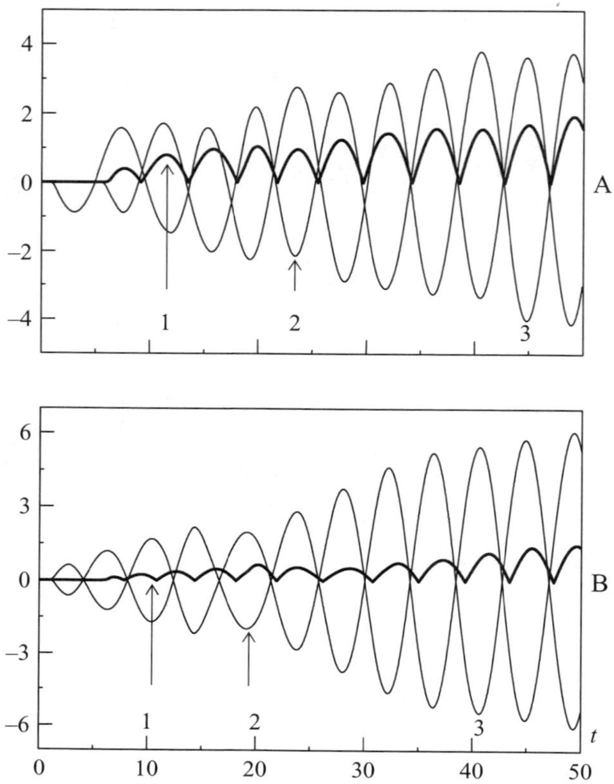

FIGURE 6.38. (See Fig. 6.37). Time–space field's characteristics of (A) H_{013}- and (B) H_{021}-eigenmodes: (A) $1 - |w_0\,(0, t)|$; $2 - U\,(\pi, -0.2, t)$; $3 - U\,(\pi, -2.6, t)$; (B) $1 - |w_{\pm 1}\,(0, t)|$; $2 - U\,(1.994, -2.5, t)$; $3 - U\,(4.286, -2.5, t)$.

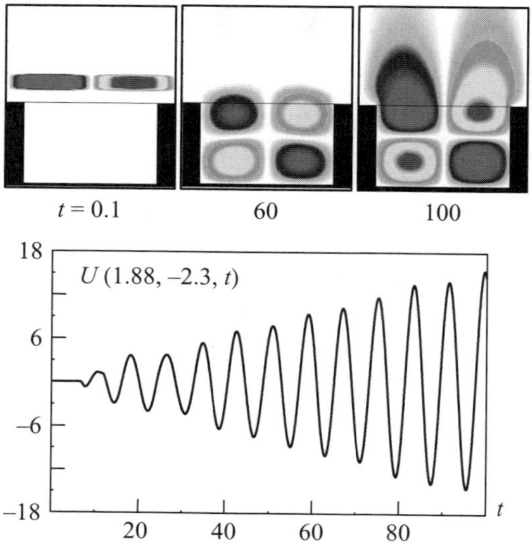

FIGURE 6.39. H_{022}-mode in lamellar grating.

FIGURE 6.40. H_{031}-mode: (1) $|w_0(1.5, t)|$; (2) $U(\pi, -1.4, t)$; (3) $U(1.47, -1.8, t)$.

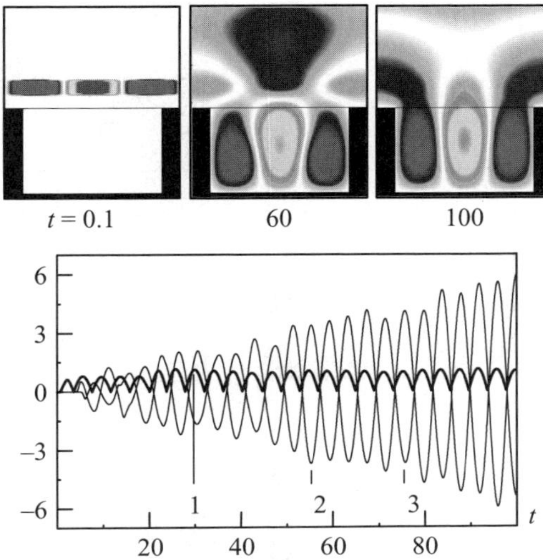

We draw such a conclusion on comparing the frequency characteristics of functions $w_{\pm 1}(0, t)$ (for sufficiently large t they oscillate at the frequency $\tilde{\kappa}$) and $U(g, t)$ in the local peak point of $\tilde{U}(g, \bar{\kappa})$ (the oscillation frequency of this function coincides with $\bar{\kappa}$). Hence, the increase of $|U(g, t)|$ is a temporary effect which will be superseded by a decrease of $|U(g, t)|$, that, in its turn, is followed by an increase and so on. All this depends on the phase at which the free oscillation field is fed by the source. The periodicity of this process with two single-frequency components is determined by the difference $\tilde{\kappa} - \bar{\kappa}$ (see Section 6.2.4).

The eigenfrequency $\bar{\kappa}$ for the oscillation $\tilde{U}(g, \bar{\kappa})$ of an H_{013}-mode is complex. The difference $\tilde{\kappa} - \operatorname{Re}\bar{\kappa}$ is here less than in the case of an H_{021}-oscillation. However, at large t the contribution of the eigenfree oscillation $\tilde{U}(g, \bar{\kappa})$ (of the single-frequency component with $\kappa = \operatorname{Re}\bar{\kappa}$) into the total field $U(g, t)$ becomes exponentially small, and the process enters the harmonic phase.

The current sources

$$F(g, t, \tilde{\kappa}) = 10\chi(z - 0.5)\chi(1 - z)\sin(y)\cos(\tilde{\kappa}t)$$

(for Fig. 6.39) and

$$F(g, t, \tilde{\kappa}) = 10\chi(z - 0.5)\chi(1 - z)\sin(1.5y)\cos(\tilde{\kappa}t)$$

(for Fig. 6.40) of a certain symmetry class, the same as in the case considered above, induce fields $U(g, t)$, whose structure resembles that of the free H_{022}- and H_{031}-oscillations. The real parts of eigenfrequencies $\bar{\kappa}$ of these oscillations almost coincide both with each other and with the central frequency $\tilde{\kappa} = 0.771$ of the sources $F(g, t, \tilde{\kappa})$. The imaginary part $\bar{\kappa}$ for an H_{022}-oscillation is zero. The parameters of the computational experiment ($\theta = 0.8$, $\varepsilon = 6.92$, and $\delta = 0.502$) derive from results presented in Figure 2.12 of the book [9].

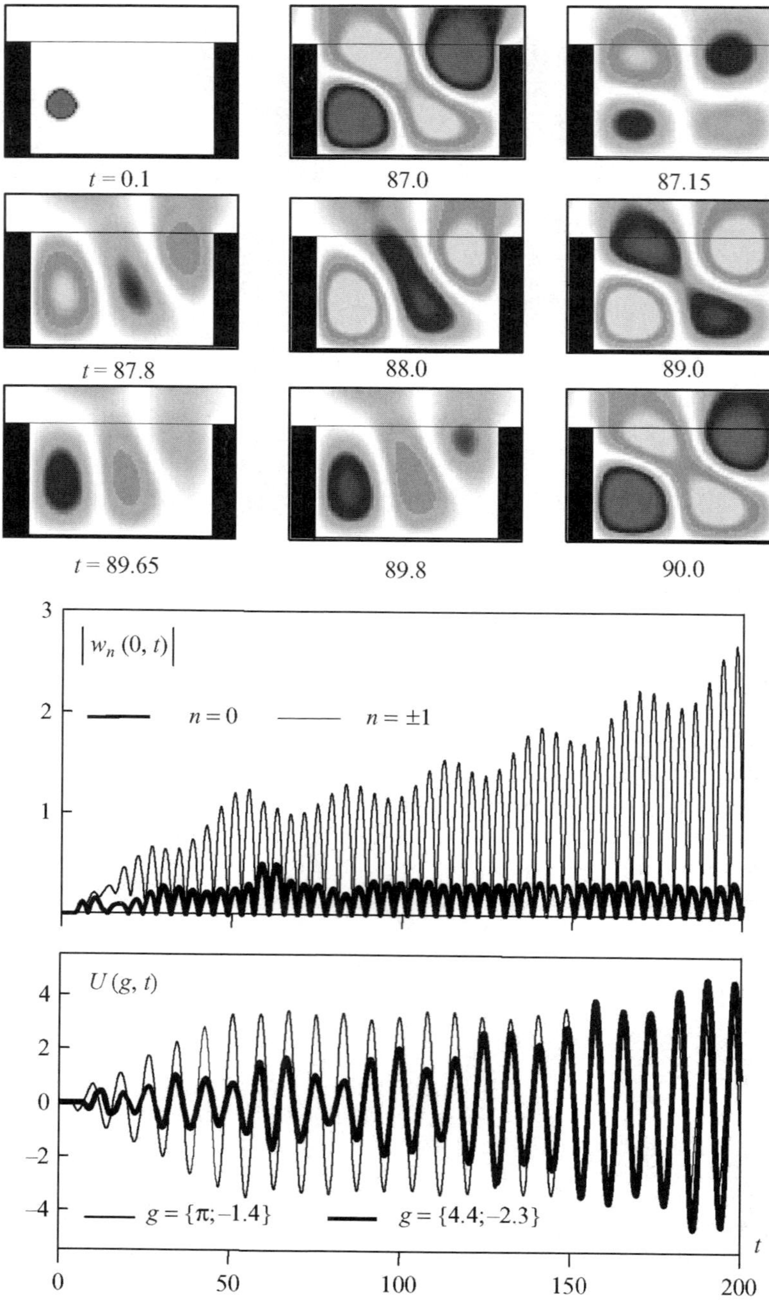

FIGURE 6.41. Forced mode coupling of eigenoscillations from different symmetry classes.

For $\Phi \neq 0$, the oscillation should be of one common symmetry class, and the frequency $\kappa = \tilde{\kappa}$ should determine the center of the interaction zone (intermode coupling zone). In the excitation problems, such interaction can be simulated also in the cases when $\Phi = 0$, by choosing the source $F(g, t, \tilde{\kappa})$ such that it contains components which are both symmetric and asymmetric with respect to the symmetry axis of the structure (and the initial boundary value problem as a whole). The result of such a simulation is presented in Figure 6.41. The field $U(g, t)$ at the initial stage of the process changes the structure dynamically, repeatedly passing the chain "quasi H_{022}-mode $\rightarrow$ hybrid mode $\rightarrow$ quasi H_{031}-mode $\rightarrow$ hybrid mode." After the source

$$F(g, t, \tilde{\kappa}) = 10\chi(0.125 - (z + 1.8)^2 - (y - 1.47)^2)\chi(50 - t)\cos(\tilde{\kappa}t);$$
$$\tilde{\kappa} = 0.771$$

is turned off, at the time $t = 50$ its residual field feeds mainly the component that is determined by the super-high-Q H_{022}-oscillation.

The general pattern of the process stays practically unchanged also at significant shifts of the central frequency $\tilde{\kappa}$ of the source $F(g, t, \tilde{\kappa})$, only if by such shifts no EFs of other high-Q free oscillations that have not taken active part in the process do not happen to be swept up. Only the field $U(g, t)$ is split into two essentially single-frequency components by the time $t = 50$, and the excitation level of the one that dominates in the grating slot ($\kappa = \mathrm{Re}\,\bar{\kappa}$), decreases according to the variations in the energy spectrum of $F(g, t, \tilde{\kappa})$. The second component, having the frequency $\kappa = \tilde{\kappa}$, dominates in the reflection zone of the structure. After the source is turned off, it leaves the near field of the structure, and all main characteristics of the process are determined here by the fields and EFs of free oscillations of H_{022}- and H_{031}-modes. Most obviously, the effect of substitution of one single-frequency component in the near field of the structure for an other, is illustrated in the concluding example of Section 6.5.2: the field of the source gives way to the field of the harmonic free oscillation $\tilde{U}(g, \bar{\kappa})$ with the real eigenfrequency $\bar{\kappa} \neq \tilde{\kappa}$.

6.5. Pulse Deformations by Free Propagation in Regular Sections of the Floquet Channels

The pulse is deformed, not only due to scattering by resonant inhomogeneities of the Floquet channel, but also during free propagation through finite regular sections of the corresponding waveguide channel. In such a case some specific factors are decisive, and they should be taken into account when designing functional units intended to operate with signals of certain shape and duration. To analyze this problem, we use well-known and frequently applied methods for studying pulse propagation in homogeneous (dispersive and absorbing) media and transmission lines. The most concentrated reviews of the relevant results are presented in [138,196,197].

6.5.1. Deformation of Narrowband Signals

Assume that the E- or H-polarized field

$$U^s(g,t) = \sum_{n=-\infty}^{\infty} w_n(z,t)\, e^{i\Phi_n y}; \quad z \geq 0$$

is formed by a grating (see formula (6.37)) and determined in the plane $z = 0$ of the cross-section of the Floquet channel $\mathbf{R}$ by the values

$$w_n(0,t) = F^{-1}[R_n(\kappa)](t) = \int_{-\infty}^{\infty} R_n(\kappa)\, e^{-i\kappa t} d\kappa \leftrightarrow R_n(\kappa)$$

of the elements $w_n(z,t)$ of its evolutionary basis (see Section 5.2.2). By studying the changes in $U^s(g,t)$ as a transient wave propagates into the region of large z, one can proceed either from the frequency

$$w_n(z,t) = \int_{-\infty}^{\infty} R_n(\kappa)\, e^{i(\Gamma_n z - \kappa t)} d\kappa \leftrightarrow R_n(\kappa)\, e^{i\Gamma_n z}, \tag{6.46}$$

or directly from the spatio–temporal representations:

$$w_n(z,t) = -\int_0 J_0\{\Phi_n[(t-\tau)^2 - z^2]^{1/2}\}\,\chi[(t-\tau)-z]\left[\left.\frac{\partial w_n(z,\tau)}{\partial z}\right|_{z=0}\right] d\tau; \tag{6.47}$$

$$w_n(z,t) = w_n(0,t-z) - \Phi_n z \int_0 w_n(0,\tau)\, \frac{J_1\left(\Phi_n\sqrt{(t-\tau)^2 - z^2}\right)}{\sqrt{(t-\tau)^2 - z^2}}\,\chi(t-\tau-z)\, d\tau. \tag{6.48}$$

The first one is obvious (see, e.g., (6.10) and (6.37)). The second is the exact radiation condition for the elements of the evolutionary basis of the signal propagating toward increasing z in a regular semi-infinite Floquet channel (see Section 2.5.2). The third follows from the first one, using the properties of the convolution of the Fourier transform:

$$\begin{aligned} w_n(z,t) &= \int_{-\infty}^{\infty} R_n(\kappa)\, e^{i\left(z\sqrt{\kappa^2-\Phi_n^2} - \kappa t\right)} d\kappa \\ &= \frac{1}{2\pi}\left(\int_{-\infty}^{\infty} R_n(\kappa)\, e^{-i\kappa t} d\kappa\right) * \left(\int_{-\infty}^{\infty} e^{iz\sqrt{\kappa^2-\Phi_n^2}}\, e^{-i\kappa t} d\kappa\right) \\ &= \frac{w_n(0,t)}{2\pi} * \left(\int_{-\infty}^{\infty} e^{i\sqrt{\kappa^2-\Phi_n^2}\, z}\, e^{-i\kappa t} d\kappa\right). \end{aligned} \tag{6.49}$$

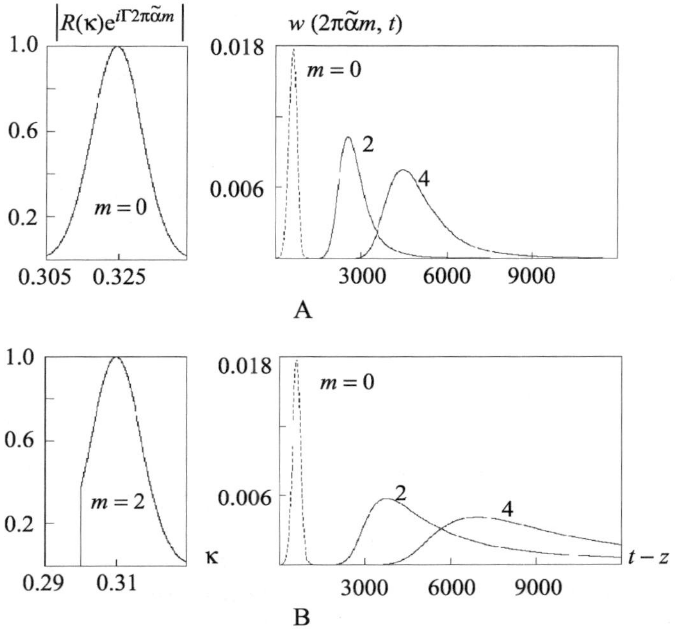

FIGURE 6.42. Propagation of narrowband pulse (6.46), (6.50) along regular Floquet channel: $\tilde{\alpha} = 100$, $\Phi = 0.3$, $\tilde{T} = 600$. (A) $\tilde{\kappa} = 0.325$; (B) $\tilde{\kappa} = 0.31$.

Substituting

$$\begin{aligned}\int_{-\infty}^{\infty} e^{iz\sqrt{\kappa^2-\Phi_n^2}} e^{-i\kappa t} d\kappa &= \frac{\partial}{\partial z} \int_{-\infty}^{\infty} \frac{e^{iz\sqrt{\kappa^2-\Phi_n^2}}}{i\sqrt{\kappa^2-\Phi_n^2}} e^{-i\kappa t} d\kappa \\ &\overset{-i\kappa=p}{=} -2\pi \frac{\partial}{\partial z}\left(\frac{1}{2\pi i}\int_{-i\infty}^{i\infty} \frac{e^{-z\sqrt{p^2+\Phi_n^2}}}{\sqrt{p^2+\Phi_n^2}} e^{pt} dp\right) \\ &= -2\pi \frac{\partial}{\partial z}\left(J_0\left(\Phi_n\sqrt{t^2-z^2}\right)\chi(t-z)\right) \\ &= 2\pi\left(\delta(t-z) - \Phi_n z \frac{J_1\left(\Phi_n\sqrt{t^2-z^2}\right)}{\sqrt{t^2-z^2}}\chi(t-z)\right)\end{aligned}$$

into (6.49) and assuming that $w_n(0,t) = 0$ for $t < 0$, we obtain (6.48). Here, as before, J_m is the Bessel function and χ is the Heaviside function. For dissipative media instead of (6.47) one can use the exact radiation conditions, similar to those derived in Section 4.5.

In the following analysis, for convenience we omit the index n in the functions w_n, Γ_n, Φ_n, and R_n. The results shown in Figures 6.42 to 6.45 (narrowband pulses),

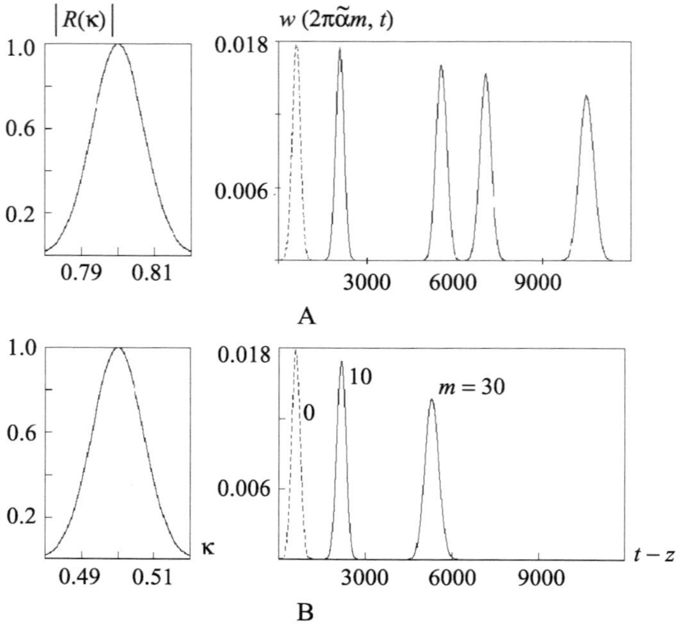

FIGURE 6.43. The influence of high frequency filling onto pulse (6.46), (6.50) propagation: $\tilde{\alpha} = 100$, $\Phi = 0.3$, $\tilde{T} = 600$. (A) $\tilde{\kappa} = 0.8$; (B) $\tilde{\kappa} = 0.5$.

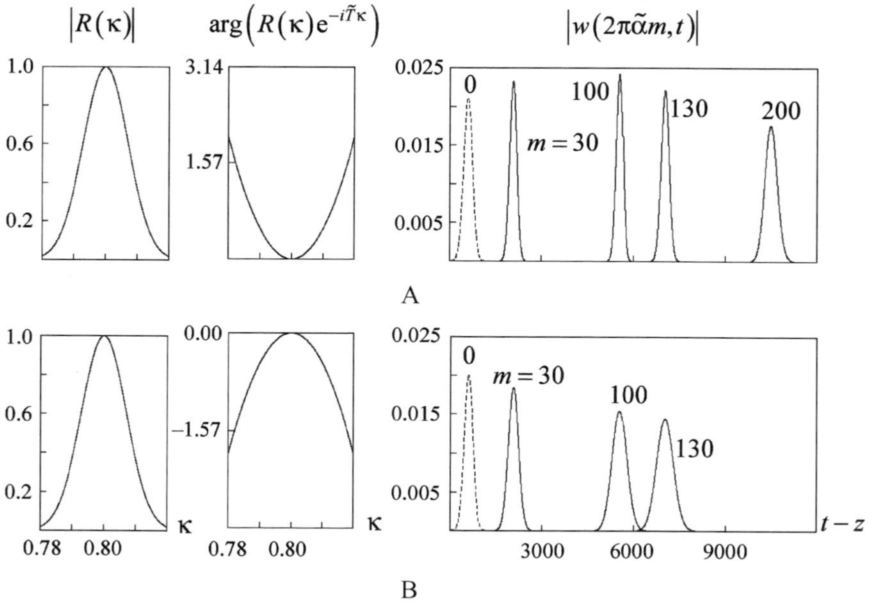

FIGURE 6.44. Propagation of narrowband Gaussian pulse (6.46), (6.51) with quadratic phase modulation: $\tilde{\alpha} = 100$, $\Phi = 0.3$, $\tilde{T} = 600$, $\tilde{\kappa} = 0.8$. (A) $\tilde{\beta} = 1/4\tilde{\alpha}^2$; (B) $\tilde{\beta} = -1/4\tilde{\alpha}^2$.

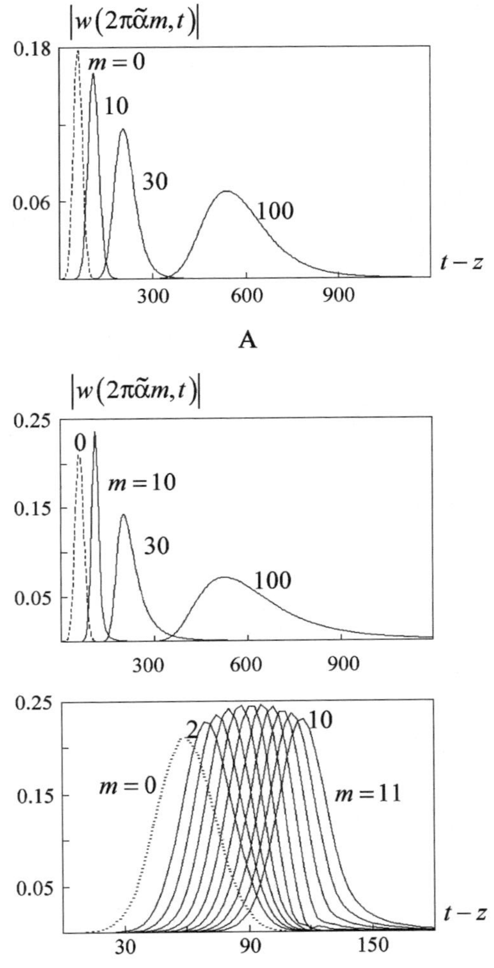

FIGURE 6.45. Time–space focusing of pulse with quadratic phase modulation: $\tilde{\alpha} = 10$, $\Phi = 0.3$, $\tilde{T} = 60$, $\tilde{\kappa} = 0.8$. (A) $\tilde{\beta} = 0$; (B) $\tilde{\beta} = 1/4\tilde{\alpha}^2$.

are obtained from formula (6.46). As $R(\kappa)$, we have chosen the functions

$$R(\kappa) = \mathrm{e}^{-\tilde{\alpha}^2(\kappa-\tilde{\kappa})^2}\mathrm{e}^{i\tilde{T}\kappa} \leftrightarrow w(0,t) = \frac{\sqrt{\pi}}{\tilde{\alpha}}\mathrm{e}^{-\left(t-\tilde{T}\right)^2/4\tilde{\alpha}^2}\mathrm{e}^{-i\tilde{\kappa}\left(t-\tilde{T}\right)}, \tag{6.50}$$

$$R(\kappa) = \mathrm{e}^{-\tilde{\alpha}^2(\kappa-\tilde{\kappa})^2/\left(1+4i\tilde{\beta}\tilde{\alpha}^2\right)}\mathrm{e}^{i\tilde{T}\kappa}$$

$$\leftrightarrow w(0,t) = \frac{\sqrt{\pi\left(1+4i\tilde{\beta}\tilde{\alpha}^2\right)}}{\tilde{\alpha}}\mathrm{e}^{-\left(t-\tilde{T}\right)^2/4\tilde{\alpha}^2}\mathrm{e}^{-i\tilde{\kappa}\left(t-\tilde{T}\right)}\mathrm{e}^{-i\tilde{\beta}\left(t-\tilde{T}\right)^2}. \tag{6.51}$$

The maximum of the function $w(0,t)$ is reached for $t = \tilde{T}$ and its significant values are concentrated on the interval $[\tilde{T} - 6\tilde{\alpha}; \tilde{T} + 6\tilde{\alpha}]$ for t. The values $w(0,t)$ outside this interval are less than $w\left(0,\tilde{T}\right) \cdot 10^{-3}$. The frequency function $R(\kappa)$

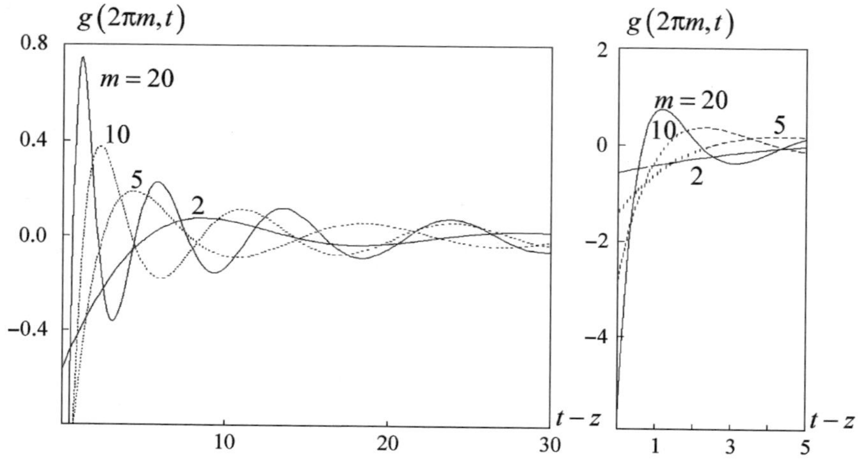

FIGURE 6.46. The tail of pulse (6.55): $\Phi = 0.3$; $m = 2, 5, 10, 20$.

has its maximum for $\kappa = \tilde{\kappa}$, and we assume it to be non-zero only for values of κ from the interval $[\tilde{\kappa} - 3/\tilde{\alpha}; \tilde{\kappa} + 3/\tilde{\alpha}]$. Outside this interval, $R(\kappa) < R(\tilde{\kappa}) \cdot 10^{-3}$.

For wideband pulses (see Figs. 6.46 to 6.49), the functions $w(z, t)$ were calculated from formula (6.48). The delay time $\tilde{T}$ was chosen so that the function

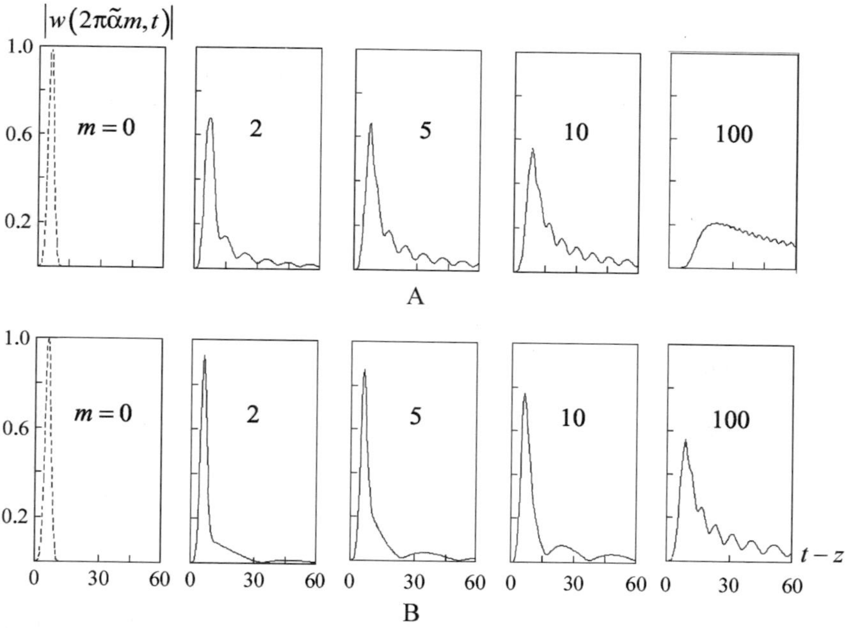

FIGURE 6.47. Propagation of the pulse with real-valued envelope function in Floquet channel for different values of parameters Φ: $\tilde{\alpha} = 1$, $\tilde{T} = 6$, $\tilde{\kappa} = 0.8$. (A) $\Phi = 0.3$; (B) $\Phi = 0.1$.

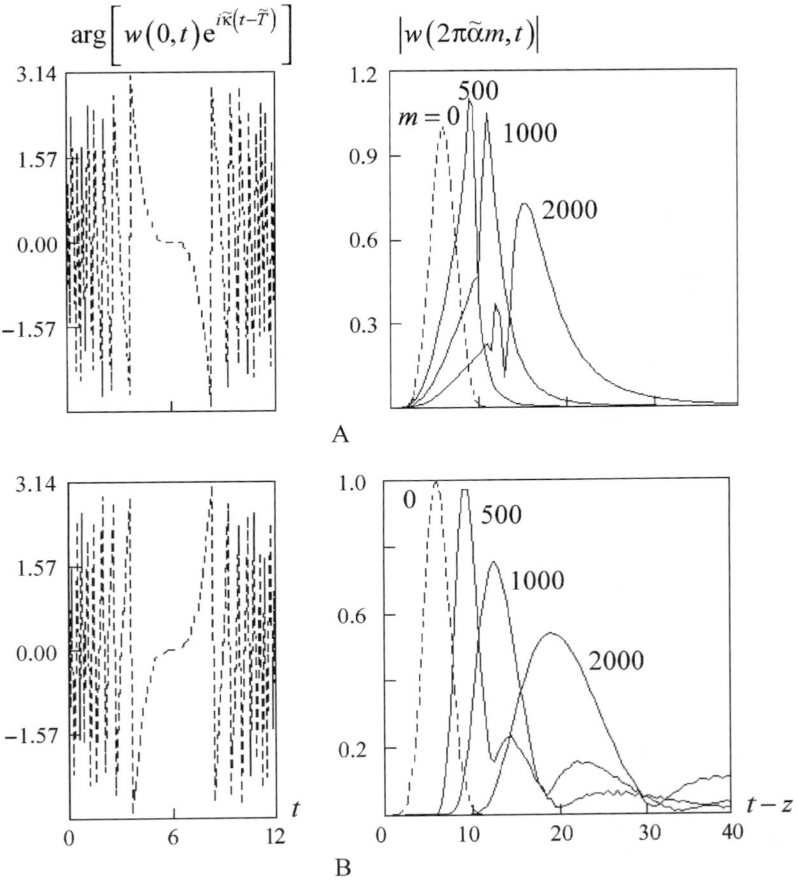

FIGURE 6.48. Propagation of the pulse with cubic phase modulation: $\tilde{\alpha} = 1$, $\Phi = 0.3$, $\tilde{T} = 6$, $\tilde{\kappa} = 7$. (A) $\tilde{\gamma} = 0.25$; (B) $\tilde{\gamma} = -0.25$.

$w(0, t)$ could be assumed to be zero for times $t < 0$. The distance scale z in all the figures is determined by the value of m in $z = 2\pi m\tilde{\alpha}$.

If the frequency band occupied by the pulses with spectral amplitudes (6.50), (6.51) does not contain the threshold point of the Floquet channel (i.e., the point κ such that $\Gamma(\kappa) = 0$), then for sufficiently large $\tilde{\alpha}$ the function $\Gamma(\kappa)$ in (6.46) can be substituted by a finite part of its Laurent series

$$\Gamma(\kappa) = \Gamma(\tilde{\kappa}) + \Gamma'(\tilde{\kappa})(\kappa - \tilde{\kappa}) + \frac{\Gamma''(\tilde{\kappa})}{2}(\kappa - \tilde{\kappa})^2 + \frac{\Gamma'''(\tilde{\kappa})}{6}(\kappa - \tilde{\kappa})^3 + \ldots.$$

Here, Γ', Γ'', and so on, are the derivatives of the function $\Gamma(\kappa)$ with respect to κ. In the case $\operatorname{Im}\Gamma(\kappa) = 0$ ($\Gamma'(\tilde{\kappa}) = \tilde{\kappa}/\Gamma(\tilde{\kappa}) > 0$, $\Gamma''(\tilde{\kappa}) = -\Phi^2/\Gamma^3(\tilde{\kappa}) < 0$), such a substitution practically does not distort the exact value of the integral in

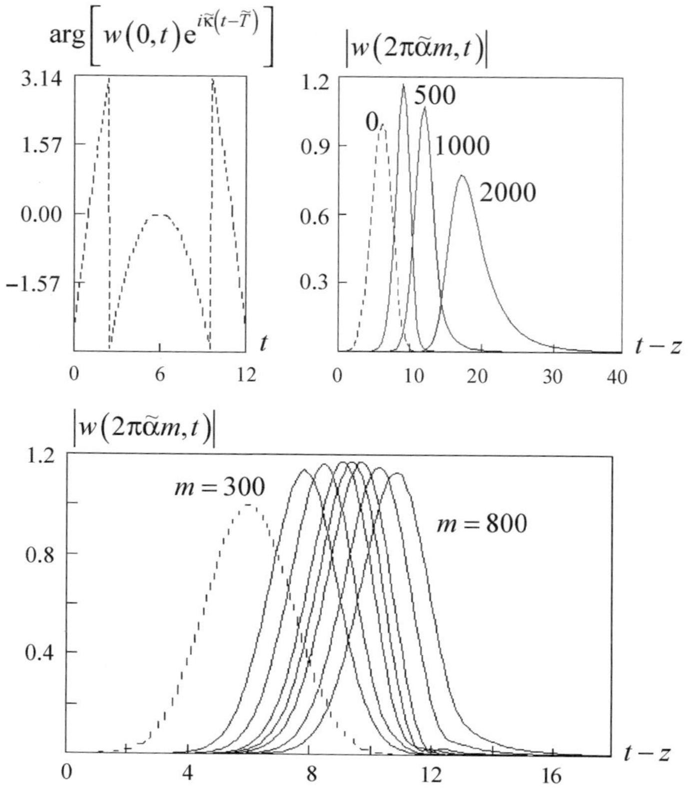

FIGURE 6.49. Propagation and focusing of wideband pulses with quadratic phase modulation: $\tilde{\alpha} = 1$, $\Phi = 0.3$, $\tilde{T} = 6$, $\tilde{\kappa} = 7$, $\tilde{\beta} = 0.25$. Results presented in lower picture are calculated for the values of $m = 0$ (dashed line), $m = 300, 400, 500, 600, 700, 800$.

(6.46) for any z and this allows us to obtain representations for $w(z, t)$ in closed form:

$$
\begin{array}{ccccc}
w(0,t) = \dfrac{\sqrt{\pi}}{\tilde{\alpha}} & \times & e^{-\frac{t^2}{4\tilde{\alpha}^2}} & \times & e^{-i\tilde{\kappa}t} \\
\Downarrow 1 & & \Downarrow 2 & & \Downarrow 3 \\
w(z,t) = \dfrac{\sqrt{\pi}}{\sqrt[4]{\tilde{\alpha}^4 + \dfrac{z^2\Phi^4}{4\Gamma^6}}} & \times & e^{-\frac{\left(t-\frac{z\tilde{\kappa}}{\Gamma}\right)^2}{4\left[\tilde{\alpha}^2 + \frac{z^2\Phi^4}{4\tilde{\alpha}^2\Gamma^6}\right]}} & \times & e^{-i\left[\tilde{\kappa}t - \frac{\left(t-\frac{z\tilde{\kappa}}{\Gamma}\right)^2\left(\frac{z\Phi^2}{2\tilde{\alpha}^2\Gamma^3}\right)}{4\left[\tilde{\alpha}^2 + \frac{z^2\Phi^4}{4\tilde{\alpha}^2\Gamma^6}\right]}\right]} \quad \Downarrow 4, \\
 & & & & \times e^{i\left[z\Gamma - \frac{1}{2}\arg\left(\tilde{\alpha}^2 + \frac{iz\Phi^2}{2\Gamma^3}\right)\right]}
\end{array}
\tag{6.52}
$$

$$
\begin{array}{ccccc}
w(0,t) = \dfrac{\sqrt{\pi(1+4i\tilde{\beta}\tilde{\alpha}^2)}}{\tilde{\alpha}} & \times & e^{-\frac{t^2}{4\tilde{\alpha}^2}} & \times & e^{-i(\tilde{\kappa}t+\tilde{\beta}t^2)} \\
\Downarrow 1 & & \Downarrow 2 & & \Downarrow 3 \\
w(z,t) = \dfrac{\sqrt{\pi(1+4i\tilde{\beta}\tilde{\alpha}^2)}}{\tilde{\alpha}\sqrt[4]{(1+2z\Gamma''\tilde{\beta})^2+\dfrac{z^2\Gamma''^2}{4\tilde{\alpha}^4}}} & \times & e^{-\frac{(t-z\Gamma')^2}{4\tilde{\alpha}^2\left[(1+2z\Gamma''\tilde{\beta})^2+\frac{z^2\Gamma''^2}{4\tilde{\alpha}^4}\right]}} & & \\
\bar{\Downarrow} 4 & & & & \\
\times\, e^{i\left[z\Gamma-\frac{1}{2}\arg\left(\tilde{\alpha}^2-\frac{i\Gamma''z}{2}+2z\Gamma''\tilde{\beta}\tilde{\alpha}^2\right)\right]} & \times & e^{-i\left(\tilde{\kappa}t+\frac{(t-z\Gamma')^2\left(4\tilde{\beta}\tilde{\alpha}^2+\frac{\Gamma''z}{2\tilde{\alpha}^2(1+2z\Gamma''\tilde{\beta})}\right)}{4\left[\tilde{\alpha}^2(1+2z\Gamma''\tilde{\beta})+\frac{z^2\Gamma''^2}{4\tilde{\alpha}^2(1+2z\Gamma''\tilde{\beta})}\right]}\right)} & &
\end{array}
\quad . \tag{6.53}
$$

To simplify the equation, we have here assumed $\tilde{T} = 0$ and omitted the value of the argument $\kappa = \tilde{\kappa}$ of the function $\Gamma(\kappa)$ and its derivatives. The cofactors in formulas (6.52), (6.53) are separated so as to make evident the changes in the main characteristics of the pulse that propagates toward increasing z.

By comparing the first two factors in the representations (6.52) (at the start in plane $z = 0$, it is a Gaussian pulse with a real envelope) we obtain information on the distortion in the amplitude envelope of the signal: its shape remains unchanged, but the duration of the signals extends and the height of its amplitude center decreases (in proportion to $\left(1 + 0.25z^2\Phi^4\Gamma^{-6}\tilde{\alpha}^{-4}\right)^{1/4}$). The second cofactor for $w(z,t)$ also determines the average envelope velocity of the pulse, that is, the velocity of the principal part of the signal:

$$
v_{group} = \Gamma/\tilde{\kappa} = \left(1 - \Phi^2/\tilde{\kappa}^2\right)^{1/2} \leq 1. \tag{6.54}
$$

For similar values of the parameter Φ of the Floquet channel **R**, the pulses with a larger central frequency $\tilde{\kappa}$ move faster. For similar high-frequency content, different partial components of the field $U^s(g,t)$ (the components with numbers $n_1 \neq n_2$) have similar envelope velocity only if their propagation constants $\Gamma_{n_1}(\tilde{\kappa})$ and $\Gamma_{n_2}(\tilde{\kappa})$ coincide. In the frequency domain, the mode with coinciding Γ is called the autocollimation mode. It is used in synthesizing plane mirrors, which are capable of concentrating the complete energy of a scattered field in harmonics propagating toward an incident wave (see [3,6,9]: the effects of complete non-mirror reflection). The controllable time separation of partial components of a scattered field that is conditioned by a mismatch of the envelope velocities of the corresponding pulses, furnishes additional possibilities for an efficient use of the selective properties of gratings.

The third cofactor is responsible for the dynamics of changes of the high-frequency pulse filling and the phase modulation that is induced during the propagation process. The fourth, additionally occurring factor is associated with the spatial tuning of the phase envelope. Note that for $\Phi = 0$, none of the main characteristics of a pulse propagating in such a case with the velocity of light is changed.

The validity of the conclusions is backed up by the results of computational experiments (see, e.g., Figs. 6.42A and 6.43: the distinctions in the envelope velocities of pulses having different high-frequency content are evident; pulses with larger $\tilde{\kappa}$ propagate longer distances without noticeable changes in their main characteristics, etc.). The computational experiments allowed us to analyze cases that cannot be studied correctly by means of the scheme that was applied above. Thus, even the slightest non-analyticity of a signal in the frequency domain results in a drastic change of its energy spectrum, a rapid blurring of the amplitude center, and the appearance of a long-lasting tail competing with the principal part (see Fig. 6.42B: the threshold point $\kappa : \Gamma(\kappa) = 0$ lies within the frequency band occupied by the pulse).

An analysis using formulas (6.53) in the case when $\tilde{\beta} < 0$ reveals no regularities that are qualitatively different from those already known (see, e.g., Fig. 6.44B; at the start in plane $z = 0$, we have a Gaussian pulse with a quadratic phase modulation).

If $\tilde{\beta} > 0$, the function $f(z) = \left(1 + 2z\Gamma''\tilde{\beta}\right)^2 + z^2\Gamma''^2/\left(4\tilde{\alpha}^4\right)$, that determines the quantitative parameters of an amplitude envelope, does not increase monotonically as z increases, but first decreases up to $f(z_{foc}) = (1 + 16\tilde{\alpha}^4\tilde{\beta}^2)^{-1}$, that is reached at the point $z_{foc} = 8\tilde{\alpha}^4\tilde{\beta}\Gamma^3[(1 + 16\tilde{\alpha}^4\tilde{\beta}^2)\Phi^2]^{-1}$, $\Phi \neq 0$, and only after that the continuous increase begins. The point z_{foc} is a point in space passed by a pulse having a maximally compressed principal part, the point of maximum focusing of a transient wave. The attainable compression rate is determined by the value $f(z_{foc})$: the larger $\tilde{\alpha}$ and $\tilde{\beta}$ are, the more efficient is the compression, and the compression rate does not depend on the parameter Φ ($\Phi \neq 0$) of the Floquet channel and the high-frequency content $\tilde{\kappa}$.

The envelope velocity of a pulse with a quadratic phase modulation remains the same as for a pulse with a real envelope (see, e.g., Figs. 6.44 and 6.45). The phase envelope of the latter varies as it propagates (see formula (6.52)), also due to the induced quadratic-in-time modulation. But this type of modulation does not cause the after-focusing of the pulse (the parameter for the coefficient $\tilde{\beta}$ is negative for any z). The signs of a coming or already realized focusing can be easily seen by computing the pulse characteristics for different distances of propagation of the pulse (see Fig. 6.44). An example of a detailed analysis of this effect is given at the bottom of Figure 6.45B.

6.5.2. *Propagation and Deformation of Wideband Pulses*

Let $w(0, t) = \delta(t)$. Then, according to (6.48),

$$w(z, t) = \delta(t - z) - \frac{\Phi z}{2} \frac{J_1\left(\Phi\sqrt{t^2 - z^2}\right)}{\sqrt{t^2 - z^2}} \chi(t - z) = \delta(t - z) + \frac{g(z, t)}{2}. \tag{6.55}$$

Thus, an undistorted pulse propagating with the velocity of light is followed by an oscillating tail. In the coordinate $t - z$, the frequency and the amplitude of these oscillations increase as the distance traveled increases. The first peak of the amplitude envelope of the tail gradually approaches the original δ-pulse, and the remaining pulses line up after it at shorter and shorter distances (see Fig. 6.46). The signal is substantially distorted, mainly because the local amplitudes' centers of the pulses forming the tail move with velocities exceeding the velocity of light. Such effects accompany the propagation of all wideband pulses with a nonanalytic envelope. Unlike the case of a δ-pulse, here the precursor can be significantly distorted at distances that are sufficient for the local amplitude centers of the tail to surpass the amplitude center of a pulse started in the plane $z = 0$.

Figure 6.47 (here $w(0, t) = \exp[-(t - \tilde{T})^2/4\tilde{\alpha}^2 - i\tilde{\kappa}(t - \tilde{T})])$ confirms this statement: the efficient frequency band occupied by the pulses lies in the region $[\tilde{\kappa} - 3; \tilde{\kappa} + 3]$, containing the threshold points $\kappa = 0.3$ (Fig. 6.47A) and $\kappa = 0.1$ (Fig. 6.47B), dividing the spectral region into subregions with essentially different wave propagation conditions.

The wideband pulses

$$w\,(0, t) = \exp[-(t - \tilde{T})^2/4\tilde{\alpha}^2 - i\tilde{\kappa}(t - \tilde{T}) - i\tilde{\gamma}(t - \tilde{T})^3]$$

with a cubic phase modulation vary during the propagation a similar way (see Fig. 6.48). For sufficiently large absolute values of the modulation coefficients $\tilde{\gamma}$, the main part splits up, and the newly formed amplitude centers surpass the principal one (for $\tilde{\gamma} > 0$), or lag behind it (for $\tilde{\gamma} < 0$).

The quadratic phase modulation of a wideband pulse with an analytic envelope in the frequency domain affects its propagation in a way that is similar to the case of narrowband signals. See, for example, Figure 6.49. Here,

$$w(0, t) = \exp[-(t - \tilde{T})^2/4\tilde{\alpha}^2 - i\tilde{\kappa}(t - \tilde{T}) - i\tilde{\beta}(t - \tilde{T})^2],$$

and at the bottom part the focusing effect is considered in detail, which takes place, as before, for positive $\tilde{\beta}$.

The canonic quasi-periodic Green's function (the Green's function of the regular Floquet channel; see Section 6.2.1)

$$\tilde{G}_0\,(g, g_0, \kappa, \Phi) = -\frac{i}{4\pi}\sum_{n=-\infty}^{\infty} \mathrm{e}^{i[\Phi_n(y-y_0)+\Gamma_n|z-z_0|]}\Gamma_n^{-1}$$

in local variables on the surface **K** has simple poles at the threshold points $\kappa_n : \Gamma_n\,(\kappa_n) = 0$. The residues $\mathrm{Res}\tilde{G}_0\,(\kappa_n) = -i\exp\,(i\,\Phi_n\,(y - y_0))\,(8\pi\kappa_n)^{-1}$ of the function $\tilde{G}_0\,(\kappa)$ in these poles determine the possible free oscillations of field $\tilde{U}_n\,(g, \bar{\kappa}_n) = \exp\,(i\,\Phi_n y)$ in the regular Floquet channel at real EFs $\bar{\kappa}_n = \kappa_n = \pm\,|\Phi_n|$. We can answer questions concerning how the elements $\bar{\kappa}_n$ of the spectral set Ω_κ influence the behavior of pulse propagation by analyzing quite simple cases. Below we give an example of such an analysis: the source $F\,(g, t)$ of transient waves $U\,(g, t)$ chooses (according to the completely transparent requirement $\int_Q \tilde{U}_n\,(g, \overline{\kappa}_n)\,F\,(g, t)\,dg \neq 0$) only one couple of free oscillations to take part

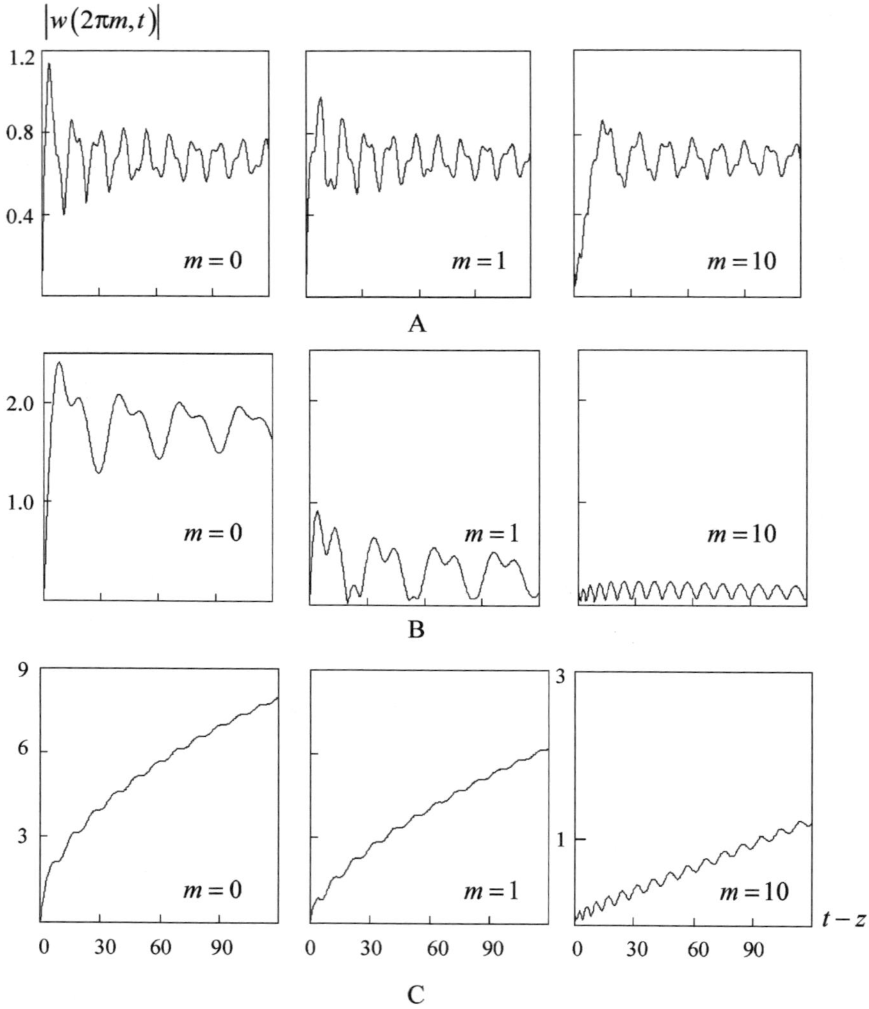

FIGURE 6.50. Propagating of wave active by continuously (starting moment $t = 0$) operating source $F(g, t) = \exp(i\Phi y)\,\delta(z)\exp(-i\tilde{\kappa}t)$, $\Phi = 0.3$. (A) $\tilde{\kappa} = 0.8$; (B) $\tilde{\kappa} = 0.1$; (C) $\tilde{\kappa} = \Phi = 0.3$.

in the process and therefore clears the pattern from the many small, irrelevant details.

Figure 6.50 illustrates the propagation of the wave $U(g, t) = w(z, t)\exp(i\Phi y)$, generated by a source $F(g, t) = \exp(i\Phi y)\,\delta(z)\exp(-i\tilde{\kappa}t)$ concentrated in the plane $z = 0$, at $\operatorname{Re}\Gamma(\tilde{\kappa}) = \operatorname{Re}\sqrt{\tilde{\kappa}^2 - \Phi^2} = \operatorname{Re}\sqrt{\tilde{\kappa}^2 - \bar{\kappa}^2} > 0$ (Fig. 6.50A), $\operatorname{Im}\Gamma(\tilde{\kappa}) > 0$ (Fig. 6.50B), and at $\Gamma(\tilde{\kappa}) = 0$ (Fig. 6.50C).

We recall (see Sections 5.2.1 and 5.5.1) that the spatio–temporal amplitudes $w_n(z,t)$ of an arbitrary signal $U(g,t) = \sum_n w_n(z,t)\exp(i\Phi_n y)$ in the regular Floquet channel **R**, that is generated by any sources $F(g,t) = \sum_n a_n(z,t)$ $\exp(i\Phi_n y)$, $\varphi(g) = \sum_n b_n(z)\exp(i\Phi_n y)$ and $\psi(g) = \sum_n c_n(z)\exp(i\Phi_n y)$ with compact supports in $\overline{\mathbf{R}}$, are given by the solutions

$$w_n(z,t) = -\frac{1}{2}\int_{-\infty}^{\infty}\int_{-\infty}^{\infty} J_0[\Phi_n((t-\tau)^2-(z-\omega)^2)^{1/2}] \times \chi(t-\tau-|z-\omega|)\, f_n(\omega,\tau)\, d\tau \tag{6.56}$$

of the generalized Cauchy problems

$$B(\Phi_n)[w_n(z,t)] \equiv \left[-\frac{\partial^2}{\partial t^2}+\frac{\partial^2}{\partial z^2}-\Phi_n^2\right] w_n(z,t) = f_n(z,t) = a_n(z,t) - \delta^{(1)}(t)\, b_n(z) - \delta(t)\, c_n(z); \quad z,t\in\mathbf{R}^1. \tag{6.57}$$

From (6.56), we obtain for $z \geq 0$,

$$\begin{aligned} w(z,t) &= -\frac{1}{2}\int_0^{t-z} J_0\left(\bar\kappa\sqrt{(t-\tau)^2-z^2}\right)e^{-i\tilde\kappa\tau}\,d\tau \\ &= -\frac{1}{2}e^{-i\tilde\kappa t}\left[\left\{\begin{array}{ll} \exp\left(-z\sqrt{\bar\kappa^2-\tilde\kappa^2}\right)\left(\bar\kappa^2-\tilde\kappa^2\right)^{-1/2}, & \tilde\kappa<\bar\kappa \\ i\exp\left(iz\sqrt{\tilde\kappa^2-\bar\kappa^2}\right)\left(\tilde\kappa^2-\bar\kappa^2\right)^{-1/2}, & \tilde\kappa>\bar\kappa \end{array}\right\}\right. \\ &\quad \left. - \int_t^{\infty} J_0(\bar\kappa\sqrt{\tau^2-z^2})e^{i\tilde\kappa\tau}\,d\tau \right] \end{aligned} \tag{6.58}$$

and

$$\begin{aligned} w(0,t) &= -\frac{1}{2}\int_0^t J_0(\bar\kappa(t-\tau))\,e^{-i\tilde\kappa\tau}\,d\tau = -\frac{1}{2}t\,[J_0(\bar\kappa t) - i J_1(\bar\kappa t)] \\ &\overset{t\gg1}{\approx} -\sqrt{\frac{t}{2\pi\bar\kappa}}\,e^{-(\bar\kappa t-\pi/4)} \end{aligned} \tag{6.59}$$

for the case $z = 0$ and $\tilde\kappa = \bar\kappa$.

The signal $w(z,t)$ (see (6.58)) is a superposition of two signals with the amplitude centers in the spectral region (central frequencies) at the points $\tilde\kappa$ and $\bar\kappa$. For $\tilde\kappa > \bar\kappa$, both signals propagate under almost equal conditions. The deformations $w(z,t)$ are conditioned mainly by the fact that the signal with central frequency $\tilde\kappa$ moves faster than the signal with central frequency $\bar\kappa$. For $\tilde\kappa < \bar\kappa$, the central frequency $\tilde\kappa$ of one of the signals is below the critical point $\kappa = \Phi = \bar\kappa$ of the Floquet channel for the corresponding spatial harmonic (see the representation (6.5)), and this results in an exponential attenuation of this signal as the distance z increases. Already for $z = 20\pi$ ($m = 10$), the energy spectrum of $w(z, t)$ is determined solely by the signal propagating without attenuation and having the amplitude center at the point $\kappa = \bar{\kappa}$. The field $U(g, t)$ is quasi-harmonic far from the source, but the period of the corresponding oscillations is determined, not by the source, but by the real EFs of the regular Floquet channel $\bar{\kappa} = \pm|\Phi|$ that is situated on both sides of the amplitude center of the function $F(g, t)$ in the frequency domain.

If $\tilde{\kappa} = \bar{\kappa}$, the frequency of the source that excites the channel coincides not only with the real EF of the regular Floquet channel, but also with its threshold point. Such a situation was not considered in Section 6.2.4. Now we can use a concrete example (see Fig. 6.50C and (6.59)) to observe the specific features of the process conditioned by this coincidence: both for small and large t, the near field of the source increases in proportion to $\sqrt{t}$ (in the case described by (6.28), it is proportional to t); the rate of increase of the field strength is reduced as the distance to the source grows, and the process gradually enters the quasi-harmonic phase of propagation of a signal with central frequency $\bar{\kappa}$ and a small amplitude.

The source $F(g, t) = \chi(z - z_1)\chi(z_2 - z)\chi(t_1 - t)\cos(\tilde{\kappa}t)$ induces, in the regular Floquet channel $\mathbf{R}$ with $\varepsilon - 1 = \sigma = 0$ and $\Phi = 0$ (see formulas (6.56) and (6.57)), the field $U(g, t) = G * F$, where $G(z, t) = -\chi(t - |z|)/2$ is the fundamental solution of the operator $B(0) \equiv [-\partial^2/\partial t^2 + \partial^2/\partial z^2]$. Expanding the convolution in the expression $U(g, t) = G * F$, for $z > z_2$ we obtain

$$U(g, t) = -\frac{1}{2}\int_0\int_{z_1}^{z_2} \chi[t - \tau - z + \omega]\chi(t_1 - \tau)\cos(\tilde{\kappa}\tau)\, d\omega d\tau$$

$$\overset{t-z+z_1>t_1}{=} -\frac{z_2 - z_1}{2}\int_0^{t_1} \cos(\tilde{\kappa}\tau)d\tau = -(z_2 - z_1)\sin(\tilde{\kappa}t_1)/2\tilde{\kappa}. \quad (6.60)$$

From (6.60), it follows that the time t_1 (the switch-off time of the source) determines a constant (viz. from $-(z_2 - z_1)/2\tilde{\kappa}$ to $+(z_2 - z_1)/2\tilde{\kappa}$; see Fig. 6.51) field-strength level in any crosssection z of channel $\mathbf{R}$ and for all observation times t, only if the inequality $t - t_1 > z - z_1$ is satisfied. The source field oscillating with the frequency $\tilde{\kappa}$ is here displaced by a stationary field $\tilde{U}(g, \bar{\kappa})$ of free oscillation (eigenfrequency $\bar{\kappa} = 0$ if $\Phi = 0$), constant along directions y and z ($\tilde{U}(g, \bar{\kappa}) =$ const).

The effect is somewhat blurred when the source is tuned to excite other ($\tilde{U}_n(g, \bar{\kappa}_n) = \exp(i\Phi_n y)$, $\bar{\kappa}_n = \pm|\Phi_n|$) free oscillations of the field in the regular Floquet channel; that is, for instance when

$$F(g, t) = \chi(z - z_1)\chi(z_2 - z)\chi(t_1 - t)\cos(\tilde{\kappa}t)\, e^{i\Phi_n y}.$$

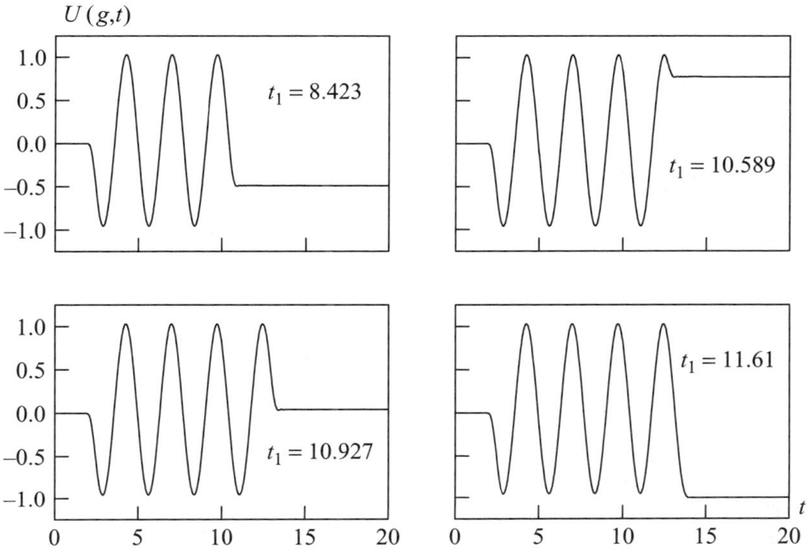

FIGURE 6.51. Field $U(g,t)$, $g = \{0 < y < 2\pi; z = 2\}$, generated by current source $F = 10\chi(-z)\chi(z+0.5)\cos(\tilde{\kappa}t)\chi(t_1 - t)$; $\tilde{\kappa} = 2.3$ in regular Floquet channel with $\Phi = 0$.

In this case, the process is very similar to that considered above for parameter values $\tilde{\kappa} < \bar{\kappa}$. The difference is only that the transition to the quasi-harmonic mode, whose frequency is determined by the eigenfrequency $\bar{\kappa}_n$, is possible already in the field $U(g,t)$ that is closer to the source. In pure form, the effect can be realized in all directing structures where time-harmonic waves can propagate for any values of the frequency parameters κ or k, that is, in structures with zero cut-off point for fundamental waves.

7
Model Synthesis of Resonance Quasi-Optical Devices: Dispersive Open Resonators, Absorbing Coatings, and Pattern-Forming Structures

7.1. Introduction

Research on resonance wave scattering has made important progress in recent years thanks to a series of profound analytical studies on some general issues and widespread implementation of a new methodology of scientific search based on adequate mathematical modeling and computational experiment. Some physical effects and phenomena discovered and studied in the course of analyzing various resonance structures have become a guide for scientific instrument-making. However, most of the results still remain within the realm of theory. One of the reasons is the complexity of the primary modeling of the problem that should take into account the most complete number of the functionally relevant electromagnetic characteristics of separate elements and allow an estimate of the effectiveness of the system or performance of the device as a whole. As for the resonance quasi-optic devices, the problem becomes even more complicated due to the substantial difference (by an order or more) in the characteristic dimensions of the scatterers that jointly create the electromagnetic field with the given parameters.

The classical approaches and methods do not allow building a sufficiently comprehensive and, at the same time, operating model of such a unit, to achieve computer-aided synthesis and optimization. One cannot avoid auxiliary theoretical calculations, including the analysis and synthesis of separate resonance elements, as well as an approximate model synthesis of the whole device. The latter is based on the achievements of classical theory and solves practically very important problems, by coordinating in the best possible way the functions of individual electromagnetic objects with the general requirements on the whole unit.

This chapter is devoted to the solution of some methodological problems of this new, both theoretically and practically important direction. Here the authors have scheduled and implemented, in a series of concrete problems, the sequence of steps that will result in bringing the theoretical model of the developed resonance quasi-optic device to the level of a laboratory specimen, whose further adjustment and optimization will not be very labor- and time-consuming. This sequence consists of the following stages.

a. Electromagnetic analysis of all the key quasi-optical and resonance elements of the unit;
b. Development of an approximate mathematical model of the quasi-optic unit, which describes the resonance elements not by geometrical and material parameters but by functions (operatorfunctions) X that determine their electromagnetic characteristics;
c. An approximate analysis and synthesis (optimization) of the developed model and definition of the possible requirements for implementation to functions X;
d. Forming, on the basis of these requirements, a standard input data set, statement, and solution of the inverse problems that reconstruct the necessary material and geometrical parameters of the resonance elements; and
e. Analysis and optimization of the model of the quasi-optic unit together with all its resonance elements that are described by actual material and geometric parameters.

At stage (a), any of the methods and results of classical electromagnetic theory can be applied. It is essential here to determine the range of functional efficiency of individual elements and coordination of these functions with the requirements for the unit as a whole. At this stage, we use results from [3,5–7,9–11,14,137,198] for building the selective elements in the model (usually represented by gratings), as well as results concerning the time-domain methods, that were considered above.

The electromagnetic representation of the modeled quasi-optic devices at stage (b) should simultaneously satisfy two requirements that are difficult to combine in practice. First, the presentation should be quite simple, suitable for further qualitative analysis and parametric optimization by analytical means. Second, the simplifications that are unavoidable in modeling should not essentially distort the idea of the real physical processes in the system and the importance of the part played by separate dispersive elements. Solutions of the problems considered in Section 7.2 generally satisfy these requirements. The key point here is an "equivalent" model substitution of the resonant physical object (grating) by its electromagnetic characteristics in the system of the generalized parameters of the device [9,173]. This method allows us to eliminate the difficulties arising due to the presence of geometrical parameters with incommensurable values, and thus can be applied in other problems of model synthesis with different dispersive elements. In the sequence of operations described above, this method allows us to split rather complicated original problems into several simpler ones, which can be solved consecutively, one after another, or independently of each other, and to examine more closely the most interesting points and effects in these problems.

The synthesis (optimization) of the simplified model of a quasi-optical device (stage (c)) is based on standard tools of computational mathematics and allows us to determine the values of geometric and material parameters of quasi-optical elements necessary for further calculations, as well as the values of functions X that describe the assumed electromagnetic characteristics of the dispersive elements. The values of X are used later on (stage (d)) to build the standard set of input data in the inverse problems (i.e., problems of visualization, synthesis, and parametric optimization). Using these results we determine the material and geometric parameters of resonance scatterers (see [173,199–206] and Section 7.3).

The stage (e), that is, the input of the real geometric and material parameters of all the device elements into the optimized (in the first approximation) model, is possible only by using time-domain methods. The final calculation of the modeled unit, including verification of the data obtained, cannot be done by means of the frequency-domain methods that are mainly oriented on quite simple objects with a limited range of values of the key geometric parameters. It should be noted that even the time-domain methods are not to be expected to provide optimal solutions in problems such as those considered in Sections 7.4 and 7.5 (examples of analysis at stages (a) and (e)), unless appropriate preparations have been made for the corresponding computational experiments (a good starting guess is required) by employing frequency-domain methods as well (stages (a) to (d) of the above sequence of operations).

7.2. Model Synthesis of Structures Including Grating Dispersive Elements (Frequency Domain)

The content of this section is devoted to consideration of the key problems emerging in electromagnetic modeling of quasi-optical devices whose functional selective elements are represented by periodic structures. Attention is mainly given to the principal question of model synthesis: the analytical description of the conditions of optimal matching of the possible operation modes of electromagnetic devices and forming, on this basis, a standard set of input data for the synthesis of the grating configuration.

7.2.1. *Optimization of the Absorbing Properties of Coatings*

Consider, as a model, a metal periodic grating embedded into a dielectric layer of the thickness $2\pi\gamma$ (Fig. 1.1B). The absorption of the coating material is determined by the part $\mathrm{Im}\,\tilde{\varepsilon} > 0$ of the value $\tilde{\varepsilon} = \mathrm{Re}\,\tilde{\varepsilon} + i\mathrm{Im}\,\tilde{\varepsilon}$, $\mathrm{Re}\,\tilde{\varepsilon} \geq 1$, which, in this case, is assumed to be constant. The direct and inverse problems of the frequency domain associated with such an object can be solved by the same methods and within the same algorithms that we use in Section 7.3 (see also papers [9,10,205,206]). Let an E-polarized monochromatic wave $\tilde{U}_0^i(g,\kappa) = \exp[i(\Phi_0 y - \Gamma_0(z - 2\pi\delta))]$ be incident on the structure shown in Figure 1.1B. The total diffraction field (see the statement of the elliptic boundary value problems in Section 6.2.1) are represented as

$$\tilde{U}(g,\kappa) = \begin{cases} \tilde{U}_0^i + \sum\limits_{n=-\infty}^{\infty} a_n e^{i(\Phi_n y + \Gamma_n(z-2\pi\delta))}; & z \geq 2\pi\delta \\ \sum\limits_{n=-\infty}^{\infty} \left(d_n e^{-i\Gamma_n^{\varepsilon}(z-2\pi\delta)} + c_n e^{i\Gamma_n^{\varepsilon}(z-2\pi(\delta-\gamma))}\right) e^{i\Phi_n y}; \\ 2\pi(\delta-\gamma) \leq z \leq 2\pi\delta \end{cases}, \tag{7.1}$$

where $\Gamma_n^\varepsilon = \left(\kappa^2\tilde{\varepsilon} - \Phi_n^2\right)^{1/2}$, and $\mathrm{Im}\,\Gamma_n^\varepsilon \geq 0$, $\mathrm{Re}\,\Gamma_n^\varepsilon \geq 0$. By applying the method of generalized scattering matrices [3,137] at the boundaries of the partial domains in the planes $z = 2\pi\delta$ and $z = 2\pi(\delta - \gamma)$, obtain

$$a = R_{1\to 1}[v] + T_{1\to\varepsilon}E[c], \quad d = T_{\varepsilon\to 1}[v] + R_{\varepsilon\to\varepsilon}E[c], \quad c = RE[d]$$

or, eliminating vector d,

$$c = RER_{\varepsilon\to\varepsilon}E[c] + RET_{\varepsilon\to 1}[v], \quad a = R_{1\to 1}[v] + T_{1\to\varepsilon}E[c]. \tag{7.2}$$

Here, $a = \{a_n\}$, $v = \{\delta_n^0\}$, $c = \{c_n\}$, and $d = \{d_n\}$ are the complex amplitudes of plane waves that form the field (7.1); $E = \{\delta_n^m \exp(i\Gamma_n^\varepsilon 2\pi\gamma)\}_{m,n}$, $R_{\varepsilon\to\varepsilon} = -R_{1\to 1}$, $T_{1\to\varepsilon} = I - R_{1\to 1}$, $T_{\varepsilon\to 1} = I + R_{1\to 1}$, and $R_{1\to 1} = \{\delta_n^m(\Gamma_n - \Gamma_n^\varepsilon)(\Gamma_n + \Gamma_n^\varepsilon)^{-1}\}_{m,n} = \{\delta_n^m r_n\}_{m,n}$ are diagonal matrices describing the transforming properties of the boundary $z = 2\pi\delta$ and of the dielectric layer $2\pi(\delta - \gamma) \leq z \leq 2\pi\delta$; I is an identity matrix; $R = \{R_{np}\}_{n,p}$ is a generalized scattering matrix of the grating in a medium having material parameters $\tilde{\varepsilon}\varepsilon_0$ and μ_0 (see [3,6] and Section 6.2.1; R_{np} is the transformation coefficient of the p th incident harmonic into the n th reflected one). The latter can be calculated by using analytical regularization methods. This matrix generate in the pair of spaces $\tilde{l}_2 \to \tilde{l}_2$ of infinite sequences a bounded operator [5,9,10]. The operators generated by the diagonal matrices $R_{1\to 1}$, $R_{\varepsilon\to\varepsilon}$, $T_{1\to\varepsilon}$, and $T_{\varepsilon\to 1}$, are bounded, too. The elements of the diagonal matrix E decrease exponentially as the index grows. Hence, the associated operator is completely continuous, and (7.2) is a Fredholm problem. Its solution exists and is unique for all real κ (see Statements 6.1 and 6.4) and can be obtained by using the truncation method that converges exponentially in the norm of the space $\tilde{l}_2$.

As for the inverse problems (the problems of visualization or synthesis of the reflecting grating in a medium with parameters $\tilde{\varepsilon}\varepsilon_0$ and μ_0, which we refer to as an ε-medium), the modification of the basic algorithms from [205,206] and Section 7.3 can be reduced to a clear operation implying the complete substitution of parameter κ^2 by $\kappa^2\tilde{\varepsilon}$ (Γ_n by Γ_n^ε for all n).

Let $\mathrm{Im}\,\Gamma_0 > 0$: the structure is excited by a homogeneous plane wave. Let us use the qualitative characteristic of the scattering process $\{N, M\}$ (see Section 6.3.2), that enables us to predict, with sufficient reliability, the possibility of certain modes, which are described by purely quantitative indices, for example, the value of $W_n^a = |a_n|^2\,\Gamma_0^{-1}\mathrm{Re}\,\Gamma_n$. Here $N = \sum |\Gamma_n|^{-1}\,\mathrm{Re}\,\Gamma_n$ is the number of channels through which the energy can be radiated into free space. M is a similar characteristic for the ε-medium by $\mathrm{Im}\,\tilde{\varepsilon} = 0$. It determines the number of harmonics in the layer that are active in the energy exchange between the radiation zone $z > 2\pi\delta$ and the grating. W_n^a is the part of the scattered energy carried away from the grating by the n th wave having the complex amplitude a_n. If $\mathrm{Im}\,\tilde{\varepsilon} = 0$, from the conservation laws (see Section 6.2.1) it follows that $W = \sum W_n^a = 1$. If $\mathrm{Im}\,\tilde{\varepsilon} > 0$, the difference $1 - W$ determines the degree of field energy absorbed in the structure, whereas for the case of $N = 1$ this role is played by the value $1 - W_0^a$.

Consider first the range of values of the parameters κ, Φ, and $\tilde{\varepsilon}$ that is described by the vector $\{N, M\}$ where $M = N = 1$. Ignoring the fields that exponentially attenuate in the layer even for $\operatorname{Im}\tilde{\varepsilon} = 0$, from (7.2) obtain

$$a_0 = (r_0 + K)(1 + r_0 K)^{-1}; \quad K = R_{00} \exp\left(i4\pi\gamma\Gamma_0^{\varepsilon}\right) \tag{7.3}$$

or

$$W_0^a = \frac{|r_0|^2 + |K|^2 + 2|r_0||K|\cos(\arg K - \arg r_0)}{1 + |r_0|^2|K|^2 + 2|r_0||K|\cos(\arg K + \arg r_0)}; \quad 0 \le \arg(\ldots) < 2\pi. \tag{7.4}$$

The error in the representation (7.3) is of order $O(\exp(-4\pi\gamma \operatorname{Im}\Gamma_{\pm1}^{\varepsilon}))$. For the case of a perfect substrate, formulas (7.3), (7.4) are exact, and $R_{00} = -1$. Minimizing (7.4) as a function of $\arg K$, we obtain the following estimation of part of the reflected wave energy: $W_0^a < (|r_0| - |K|)^2(1 - |r_0||K|)^{-2}$ when $\cos(\arg K - \arg r_0) = -1$. The last condition in terms of electromagnetic and geometric parameters of the structure is written as

$$4\pi\gamma \operatorname{Re}\Gamma_0^{\varepsilon} + \arg R_{00} - \arg r_0 = -\pi + 2\pi m; \quad m = 0, 1, 2, \ldots \tag{7.5}$$

and for $\arg r_0$ close to π, which is obviously true for not very large $\operatorname{Im}\tilde{\varepsilon}$, almost coincides with the exact condition of a transverse resonance at the zero harmonic in the layer between free space and the grating. In the interval

$$0 < |K| = |R_{00}| \exp\left(-4\pi\gamma \operatorname{Im}\Gamma_0^{\varepsilon}\right) < 2|r_0|(1 + |r_0|^2)^{-1} \tag{7.6}$$

of possible values of $|R_{00}|$, the absorption capability of the structure is higher than that of the ε-half-space, that is, $W_0^a < |r_0|^2$, if condition (7.5) is satisfied. The absolute minimum of the reflected energy under similar conditions is reached in the case of $|R_{00}| = |r_0| \exp(4\pi\gamma \operatorname{Im}\Gamma_0^{\varepsilon})$. Ignoring condition (7.5), we obtain from (7.4) $W_0^a < |r_0|^2$, if

$$|K| = |R_{00}| \exp\left(-4\pi\gamma \operatorname{Im}\Gamma_0^{\varepsilon}\right) < 2|r_0|[-\cos(\arg K - \arg r_0)](1 + |r_0|^2)^{-1}. \tag{7.7}$$

In the considered parameter range, the schemes of behavior of a layer with a corrugated and even substrate are similar. Still, in the trivial case, $|K|$ can be reduced up to sufficient values only at the expense of increasing the layer thickness ($|K| = \exp(-4\pi\gamma \operatorname{Im}\Gamma_0^{\varepsilon})$), and condition (7.5) is satisfied if this thickness corresponds to odd numbers of a quarter of the wavelength λ_ε in the ε-medium ($2\pi\gamma/\lambda_\varepsilon = (2m - 1)/4; m = 1, 2, \ldots$). The use of a grating allows us to solve the problem of optimizing of the absorbing capability of the coating more efficiently. First, due to the optimal choice of the wave accelerating values of $\arg R_{00}$ in (7.5) the required layer thickness can be substantially reduced, and hence also the dimension and the weight of the construction as a whole, and second, by selecting the parameters and operation mode of the gratings for which the value $|R_{00}|$ lies within the required value range with the minimum height of the layer being $2\pi\gamma$.

It is also important that such a choice that is sufficient in all the parameters can be done not only for separate sets of values κ and Φ, but also in quite wide ranges of variation of the length and the angle of the incident wave [3,6,9]. Thus, for instance, from the energy conservation laws it follows that the efficiency of a grating, placed in an absorbing medium, in the zero order of the spectrum (the value $|R_{00}|^2$) is mainly determined by the near field strength. Papers [6,9] suggest some methods that allow this characteristic to be controlled. Here we only mention two of them: excitation of the grating at the frequencies that are close to the eigenfrequencies (effects with the different bandwidth), and the selection of the nature of singularities in the geometry of the structure (super-wideband effects).

Let us make one of the possible formulations of the problems of synthesis of gratings, that will provide the required characteristics of the coating in the region $\{1, 1\}$. Here we fix the physical values: l is the period of grating, h is the thickness of the layer, and $\tilde{\varepsilon}$ is the relative dielectric permittivity of the layer material (see Fig. 1.1B). We specify the intervals of variation of wavelength $\lambda_1 < \lambda < \lambda_2$ and the direction wave arrives from $\alpha_1 < \alpha < \alpha_2$, as well as the maximum, with respect to all values λ and α, level of the reflected energy $\max W_0^a = W$. According to the values

$$
\begin{aligned}
&2A|K|_{\max} = B + (B^2 - 4AC)^{1/2},\\
&2A\,|K|_{\min} = \begin{cases} 0; & C \le 0 \\ B - (B^2 - 4AC)^{1/2}; & C > 0 \end{cases};\\
&B = 2|r_0|[W\cos(\arg K + \arg r_0) - \cos(\arg K - \arg r_0)],\\
&A = 1 - W\,|r_0|^2, \quad C = |r_0|^2 - W, \quad B^2 \ge 4AC,
\end{aligned}
\tag{7.8}
$$

obtained from (7.4), a layer of parameters $l/\lambda_1 \le \kappa \le l/\lambda_2$, $\alpha_1 < \alpha < \alpha_2$, $|K|_{\min}\exp(4\pi\gamma\,\mathrm{Im}\,\Gamma_0^{\varepsilon}) \le |\,R_{00}\,(\kappa, \Phi, \arg\, R_{00})| \le |\,K|_{\max}\exp(4\pi\gamma\,\mathrm{Im}\,\Gamma_0^{\varepsilon})$, $\cos\left(\arg R_{00} + 4\pi\gamma\mathrm{Re}\,\Gamma_0^{\varepsilon} - \arg r_0\right) \le -0.5$ is built, that determines $R_{00}\,(\kappa, \Phi)$ $(\Phi = \kappa\sin\alpha)$. The data on function $R_{00}\,(\kappa, \Phi)$ (on the amplitude $a_0\,(\kappa, \Phi)$ of the principal harmonic of the spatial spectrum of the grating in the ε-medium) are further used to set up the input data set in the problem of synthesis of a periodic corrugated surface (see Section 7.3).

Passing into the region $\{1, 2\}$, the physics of the investigated processes becomes more diversified; there are more ways to achieve the required electromagnetic characteristics of the coating (see [3,6,9] and Section 6.3.2: the effects of the waves trapped in a layer, $N < M$). Ignoring the influence of the harmonics of higher order ($d_n = c_n = 0, n = 1, \pm 2, \pm 3, \ldots$) that are evanescent even in an ideal ($\mathrm{Im}\,\tilde{\varepsilon} = 0$) layer, we obtain from (7.2),

$$
\begin{aligned}
&a_0 = \frac{r_0 + K}{1 + r_0 K};\\
&K = R_{00}\exp\left(i4\pi\gamma\Gamma_0^{\varepsilon}\right) - \frac{R_{0,-1}R_{-1,0}r_{-1}\exp\left(i4\pi\gamma\left(\Gamma_0^{\varepsilon} - \Gamma_{-1}^{\varepsilon}\right)\right)}{1 + R_{-1,-1}r_{-1}\exp\left(i4\pi\gamma\Gamma_{-1}^{\varepsilon}\right)}.
\end{aligned}
\tag{7.9}
$$

The order of the error in the representation (7.9) is $\mathrm{O}(\exp(-4\pi\gamma\mathrm{Im}\,\Gamma_m^{\varepsilon}); m = 1, \pm 2)$. Here, as before, R_{np} are the elements of a generalized scattering matrix of grating in the ε-medium, and r_n is the reflection coefficient of the nth harmonic from a flat boundary $z = 2\pi\delta$.

The value $W_0^a(K)$ is minimized under the same conditions as in the case above (see the analysis after formulas (7.3) to (7.8)). However, because a minus first harmonic trapped in the layer is involved in the process now, there are many more methods available to reach the extremes. Thus, if the essential contribution to the formation of K is made by the first component (low degree of excitation of the wave with the number $n = -1$), a satisfactory solution can be obtained by synthesizing the grating after the scheme based on formulas (7.8). Otherwise, if the main contribution is made by the second component ($|R_{00}| \ll 1$), the result $W_0^a < |r_0|^2$ is obtained when the electromagnetic characteristics of the periodic structure satisfy the following conditions.

$$4\pi\gamma\left(\operatorname{Re}\Gamma_0^\varepsilon + \operatorname{Re}\Gamma_{-1}^\varepsilon\right) + \arg R_{0,-1} + \arg R_{-1,0} + \arg r_{-1} + \pi - \arg r_0 = 2\pi m - \pi, \quad (7.10)$$

$$4\pi\gamma\operatorname{Re}\Gamma_{-1}^\varepsilon + \arg R_{-1,-1} + \arg r_{-1} = 2\pi m; \quad m = 0, 1, 2, \ldots; \quad (7.11)$$

$$\frac{|R_{0,-1}||R_{-1,0}||r_{-1}|\exp\left[-4\pi\gamma(\operatorname{Im}\Gamma_0^\varepsilon + \operatorname{Im}\Gamma_{-1}^\varepsilon)\right]}{1 + |R_{-1,-1}||r_{-1}|\exp\left(-4\pi\gamma\operatorname{Im}\Gamma_{-1}^\varepsilon\right)} < \frac{2|r_0|}{1 + |r_0|^2}. \quad (7.12)$$

For small $\operatorname{Im}\tilde{\varepsilon}$, that is, if $|\arg r_0 - \pi| \ll 1$, the first one is the condition of a transverse crossed resonance in the layer (at the zero and the minus first harmonics). The zero harmonic propagates away from the boundary $z = 2\pi\delta$ toward the grating and transforms (with the coefficient $R_{-1,0}$) into the minus first harmonic. The latter double-crosses the layer, reflecting from the boundary $z = 2\pi\delta$, and transforms (with the coefficient $R_{0,-1}$) again into the harmonic with $n = 0$. This harmonic terminates the cycle while reflecting from the surface $z = 2\pi\delta$ with the common phase incursion divisible by 2π. A wave trapped in the layer should go the same way to reconstruct the phase characteristic (condition (7.11)).

For achieving the result $W_0^a < |r_0|^2$ (i.e., the same as in the region $\{1, 1\}$; see (7.7), (7.8)), the requirement $\cos(\arg K - \arg r_0) = -1$, from which in the considered case follow the conditions (7.10), (7.11), is not necessary. It is enough, for instance, to set the following restriction: $\cos(\arg K - \arg r_0) \leq -0.5$. However, in this case the search for the optimal absolute values of the elements of the generalized scattering matrix R is substantially complicated as they refer to corresponding phase characteristics. Conditions (7.10), (7.11) are the resonance conditions. Their influence in the search of extrema of characteristics (in the absolute values, in the bandwidth they are stored in, etc.) increases many times if they act together. All said is true also for (7.14), (7.15) (see below), that allow us, in combination with (7.10), (7.11) to optimize the absorbing capacities of the structure in the region $\{1, 2\}$ on any level of excitation of the principal (0th) and the minus first harmonics in the layer. This conclusion, important for model synthesis, is based on the results of analysis of various resonance and anomalous effects occurring due to the interaction (mode coupling) of free field oscillations in periodical structures (see [9] and Section 6.3). It is fit for solving the optimization problems on the basis of a one-frequency (central frequency), one directional (the average wave incidence angle is α) periodic structure synthesis, the input data for which, for the harmonics propagating in the ε-layer, are formed by the conditions such as (7.10) to (7.15). The efficiency of such an approach is estimated then after the solution to the direct problem for a layer with a synthesized substrate in the complete variation range of κ and α (stage (e)).

We return again to expression (7.9) and derive from it the following conditions of the global minimum of W_0^a. If the phase characteristics simultaneously satisfy the requirements (7.5), (7.10), and (7.11), in such a case for the energy characteristics we obtain

$$|R_{00}|\exp\left(-4\pi\gamma\,\mathrm{Im}\,\Gamma_0^\varepsilon\right)+\frac{|R_{0,-1}||R_{-1,0}|\,|r_{-1}|\exp\left[-4\pi\gamma\left(\mathrm{Im}\,\Gamma_0^\varepsilon+\mathrm{Im}\,\Gamma_{-1}^\varepsilon\right)\right]}{1+\left|R_{-1,-1}\right|\,|r_{-1}|\exp\left(-4\pi\gamma\,\mathrm{Im}\,\Gamma_{-1}^\varepsilon\right)}$$
$$=|r_0|\,. \tag{7.13}$$

If conditions (7.5) and (7.11) are satisfied, and

$$4\pi\gamma\left(\mathrm{Re}\,\Gamma_0^\varepsilon+\mathrm{Re}\,\Gamma_{-1}^\varepsilon\right)+\arg R_{0,-1}+\arg R_{-1,0}+\arg r_{-1}-\arg r_0=2\pi m-\pi;$$
$$m=0,1,\ldots, \tag{7.14}$$

then the plus sign of the second component in (7.13) should be changed to a minus sign. If conditions (7.5), (7.10) are satisfied, and

$$4\pi\gamma\,\mathrm{Re}\,\Gamma_{-1}^\varepsilon+\arg R_{-1,-1}+\arg r_{-1}=\pi+2\pi m;\quad m=0,1,2,\ldots, \tag{7.15}$$

the same changes should be undertaken in the denominator of the second component. The possibilities to extend the chain of variants are obvious. All of them are associated with the implementation of a certain combination of resonances at the zero and the minus first harmonics in the layer, and imply restrictions of similar type on the absolute values of the elements of the generalized scattering matrix of the grating in the ε-medium.

While opening the boundary $z=2\pi\delta$ for a minus first harmonic propagating in the layer, we enter the region $\{2,2\}$, where the energy is now transferred from the structure away into the far zone by two homogeneous plane waves having the amplitudes a_0 and a_{-1} ($\mathrm{Im}\,\Gamma_{-1}=0$). For the first of these amplitudes, expression (7.9), as well as the conclusion made in the preceding item, are valid. However, even by $W_0^a=0$ the energy goes away through the channel associated with the minus first harmonics. From the expression

$$\begin{aligned}a_{-1}&=(1+r_0)(1-r_{-1})\,R_{-1,0}\exp\left(i2\pi\gamma(\Gamma_0^\varepsilon-\Gamma_{-1}^\varepsilon)\right)\\&\quad\times[(1+R_{-1,-1}r_{-1}\exp\left(i4\pi\gamma\Gamma_{-1}^\varepsilon\right))(1+R_{00}r_0\exp\left(i4\pi\gamma\Gamma_0^\varepsilon\right))\\&\quad-R_{0,-1}R_{-1,0}r_0r_{-1}\exp\left(i4\pi\gamma(\Gamma_0^\varepsilon+\Gamma_{-1}^\varepsilon)\right)]^{-1},\end{aligned}$$

it follows that $W_{-1}^a=0$ only in the trivial case $R_{-1,0}=0$. In the autocollimation mode ($\Phi=0.5$, $\Gamma_0=\Gamma_{-1}$, $\Gamma_0^\varepsilon=\Gamma_{-1}^\varepsilon$, $r_0=r_{-1}$, $R_{00}=R_{-1,-1}$), when the minus first harmonic propagates toward the wave incident on the structure,

$$\begin{aligned}a_{-1}&=(1-r_0^2)R_{-1,0}\exp\left(i4\pi\gamma\Gamma_0^\varepsilon\right)\\&\quad\times[(1+r_0R_{00}\exp\left(i4\pi\gamma\Gamma_0^\varepsilon\right))^2-R_{0,-1}R_{-1,0}r_0^2\exp\left(i8\pi\theta\Gamma_0^\varepsilon\right)]^{-1},\end{aligned}$$

and the values of W_{-1}^{a} have a minimum, if

$$4\pi\gamma \operatorname{Re}\Gamma_0^{\varepsilon} + \arg R_{00} + \arg r_0 = 2\pi m; \quad m = 0, 1, 2, \ldots, \tag{7.16}$$

$$8\pi\gamma \operatorname{Re}\Gamma_0^{\varepsilon} + \arg R_{0,-1} + \arg R_{-1,0} + 2\arg r_0 + \pi = 2\pi m; \quad m = 0, 1, 2, \ldots. \tag{7.17}$$

It is easy to see that conditions (7.16), (7.17) are opposite to those under which a substantial reduction of the structure's efficiency in the zero order of the spectrum can take place.

Thus, the most favorable region for optimizing the absorbing capacity of the coatings is the region $\{1, M\}$ with $M \geq 2$, where a reasonable use of resonances at the radiating and trapped in the layer waves allows us to control the value of W within value ranges of λ and α of different widths, with the layer thickness not being very large and the matching at the boundary $z = 2\pi\delta$ being weak. The growth of M, which may be not only formal, but actual (due to the harmonics that are evanescent in the layer, although still participating in the energy transfer), can prompt reaching the required characteristics of the synthesized coating. The growth of N reduces the general reachable level $1 - W$. However, it allows us to state and solve (by using tools similar to those described above) the problems of optimal spatial reorientation of energy flows.

7.2.2. Pattern-Forming Grating Structures

Periodical pattern-forming structures (PFSs) are widely applied in antenna engineering and in scientific instruments. Statements of the model problems (in the theory of phased antenna gratings, diffraction radiation theory, etc.), whose analysis is the main source of knowledge relevant for applications, as a rule do not go beyond the frameworks of idealizations that are traditional for classical theory (an infinite grating of a canonical geometry in a plane wave field) and are not oriented to solving the problems of synthesis or optimization. Thus, for instance, modeling of a system "planar dielectric waveguide—reflecting grating", utilizing the effect of transforming the exponentially decreasing part of the field of an eigen E-polarized wave of a planar waveguide into propagating harmonics of the spatial spectrum of a periodic structure, is usually (in the approximation of the specified field) reduced to solving a standard boundary-value problem (6.2) to (6.5) with $\tilde{U}_p^i(g, \kappa) = \exp[i(\Phi_p y - \Gamma_p(z - 2\pi\delta))]$, $\operatorname{Re}\Gamma_p = 0$, $\operatorname{Im}\Gamma_p > 0$ (see Section 6.3.2 for an example concerning the synthesis of reflecting gratings for the devices of relativistic diffraction electronics).

Adequate electromagnetic modeling and synthesis of PFSs are more complicated problems: one has to take into account the presence of compact inhomogeneities near the finite or infinite periodic structures; to take into consideration the spatially limited wave sources and sources generating a limited spot of the excitation field, to predict the influence of random deviations in the parameter values, and so on. The choice of method to be applied to solve these problems should be decided in every specific case taking into account the actual conditions.

The common point here is only that the time domain methods are more efficient at stages (a) and (e), and the frequency domain methods are more suitable for the preparatory analytical work. Below we cite some results of such work, stating and solving nonstandard problems of the model synthesis of grating PFSs.

The values of the field pattern

$$D(\alpha) = \cos\alpha \sum_{n:|\kappa \sin\alpha + n| \le 0.5} \eta_n R_n(g_0, -\kappa \sin\alpha - n) \tag{7.18}$$

shaped by the reflecting periodic structure that is excited by a point source (6.21) (see Fig. 1.1B and Statement 6.11) are determined by the amplitudes $R_n(g_0, \Phi)$ of the propagating harmonics of the Green function $\tilde{G}(g, g_0, \Phi)$ of the grating in the field of quasi-periodic point sources (6.7). For $\tilde{G}(g, g_0, \Phi)$, according to condition (6.10), in the radiation zone $z \ge 2\pi\delta$ ($g_0 = \{y_0, z_0\}, 0 \le y_0 \le 2\pi, 0 < z_0 < 2\pi\delta$) the following expression is true.

$$\tilde{G}(g, g_0, \Phi) = \sum_{n=-\infty}^{\infty} R_n(g_0, \Phi)\, e^{i[\Phi_n y + \Gamma_n(z - 2\pi\delta)]}. \tag{7.19}$$

The diffraction characteristics of the grating in the field of a plane wave are easy to determine if the fundamental solution $\tilde{G}(g, g_0, \Phi)$ to the boundary value problem is known. In numerical analysis the respective relations are rarely used; simpler and more efficient algorithms have been developed, in fact, within the framework of plane wave diffraction theory (see, e.g., [3,5,6,9,11]). The result given below allows us to adapt these efficient algorithms for solving an inverse problem in the following sense: to restore function $\tilde{G}(g, g_0, \Phi)$ after the scattering characteristics of plane waves.

Statement 7.1 [9]. *In the case of an E-polarized field, amplitudes $R_n(g_0, \Phi)$ of the Green function (7.19) are determined by the relation*

$$4\pi R_{-p}(g_0, \Phi)\,\Gamma_{-p} = -i\tilde{U}(g_0, -\Phi). \tag{7.20}$$

Here, $\tilde{U}(g, -\Phi)$ is the value (at the point $g \in \mathbf{Q}$) of the unique nonzero component of the total electric field generated by the grating (see problem (6.2) to (6.5)) that is being excited by a plane homogeneous or inhomogeneous harmonic wave

$$\tilde{U}_p^i(g, -\Phi) = \exp[i(\Phi_p(-\Phi)y - \Gamma_p(-\Phi)(z - 2\pi\delta))];$$

$$\Phi_p(-\Phi) = p - \Phi, \quad \Gamma_p(-\Phi) = (\kappa^2 - [\Phi_p(-\Phi)]^2)^{1/2}.$$

Thus, an infinite grating in a field of a point source turns out to be quite a simple object to analyze and synthesize (see Section 7.3). Being incorporated into one computational scheme, the solution algorithms of the direct and inverse problems of the plane wave diffraction and formulas (7.18), (7.20) provide exhaustive information on the function $D(\alpha)$ and enable one to operate, for searching the optimal variants, with a wide range of means available using the specific features of the distribution of values $\tilde{U}(g_0, -\Phi)$ in the vicinity of the efficient scatterers. Examples of successful implementations of this approach are reported in [9,173].

The notions of a point source or a multitude of point sources composing a compact set in the plane $\mathbf{R}^2$ (in this case the above results can obviously be generalized), that are used in theoretical studies, often lose their initial sense when being implemented in practice. In the quasi-optical engineering both millimeter and submillimeter wavelength ranges, more convenient and widely accepted both in theoretical and applied studies, is the notion of a wave beam generated by a finite-dimension aperture with the specified primary field pattern (excitation field). Below we will consider only some specific features of the analysis and synthesis methods applied for gratings that are caused by the fact that the exciting field is a wave beam field. A more detailed consideration of the general questions of the analysis is presented in [14].

Suppose that a beam of plane harmonic waves

$$\tilde{P}^i(g) = \int_{-\infty}^{\infty} q(\Phi) e^{i(\Phi_0 y - \Gamma_0(z - 2\pi\delta))} d\Phi; \quad g = \{y, z\}, \quad z \geq 2\pi\delta \tag{7.21}$$

is incident to the grating (Fig. 1.1B). Here, Φ has the meaning of a product $\kappa \sin\alpha$, where α is the incidence angle of a beam component, and

$$q(\Phi) = \frac{1}{\kappa \cos\alpha_0} \exp\left\{-\left[\frac{w(\kappa \sin\alpha_0 - \Phi)}{2\cos\alpha_0}\right]^2 - i\Phi_0 y_0 + i\Gamma_0(z_0 - 2\pi\delta)\right\}.$$

The efficient beam width in the aperture plane that is normal to the beam axis (determined by the angle α_0) and crossing the point $\{y_0, z_0\}$ of this axis, is w. For every Φ, the beam component (a plane homogeneous or inhomogeneous wave with a unit amplitude) generates, in the region $z \geq 2\pi\delta$, a scattered field $\tilde{U}^s(g, \Phi) = \sum_{n=-\infty}^{\infty} R_{n0}(\Phi) \exp[i(\Phi_n y + \Gamma_n(z - 2\pi\delta))]$. Applying the superposition principle, for the scattered field generated by the periodic structure that is being irradiated by the beam $\tilde{P}^i(g)$, we obtain

$$\tilde{P}^s(g) = \int_{-\infty}^{\infty} q(\Phi)\, \tilde{U}^s(g, \Phi)\, d\Phi. \tag{7.22}$$

While implementing (7.22), similarly to the case of a point source, a problem of determining the integration path arises: the real axis Φ contains branch points of the surface $\mathbf{F}$ and, maybe, the poles of the function $\tilde{U}^s(g, \Phi)$ (see Section 6.2.3 and Statement 4.11). The solution suggested in [173] yields a physically correct result that is formulated below.

Statement 7.2. *Suppose that for the given frequency $\kappa > 0$ the poles $\bar{\Phi}_n$ of the function $\tilde{U}^s(g, \Phi)$ at the real axis of the first sheet of the surface* $\mathbf{F}$ *can only be simple, and among the elements of the set $\{\bar{\Phi}_n \in \Omega_\Phi : \mathrm{Im}\bar{\Phi}_n = 0\}$ there are no branch points of the surface* $\mathbf{F}$. *In such a case, for $r \to \infty$ ($|\alpha| < \pi/2$) and*

$|y| \to \infty$ $(|\alpha| = \pi/2)$,

$$\tilde{P}^s(g) = \left(\frac{2\pi\kappa}{r}\right)^{1/2} e^{i(r\kappa - \pi/4)} \cos\alpha \sum_{n=-\infty}^{\infty} q(-\kappa\sin\alpha - n) R_{n0}(-\kappa\sin\alpha - n)$$

$$+ \begin{cases} \pi i \sum\limits_{\bar{\Phi}_m \in M^+} \tilde{U}_m(g, \bar{\Phi}_m) \operatorname{Res} R_{00}(\bar{\Phi}_m) \sum\limits_{k=-\infty}^{\infty} q(\bar{\Phi}_m + k); & y > 0 \\ -\pi i \sum\limits_{\bar{\Phi}_m \in M^-} \tilde{U}_m(g, \bar{\Phi}_m) \operatorname{Res} R_{00}(\bar{\Phi}_m) \sum\limits_{k=-\infty}^{\infty} q(\bar{\Phi}_m + k); & y < 0 \end{cases}$$

$$+ O(r^{-1}) \tag{7.23}$$

The sets M^+ and M^- consist of all real poles of the function $\tilde{U}^s(g, \Phi)$ in the interval $|\Phi| \leq 0.5$, which corresponds to slow surface waves of the grating $\tilde{U}_m(g, \bar{\Phi}_m)$ that transfer the energy into the positive and negative directions along the axis y; $\tilde{U}_m(g, \bar{\Phi}_m)$ are normalized so that the amplitudes of their principal waves in the region $z > 2\pi\delta$ are equal to unity; r and α are polar coordinates (see Fig. 1.1B).

The radiation field far away from the zone of interaction between the beam $\tilde{P}^i(g)$ and the grating is formed by a diverging cylindrical wave having the amplitude (field pattern)

$$D(\alpha) = \cos\alpha \sum_{n=-\infty}^{\infty} q(-\kappa\sin\alpha - n) R_{n0}(-\kappa\sin\alpha - n); \quad |\alpha| < \pi/2, \tag{7.24}$$

and by eigensurface waves of an open periodic waveguide transferring the energy toward $|y| \to \infty$. The fact that the field $\tilde{P}^s(g)$ is free from partial components arriving in this region proves the correctness of the choice made in [173] when specifying the integration contour in (7.22).

By constructing $D(\alpha)$, the direct implementation of formula (7.24) requires a repeated solution to the problem of the type (6.2) to (6.5) for varying values of $\Phi = -\kappa\sin\alpha - n$, $|\alpha| < \pi/2$, $n = 0, \pm 1, \ldots$. If the efficient width w of the beam $\tilde{P}^i(g)$ is large, it hardly can be used also in forming the input data set in the synthesis problem for periodic PFSs. The result following from [173] considerably improves the situation.

Statement 7.3. *The following equality is true:*

$$R_{m0}(-\kappa\sin\alpha - m) = R_{m0}(\kappa\sin\alpha)\,\Gamma_m(\kappa\sin\alpha)\,\Gamma_0^{-1}(\kappa\sin\alpha). \tag{7.25}$$

Thus, along with (7.24), we can use the equivalent representation

$$D(\alpha) = \cos\alpha \sum_{n=-\infty}^{\infty} q(-\kappa\sin\alpha - n) R_{n0}(\kappa\sin\alpha) \frac{\Gamma_n(\kappa\sin\alpha)}{\Gamma_0(\kappa\sin\alpha)}; \quad |\alpha| < \frac{\pi}{2}, \tag{7.26}$$

which not only essentially simplifies the solution to the analysis problems, but allows us to apply already tested approaches and methods to the synthesis problems as well (see Section 7.3 and [173]).

7.2.3. Mode Selection in Open Resonators with Mirrors Made of Gratings

In this section, we consider more closely the modes-electing mechanism in open resonance systems. Such studies were started in [9], where a practically realizable approach to the problem of synthesizing a resonator with a very rarified spectrum and given electromagnetic characteristics at the working wavelength was suggested. The essence of the approach is as follows. A prototype (open or closed resonator) is chosen having a simple spectrum Ω_k^{prot}. At one of the eigenfrequencies $\overline{k} \in \Omega_k^{prot}$, the nonselective mirrors are replaced by electromagnetically equivalent grating mirrors, the spectrum Ω_k^{mod} of the modified resonator still having the oscillation $\tilde{U}\left(g, \overline{k}\right)$ of the prototype resonator that corresponds to the eigenfrequency $\overline{k}$. The other components of Ω_k^{mod} should satisfy the overdetermined set of dispersion equations including the modified dispersion equation of the prototype and relations that should be satisfied in order to maintain stable oscillations in open resonators (ORs) with grating mirrors. The complete set of dispersion equations includes the complex amplitudes R_{n0} and T_{n0} of the grating field (6.5) as a functions of frequency and incidence angle of plane waves (the plane waves that excite the grating and those composing its secondary field are considered to be partial components of the free oscillation field in OR). The set is overdetermined, hence, the spectrum Ω_k^{mod} will be definitely rarefied to a large extent (a part of the spectrum will be removed).

In fact, the spectrum Ω_k^{mod} is rarefied also by the significant increase in the Q-factor of the free oscillation remaining in the spectrum of the dispersive resonator. Whether the corresponding options will be realized depends on the values of the amplitudes R_{n0}, T_{n0} in the considered key parameter ranges. All together, these factors, causing the rarefication of Ω_k^{mod}, form the oscillation-selecting mechanism in the synthesized OR. This mechanism can prove very efficient when it comes to properly operating with various resonance scattering modes in the gratings (see Section 6.3.2 and papers [3,6,9,11]).

The approach proposed in [9] cannot be defined as a rigorous one. However, it satisfies the requirements and problems of the model synthesis and is based on the approximations that are accepted in quasi-optics. The analysis and synthesis of the dispersive elements (gratings) can be performed by mathematically valid methods; their effect on the characteristics of the quasi-optical system is demonstrated quite comprehensively and correctly by dispersion equations. What is the novelty of the approach that is based, similar to those from [207,208], on the electromagnetically equivalent replacement of mirrors? In [207,208], out of the complete set of electromagnetic characteristics of gratings, that is, complex amplitudes and propagation directions of the harmonics of the scattered field as functions of

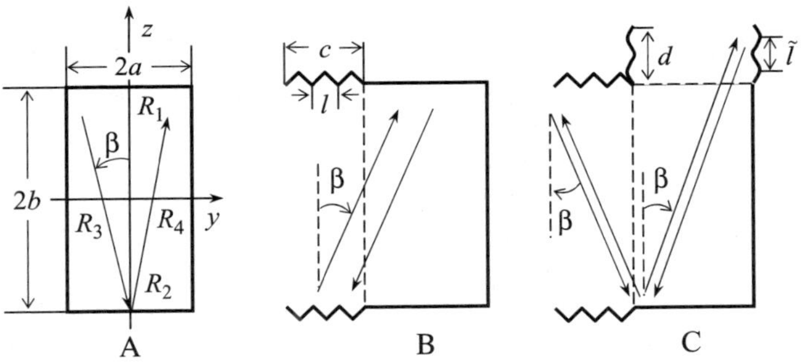

FIGURE 7.1. (A) Schematic presentation of prototype resonator and (B, C) modified resonators.

frequency, conditions of excitation, geometrical and material parameters, only the angle of divergence and, at most, experimentally determined diffraction efficiency are chosen to be used in the resonator model. It simplifies the analysis; still the conclusions can hardly be guaranteed to be correct.

Below, in a way similar to [9], the equivalence is used only to select the parameters of the synthesized resonator and to determine the possible ways of solving the problem (the prototype principle). It suffices to establish the equivalence at one certain point of the spectrum Ω_k^{prot}, and this essentially simplifies the synthesis at all stages. The distinctions in other spectral characteristics of the prototype and the resonator having a modified geometry are determined by the dispersion equations, whose analysis leads to solving the direct problems of the theory of wave diffraction by gratings. The result can be refined at stage (e), which is based on the time-domain methods allowing us to accurately calculate and optimize the key characteristics and parameters of the synthesized OR (see Section 7.4).

As a prototype, we choose a simple two-dimensional resonator (Fig. 7.1A). Its boundaries are not transforming (their operation regime is "incident wave → mirrored wave"), however, corresponding reflection coefficients (R_j, $j = 1, 2, 3, 4$) can be functions of the frequency and plane wave incidence angle. In the approximation of partial components of the free oscillation fields, the set of modes of such a structure

$$\tilde{U}_{nm}(g, \bar{k}_{nm}) = \left\{ e^{-i\bar{k}_{nm}(z-b)\cos\bar{\beta}_{nm}} + A_{nm} e^{i\bar{k}_{nm}(z+b)\cos\bar{\beta}_{nm}} \right\}$$
$$\times \left\{ e^{i\bar{k}_{nm}(y+a)\sin\bar{\beta}_{nm}} + B_{nm} e^{-i\bar{k}_{nm}(y-a)\sin\bar{\beta}_{nm}} \right\};$$
$$A_{nm} = R_1^{-1} e^{-2ib\bar{k}_{nm}\cos\bar{\beta}_{nm}}, \quad B_{nm} = R_3^{-1} e^{-2ia\bar{k}_{nm}\sin\bar{\beta}_{nm}} \tag{7.27}$$

and its eigenparameters $\bar{k}_{nm}$ and $\bar{\beta}_{nm}$ are determined from a simple set of dispersion equations

$$R_1 R_2 = e^{-i4bk\cos\beta}, \tag{7.28}$$
$$R_3 R_4 = e^{-i4ak\sin\beta}; \quad 0 < |\beta| < \pi/2. \tag{7.29}$$

Here, as often before, we consider oscillations of an E-polarized field ($E_x = \tilde{U}_{nm}$, $E_y = E_z = H_x = 0$ and $R_{n0} = a_n$); still, unlike, for example, Chapter 6, the coordinates y, z and the magnitudes $a, b, \lambda = 2\pi/k$ are dimensional quantities, as is convenient for a system having a large number of characteristic parameters. If, while modifying the resonator, a grating of the period length l is used, then the relation between the diffraction characteristics, being the functions of parameters k, β, and the functions of the parameters κ, Φ (see Chapter 6), on the other hand, is determined as $\kappa = l/\lambda = lk/2\pi$, $\Phi = \kappa \sin\alpha$; α (the incidence angle) is determined via β, while taking into account the mutual orientation of the longitudinal axis of the resonator and the normal to the plane of the periodic structure. The dimensionless geometric parameters of the grating $2\pi\delta$ (with the period 2π) are converted into the dimensional h (period length l) through the relation $\delta = h/l$.

Assume now that the synthesized resonator should maintain an oscillation of the prototype resonator having $\bar{k} = \bar{k}_{NM}$, which implies an additional substantial rarefication of the spectrum in the frequency band enclosing $\bar{k}_{NM}$. Here, in analogy with [9], we distinguish between the pure rarefication due to the decreased number of stable oscillations, and the actual rarefication, accompanied by a substantial decay of the Q-factor of the possible satellite stable oscillations. There are many ways to solve this problem (see [9]). Here we restrict ourselves only to one of them, namely, the simplest one. Our goal is to clearly demonstrate the scheme and emphasize the key factors that are relevant to the oscillation selection in the dispersive OR.

Let us make the following changes in the prototype geometry: we remove the specular reflector in the plane $y = -a$, and apply reflective periodic gratings having $l\mathrm{Re}(\bar{k}_{NM} \sin\bar{\beta}_{NM}) = \pi$ and $c > 2b\mathrm{Re}(\mathrm{tg}\bar{\beta}_{NM})$ onto the extension of the mirrors $z = \pm b$ (see Fig. 7.1B). The period structure (profile) of the gratings is determined by the condition $a_{-1}(\bar{k}_{NM}, \bar{\beta}_{NM}) = R_3$, that provides the equivalence of the dispersive resonator to the prototype resonator at the frequency $k = \bar{k}_{NM}$. Actually, being exposed to such changes, the gratings that are excited in the auto collimation reflection mode at the minus first harmonic of the spatial spectrum (see [6]: effects of the nonspecular wave reflection) produce an equivalent reflection coefficient at the conditional boundary $y = -a$, which is equal to R_3. In order to confirm this statement, it is enough to perform a quite simple but cumbersome procedure consisting in the description of the complete fields, that is, the incident and the reflected ones, in the partial domains $y > -a$ (taking into account the reflection coefficients R_1 and R_2 from the waveguide walls) and $y < -a$ (using the conditions of the autocollimation reflection at the minus pth harmonic, which is necessary for obtaining stable oscillations) with these fields being then joined at the boundary $y = -a$. The complete spectrum of the synthesized resonator is determined from a set of dispersion equations

$$R_1 R_2 = \mathrm{e}^{-i4bk\cos\beta}, \tag{7.30}$$

$$a_{-p}(k, \beta) R_4 = \mathrm{e}^{-i4ak\sin\beta}, \tag{7.31}$$

$$l\mathrm{Re}(k\sin\beta) = \pi p; \quad p = 1, 2, 3, \ldots. \tag{7.32}$$

Here, it is taken into account that stable oscillations, that is, those not changing the field geometry, can be generated in such a resonator also during the autocollimation reflection mode at the upper harmonics of the spatial spectrum ($p = 2, 3, \ldots$). The first equation (if compared to the set (7.28), (7.29)) remains unchanged. The changes made in the second equation have already been considered above. The role of the third, additional, equation is the filtration of possible stable oscillations; the plane wave reflected by the grating (a harmonic of the diffraction spectrum having the number $-p$) should be moving straight toward the arriving wave. A simple analysis after the formulas $\alpha_n = -\arcsin[(n + \Phi)\kappa^{-1}]$ for the angles α_n at which the propagating secondary field harmonics going out of the grating (see Section 6.3.2), shows that the oscillations, whose partial components do not satisfy (7.32), cannot be maintained in an open resonator, as the magnitude of deviation in the propagation directions of the corresponding plane waves from the longitudinal axis of the resonator grows with every complete passing.

If the eigenfrequency for the dispersive OR is chosen to be equal to EF $\bar{k}_{NM}$ of the prototype resonator, then in view of (7.32), the values $\mathrm{Re}\,(k \sin\beta)$ for the oscillations suitable for forming the spectrum Ω_k^{mod} are determined as follows (necessary condition).

$$\mathrm{Re}\,(k \sin\beta) = \mathrm{Re}(\bar{k}_{N,M\cdot p} \cdot \sin\bar{\beta}_{N,M\cdot p}) - \frac{p-1}{4a}\,(\arg R_3 + \arg R_4)\,. \qquad (7.33)$$

Here, $\bar{k}_{nm}$, $\bar{\beta}_{nm}$ are the eigencharacteristics of the prototype resonator and $p = 1, 2, \ldots$. The requirement of satisfying (7.33) results in a substantial rarefication of the spectral set and guarantees, in the synthesized resonator, an efficient selection according to the groups of eigenoscillations having different transversal indices. Thus, for instance, if $\arg R_3 + \arg R_4 = 0$, then the possible free oscillation is freed from the groups of eigenoscillations of the prototype resonator having frequencies $\bar{k}_{nm}$, with $m < M$ and $m \neq M \cdot p$; $p = 1, 2, \ldots$. The geometry factor (changes in the mirrors, locations and their operation in the nonspecular reflection mode) is decisive here. This factor is considered to be the first component of the oscillation-selecting mechanism in a dispersive OR.

To implement the values of $\mathrm{Re}\,(k \sin\beta)$, determined by (7.33), as eigenvalues, the equation (7.31) should be satisfied first, which, together with (7.33), leads to

$$\arg a_{-p}\,(k, \beta) = (p-1)\arg R_4 + p \arg R_3; \quad p = 1, 2, \ldots, \qquad (7.34)$$

$$\mathrm{Im}\,(k \sin\beta) = (4a)^{-1}\,\{\ln|R_4| + \ln|a_{-p}(k, \beta)|\}. \qquad (7.35)$$

Condition (7.34) causes an additional spectrum rarefication, as it implies strict requirements for the argument value of the coefficient of reflection into the autocollimating minus pth harmonic at the points k and β, that correspond to the free field oscillations: $\arg a_{-p}\,(k, \beta)$ is a complex function of parameters k, β, and so the space of the values of k and β, satisfying both (7.33) and (7.34), is narrower than the similar region for values satisfying only (7.33). The key role belongs here to the dependence of $\arg a_{-p}$ on k and β, which is considered to be the second component of the selecting mechanism.

Assume that the point $\{\bar{k}, \bar{\beta}\}$ is the eigenvalue of the dispersive resonator whose spectrum is already rarefied due to the first two components of the selecting mechanism (pure rarefication; $\bar{k}$ and $\bar{\beta}$ satisfy to (7.30), (7.33), (7.34)). From (7.35) follows that the Q-factors of the remaining satellite oscillations are determined by the value of $|a_{-p}(\bar{k}, \bar{\beta})|$. As the nonspecular reflection mode has a resonance nature, it is unlikely that for $\bar{k} \neq \bar{k}_{NM}$ and $\bar{\beta} \neq \bar{\beta}_{NM}$, the value of $|a_{-p}(\bar{k}, \bar{\beta})|$ will be large enough. Hence, the Q-factors of the satellite oscillations in the dispersive resonator will not be high, either. This is the third component of the selective mechanism: the dependence between the absolute values of the reflection coefficient of the grating and the excitation parameters. This component ensures the actual additional spectrum rarefication. The second and third components are easily controllable, as the geometric and material parameters of the grating can be varied, as well as the scheme of its equivalent integration into the prototype geometry.

The distinguishing feature of the considered approach is that the key characteristics of the chosen eigenoscillation remain unchanged when the prototype geometry undergoes a modification. From here follows the condition of a successful solution to the problem of model synthesis of a dispersive resonator: one of the oscillations generated by the prototype resonator should possess the required characteristics, that is, the wavelength of operation, Q-factor, field geometry. Thus, the classical Fabry–Perot resonator with plane specular reflectors (i.e., an open-ended section of a plane-parallel waveguide) cannot be taken as the basis for synthesizing a structure maintaining a high-quality eigenoscillation for the values of the transverse index m that are larger than several units. The most useful geometry for this case is the prototype geometry shown in Figure 7.1A. In the considered modification, its longitudinal oscillation, having experienced some minor disturbances, remains in the spectrum of the dispersive OR. Further changes in the prototype specular reflector geometry can make the selection procedure more efficient. Thus, for instance, while replacing the plane specular reflector $z = b$ with the grating sections within the interval $z > b$ of planes $y = \pm a$ (see Fig. 7.1C; the grating parameters are subject to the conditions of small disturbance of the prototype oscillation at $\bar{k} = \bar{k}_{NM}$: $\tilde{l}\mathrm{Re}\left(\bar{k}_{NM}\cos\bar{\beta}_{NM}\right) = \pi$, $d > 2a\mathrm{Re}\left(\mathrm{ctg}\bar{\beta}_{NM}\right)$, $\tilde{a}_{-1}\left(\bar{k}_{NM}, \bar{\beta}_{NM}\right) = R_1$) we enlarge the total amount of conditions imposed on the free oscillations; the spectral set is being further considerably rarefied, with the eigenfrequencies of longitudinal oscillations being withdrawn. The complete set of dispersion equations of the synthesized OR will then consist of (7.31), (7.32) and

$$\begin{cases} \tilde{a}_{-\tilde{p}}(k, \beta) R_2 = \mathrm{e}^{-i4bk\cos\beta} \\ \tilde{l}\,\mathrm{Re}(k\cos\beta) = \pi\tilde{p}; \quad \tilde{p} = 1, 2, \ldots \end{cases} . \tag{7.36}$$

Conditions (7.33), (7.34), that provide for the pure rarefying of the spectrum, are complemented by the following.

$$\begin{cases} \mathrm{Re}(k\cos\beta) = \mathrm{Re}(\bar{k}_{N\cdot\tilde{p},M}\cos\bar{\beta}_{N\cdot\tilde{p},M}) - \dfrac{\tilde{p}-1}{4b}(\arg R_1 + \arg R_2) \\ \arg\tilde{a}_{-\tilde{p}}(k, \beta) = (\tilde{p}-1)\arg R_2 + \tilde{p}\arg R_1 \end{cases} . \tag{7.37}$$

The third component of the selection mechanism also becomes evident, which, being put into action, causes the reduction of the Q-factor of the accompanying stable oscillations of the dispersive OR. Formally, it is expressed by adding to (7.35) the condition

$$\operatorname{Im}(k\cos\beta) = (4b)^{-1}\{\ln|R_2| + \ln|\tilde{a}_{-\tilde{p}}(k,\beta)|\}.$$

We do not go deep into the details of the changes caused by the next step of modification, we just mention that satisfying conditions (7.33), (7.37) results in a significant rarefying of the spectral set. Thus, for example, when $\arg R_1 + \arg R_2 = 0$ and $\arg R_3 + \arg R_4 = 0$ only oscillations at the frequencies close to $\bar{k}_{N\cdot\tilde{p},M\cdot p}$; $p, \tilde{p} = 1, 2, \ldots$ are possible. The examples of implementation of this and similar approaches (selection of gratings that are able to provide the required diffraction characteristics, experimental and theoretical study of dispersive resonators with grating mirrors in the frequency domain, etc.) for diverse modifications of the geometry of the prototype resonator and the modification methods can be found in [9,209–211]. It is obvious that in the frequency domain, the theoretical study of the structures synthesized after this method cannot provide comprehensive and precise information. It can be well complemented by the results of analysis in the time domain. The analysis of quasi-optical systems containing grating dispersive elements is the subject of Sections 7.4 and 7.5.

7.3. Inverse Problems in Electromagnetic Theory of Gratings (Frequency Domain)

7.3.1. General Statements

Let us come back to the problem (6.1) to (6.5), assuming now that the field $\tilde{U}(g,\kappa)$ in the form (6.5) is defined partially or completely by its complex amplitudes $\{R_{np}\}$, $\{T_{np}\}$; p is a fixed number. We now describe, without going into details of problems associated with the complexity and level of errors in input data, the inverse problems of interest as problems of the reconstruction of profiles **S** and (or) relative permittivity $\tilde{\varepsilon}(y,z)$ of periodic structures, compatible (or close to compatible) with the required characteristics X.

In reconstruction problems in the case when input data are exactly known the question about existence of the solution does not arise. The main problems are connected with the definition of a sufficient body of input data that may guarantee the uniqueness of the solution. In synthesis problems, when the problem of possible achievement of required (or close to required) characteristics is under consideration, the main problem is the existence of the solution. Here the exact and complete input data (data, providing a unique solution to the inverse problem) are only abstract notions, useful for an analytic study. The use of incomplete and (or) approximate input data (as is typical in most applications) gives rise to the following problems. First, it is necessary to define more precisely the notion "solution to the problem". Here the solution in a conventional meaning (the object with really exciting parameters of geometry and materials) may not exist. Second, the question about the existence of the newly defined meaning of "solution" and its stability in respect to small variations in the input data arises.

Consider several possible solutions of such problems, applying the theory of ill-posed problems [56]. To be specific, consider the synthesis problem for a reflective grating (the profile is defined by the contour $\mathbf{S} \in \mathbf{L}$, $\tilde{\varepsilon}(y,z) \equiv 1$) forming the required (or close to the required) diffraction characteristic $X \in \mathbf{M}$. Here $\mathbf{L}$ and $\mathbf{M}$ are given by parametric descriptions of $\mathbf{S}$ and X as elements in Banach spaces. In the case when $X = \{R_{np}\} \cup \{T_{np}\}$, $\mathbf{M}$ is a space of infinite sequences, and a possible structure of operator $D : \mathbf{L} \to \mathbf{M}$, $X = D[\mathbf{S}]$ is obtained by means of the algorithm of solving the direct problem. If we use a characteristic $\tilde{X}$ that is in a certain sense close to X, it is impossible to guarantee in the general case that among the elements of the space $\mathbf{L}$ there is at least one ($\mathbf{S}$) such that $\tilde{X} = D[\mathbf{S}]$. Preliminary selection of data $\tilde{X}$ by results of the verification of all necessary conditions and requirements (conservation laws, reciprocity relations, analytical properties) is not able to solve the problem of existence of the solution, but in several cases the analysis connected with such selection turns out to be rather useful, inasmuch as taking into consideration the properties of the operator D, one can reduce the domain of search $\mathbf{S}$. Assume that the solution to the inverse problem has to be in a closed set $\mathbf{K}$ that is compact in $\mathbf{L}$. Then it can be defined as an element $\tilde{\mathbf{S}} \in \mathbf{L}$, and the functional $\psi(\mathbf{S}) = \|\tilde{X} - X\|$ (norm of discrepancy), $X = D[\mathbf{S}]$, $\mathbf{S} \in \mathbf{K}$ then has a minimum.

The solution $\tilde{\mathbf{S}}$ defined in such a way does exist. Continuous dependence of $\tilde{\mathbf{S}}$ from $\tilde{X}$ can be realized by changing the volume of input data: if for exact input data the inverse problem has not more than one solution and the projection $\tilde{X} \in \mathbf{M}$ to the set $D[\mathbf{K}]$ is unique, then the solution $\tilde{\mathbf{S}}$ for given $\tilde{X}$ is also unique and depends continuously (within the metric of the space $\mathbf{L}$) on $\tilde{X}$ [56].

Problems for which it is impossible to reduce the domain of search for the solution to a compact set $\mathbf{K}$ and with input data $\tilde{X}$ such that $\tilde{X} \neq D[\mathbf{S}]$ for all $\mathbf{S} \in \mathbf{L}$, are called essentially ill-posed problems. Finding approximate solutions that are stable with respect to changes in input data may be reduced here to the construction of the regularizing operators R_ξ and to the definition of parameter of regularization ξ, using additional information about $\tilde{X}$. The requirement for the solution to be stable is part of the property of the fundamental notion of a regularizing operator that is defined over all $X \in \mathbf{M}$ and approaches the operator D^{-1} over elements $X = D[\mathbf{S}]$; $\mathbf{S} \in \mathbf{L}$. There are numerous schemes for the construction of R_ξ. These schemes, as are those for the regularization parameters, are usually defined in an ambiguous way. The most conventional scheme is connected with minimization of the functional

$$\psi(\mathbf{S}, \xi) = \|\tilde{X} - X\|^2 + \xi\varphi(\mathbf{S}); \quad X = D[\mathbf{S}], \quad \mathbf{S} \in \mathbf{L}_1, \tag{7.38}$$

where $\varphi(\mathbf{S})$ is a stabilizing functional and $\mathbf{L}_1$ is everywhere dense in $\mathbf{L}$. For any $\tilde{X} \in \mathbf{M}$ and any $\xi > 0$ there is the element $\tilde{\mathbf{S}} \in \mathbf{L}_1$, for which (7.38) reaches its exact lower bound. The choice of the value of ξ defines if the operator $R_\xi : \tilde{\mathbf{S}} = R_\xi[\tilde{X}]$ is the regularizing operator for the inverse problem under consideration. That is, if the solution $\tilde{\mathbf{S}}$ to this inverse problem is stable. If for exact input data X there is a unique solution $\mathbf{S}$ to the inverse problem, then for any $\eta > 0$ there are β and ξ (depending on β) such that from $\|X - \tilde{X}\| < \beta$ follows the inequality $\|\mathbf{S} - \tilde{\mathbf{S}}\| < \eta$.

The solution $\tilde{\mathbf{S}}$ naturally depends on the value of ξ and on the form of the functional $\varphi(\mathbf{S})$. But even for fixed ξ and $\varphi(\mathbf{S})$ it may not be unique. In [56] sufficient conditions providing uniqueness of $\tilde{\mathbf{S}}$ are given. The following requirements are sufficient: D is a linear operator and $\varphi(\mathbf{S})$ is a quadratic functional. The linearity property of D essentially simplifies the routine of the determination of ξ as a function of β. The relevant resolvent operator is the regularizing operator for the corresponding inverse problem. This problem (under unimportant restrictions for $\varphi(\mathbf{S})$) may be used to search for the element $\tilde{\mathbf{S}}$, for which the functional (7.38) has a minimum (under restrictions of classical type). That is, the inverse problem may be reduced to solving the Euler equation.

Thus, we can conclude that for linear operators D, if the exact input data provide the unique solution to the inverse problem, the latter one may be correctly formulated and correctly resolved numerically. In order to utilize such possibilities it is necessary to complement data $\tilde{X}$ up to the volume that in the case of exact data X provides the inverse problem with a unique solution. If the operator D is not linear (the predominant situation for problems of interest), it is necessary to carry out additional work and linearize the relations between the parameters of the unknown object and its diffraction characteristics that play the role of input data.

One of the ways of resolving the problem of linearization is connected with substituting the equation $X = D[\mathbf{S}]$ by the equivalent sequence

$$X = D_1[\mathbf{S}_1], \quad X_1 = B_1[\mathbf{S}_1]; \quad X_1 = D_2[\mathbf{S}_2], \quad X_2 = B_2[\mathbf{S}_2]; \quad \ldots;$$
$$X_{n-1} = D_n[\mathbf{S}], \tag{7.39}$$

where D_j are linear operators and X_j are auxiliary data.

The inversion of n problems (7.39) results in the determination of profiles $\mathbf{S}$ of the gratings that give the required diffraction characteristic X. Such an approach has been implemented in [15] for reconstruction of impedance values on the surface of scatterers by a given pattern. If the problem is linearized accordingly (7.39), then inaccuracies in the specification of X result in inaccuracies in the determination of the kernel of the operator D_n. The drawbacks arising here are of a nonprincipal character. The changes that have to be introduced into the code, implemented according to the algorithm of the regularized solution of the corresponding linear operator equation of the first kind, are rather formal and insignificant [56]. Alas, in most situations this simple schema cannot be implemented or turns out to be numerically inefficient (too time consuming, etc.).

The second frequently used version of the linearization of inverse problems (see, e.g., [212–215]) is based on the classic Newton–Kantorovitch method. The following factors define the efficiency of the algorithm for construction of the regularized solution $\tilde{\mathbf{S}}$ in this case: having a good initial guess (there are no general recipes for how to find this), the available analytic or numerical routines for calculation of Frechét derivatives of the operator D, and the parameters of the direct problem-solving algorithms (such as guaranteed required accuracy of the numerical solution, efficiency, and speed). In Section 7.3.3 we consider several more versions of approximate linearization used for solving the inverse problem for reflective gratings with grooves of arbitrary profile. We take up in that section the analysis of the problems connected with those versions. Just one remark here: a variety of applied inverse problems of acoustics, optics, nondestructive testing, and remote sensing can be treated within the framework of similar techniques.

7.3.2. Uniqueness Theorems

We have already pointed out the significance of definition of a sufficiently rich set of input data that provide the uniqueness of the solution to the inverse problem so that these problems can be analyzed in formally correct way. The relevant results have to be accumulated for various objects, parameters, and characteristics, obtainable by measurements. The statements formulated below concerning uniqueness of the solution to inverse problems are obtained assuming that all input data are defined exactly and are relevant to objects existing in reality. The methods and techniques used in the derivation of the proofs of these statements are based on the results published in [9,199,201,216–219]. Here we outline only the schemes of the proofs. We start with one rather general result [9,15], describing analytic properties of the solutions to the Helmholtz equation as functions of spatial coordinates. This result is frequently used in the study of the problem of uniqueness of the solutions.

Statement 7.4 (about analyticity). *Let $\tilde{\mathbf{R}}$ be a set of points $g = \{y, z\}$ of the strip $\mathbf{R}$ (Floquet channel), where the grating scatterers and sources of excitation $\tilde{f}(g)$ are localized. Then the solution $\tilde{U}(g, \kappa)$, $\kappa \notin \{\bar{\kappa}_n\} \cup \{\kappa_n\}$ to the problem (6.8) to (6.10) is an analytic function of the variables y u z everywhere in $\mathbf{R}\backslash\tilde{\mathbf{R}}$ (representable by convergent power series). The property of uniqueness of continuation in the weak sense follows from this: if a solution to the problem (6.8) to (6.10) is equal to zero in an open set in $\mathbf{R}\backslash\tilde{\mathbf{R}}$, then it is equal to zero identically. This result is also valid for the solutions $\tilde{U}(g, \kappa)$ to the diffraction problems (6.1) to (6.5). The uniqueness of the solution to the Cauchy problem for the Helmholtz equation is closely connected with the analyticity too: if this solution is subject to zero Cauchy conditions on a curve* **S**, *then it is identically equal to zero.*

Let the fields incident on a periodic grating (Fig. 1.1) be given or be absent (spectral boundary value problems for the eigenfields $\tilde{U}_n(g, \bar{\kappa}_n)$). Let us take as input data of inverse problems of reconstruction of **S** and (or) $\tilde{\varepsilon}(y, z)$ the values X of the complete field along the coordinate lines bounding the domain, containing the scatterers and sources. Thus, for example, for a semi-transparent structure (see Fig 1.1A)

$$X = \left. \begin{cases} \tilde{U}(g, \kappa), & \kappa > 0 \\ \tilde{G}(g, g_0, \kappa), & g_0 \in \mathbf{Q}_\delta, \kappa > 0 \\ \tilde{U}_n(g, \bar{\kappa}_n), & \bar{\kappa}_n \in \Omega_\kappa \subset \mathbf{K} \end{cases} \right|_{z=\pm 2\pi\delta, \qquad 0 \le y \le 2\pi}$$

in the case of excitation by a plane wave, by a wave from a point source, or in the regime of eigenoscillations. Note that X given thus defines in a unique way the field scattered by the grating (or eigenfield) everywhere in $\mathbf{Q}\backslash\mathbf{Q}_\delta$ and everywhere in $\mathbf{Q}$ in the case $\tilde{\varepsilon}(y,z) \equiv 1$. It follows the radiation conditions (6.5), (6.10) and from Statement 7.4. It is clear that the same amount of information can be obtained by specifying X in any 2-D open set in $\mathbf{Q}\backslash\mathbf{Q}_\delta$ above and below the grating.

Statement 7.5 (about uniqueness in the case $\tilde{\varepsilon}(y,z) \equiv 1$ for E-polarization). *The contours* $\mathbf{S}$ *of perfectly conducting gratings are defined uniquely by specifying the total diffraction field* $\tilde{U}(g,\kappa)$ *(eigenmode field* $\tilde{U}_n(g,\bar{\kappa}_n)$*) on any interval of variation of frequency parameter* κ *(at single complex eigenfrequency* $\bar{\kappa} \in \Omega_\kappa$, $\mathrm{Im}\bar{\kappa} \neq 0$*). For any finite* $\delta > 0$ *there is a finite composition of modes of excitation of the structure that differ from each other by the collections of sounding frequency, excitation angles, or the numbers of incident harmonic, for which the specification of diffraction characteristics* $X = \tilde{U}(g,\kappa)$*;* $\kappa > 0$, $z = \pm 2\pi\delta$, $0 \le y \le 2\pi$ *unambiguously define the* $\mathbf{S}$.

The major idea of the proofs of all the facts of Statement 7.5 has been formulated for the first time by M. Schiffer (see [220]). The key point consists in creating a contradiction from the assumption that there are two different solutions to the inverse problem, with obvious consequences from Statement 7.4. For example, if we assume that at a certain frequency κ two different contours $\mathbf{S}_1$ and $\mathbf{S}_2$ such that $\overline{\mathrm{int}\,\mathbf{S}_1} \cap \overline{\mathrm{int}\,\mathbf{S}_2} = \varnothing$ generate the same data $X = \tilde{U}$, then the field $\tilde{U} - \tilde{U}_p^i$, according to Statement 7.4, has to be an entire function of the coordinates y and z. A solution to the Helmholtz equation, satisfying the quasi periodic boundary conditions (6.3), in this case has to be either a packet of plane waves or a function that is identically equal to zero. The first possibility gives a contradiction to the radiation conditions (6.5): waves coming from infinity must not be present in the scattered field. The second one leads to the obviously unrealizable requirement that $\tilde{U}_p^i(g) = 0$ on $\mathbf{S}_1$ and $\mathbf{S}_2$.

Following the terminology of references [217,218], let us call two gratings defined by contours $\mathbf{S}_1$ and $\mathbf{S}_2$ equivalent at given frequency κ with respect to coinciding characteristics X, if the set of points g in the Floquet channel $\mathbf{R}$ that are not common to $\overline{\mathrm{int}\mathbf{S}_1}$ and $\overline{\mathrm{int}\mathbf{S}_2}$ is not empty. Equivalent gratings (scatterers) naturally do exist. The simplest example here may be half-spaces with plane perfectly reflecting boundaries at $z = \mathrm{const}$, providing a change of the phase of the field $\tilde{U}(g,\kappa)$ by a multiple of 2π. The arguments mentioned above confirm that equivalent scatterers give with necessity a nonempty intersection of $\overline{\mathrm{int}\mathbf{S}_1}$ and $\overline{\mathrm{int}\mathbf{S}_2}$. The domains $\mathrm{int}\mathbf{S}_1\backslash(\mathrm{int}\mathbf{S}_1 \cap \mathrm{int}\mathbf{S}_2)$ and $\mathrm{int}\mathbf{S}_2\backslash(\mathrm{int}\mathbf{S}_1 \cap \mathrm{int}S_2)$ in $\mathbf{R}$ are resonant for the sounding frequency κ. The total diffraction field $\tilde{U}(g,\kappa)$ or field of eigenoscillation $\tilde{U}_n(g,\bar{\kappa}_n)$ here plays simultaneously the role of eigenfields of closed resonators (with generalized boundary conditions (6.3), (6.4)) with a discrete, real-valued spectrum of finite multiplicity. This fact, which is a direct consequence of Statement 7.4, leads to the step that is natural for providing inverse problems with unique solutions: the sounding within a certain frequency range or utilization of complex eigenfrequencies of free oscillations of the field. In the limit case of two unclosed thin screens $\mathbf{S}_1$ and $\mathbf{S}_2$ the above statements partially lose their validity and require a more detailed consideration. The boundaries of the resonant range for sounding frequency are defined here as the union of $\mathbf{S}_1$ and $\mathbf{S}_2$. None of the perfectly sharp edges of $\mathbf{S}_1$ and $\mathbf{S}_2$ is "hanging" in $\mathbf{R}$, but closing at the point g of dual contour is necessary. If we assume the opposite, then we arrive at a contradiction to Statement 7.4, as geometrical singularities of boundaries $\mathbf{S}$ (edges) cause isolated singular points of the functions $\tilde{U}(g,\kappa) - \tilde{U}_p^i(g,\kappa)$, $\tilde{U}_n(g,\bar{\kappa}_n)$, or $\tilde{G}(g,g_0,\kappa)$ to appear when they are analytically continued from the domain $\mathbf{R}\backslash\overline{\text{int}\mathbf{S}}$ (see, e.g., reference [221]).

For any finite $\delta > 0$ and any fixed mode of excitation $\tilde{U}_p^i(g,\kappa,\Phi)$, the number of equivalent gratings is finite (the mode of excitation is uniquely defined by the parameters κ, Φ, and number p of the incident plane homogeneous or inhomogeneous E-polarized wave). If this statement were not true, then finite frequency κ would turn out to be resonant for a sequence of domains with unboundedly decreasing volumes and we would arrive at a contradiction to known [220,222] properties of eigenvalues of the relevant homogeneous problems. Considering the arbitrary interval $[\kappa_1;\kappa_2]$ of "allowed" sounding frequencies, we can be sure that in this interval there are a finite number of values κ (this number depends on value of δ), such that if we use them for obtaining characteristics X, then we can define $\mathbf{S}$ in a unique way. Actually, starting the visualization procedure from any $\kappa \in [\kappa_1;\kappa_2]$ and having defined the relevant number of equivalent scatterers, even in most unfavorable situations (such as getting exactly at the resonant frequencies of already clarified resonant regions), we can arrive at the true profile $\mathbf{S}$ through a finite number of steps. The maximum number of such steps exceeds by unity the necessarily finite number (in the finite interval $[\kappa_1;\kappa_2]$) of all eigenfrequencies of all resonant regions, formed by the gratings that are equivalent at the frequency κ. It is clear that similar results can be achieved by stepwise changing other parameters that define the excitation mode of the structure. The methods and techniques of sounding using multiple frequencies and positions, widely used in applications, are verified in such a way.

Statement 7.6 [9,201] (about uniqueness in the case $\mathbf{S} = \emptyset$ for E-polarization). *The function* $\tilde{\varepsilon}(y,z) \in \mathbf{C}\left(\bar{\mathbf{Q}}_\delta\right)$, *defining the constitutive parameters of dielectric periodic structure, is defined in a unique way by values of* $X = \tilde{U}(g,\kappa)$, $z = \pm 2\pi\delta$, $0 \leq y \leq 2\pi$ *of the distribution of the total diffraction field within any interval of variation of the frequency parameter* κ.

Every inverse problem with specified input data formulates, as a rule, in a new fashion the questions of existence and uniqueness of the solution. An analysis of all possible situations can hardly be supplied with the results listed above. For example, the information about phase characteristics of diffraction processes is not always available, and energy characteristics have analytic properties that are not essential for the analysis. Periodic structures at any frequency form only a finite number of channels that radiate energy into the free space (in (6.5) only harmonics of numbers n such that $\mathrm{Im}\Gamma_n = 0$ can propagate without attenuation). That means that true information about total field $\tilde{U}(g, \kappa)$ may be obtained only by measurements close to the grating.

The information about amplitudes of propagating harmonics is more easily accessible. The numerical algorithms should be oriented toward using this information. But here we also meet the requirements to apply nonconventional analysis: there are plenty of examples [6] showing that even such measurements, performed within the frequency range, may not be enough for an unambiguous definition of the surface profile. Thus, in scattering of a plane wave by an arbitrary semi-transparent (see Fig. 1.1A) or reflective (see Fig. 1.1B) periodic structure, the reflection coefficient R_{00} does not depend on the sign of the incident angle ($R_{00}(\Phi) = R_{00}(-\Phi)$). In the language of optics this fact, which follows from the reciprocity relations (6.12), means that the efficiency of any grating in zero order of the reflected spatial spectrum does not change when the structure is rotated by 180° around the axis normal to the plane of the grating. In a single-mode range ($\mathrm{Im}\Gamma_n > 0$ for $n \neq 0$) the value $|T_{00}|$ also does not depend on the sign of Φ (see [3,6,9]).

The following example, rather unexpected against the background of Statement 7.5, shows the necessity of a rather careful approach to the analysis of the problem of uniqueness of the solution to inverse problems. In the case of H-polarization of the field with zero values of Φ and p, for any finite δ, there are arbitrarily many equivalent gratings of knife type (gratings composed with infinitely thin metal stripes of finite width, place in planes $y = \mathrm{const}$) in any sounding frequency range. Such structures are practically unable to disturb the incident wave.

Concluding this section, it is worth noting that the questions of uniqueness and existence of solutions to inverse problems will still be on the agenda of researchers for many years to come. Even a partial resolution of some of these problems is likely to promote progress in this scientific field.

7.3.3. Arbitrary Profile Reflective Grating: Visualization Problem

A computational scheme, which in the case of single-frequency and single-position sounding allows us to define with sufficient accuracy a profile of a perturbation of a regular perfectly conducting boundary in two-dimensional space $\mathbf{R}^2$ that is of finite length and small in depth, has been suggested in [223]. This scheme may be considered as one particular version of a more general approach that is based on the idea of quasi-linearization of integral representations of potential theory. This idea is rather universal with respect to the classes of inverse boundary value problems that might be resolved numerically in an efficient way. It may also stimulate the appearance of many numerical algorithms that are simple in implementation and which can be used both in the long wave range and in the resonant range. In the present section we study several of these possibilities.

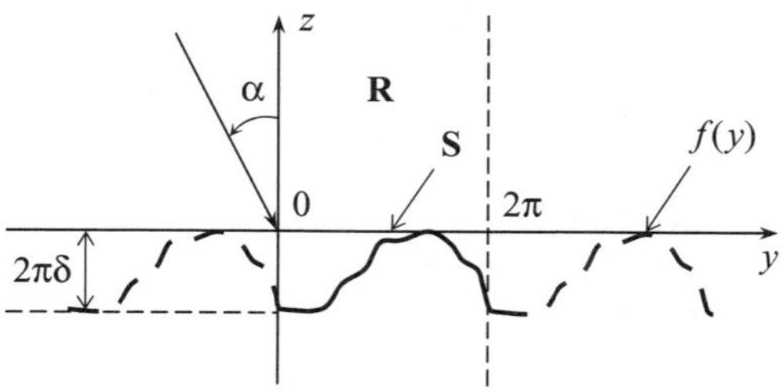

FIGURE 7.2. Presentation of geometry of the problem.

The arbitrary profile reflective grating has been chosen as a model structure. The results we have obtained may be useful in solving diagnostic problems, in numerical visualization of compact dielectric and metal discontinuities in dielectric half-space, in problems of reconstruction of periodic and nonperiodic boundaries among media, profiles of metal substrates of frequency-selective magnetodielectric layers, and so on. The universality of the algorithms and their ability to be adapted without considerable changes to essentially different situations and the transparency of the final computational schemes should make this approach useful for solving numerous applied problems of acoustics, optics, and microwave engineering.

Consider a perfectly conducting grating (see Fig. 7.2) placed in the field of the E-polarized plane wave $\tilde{U}_0^i = \exp\left[i\left(\Phi_0 y - \Gamma_0 z\right)\right]$, $\Phi = \kappa \sin\alpha$. The solution to the direct problem of the kind (6.2) to (6.5) can be represented in this case as a single-layer potential [9,15]

$$\tilde{U}(g) - e^{i(\Phi_0 y - \Gamma_0 z)} = -\int_{\mathbf{S}} \mu(g_0)\tilde{G}_0(g, g_0)\, ds_{g_0} \tag{7.40}$$

everywhere in $\mathbf{R}\backslash\mathbf{S}$. Here $\mathbf{S}$ is part of the periodic contour belonging to $\mathbf{R}$, the continuous density $\mu(g_0)$ is the solution to the singular boundary equation

$$\int_{\mathbf{S}} \mu(g_0)\tilde{G}_0(g, g_0)\, ds_{g_0} = \tilde{U}^i(g); \quad g = \{y, z\} \in \mathbf{S}, \tag{7.41}$$

and $\tilde{G}_0$ is the canonical Green function (Green function of the regular Floquet channel; see Sections 6.2.1 and 6.5.2).

The inverse problem (problem of visualization or problem of the reconstruction of groove profile) is to define the contour $\mathbf{S}$ by the complex amplitudes $\{a_n\} = \{R_{n0}\}$ of the total field $\tilde{U}(g, \kappa)$, that is given in the domain $z > 0$. Let the angle α (the angle of incidence of the wave $\tilde{U}_0^i$) and the frequency κ be fixed. The contour $\mathbf{S}$ is defined by the single-valued function $f(y) : f(y) \le 0$, which is 2π periodic in $\mathbf{R}^2$. The assumption that the function $f(y)$ is single-valued does not bring any principal restriction into the solution to direct and inverse problems or to the corresponding algorithms. The relevant parameterization of the contour $\mathbf{S}$ can efficiently remove such restrictions [9,224].

Taking into account the radiation condition of the type (6.5) for the reflection zone $z \geq 0$ and the electromagnetically evident requirement for the field in perfect conductor ($\tilde{U}(g) = 0$ for $z < f(y)$), we derive from (7.40),

$$\left\{ \begin{array}{c} a_n \\ -\delta_n^0 \end{array} \right\} = \frac{i}{4\pi\Gamma_n} \int_0^{2\pi} \eta(y_0) e^{\{\mp\} i\Gamma_n f(y_0)} e^{-iny_0} dy_0; \quad n = 0, \pm 1, \ldots,$$

$$\eta(y_0) = \mu(y_0, f(y_0)) \exp(-i\Phi_0 y_0) \left[1 + \left(\frac{d}{dy_0} f(y_0) \right)^2 \right]^{1/2}. \quad (7.42)$$

After substitution of the expansion of the function $\exp[\pm i\Gamma_n f(y_0)]$ in a power series into (7.42), we arrive at the relations

$$c_n = -2\pi i \Gamma_n \left(a_n - \delta_0^n \right)$$
$$\approx \int_0^{2\pi} \eta(y_0) \left\{ 1 - \frac{\Gamma_n^2 f^2}{2!} + \cdots + (-1)^{N-1} \frac{(\Gamma_n f)^{2(N-1)}}{[2(N-1)]!} \right\} e^{-iny_0} dy_0,$$

$$d_n = 2\pi\Gamma_n \left(a_n + \delta_n^0 \right) \approx \int_0^{2\pi} \eta(y_0) \left\{ \Gamma_n f + \cdots + (-1)^{N+1} \frac{(\Gamma_n f)^{2N-1}}{(2N-1)!} \right\} e^{-iny_0} dy_0;$$

$$n = 0, \pm 1, \ldots. \quad (7.43)$$

The accuracy of (7.43) is defined by number of terms retained in the power series (in our case it is equal to $2N$).

We introduce new variables $a_n^{(m)}$ according to

$$a_n^{(m)} = \int_0^{2\pi} f^m(y_0) \eta(y_0) e^{-iny_0} dy_0; \quad m = 0, 1, \ldots, 2N-1, \quad n = 0, \pm 1, \ldots. \quad (7.44)$$

That allows us to reduce (7.43) to the equivalent quasi-linear problem

$$\left\{ \begin{array}{l} a^{(m)} = \left(a^{(m-1)} * f \right); \quad m = 1, 2, \ldots, 2N-1 \\ d \approx D^{(1)} \left[a^{(1)} \right] - D^{(3)} \left[a^{(3)} \right] + \cdots + (-1)^{N+1} D^{(2N-1)} \left[a^{(2N-1)} \right]. \\ c \approx a^{(0)} - D^{(2)} \left[a^{(2)} \right] + \cdots + (-1)^{N-1} D^{(2N-2)} \left[a^{(2N-2)} \right] \end{array} \right. \quad (7.45)$$

Here, $a^{(m)} = \left\{ a_n^{(m)} \right\}$, $d = \{d_n\}$, $c = \{c_n\}$, $f = \{f_n\}$; f_n are Fourier coefficients of the function $f(y)$; $D^{(m)} = \left\{ \delta_n^p (\Gamma_n)^m / m! \right\}_{n,p}$ are the diagonal type matrix operators. The asterisk stands for the convolution in the space of infinite sequences; that is,

$$(a * b) = \left\{ \sum_p a_{n-p} b_p \right\} = \left\{ \sum_p b_{n-p} a_p \right\}.$$

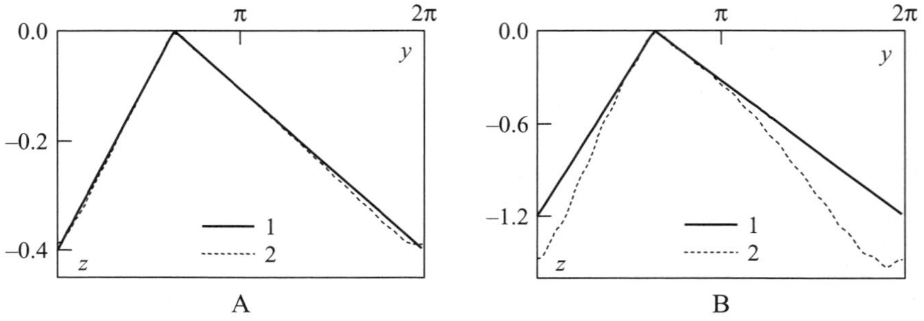

FIGURE 7.3. Solution to inverse problem for echelette gratings with different depth of grooves (A and B) for $\alpha = 0$ and $\kappa = 1.2$: 1, true profile f; 2, $\widehat{f}$.

The complete system (7.45), in which $2N + 1$ unknown vectors $f, a^{(0)}, \ldots, a^{(2N-1)}$ are connected by $2N + 1$ equations, provides a basis for the numerical solution of the inverse boundary value problem.

For $N = 1$ (this approximation has been used in [223]), the system (7.45) has the explicit solution

$$f(y) \approx \widehat{f}(y) = \operatorname{Re}\left\{ i \sum_n \left(a_n + \delta_n^0\right) e^{iny} \Big/ \sum_n \Gamma_n \left(a_n - \delta_0^n\right) e^{iny} \right\}. \qquad (7.46)$$

Numerical implementation of (7.46) allows us to perform preliminary estimations about the capacity of numerical algorithms, constructed according to (7.45).

For numerical experiments, the initial data (complex amplitudes $\{a_n\}$) (see Figs. 7.3 to 7.8; the true profile $f(y)$ is shown in all the figures as a solid line) have been provided by the numerical solution to direct problems that has been obtained by means of the method of analytical regularization of the singular integral equations (7.41) [9,10,224]. As we have mentioned earlier, this method provides the required accuracy of the results within the considered range of parameters κ, α, δ. That is why we assume that the initial data correspond to a real scatterer. As a result the question about the existence of the solution to the inverse problem that in general case for fixed κ and α is not unique (see Section 7.3.2) can be withdrawn.

The accuracy of the solutions to inverse problems (the reconstructed profiles $\widehat{f}$ are represented by dashed lines) within the framework of the present approximation is affected considerably by the relative depth of the grooves (with respect to wavelength, viz. the parameter $\kappa\delta$) and length of the period (parameter κ). The relative error in the determination of $f(y)$ in a uniform metric (by maximal deviation on the interval $0 < y < 2\pi$) for $\kappa < 2$, $\alpha < 80^\circ$, and δ up to about 0.15 does not exceed 6% for any of the profiles (see Fig. 7.3A and Fig. 7.4B). The growth of the error through the increase of κ is faster than its decrease for decreasing values of δ. This is observed also in the cases when $\kappa\delta$ does not increase (see Fig. 7.3B and Fig. 7.4). The decrease of the grating depth in an integral metric (in the metric

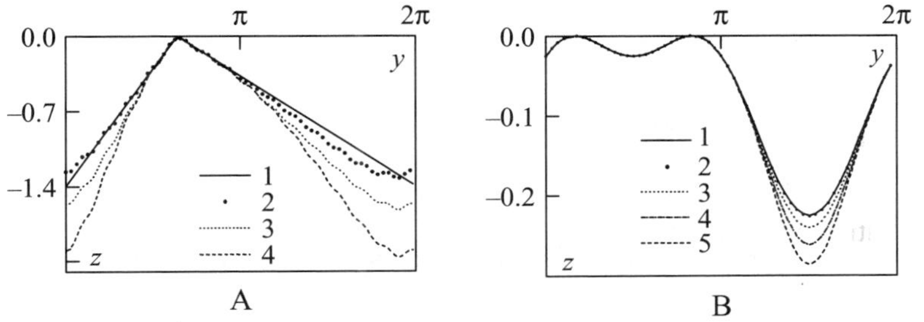

FIGURE 7.4. Results of one step routine of profile reconstruction for $\alpha = 0$ and for various values of κ: 1, true profile; 2, computed profile for (A) $\kappa = 0.8$ and (B) $\kappa = 1.4$; 3, for (A) $\kappa = 1.05$ and (B) $\kappa = 2.4$; 4, for (A) $\kappa = 1.2$ and (B) $\kappa = 3.2$; 5, for $\kappa = 3.8$.

of the space $\mathbf{L}_1\,[(0; 2\pi)]$ of real-valued functions with integrable absolute value) allows one to move the limit of acceptable accuracy of the reconstruction toward greater values of δ and κ. It is significant that the error of the results increases slightly when a larger part of a grating surface is shadowed. The curves in Figures 7.5B and 7.4B, where the integral norm of the true profile $f(y)$ is sufficiently large, confirm this conclusion: the reconstruction error both in a uniform and in an integral metric is distinctly higher than in the case presented in Figure 7.5A, in spite of a considerable decrease of δ.

The simple version when $N = 1$ can also be used efficiently for large δ (or in a more general situation: for a high integral norm of the reconstructing function $f(y)$) in the framework of the following iteration procedure. The function $\widehat{f}(y)$ given by (7.46) acts as a starting approximation allowing us to define more accurately the values of vectors $a^{(0)}$ and $a^{(1)}$ in the first equation of the system (7.45), which then will be inverted following the same scheme. Thus, when constructing

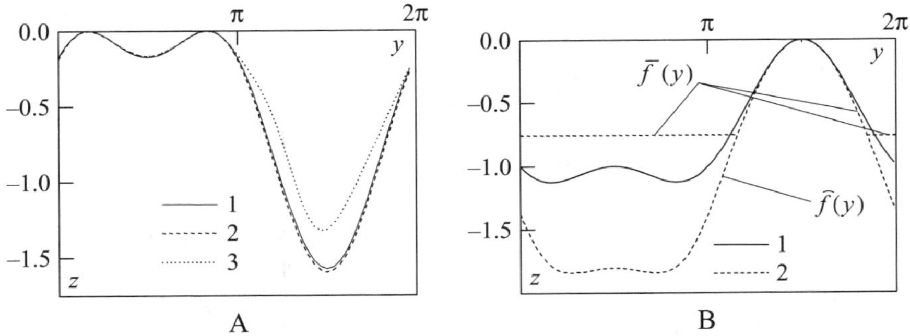

FIGURE 7.5. Profile reconstruction for the gratings with various depth of grooves for various values of α and $\kappa = 1.2$. (A) 1, true profile f; 2, $\widehat{f}$ for $\alpha = 0$; 3, $\widehat{f}$ for $\alpha = 80°$. (B) $\alpha = 0$; 1, f; 2, $\widehat{f}$ and $\bar{f}$.

$\widehat{f}(y)$, the error

$$r_n^{(0)} = \int_0^{2\pi} \eta(y_0)\left[\cos\left(\Gamma_n f(y_0)\right) - 1\right] e^{-iny_0} dy_0; \quad n = 0, \pm 1, \ldots$$

in the determination of true values of $a_n^{(0)}$ is given by the value of the rejected remainder of a power series (in the lower equation of the system (7.45) that are all the members, starting from $D^{(2)}[a^{(2)}]$ and so on); then in the second step, while obtaining the first approximation $\widehat{f}1$, it is assumed that, not rejection, but substitution of the remainder with the similar one that relevant to the already defined function $\widehat{f}$. As a result, the error in the definition of $a_n^{(0)}$ with a sufficiently good initial guess $\widehat{f}$ reduces to

$$\widehat{r}_n^{(0)} = \int_0^{2\pi} \{[\eta(y_0) - \widehat{\eta}(y_0)][\cos(\Gamma_n f(y_0)) - 1] + \widehat{\eta}(y_0)[\cos(\Gamma_n f(y_0)) - \cos(\Gamma_n \widehat{f}(y_0))]\} e^{-iny_0} dy_0,$$

where $\widehat{\eta}(y)$ is the potential relevant to the profile $\widehat{f}(y)$. To construct it, it is necessary to solve the direct problem. The properties of the resolvents of the corresponding boundary value problems [9,10,224] allow us to state that numerical routines that implement the scheme suggested above are convergent if the restrictions of the following type

$$\left|\frac{\eta(y) - \widehat{\eta}(y)}{\eta(y)}\right| < 0.5, \quad \left|\frac{\cos[\Gamma_n f(y)] - \cos[\Gamma_n \widehat{f}(y)]}{\cos[\Gamma_n f(y)]}\right| \leq 0.5;$$
$$0 \leq y \leq 2\pi, \quad n = 0, \pm 1, \ldots \tag{7.47}$$

are valid. Numerical tests confirm this conclusion. Already in the second step (the determination of $\widehat{f}1(y)$ corrected with the help of $\widehat{f}(y)$ values of $a_n^{(0)}$ and $a_n^{(1)}$; further more we present this result as $\widehat{f}1[\widehat{f}]$, similar for $\widehat{f}2[\widehat{f}1]$ etc.) we reconstruct the profile for the gratings with maximal groove depth that is two times larger (see Fig. 7.6). The error in a uniform metric does not exceed 6%.

For larger δ and (or) κ, the initial guess $\widehat{f}$ does not always provide a starting point for a convergent iteration procedure (see Fig. 7.7): the inequality (7.47) does not hold for all values of y and (or) n. In such a situation we must not take the functions $\widehat{f}$, $\widehat{f}1[\widehat{f}]$, and so on, but instead their corrected analogues $\bar{f}$, $\bar{f}1[\widehat{f}]$, and so on, to obtain successive approximations. There are several ways of introducing such corrections; one of the simplest is presented in Figure 7.5B. We describe here in more detail a method that takes into account the peculiar features of the basic algorithm and the behavior of $\widehat{f}(y)$ and its successive approximations, which have been revealed in the course of numerical experiments.

It should be emphasized that the first artificial adjustment of the starting approximation $\widehat{f}$ has already been performed in the framework of the representation

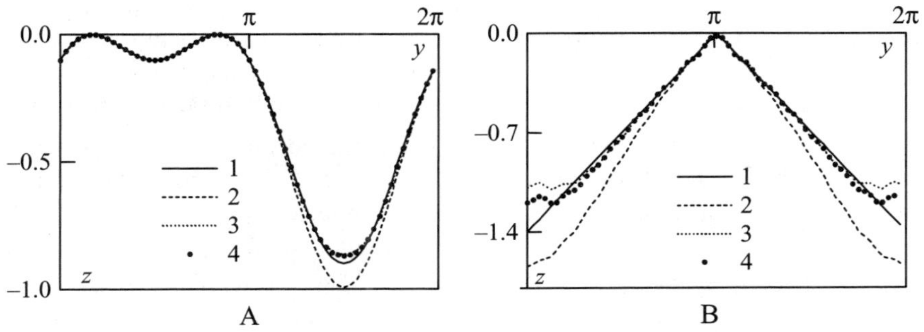

FIGURE 7.6. Illustration of using $\widehat{f}$ as initial guess in iteration schema for reconstructing f for $\alpha = 0$, (A) $\kappa = 1.4$, and (B) $\kappa = 1.05$: 1, f; 2, $\widehat{f}$; 3, $\widehat{f}1[\widehat{f}]$; 4, $\widehat{f}2[\widehat{f}1]$.

(7.46). A rather natural step for seeking a real-valued function has been taken: we discarded the imaginary part of the solution as it causes an accumulation of errors and the stability of the schema can be destroyed. The same principles (the illumination of factors that evidently disturb the scheme) are put into the routine of the construction of $\bar{f}(y)$ (adjusted initial guess) and if necessary $\bar{f}1[\widehat{f}]$, $\bar{f}1[\bar{f}]$, and so on. The following facts provide the background for this action. First, for each iteration within a certain κ range there is the upper limit of the groove depth, such that for smoother gratings the profile $f(y)$ can be reconstructed with geometrical accuracy. For example, for $\widehat{f}(y)$ at $\kappa < 2$ this depth is given by the inequality $z \geq -0.5$. Second, the greatest discrepancy between $\widehat{f}(y)$ and the true profile $f(y)$ in the interval of y where $|f(y)|$ has a maximum is caused by errors in the determination of the amplitudes f_n of the higher order components of the Fourier expansion of the function $\widehat{f}(y)$.

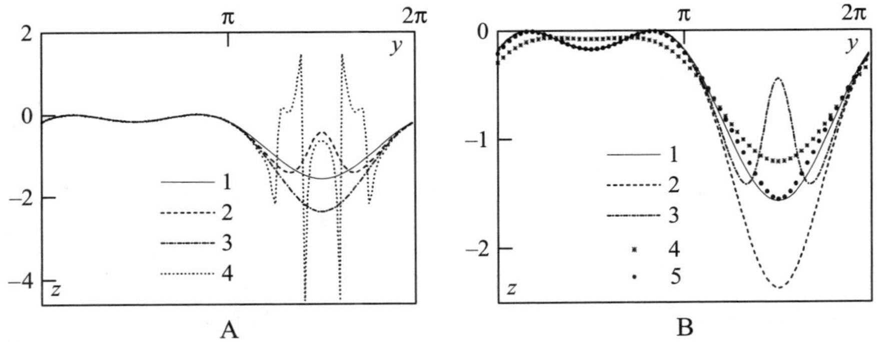

FIGURE 7.7. Regularization of iterative procedure for $\alpha = 0$ and $\kappa = 1.4$. (A) Direct schema: 1, f; 2, $\widehat{f}$; 3, $\widehat{f}1[\widehat{f}]$; 4, $\widehat{f}2[\widehat{f}1]$. (B) Modified schema: 1, f; 2, $\widehat{f}$; 3, $\widehat{f}1[\widehat{f}]$; 4, $\overline{\widehat{f}1}[\widehat{f}]$; 5, $\widehat{f}2[\overline{\widehat{f}1}]$.

Consider an example (see Fig. 7.7) that suggests the correct solution to the problem. The direct iteration procedure in the third step in the determination of $\widehat{f}2[\widehat{f}1]$ is starting to deviate from the desired solution in the interval $4 < y < 5.5$ (Fig. 7.7A). At this $f_n = 0$ for $|n| > 2$, and amplitudes of the same numbers for $\widehat{f}1[\widehat{f}]$ reach values of about $0.5 \cdot 10^{-2}$. The amplitudes of the principal components ($n = 0, \pm 1, \pm 2$) change moderately with each iteration. Having interrupted the process after arriving at $\widehat{f}1[\widehat{f}]$ and assuming $(\widehat{f}1[\widehat{f}])_n = 0$ for $|n| > 2$, we construct instead of $\widehat{f}1[\widehat{f}]$ the corrected approximation $\bar{f}1[\widehat{f}]$. The function $\widehat{f}2[\bar{f}1]$ obtained by the next iteration practically coincides with the desired function f (see Fig. 7.7B). In other natural model situations the determination of basic Fourier components is not a complicated computational problem, because Fourier amplitudes remain relatively stable in the present type of step-by-step procedure.

The vectors $a^{(0)}$ and $a^{(1)}$ can also be adjusted by means of the following scheme. The subsystem of $2N$ linear operator equations from the system (7.45) (with the exception of the first one) is inverted. The matrix operators of the convolution equations $a^{(m)} = (a^{(m-1)} * \widehat{f})$; $m = 2, \ldots, 2N - 1$ are given by the Fourier coefficients of the already obtained function $\widehat{f}(y)$. The next approximation $\widehat{f}1[\widehat{f}]$ is determined by the formula $a^{(1)} = (a^{(0)} * \widehat{f}1)$, where the vector $\widehat{f}1$ is composed of the Fourier coefficients of function $\widehat{f}1[\widehat{f}]$. The process is repeated until the required accuracy of the reconstruction of $f(y)$ is achieved. The external accuracy can be checked against the error in the reconstruction of the input data $\{a_n\}$. In any step of the scheme the next approximation can be purposefully adjusted by artificial means, specifically by using the method suggested above. For considerably larger values of δ, the suggested scheme turns out to be not as efficient as the previous one. However, it gives the desired result more rapidly if the integral norms of $f(y)$ and κ are large (δ is not too large) even for minimal values of $N = 2$ (see Fig. 7.8). Apparently both of these schemes can be used within the same algorithm: one can turn from one scheme to the other taking into account the results of earlier steps.

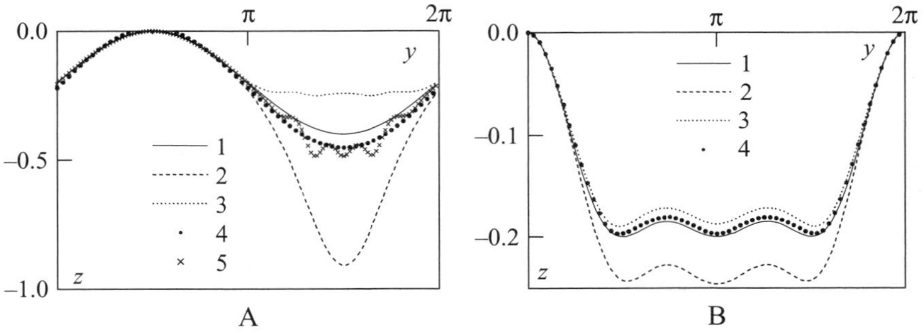

FIGURE 7.8. Results of using of iterative procedure with $N = 2$ for $\alpha = 0$ and (A) $\kappa = 3.2$, (B) $\kappa = 3.8$: 1, f; 2, $\widehat{f}$; 3, $\widehat{f}1[\widehat{f}]$; 4, $\widehat{f}2[\widehat{f}1]$.

Presently, we cannot indicate the boundary values of δ and κ for which the starting approximation $\widehat{f}(y)$ for $N = 1$ becomes invalid. We have got satisfactory results even when the period of $f(y)$ was five times larger than the wavelength and the grating depth reached the value of two wavelengths. However, it is clear that the range of δ and k where the starting approximation $\widehat{f}(y)$ matches our requirements is bounded. So we have to seek alternative procedures for constructing $\widehat{f}(y)$ for increasing δ and κ. One suggested procedure is associated with direct inversion of the quasi-linear system (7.45) for $N > 1$. The idea of quasi-linearization, embodied in this system, is rather fruitful and may serve as a basis for several other computational schemes. Such schemes undoubtedly deserve attention because they are closed (i.e., not utilizing any additional input data even in such important issues as the starting approximation); they rely only on basic numerical techniques that are flexible and easily adjustable for the required parameter range. Their capabilities allow us to solve a number of important applied problems in optics, solid-state electronics, diffraction electronics, and antenna engineering. Rather often (see Sections 6.3.2 and 7.2) in these scientific areas the technology requires the use of gratings of small groove depth (the value of δ is limited). Besides, the gratings must not generate a large number of propagating spatial harmonics (a limitation on the values of κ).

We also point out another rather important special feature of the algorithms considered. The choice of initial guess $\widehat{f}(y)$ as made for $N = 1$ regularizes in essence the ill-posed (for regime of single-frequency and single-position sounding) inverse problem of profile $f(y)$ reconstruction. Practically exact coincidence of functions $\widehat{f}(y)$ and $f(y)$ within the variation intervals of y, where values of $|f(y)|$ are not large, immediately reject the alternative versions, and if the iteration procedure converges, it converges to the true profile $f(y)$.

7.3.4. Synthesis of Reflective Gratings

The possibility of resolving the inverse boundary value problem within the framework of single-position and single-frequency schemes, that is, by response of the grating to the wave of fixed length and propagation along a fixed direction, may be considered as an advantage of the corresponding method or approach only in a case of problem visualization or diagnostic of the structure. More often the synthesis problem is required to be solved within a frequency range and (or) within an interval of angles of incidence of a plane wave. These give rise to a set of specific problems that may be resolved by reasonably balanced formation of a packet of input data or specification for the synthesis problem. Not all the components of this packet (in our case that are the absolute values and phases of complex amplitudes of a scattered by grating electromagnetic field) are firmly connected to the required electromagnetic characteristics (see Section 7.2), that the synthesized structure must possess. Part of them, by their content, may be reoriented to overcome purely methodological difficulties. In the following we consider the realization of this possibility which will lead to modifications of the already approved (in the case of fixed κ and α) computational scheme.

Functionally meaningful sets of input data for the formulation of and solution to synthesis problems for gratings is obtained conventionally as a result of parametric optimization of electromagnetic models corresponding to the device of interest. The grating plays the role of a dispersive element in this device. These data have to be checked for not being in contradiction to requirements following from basic electromagnetic principles (energy conservation law, reciprocity theorems, etc.) as well as others that are not equally general, but also important for the correct formulation of the question about the existence of the solution. Let us consider the result from [7] as an example: in one mode frequency range ($\text{Im}\,\Gamma_n > 0, n \neq 0$) the arguments of all amplitudes R_{n0} of higher-order evanescent harmonics ($n \neq 0$) of the spatial spectrum of the reflective grating satisfy the equation $\arg(R_{00}) = \arg(R_{n0})/2 + m\pi$, where m is equal to zero or one. Such a verification has to be made for the absolute values and phases of the amplitudes that are used for completing the input data to the extent necessary for the computational algorithm.

Consider now the formulation of the synthesis problem for a grating, exposed to an incident E-polarized wave $\tilde{U}_0^i(g, \kappa)$, and that has to conform to required, or close to required, values of amplitudes $a_n(\kappa, \Phi)$ within the given intervals of κ and (or) Φ. The schemes outlined below are the same in principal, with a change of the number of meaningful amplitudes (finite number) and variable parameters (κ and Φ, κ or Φ). Only the dimension of the packet of input data varies. Also the dimension (single or dual, etc.) of the integral equations where these data play a role of known parameters changes. Naturally, it is necessary to introduce the formal changes into the description of the domain of definition for the functions $a_n(\kappa, \Phi)$ (separate intervals of variation of κ and Φ, or their direct product), integral operators and so on. Granting this, we do not consider the most general situation; rather we confine ourselves to the analysis of the situation when n and Φ are fixed and the goal is to synthesize a reflective grating with a profile $z = f(y)$ such that the nth spatial spectral amplitude is equal or close to the value $a_n(\kappa)$, $\text{Re}\,\Gamma_n(\kappa) > 0$ in the frequency range $[\kappa_1; \kappa_2]$.

The first step, common to both computational schemas considered below, consists in the following. Keeping only the highest terms in the expansion of the function $\exp[i\Gamma_n f(y)]$ into a power series, from the second equation of the system (7.42) (null field equation) we derive

$$\eta(y, \kappa) \approx 2i\Gamma_0. \tag{7.48}$$

In essence, we repeat the procedure used in the construction of the function $\widehat{f}(y)$ in the problem of single-frequency and single-position reconstruction of the grating profile. Then we have inverted the first equation of the system (7.43) for $N = 1$. The important difference for the present case is that the first approximation by a current can be obtained without using the information about the amplitudes of the diffracted field. It is also important that this approximation, having the same order of error as a previous one, can be incorporated into the sequence of computational routines for a shape reconstruction and synthesis of a reflective periodic structure.

Substituting (7.48) into the first of equations (7.42), we arrive at the following classical nonlinear problem (n is fixed)

$$a_n(\kappa) = -\frac{\Gamma_0}{2\pi\Gamma_n}\int\limits_0^{2\pi} \mathrm{e}^{-i\left[\Gamma_n \widehat{f}(y)+ny\right]}dy; \quad \kappa \in [\kappa_1;\kappa_2]. \tag{7.49}$$

To solve it and to find the unknown function $\widehat{f}(y)$ we have to invert the Urysohn operator

$$A[f] = -\frac{\Gamma_0}{2\pi\Gamma_n}\int\limits_0^{2\pi} \mathrm{e}^{-i[\Gamma_n f(y)+ny]}dy, \tag{7.50}$$

which is continuous from $\mathbf{C}([0;2\pi])$ into $\mathbf{C}([\kappa_1;\kappa_2])$. The operator (7.50) is Fréchet-differentiable. Its derivative in the point $g(y) \in \mathbf{C}([0;2\pi])$, which is bounded linear operator $B_g : \mathbf{C}([0;2\pi]) \to \mathbf{C}([\kappa_1;\kappa_2])$, can be represented in the form [225],

$$B_g[f] = \frac{i\Gamma_0}{2\pi}\int\limits_0^{2\pi} \mathrm{e}^{-i[\Gamma_n g(y)+ny]} f(y)\,dy.$$

Using the Newton–Kantorovitch algorithm, we reduce problem (7.49) to the determination of $\widehat{f}(y) = \lim\limits_{m\to\infty} f_m(y)$, where $\{f_m(y)\}$ is the sequence of solutions of the linear problems

$$B_{f_m}[f_{m+1}(y) - f_m(y)] = -A[f_m(y)] + a_n(\kappa); \quad m = 0, 1, 2, \ldots \tag{7.51}$$

or, in simplified form,

$$B_{f_0}[f_{m+1}(y) - f_m(y)] = -A[f_m(y)] + a_n(\kappa); \quad m = 0, 1, 2, \ldots. \tag{7.52}$$

The bounded invertibility of the operators B_{f_m} and an appropriate choice of initial estimate $f_0(y)$ in iterative procedures are fundamental requirements to provide the convergence $f_m(y) \to \widehat{f}(y)$ and the computational efficiency of the algorithms based on the inversion of equations (7.51) or (7.52) [225]. Because the required properties are not shared by the operators B_{f_m} (whose kernels are sufficiently smooth functions), it is necessary to regularize the integral equations of the first kind (7.51) or (7.52). Consider together with the integral operator $B_{f_m} : \mathbf{L}_2([0;2\pi]) \to \mathbf{L}_2([\kappa_1;\kappa_2])$ the integral operator

$$B^*_{f_m}[a(\kappa)] = \frac{i}{2\pi}\mathrm{e}^{iny}\int\limits_{\kappa_1}^{\kappa_2} \Gamma_0 \mathrm{e}^{i\Gamma_n f_m(y)} a(\kappa)\,d\kappa,$$

that is the adjoint of B_{f_m}. Assuming that right-hand side part in (7.51) does not belong to the null space of the operator $B^*_{f_m}$, we arrive at the problem

$$B^*_{f_m} B_{f_m}[f_{m+1}(y) - f_m(y)] = -B^*_{f_m} A[f_m(y)] + B^*_{f_m}[a_n(\kappa)]; \\ 0 \le y \le 2\pi, \quad m = 0, 1, 2, \ldots, \tag{7.53}$$

that is equivalent to (7.51). The operator of this problem (7.53) is self-adjoint, positive, and compact in the pair of spaces $\mathbf{L}_2 \to \mathbf{L}_2$. This allows us [142] to introduce instead of the ill-posed operator equations of the first kind (7.53) the operator equations of the second kind by implementing the well-known Lavrentev's ξ-regularization:

$$[\xi_m + B^*_{f_{m,\xi}} B_{f_{m,\xi}}][f_{m+1,\xi}(y) - f_{m,\xi}(y)] = -B^*_{f_{m,\xi}} A[f_{m,\xi}(y)] + B^*_{f_{m,\xi}}[a_n(\kappa)];$$
$$\xi_m > 0, \quad 0 \le y \le 2\pi, \quad m = 0, 1, 2, \ldots. \tag{7.54}$$

Their solutions $f_{m,\xi}(y)$ converge to $f_m(y)$ for $\xi_m \to 0$ in an appropriate space norm. The sequence $\{f_m(y)\}$, defined in such way, gives the desired profile $\widehat{f}(y)$ as a limit when $m \to \infty$.

The equivalent reformulation and regularization of problems (7.52) can be carried out similarly. The rate of convergence $f_m(y) \to \widehat{f}(y)$ is lower in this case [225]. But we have to note the advantages provided by the static character of operator B_{f_0}. In the simplified form (7.52) it is necessary to invert B_{f_0} only once. Not going into details and assuming $f_0(y) \equiv 0$, in analogy with (7.54) we can write

$$\xi\left[f_{m+1,\xi}(y) - f_{m,\xi}(y)\right]$$
$$+\gamma_1 e^{iny} \int_0^{2\pi} e^{-iny_0}\left[f_{m+1,\xi}(y_0) - f_{m,\xi}(y_0)\right] dy_0 = \gamma_2(m, \xi, y);$$

$$\gamma_2(m, \xi, y) = \frac{i}{4\pi^2} e^{iny} \int_{\kappa_1}^{\kappa_2} \frac{\Gamma_0^2}{\Gamma_n}\left\{\int_0^{2\pi} e^{-i[\Gamma_n f_{m,\xi}(y_0)+ny_0]} dy_0\right\} d\kappa + \frac{i}{2\pi} e^{iny} \int_{\kappa_1}^{\kappa_2} \Gamma_0 a_n(\kappa)\, d\kappa,$$

$$\gamma_1 = \frac{1}{4\pi^2} \int_{\kappa_1}^{\kappa_2} \Gamma_0^2 d\kappa, \quad m = 0, 1, 2, \ldots, \quad \xi > 0, \quad 0 \le y \le 2\pi. \tag{7.55}$$

The kernels of the integral equations of the second kind (7.55) are degenerate, hence their solutions can be found in the explicit form:

$$\xi\left[f_{m+1,\xi}(y) - f_{m,\xi}(y)\right] = \gamma_2(m, \xi, y) - \gamma_1\gamma_3(m, \xi)(\xi + 2\pi\gamma_1)^{-1} e^{iny};$$
$$\gamma_3(m, \xi) = \int_0^{2\pi} e^{-iny}\gamma_2(m, \xi, y)\, dy, \quad \xi > 0. \tag{7.56}$$

Letting ξ tend to zero and calculating the sequence $\{f_m(y)\}$ for increasing values of m, we arrive at the desired function $\widehat{f}(y)$.

One of the possible and simplest versions of the choice of initial guess $f_0(y)$ we have already realized in (7.52), (7.55), (7.56). The rough solution to the problem suggested below can not be called the final solution, but the approximation we get due to it provides a rather steady start for an iterative process. For several cases the appropriate clear and simple scheme for the search of $f_0(y)$ gives us directly the solution $f(y)$ to the inverse problem of synthesis or reconstruction. We can describe the scheme, in short, in the following way (n and Φ are fixed again, the formulation of the inverse problem is the same). The representation (7.48) is used in the second equation of the system (7.43), $N = 1$. The integral equation of the first kind

$$a_n(\kappa) + \delta_n^0 = \frac{i\Gamma_0}{\pi} \int_0^{2\pi} \widehat{f}(y)\, e^{-iny} dy; \quad \kappa \in [\kappa_1; \kappa_2],$$

derived by means of Lavrentevs ξ-regularization, is reduced to the integral equation of the second kind

$$\xi \widehat{f}_\xi(y) + e^{iny}(\kappa_2 - \kappa_1) \int_0^{2\pi} e^{-iny_0} \widehat{f}_\xi(y_0)\, dy_0 = e^{iny} \frac{\pi}{i} \int_{\kappa_1}^{\kappa_2} \frac{a_n(\kappa) + \delta_n^0}{\Gamma_0} d\kappa;$$
$$0 \le y \le 2\pi. \qquad (7.57)$$

The solution to this equation $\widehat{f}_\xi(y)$ does exist, is unique, and in the limit ($\xi \to 0$) gives $\widehat{f}(y)$: $\|\widehat{f}_\xi(y) - \widehat{f}(y)\|_{L_2} \to 0$ when $\xi \to 0$. The kernel of the integral operator in (7.57) is degenerate and the function $\widehat{f}_\xi(y)$ has an explicit form that coincides with (7.56) up to notations.

The solution $\widehat{f}(y)$, relevant to M given within the frequency interval $[\kappa_1; \kappa_2]$ amplitudes $a_{n_1}, a_{n_2}, \ldots, a_{n_M}$, may be required more often in the applied problems. The technique used in this situation is identical to that for $M = 1$. Without going into detail let us write the final formula

$$\widehat{f}(y) = \text{Re}\left\{ -\frac{i}{2\Gamma_0(\kappa_2 - \kappa_1)} \int_{\kappa_1}^{\kappa_2} \sum_{m=1}^{M} \left(a_{n_m} + \delta_{n_m}^0\right) e^{imy} d\kappa \right\}. \qquad (7.58)$$

The results of numerical experiments based on implementation of (7.58), for synthesis of a structure providing amplitudes of the scattered field close to given values $a_{n_m}(\kappa)$ within the range $0.652 \le \kappa \le 0.952$, for $\alpha = 10^\circ$ are presented in Figures 7.9 to 7.11. The synthesis task has been formulated from data obtained from the solution to direct diffraction problem (6.1) to (6.5) ($p = 0$; an E-polarized field) for the grating formed by perfectly conducting semi cylinders. The corresponding profile (see Fig. 7.9) and the relevant diffraction characteristics (see Figs. 7.10 and 7.11) are shown by solid lines 1. The profiles of synthesized gratings and their diffraction characteristics are shown by dashed lines 2 to 6. The set of input data $\{a_0(\kappa); M = 1\}$ corresponds to the curves 2; the set $\{a_{-1}(\kappa), a_0(\kappa); M = 2\}$ corresponds to curves 3, curves 4 correspond to the input data set $\{a_{\pm 1}(\kappa), a_0(\kappa); M = 3\}$, curves 5, to the set $\{a_{-5}(\kappa) \div a_5(\kappa); M = 11\}$, and for curves 6, to the set $\{a_{-15}(\kappa) \div a_{15}(\kappa); M = 31\}$. The result may be estimated as a satisfactory one: the error in the fulfillment of the requirements is not large; the dynamics of profile transformation with M increasing is predictable and consistent with the Statement 7.5 about the uniqueness of the solution to the inverse problem for complete and exact input data.

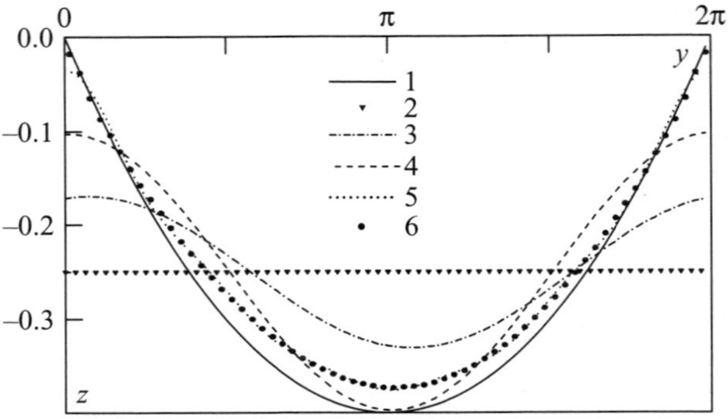

FIGURE 7.9. Profiles of control (solid line 1) and synthesized for different input data set (dashed lines 2 to 6) configurations.

Note that M in the numerical example considered is not a variable in the iterative process. Each M determines its own problem with its distinctive set of features required of the synthesized structure. Figures 7.9 to 7.11 allow us to judge the accuracy of fulfillment of these requirements in the framework of the implemented algorithm. For a correct comparative estimation of the accuracy we must take into account the arguments and the absolute values of the amplitudes $a_n(\kappa)$ on the whole interval $[\kappa_1; \kappa_2]$ for all values of n. An examination of just one picture (e.g., the picture from Figure 7.10, where the characteristic $\arg a_0(\kappa)$ for the synthesized structures deviates from the desired one) is insufficient to formulate the conclusions.

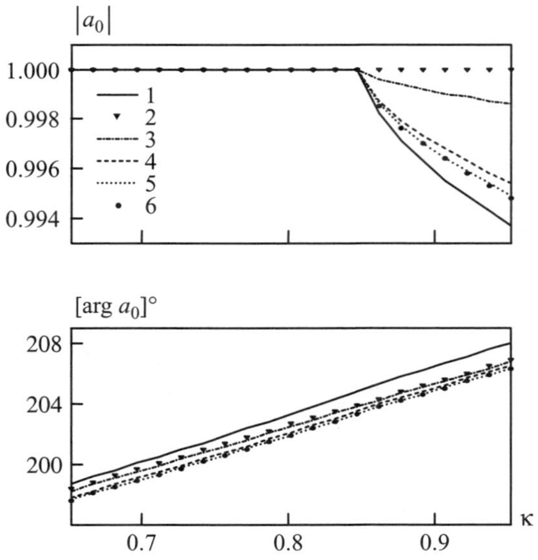

FIGURE 7.10. Amplitude $a_0(\kappa)$ of spatial spectrum of control (solid line 1) and synthesized for different input data set (dashed lines 2 to 6) structures.

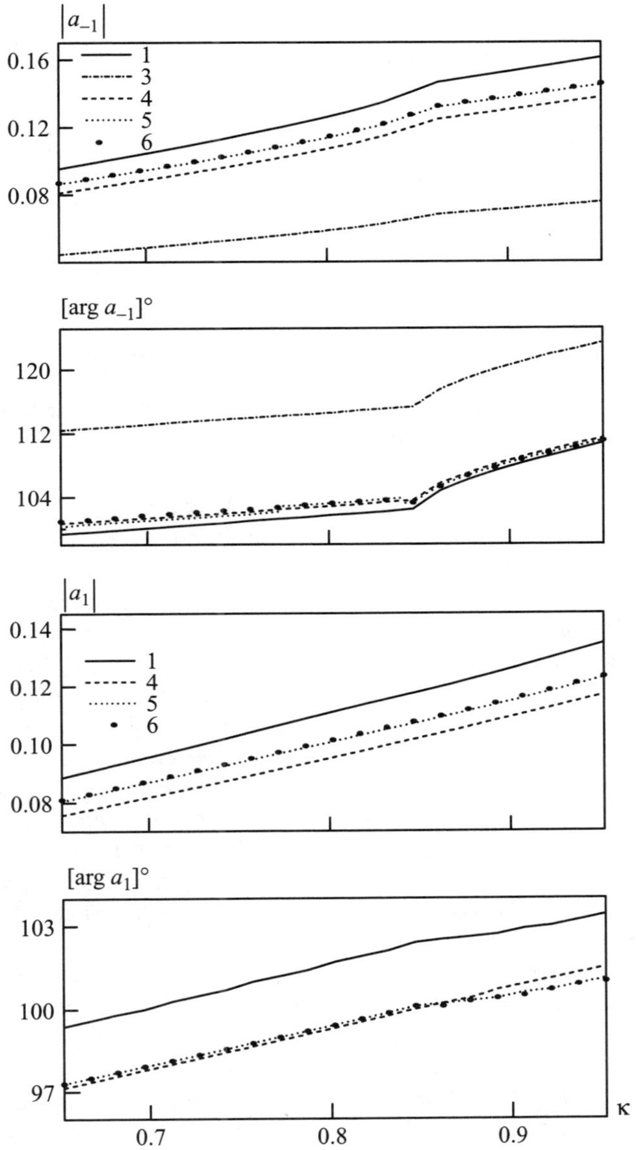

FIGURE 7.11. Amplitudes a_{-1} (κ) and a_1 (κ) of spatial spectrum of control (solid line 1) and synthesized for different input data set (dashed lines 2 to 6) structures.

The sequence of the resolved synthesis problems can be considered also as the solution of the profile reconstruction problem with an incomplete set of exact input data (the completion is achieved by a growth of M). Because the reconstruction problem is performed in a frequency range, the sequence of the obtained solutions

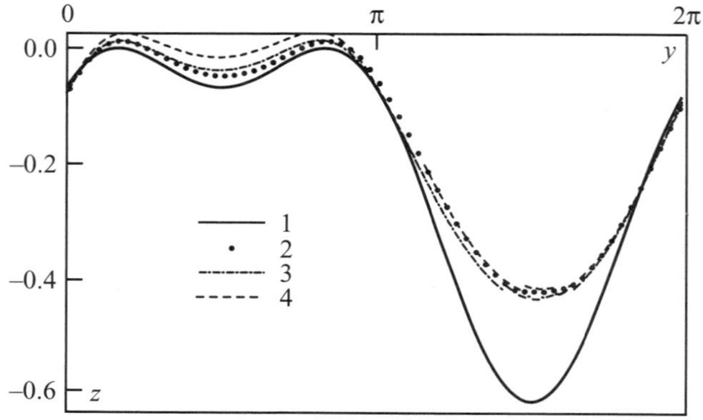

FIGURE 7.12. Control profile (solid line 1) and synthesized profiles (dashed lines 2 to 4) for different levels of input data distortion.

$\widehat{f}(y)$ must converge to the real profile $f(y)$ as M is sufficiently large (see Statement 7.5). Figures 7.9 to 7.11 confirm the intrinsic convergence of the method (curves 5 and 6 merge together almost everywhere), yet show no large discrepancy between the reconstructed (the result for $M = 31$) and the true profile (the error is up to 5 in a uniform metric). This rather regular error is caused by the approximation of the algorithm. The smaller the relative grating depth, the smaller is the error. It can be reduced substantially by the use of the two basic algorithms described above. The numerical examples presented here concern in essence the problem of choosing the initial guess in these schemes. However, the characteristics of the obtained solution allow us to use it not only as a good initial estimation in nonlinear problems (7.49), but also as a final or auxiliary result in the relevant synthesis problems or in reconstruction problems with an incomplete set of exact input data.

Several words have to be said about the robustness of the algorithm with respect to noise in the input data. Limited random deviations of the values $a_n(\kappa)$ from the exact values do not introduce large errors into the synthesis or reconstruction problem solution. This is confirmed by the results of the numerical experiments presented in Figures 7.12 and 7.13. The problems are solved in the frequency range $0.702 \leq \kappa \leq 1.202$ with $\alpha = 15^\circ$ for the test profile $f(y) = 0.27\left(\sin y + \cos^2 y - 1.25\right)$. The set of the input data is $\{a_{-10}(\kappa) \div a_{10}(\kappa);\, M = 21\}$. Curves 1 are the test grating and its scattering characteristics; curves 2 to 4 are the synthesized structures and their scattering characteristics. Curve 2 corresponds to the exact input data; curves 3 and 4 correspond to the input data distorted randomly within 5% and 10%, respectively.

7.4. Open Dispersive Resonators

In quasi-optics, the dimension of mirrors and lenses are by far much larger than the wavelength, that is why many things become clear if considered from the standpoint of geometrical optics. However not all of them: radiation is formed in the diffraction

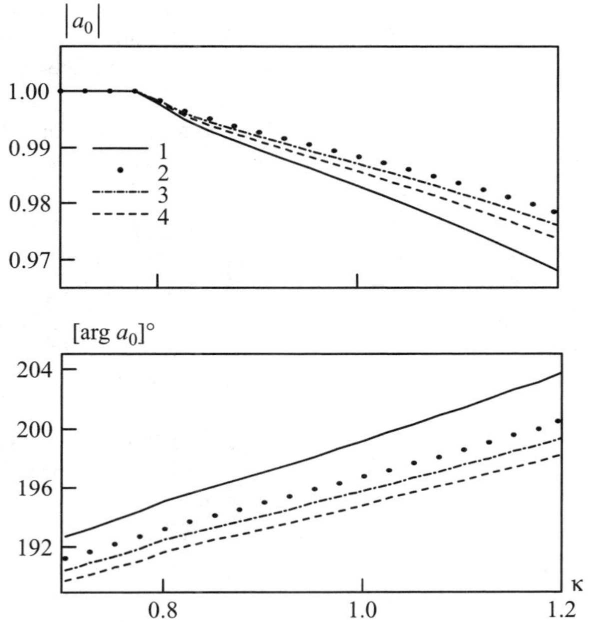

FIGURE 7.13. Scattering characteristics of control (solid line 1) and synthesized (dashed lines 2 to 4) structures.

process, hence, diffraction is more important here than in ordinary optics, and makes the quasi-optical computations more complicated. In many cases, besides the use of larges-cale calculations, there is no other way but to evaluate the diffraction effects in quasi-optical devices approximately or one may even be confined to a qualitative estimate.

We can agree with almost everything in this statement of L.A. Weinstein [226] concerning the problems of analyzing quasi-optical OR with only one reservation: "large-scale calculations" cannot be considered as an alternative to a qualitative analysis. This is the only method, however, not such a simple one to keep in mind as a reserve, just in case, of obtaining really reliable data on complicated quasi-optical devices. This also pertains to data concerning wavelengths at which it generally makes sense to employ various asymptotic and heuristic approaches and methods. Naturally, the modeling of resonance quasioptical devices only adds more problems. The qualitative analysis plays here a very specific and important role while providing good model approximations for "large-scale calculations."

7.4.1. Open Resonators of Classical Configurations: Computational Experiments in the Time Domain

Over the frequency ranges, where the use of the asymptotic method, even if it could be justified in certain cases and then only after verifying corresponding data by the results of rigorous approaches, the major part of essential questions of the theory of OR composed of thin cylindrical and spherical screens has been solved by V.P. Shestopalov and his students S.S. Vinogradov, V.N. Koshparenok, P.N. Melezhik, A.E. Poyedinchuk, Y.A. Tuchkin, and others (see in the references their papers and books) by using the analytical regularization methods. A lot of interesting physically clear data have been revealed due to the waveguide concept of the processes taking place in open oscillating systems, which has been suggested and developed by L.A. Weinstein [198,226]. However, if the mirrors have a complicated configuration, even the most powerful approaches of the frequency domain allow just an analytic description of the spectrum (the discreteness, finite multiplicity factor, and other general features that are subject to the Fredholm theorem for compact finite-meromorphic operatorfunctions), as well as a qualitative characterization of its separate components and approximately estimate their influence on the transient characteristics of OR (see Statement 1.2 and papers [10,72,198,227–229]). A sufficiently reliable calculation, (e.g., using the integral equation method) is highly timeconsuming and is a huge computational burden, so it hardly can be implemented to the extent necessary for a profound analysis. The most suitable in such a case are the direct (finite-difference) time-domain methods. It is difficult to find some objective reasons to explain such a few number of attempts to realize in the time domain those things that can hardly be done in the frequency domain. We can refer only to one example of a well-considered and realized approach to this problem, whose results are summed up in the last, sixteenth chapter ("Microcavity Ring Resonators") of the book [27].

The computational experiments, whose results are discussed in this and the next sections, involve the algorithm suggested in Section 3.3. The initial boundary value problems (3.12) (in the case of an E-polarized field) have been discredited by the finite-difference method at the rectangular coordinate grid y and z (Fig. 3.1). The rectangular analysis domain $\mathbf{Q}_L$ coincides with the field of continuous-tone images illustrating the distribution of the electric field intensity $U(g,t) = E_x$ at the points $g = \{y, z\} \in \mathbf{Q}_L$ for different observation times t. The problem of limitation of the computational space and the problem of corner points on the imaginary boundary $\mathbf{L} = \bar{\mathbf{Q}}_L \backslash \mathbf{Q}_L$ are rigorously solved within the set of equations (3.28) to (3.30) (see Statement 3.1).

The scatterer's parameters (functions $\varepsilon(g)$ and $\sigma(g)$), the current ($F(g,t)$) and momentary ($\varphi(g)$ and $\psi(g)$) sources are determined by step functions of the type $\chi[f_1(g)]\chi[f_2(g)]\ldots\chi[f_m(g)]$ (see the appendix). It is assumed that the metal details of the resonators, radiators, and the like are made of copper, and all the dimensions are set in centimeters. Thus, for instance, the mirrors of OR, whose spectral characteristics are given in Figures 7.14 to 7.16, are described by the functions

$$\sigma(g) = 2.19 \cdot 10^8 \chi[5 - |y|]\chi[4 - |z|]\chi\left[z^2 + (|y| + 4.5)^2 - 9^2\right]$$

(OR with a confocal geometry: radius R of the mirror surface curvature and spacing L between the mirrors along the main axis of the resonator are 9.0),

$$\sigma(g) = 2.19 \cdot 10^8 \chi[5 - |y|]\chi[4 - |z|]\chi\left[z^2 + (|y| + 6.5)^2 - 11^2\right]$$

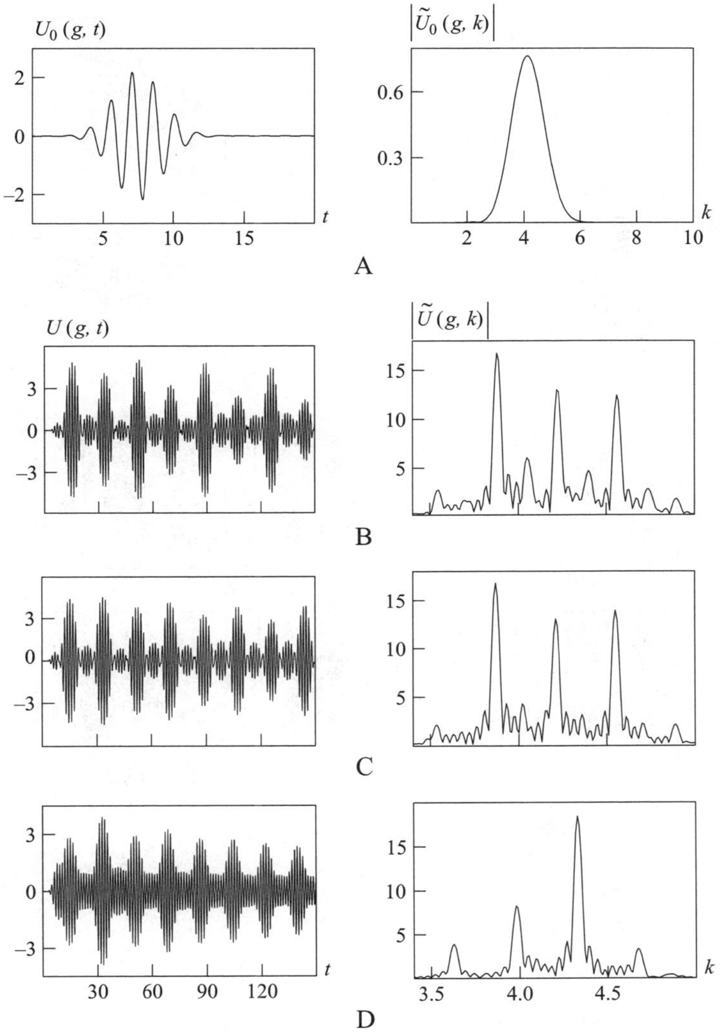

FIGURE 7.14. Spectral characteristics of (A) source, resonators of (B) confocal and (C) preconfocal geometry, and (D) resonator with a break of principal axis: $\tilde{k} = 4.2$; $g = \{0, 0\}$ (A) and $g = \{0.2, 0.2\}$ (B to D).

(OR with a preconfocal geometry: $R = 11.0$, $L = 9.0$), and

$$\sigma(g) = 2.19 \cdot 10^8 \chi[5 - |y|]\{\chi[4 - |z|]\chi[z^2 + (y - 4.5)^2 - 9^2] + \chi[4 - |z + 2|]\chi[(z + 2)^2 + (y + 4.5)^2 - 9^2]\}$$

(a break of the axis in the confocal OR: see Fig. 7.16B; the spacing between the mirror centers in the vertical direction is 2.0).

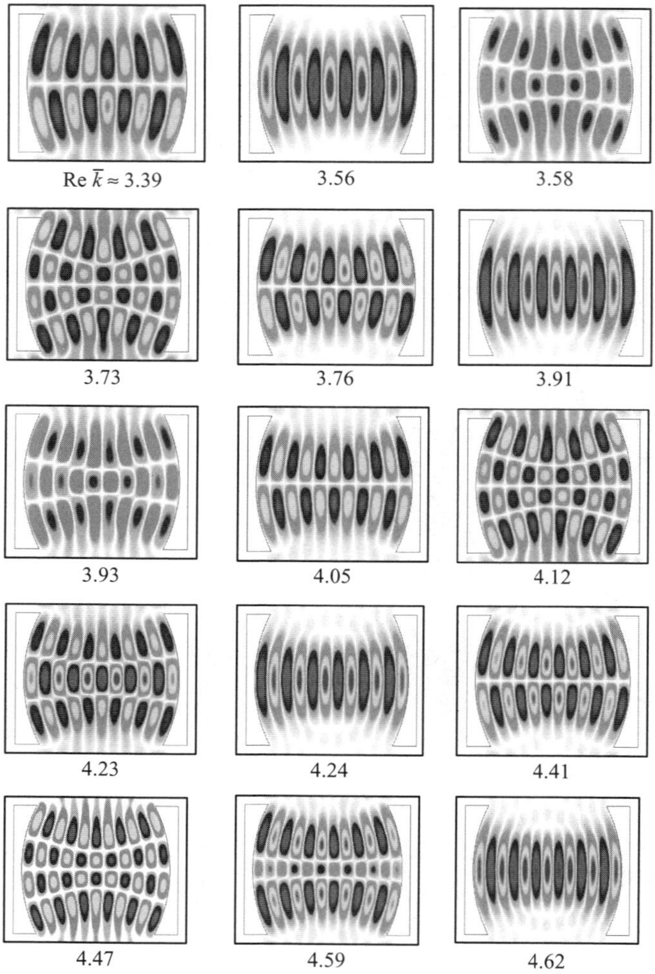

FIGURE 7.15. Configuration, sequence of steps (over frequency), and approximate values of Re $\bar{k}$ for high-Q eigenmodes of resonator with confocal geometry.

The choice of the exciting source for OR and the value of the interval $0 \leq t \leq T$ of the observation time t are relevant for the result of the experiment and should be strictly coordinated with the objectives to be reached while solving certain problems. Let us explain this by an example. A current source

$$F(g, t) = 10\chi[3.5 - |y|]\chi[1.5 - |z + 1|]\cos(\beta_1\tilde{k}y + \beta_2)\cos(\beta_3\tilde{k}z + \beta_4) \times \exp[-(t - \tilde{T})^2/4\tilde{\alpha}^2]\cos[\tilde{k}(t - \tilde{T})], \tag{7.59}$$

that was used in the considered series of experiments, has seven free parameters ($\tilde{k}$, $\tilde{\alpha}$, $\tilde{T}$ and β_j, $j = 1, 2, 3, 4$). The first one, $\tilde{k}$, determines the amplitude center of

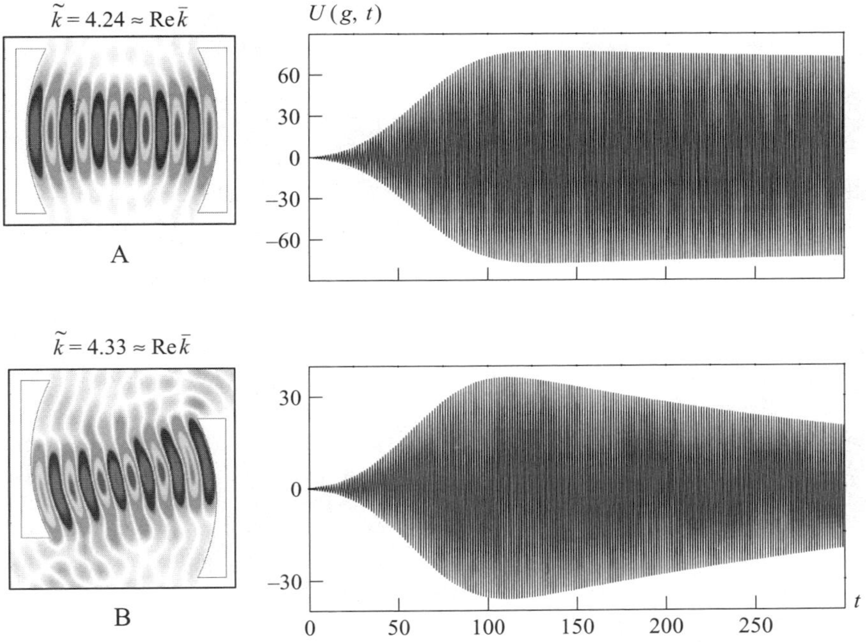

FIGURE 7.16. Modes of the same type (A) in confocal resonator ($g = \{0.36, 0\}$) and (B) in resonator with a break of principal axis ($g = \{0.5, -1\}$).

the primary signal $U_0\,(g, t)$ in the spectral domain: the maximum of the absolute value of the function

$$\tilde{U}_0\,(g, k) = \frac{1}{2\pi} \int\limits_{-\infty}^{\infty} U_0\,(g, t) e^{ikt} dt \equiv F\,[U_0\,(g, t)]\,(k) \leftrightarrow U_0\,(g, t) \qquad (7.60)$$

(here, $U_0\,(g, t)$ is the field of the source $F\,(g, t)$ in free space). Together with the parameter $\tilde{\alpha}$, it determines the band $[\tilde{k} - b/\tilde{\alpha}; \tilde{k} + b/\tilde{\alpha}]$ of frequencies k, where the normalized spectral amplitudes of the pulse $U_0\,(g, t)$ (the values of $|\tilde{U}_0(g, k)|/|\tilde{U}_0(g, \tilde{k})|$) are never less than γ. On the t-axis, the signal $U_0(g, t)$ occupies the interval $\tilde{T} - c\tilde{\alpha} \leq t \leq \tilde{T} + c\tilde{\alpha}$, outside which the values $|U_0(g, t)|/|U_0(g, \tilde{T})|$ never exceed γ. The approximate values of the coefficients b and c, corresponding to some fixed levels of γ and obtained from the known relation

$$\frac{1}{2} e^{i\tilde{T}k} \left[e^{-\tilde{\alpha}^2 (k-\tilde{k})^2} + e^{-\tilde{\alpha}^2 (k+\tilde{k})^2} \right] \leftrightarrow \frac{\pi^{1/2}}{\tilde{\alpha}} e^{-(t-\tilde{T})^2 / 4\tilde{\alpha}^2} \cos\left[\tilde{k}\left(t - \tilde{T}\right)\right],$$

are listed in the table below.

	$\gamma = 0.001$	$\gamma = 0.01$	$\gamma = 0.1$	$\gamma = 0.5$
$b \approx$	2.63	2.14	1.52	0.83
$c \approx$	5.25	4.29	3.04	1.66

The parameters β_j allow tuning the source so that it excites oscillations of a definite symmetry class or, v.v., all types of oscillations. These parameters can also be used to fix the frequency characteristics of the field $U_0(g,t)$, by imposing on it the structure of a plane wave of length $2\pi/\tilde{k}$.

So, what requirements on the above parameters should be satisfied in an analysis of the general spectral characteristics of OR (see Fig. 7.14)? It is obvious that $\tilde{k}$ should be matched with the center of the frequency interval that is being analyzed, and that $\tilde{\alpha}$ should be selected such that the level of the normalized spectral amplitudes of the signal $U_0(g,t)$ in this interval is not under 0.1 (indeed, it is desirable that it is not less than 0.5, because this leads to the minimum distortion of some ideal image that we could observe if the frequency level $|\tilde{U}_0(g,k)|$ were similar over the whole interval). The left boundary of the interval $\tilde{T} - c\tilde{\alpha} \leq t \leq \tilde{T} + c\tilde{\alpha}$ is plotted at the point $t = 0$, in order to reduce the overall computation time T, and, to retain the expected spectral characteristics of the source, the level $|U_0(g,0)|$ ($U_0(g,t) = 0, t < 0$) should be negligible ($0.001 \leq \gamma \leq 0.01$). This condition, together with the chosen $\tilde{\alpha}$, determines the efficient time duration $0 \leq t \leq 2\tilde{T}$ of the signal, and, hence, the delay time $\tilde{T}$.

In the considered stage, all types of oscillations are getting usually equal attention. That is why the field $U_0(g,t)$ of the source should not be orthogonal (see Section 6.2.4) to any of the fields of possible free oscillations in the structure (the requirement concerning the choice of value of parameters β_j; the symmetry classes of the sources and those of the excited oscillations are the same). The excitation efficiency of all the spectral components of the total field $U(g,t)$ depends strongly on the spatial dimension of the source (see Section 6.2.4), whereas by enlarging the volume of $F(g,t)$ (the range of values g, for which $F(g,t) \neq 0$), we can significantly reduce the total computation time T. The lower limit $T = 5\tilde{T}$ of possible values of T is determined by the following apparent requirement: within the time interval $0 \leq t \leq T$, OR should operate, for a while, in the forced oscillations mode ($0 \leq t \leq 2\tilde{T}$); to get rid of the spectral components in the field $U(g,t)$ (due to radiation into free space) that are not capable of building stable wave formations inside the resonator ($2\tilde{T} \leq t \leq 3\tilde{T}$), to let the high-Q oscillations manifest themselves on the background of the oscillations with a lower Q-factor ($3\tilde{T} \leq t \leq 5\tilde{T}$).

If all these requirements are satisfied, the analysis of the spectral characteristics of OR in the frequency band is reduced to the determination of the total field $U(\tilde{g},t)$ at one of the points $\tilde{g}$ inside the resonator as a function of time $t \in [0;T]$ and the analysis of its Fourier image $\tilde{U}(\tilde{g},k) \leftrightarrow U(\tilde{g},t)$ (see formula (7.60): outside the interval $t \in [0;T]$ the function $U(\tilde{g},t)$ is assumed to be zero). Certainly, these characteristics will be distorted if the point $\tilde{g}$ is located at a field node of some high-Q free oscillation $\tilde{U}_n(g,\bar{k}_n)$ ($\tilde{U}_n(\tilde{g},\bar{k}_n) = 0$): the presence of this oscillation in the spectrum Ω_k of OR will not be reflected in the spectrum determined by the function $\tilde{U}(\tilde{g}, k)$. From this follows a rule that usually applies if no more precise references are available: the point $\tilde{g}$ should be shifted, with respect to various symmetry axes and mirrors, by, respectively, one eighth and one fourth of the wavelength corresponding to the frequency $k = \tilde{k}$.

While studying separate oscillations $\tilde{U}_n(g, \bar{k}_n)$, taken from the general spectrum of OR, the interval $[\tilde{k} - b/\tilde{\alpha}; \tilde{k} + b/\tilde{\alpha}]$ ($\tilde{k} \approx \mathrm{Re}\,\bar{k}_n$) should not include resonance points in the neighborhood of $\mathrm{Re}\,\bar{k}_n$, and the level of spectral amplitudes of signal $U_0(g, t)$ at the ends of this interval should be negligible. This requirement can surely be substantially weakened if one can efficiently use sources of a certain symmetry class, oriented to symmetry classes of the oscillation being analyzed and the neighboring ones. In all other respects, the principles formulated above for selecting parameters hold also for this stage of investigation.

Within the interval [3.4; 5.0] of frequencies k, the normalized spectral amplitudes of the pulse $U_0(g, t)$ (see Fig. 7.14A), generated by the source (7.59) with $\tilde{\alpha} = 1$, $\tilde{T} = 6$, $\beta_1 = 1$, $\beta_2 = \pi/4$, and $\beta_3 = \beta_4 = 0$, are never below 0.4. For the time $T = 150$ the resonators of all three considered classic configurations efficiently select from the spectral components of field $U(g, t)$ those that are associated with the high-Q quasi-harmonic oscillations. In the plots in Figures 7.14B–D, these components correspond to rapid peaks in the characteristics $|\tilde{U}(g, k)|$. The parameters of the amplitude centers that determine the eigenfrequencies of H_{0n1}-oscillations, in the spectrum of resonators with the confocal, preconfocal, and postconfocal ($R < L$) geometry are almost identical even if the value R/L deviates significantly from unity. Only the distinctions in the capability to maintain oscillations with two and more variations of the field along the vertical axis are clearly pronounced: this capacity is much higher in a resonator with a confocal geometry. The spectrum of the resonator with a break of the axis, as anticipated, shows more distinctions: the H_{0nm}-oscillations, corresponding to the index values $m \geq 2$, become unstable, and the corresponding spectral amplitudes of the field $U(g, t)$ go down to the general background level. Still even at small squares where the mirror projections intersect, such a resonator can maintain high-Q free H_{0n1}-oscillations (see, e.g., Fig. 7.16B).

The peaks in the characteristics $|\tilde{U}(g, k)|$ allow us to easily determine the approximate values of $\mathrm{Re}\,\bar{k}$ of eigenfrequencies $\bar{k}$ for all sufficiently high-Q free oscillations maintained in a given OR. These values can be specified and the configuration of the corresponding free oscillation fields can be visualized by using a source $F(g, t)$ that is specially adjusted for such a situation. The results shown in Figure 7.15 have been obtained by analyzing fields $U(g, t)$, generated by OR of a confocal geometry. The source parameters have been chosen taking into account the respective oscillation mode and its Q-factor. In some cases two or three such iterations were needed.

The changes in the field strength $U(g, t)$, for which, after switching off the source, the quasi-mono-frequency component associated with the free oscillation field $\tilde{U}_n(g, \bar{k}_n)$ is dominates, are determined by the factor $\exp(-|\mathrm{Im}\,\bar{k}_n| t)$ (see Section 6.2.4). This fact allows us to calculate $\mathrm{Im}\,\bar{k}_n$ and the Q-factor

$Q_n = \text{Re}\,\bar{k}_n/2|\text{Im}\,\bar{k}_n|$ of the corresponding free oscillation when results similar to those presented in Figure 7.16 are used. The source (7.59) having parameters $\tilde{\alpha} = 20$, $\tilde{T} = 60$, $\beta_1 = 1$, $\beta_3 = \beta_4 = 0$, $\beta_2 = \pi/2$ (Fig. 7.16A), and $\beta_2 = \pi/4$ (Fig. 7.16B) furnishes all necessary conditions for this calculation. The total observation time T can surely be prolonged if a higher accuracy of definition for $\text{Im}\,\bar{k}_n$ is required.

7.4.2. *Dispersive Open Resonators with Grating Mirrors*

The approach whose stages have been considered in the previous section can well be applied to analyze OR with grating mirrors. Note that any experiment which is not based on a clear idea concerning the functioning of the dispersive element in the generation of high-Q field oscillations in the structure is doomed to be a failure.

The analysis is started by considering the spectral characteristics of the Fabry–Perot resonator (see Figs. 7.17 and 7.18), a resonator with plane parallel mirrors

$$\sigma(g) = 2.19 \cdot 10^8 \chi\,[4 - |y|]\,\{\chi\,(z + 0.3)\,\chi\,(-z) + \chi\,(10.3 - z)\,\chi\,(z - 10)\} \tag{7.61}$$

(the width of the resonator is 8.0, the spacing between mirrors is 10.0). The source

$$F(g,t) = 10\chi\,[3 - |y|]\,\chi\,(z-2)\,\chi\,(8-z)\cos(\tilde{k}z + \beta)\exp[-(t-\tilde{T})^2/4\tilde{\alpha}^2] \times \cos[\tilde{k}(t-\tilde{T})] \tag{7.62}$$

excites only oscillations symmetric along the main axis $y = 0$ in such a structure. The normalized spectral amplitudes of the corresponding field $U_0(g,t)$ for $\tilde{\alpha} = 1$ and $\tilde{T} = 6$ in the frequency range $[\tilde{k} - 1; \tilde{k} + 1]$ are never beneath $\gamma = 0.23$ (see Fig. 7.17A). When calculating the range characteristics $\tilde{U}(g,k) \leftrightarrow U(g,t)$, $0 \le t \le T = 150$ (see, e.g., Figs. 7.17B and 7.18B) for all considered OR these parameter values of the function $F(g,t)$ ($\beta = 0.08\tilde{k}$ for $\tilde{k} = 8$ and $\tilde{k} = 14$; $\beta = 0.08\tilde{k} + \pi/4$ for $\tilde{k} = 10$ and $\tilde{k} = 12$) remain unchanged. The source is adjusted to be suitable for analyzing the derived high-Q oscillations (examples in Fig. 7.17C: $H_{0,1,38}$-oscillation and Fig. 7.18A: $H_{0,1,26}$-oscillations) and the values of $\tilde{\alpha}$, $\tilde{T}$, and T are increased up to 20, 80, and 300, respectively. The coordinates $y = 0$, $z = 10 - \pi/2\tilde{k}$ of the points g (the resonator axis, a distance equal to a quarter sounding wavelength from the upper mirror), where the time characteristics of the signal $U(g,t)$ are determined, remain constant and will be changed only when we analyze a resonator with two selective mirrors.

Replace the bottom mirror of the resonator (7.61) by a finite metal grating

$$\sigma(g) = 2.19 \cdot 10^8 \chi\,[4 - |y|]\,\{\chi\,(-z)\,\chi\,(z + 0.235) \times \chi\,[\cos(2\pi y/0.543) - \cos(\pi/2)] + \chi\,(-z - 0.235)\,\chi\,(z + 0.535)\}.$$

Its period is $l = 0.543$; the jut height is $h = 0.235$; the slot width $d = 0.5l$; the reflection zone coincides with the half plane $z \ge 0$. The upper boundary of a single-wave range for the corresponding perfectly conducting infinite structure excited by a normally incident plane wave is determined by the value $k_{\pm 1} = 11.57$ (i.e., the

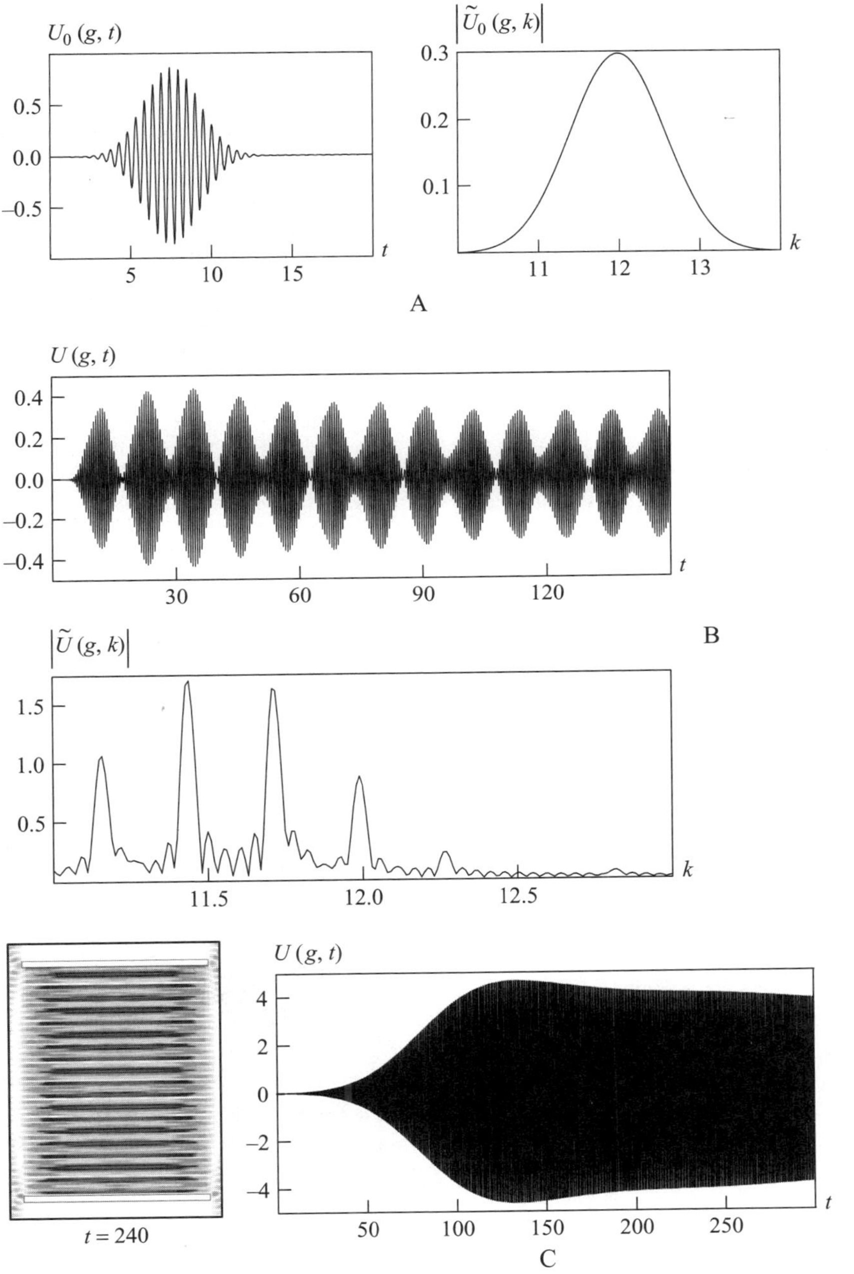

FIGURE 7.17. Analysis of dispersive open resonators: (A) $\tilde{k} = 12$, $g = \{0, 4.72\}$; (B) $\tilde{k} = 12$, $g = \{0, 9.87\}$; (C) $\tilde{k} = 1.44 \approx \mathrm{Re}\bar{k}$ (left picture) and $g = \{0, 9.87\}$ (right picture).

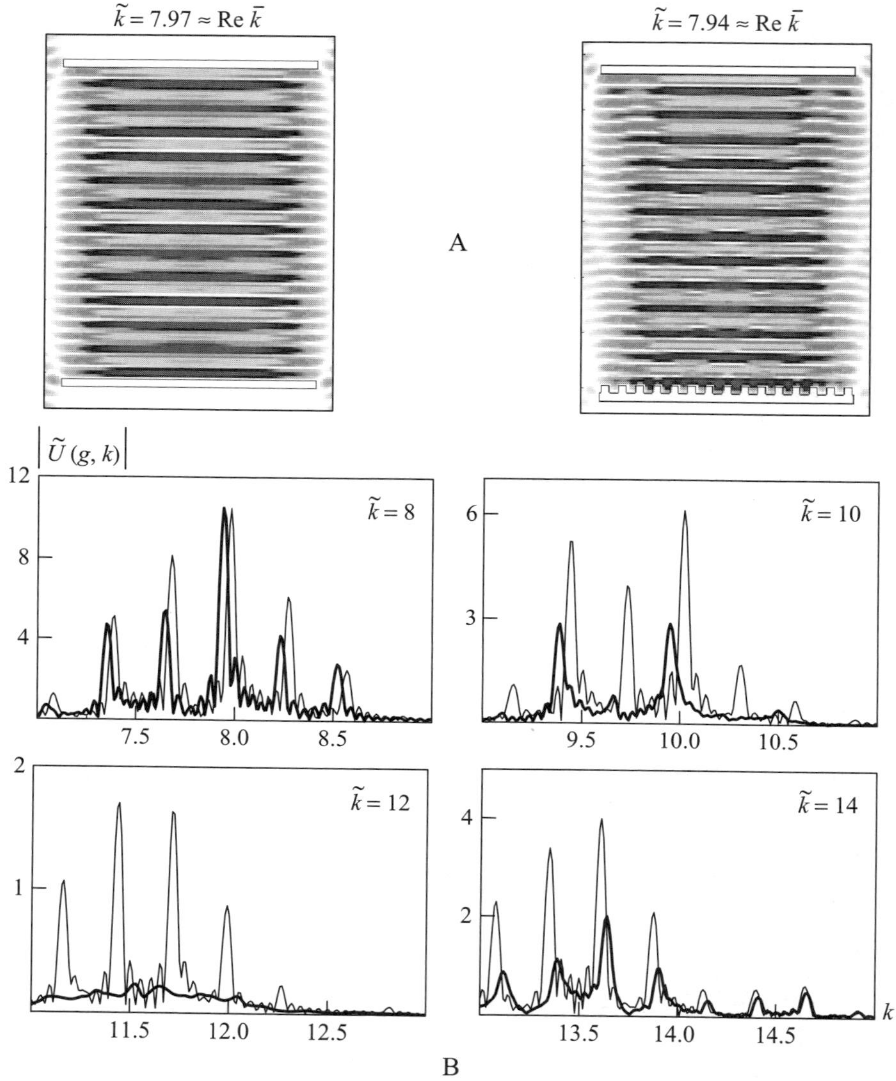

FIGURE 7.18. Comparison of spectral characteristics of Fabry–Perot resonator (fine lines) and dispersive open resonator with grating mirror (thick lines): (A) $t = 300$; (B) $g = \{0, 10 - \pi/2\tilde{k}\}$.

value $\kappa_{\pm 1} = 1$ of a dimensionless frequency parameter $\kappa = l/\lambda = lk/2\pi$). Over this range $|a_0(k)| = 1$, and the value $\arg a_0(k)$ varies within 180 to 260 degrees (the amplitude–frequency characteristics of the grating are presented in Figures 135 and 158 in the book [3]), and from an observer's standpoint in the far zone, there is no difference between the operation of a grating and a perfectly conducting plane

$z = (\pm\pi - \arg a_0)/2k$. Within the frequency range $k_{\pm 1} \le k \le 13.9$ the reflection efficiency in the zero order of the spectrum is drastically decreased (the energy is distributed among the zero, plus, and minus first spatial harmonics), then it increases up to the unit value again ($k = 17.3$) and so on.

If one wishes to forecast the nature of variations in the OR spectrum on the energy balance ratios for perfectly conducting infinite structures, the result will be as described below (see Section 6.2.1 and papers [3,9,11]; the case is quite simple, and such an approach seems to be reasonable). Up to the threshold point $k_{\pm 1}$ the distribution of the spectral amplitudes $|\tilde{U}(g,k)|$ of the field $U(g,t)$ will show no essential changes: the resonance frequencies will be tuned up to the values of $\arg a_0 \neq \pi$, when being shifted along the k axis, and the Q-factors of corresponding free oscillations will change slightly. Past the point $k_{\pm 1}$, the ratio between the levels of the local amplitude centers of functions $|\tilde{U}(g,k)|$ for the dispersive resonator and the Fabry–Perot resonator will coincide with $|a_0(k)|$, however, with minor deviations.

In fact, the uttered suggestion proves to be right, although only in part (see Fig. 7.18B): the sharp boundary by the point $k_{\pm 1}$ blurs over the value range of k from 10.0 to 13.0 ($0.86 < \kappa < 1.12$), where the amplitudes $|\tilde{U}(g,k)|$ go down to the level of the nonresonance background. The main reason (compare the data presented in Figs. 7.19A and 7.19B) is the abnormal redistribution of

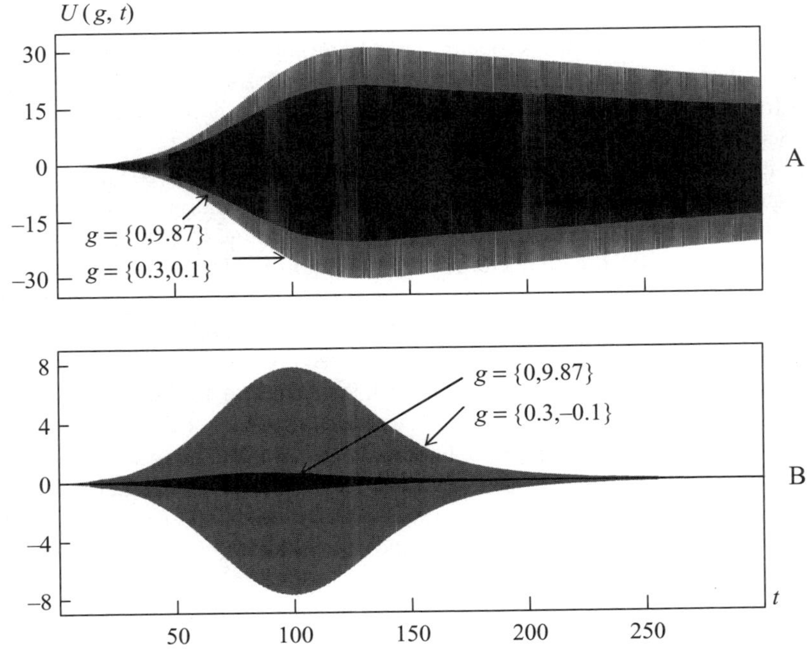

FIGURE 7.19. Field strength in the vicinity of the specular reflectors of dispersive resonator: (A) $\tilde{k} = 7.94$; (B) $\tilde{k} = 11.52$.

the energy of the field $U(g, t)$ of forced oscillations between the regions adjacent to the top (nonselective) and bottom (grating) mirrors of OR. The relative growth of the strength of the near field of the periodical structure and its efficient radiation into free space is possible due to sliding harmonics which, when approaching the threshold point $k_{\pm 1}$, respond in a way typical for the spectral domain.

Replace the bottom plane mirror of the prototype resonator (7.61) by a finite grating

$$\begin{aligned}\sigma(g) = 2.19 \cdot 10^8 \chi\,[4 - |y|]\,\{\chi\,(-y \sin\eta - z\cos\eta)\,\chi\,(y\sin\eta + z\cos\eta + 0.26)\\ \times\chi\,[\cos(2\pi(y\cos\eta - z\sin\eta)/0.75) - \cos(0.15\pi)]\\ +\chi\,(-y\sin\eta - z\cos\eta - 0.26)\,\chi\,(y\sin\eta + z\cos\eta + 0.56)\}\qquad(7.63)\end{aligned}$$

(see Fig. 7.20A; the rotation angle $\eta = 24.06^\circ$ of the structure around the point $g = \{0; 0\}$ is counted clockwise).

An infinite periodic structure having the parameters $l = 0.75$, $h = 0.26$, and $d = 0.85l$ of the grating (7.63) in the frequency range $8.7 < k < 11.9$ ($1.04 < \kappa < 1.43$) operating in the autocollimation mode at the minus first spatial harmonic (see Section 6.3.2: $2\kappa \sin\alpha = 1$; $20.5^\circ < \alpha < 28.7^\circ$) concentrates almost all of its second field energy in the plane wave that is propagating toward the incident one. On the basis of the data presented in Figure 133 in the book [6], it can be assumed that if one of the plane mirrors of the Fabry–Perot resonator (7.61) is replaced by a grating (7.63) that is properly oriented in space (for this reason we have chosen the value of η), the modified resonator will still be capable of maintaining high-Q field oscillations. However, it is true for the values $8.7 < k < 11.9$, whereas in other frequency ranges the spectrum of the dispersive OR will be heavily rarefied (see Section 7.2.3).

All that is said above is confirmed by Figures 7.20 and 7.21. The angle $\alpha = \eta$ corresponds to the frequency $k = 10.27$ of the autocollimation reflection at the minus first harmonic of the spatial spectrum of the grating. Hence, the dispersive resonator, whose spectral characteristics are given in Figure 7.20A, in the corresponding narrow frequency band resembles the characteristics of the prototype resonator, practically without distortions. The excitation level of one of the oscillations belonging to this band has slightly decreased, when that of the other, quite to the contrary, has grown. Beyond this band, the spectrum gradually takes on the form of a nonresonance background, as the plane static mirror does not allow the full implementation of the features inherent for this range of the periodic structure toward almost complete autocollimation reflection. This situation can be changed in two ways. The first one implies a permanent spatial reorientation of the mirrors according to the condition $2\kappa \sin\alpha = 1$, and the second one consists in replacing the plane mirror by a cylindrical one

$$\sigma(g) = 2.19 \cdot 10^8 \chi\,[4 - |y|]\,\chi\,(10.3 - z)\,\chi\,[y^2 + (z + R - 10)^2 - R^2]\qquad(7.64)$$

(see Figs. 7.20B to 7.23; R is the curvature radius of the mirror surface). The comparative characteristics of the high-Q oscillations that are still maintained in

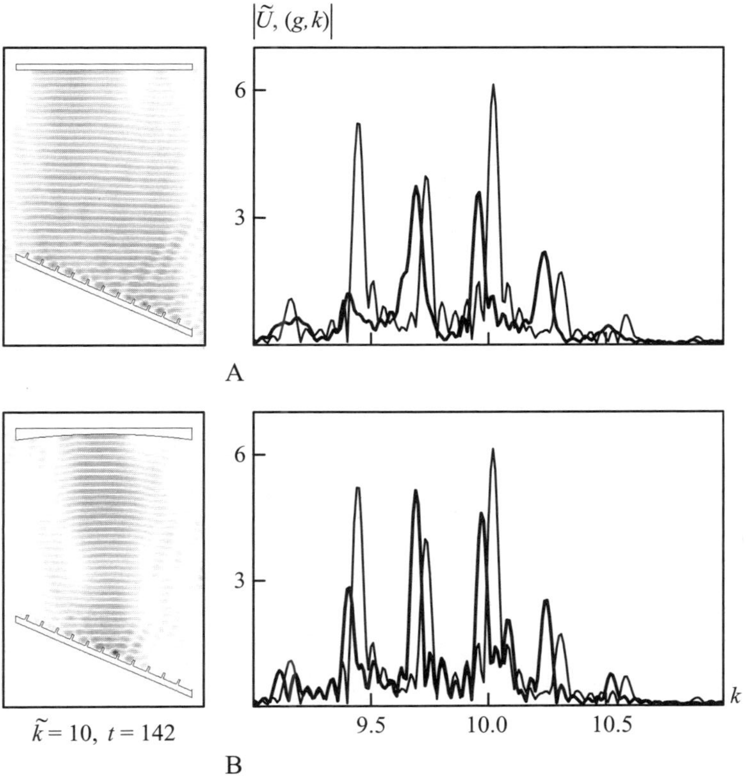

FIGURE 7.20. Changes of spectral characteristics of Fabry–Perot resonator (fine lines) when the lower specular reflector is replaced with grating, operating in auto collimation regime ($g = \{0, 9.84\}$): (A) plane upper mirror; (B) cylindrical upper mirror ($R = 30$).

the OR spectrum (i.e., the approximate values of Re $\bar{k}$, the field geometry, and the Q-factor of oscillations) are presented in Figures 7.22 and 7.23. As for the rarefication of the spectrum beyond the main band $9 < k < 11$, the effect apparently takes place (see Fig. 7.21) and entirely meets the expectations.

In conclusion, we would like to dwell on one more example (see Fig. 7.24), where both the plane mirrors of the prototype resonator (7.61) are replaced by the grating mirrors. The spacing along the $y = 0$ axis between the bottom mirror (7.63) and its replica being appropriately situated in the upper part of the space, is still equal to 10.0. A comparison of the plots in Figure 7.24C (the dots in the middle figure duplicate the characteristics of the dispersive resonator (7.63), (7.64), $R = 50$) and Figure 7.21, allows us to conclude that the focusing capability of the two grating mirrors in the autocollimation reflection mode can, in some cases, exceed by far that of two parallel plane mirrors and that of a pair "cylindrical mirror—grating". This can be explained by the fact that in such a system, the condition $2\kappa \sin \alpha = p$; $p = 1, 2, \ldots$ finds its most natural implementation, which is necessary to generate stable

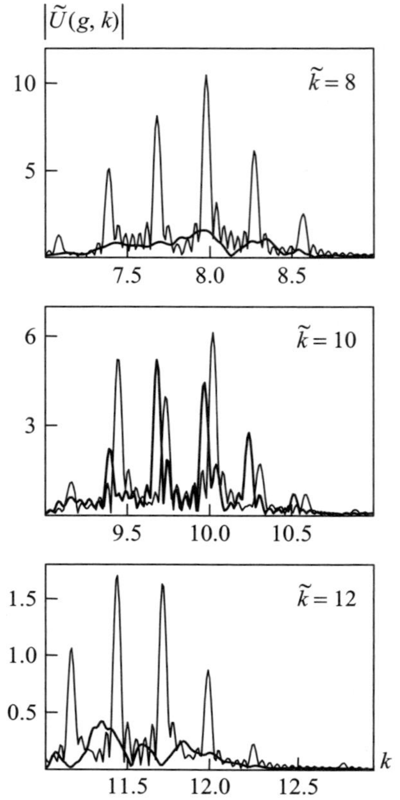

FIGURE 7.21. Spectral characteristics of dispersive open resonator (thick lines) and Fabry–Perot resonator (fine lines) in the frequency range $7 \leq k \leq 13$: $R = 50$, $g = \{0, 10 - \pi/2\tilde{k}\}$.

oscillation according to the scheme "incident wave—reflected wave propagating toward the incident wave". Two grating mirrors constitute a solution to the problem of using narrowband effects of complete nonmirror reflection in the synthesis of considerable single-mode resonance structures. A set of such mirrors (two, three, and more) makes it possible to vary, almost without any restrictions, the general geometry and the volume of OR, the Q-factor, the geometry of the oscillation field, and the spatial distribution of its intensity.

In accordance with the conclusions from [6,9] (see also Section 6.3.2), the implementation of the modes of complete or almost complete nonmirror wave reflection is associated with the excitation of eigenlike oscillations in periodical structures. Hence, for the efficient selection of oscillations in an OR with grating mirrors, the resonance conditions for the original open quasi-optic system should be coordinated with those for the dispersive element employed. It is quite a challenging problem, and its satisfactory solution can be obtained only by using a complex approach encompassing a wide range of frequency and time-domain methods. We hope that the matter presented in this section will be a strong enough affirmation of all the above statements, and will be of good assistance for beginners

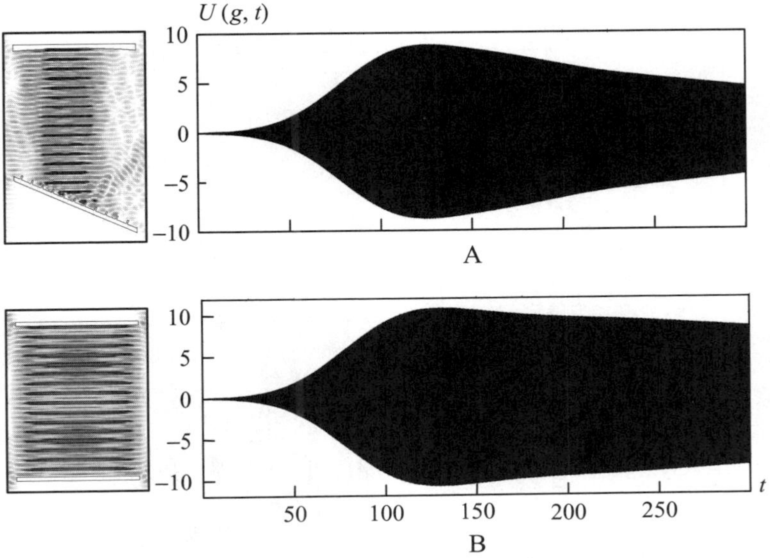

FIGURE 7.22. High-Q oscillations of $H_{0,1,32}$-type (A) in dispersive resonator ($t = 240$, $\tilde{k} = 9.68 \approx \mathrm{Re}\bar{k}$, $R = 50$, and $g = \{0, 9.85\}$) and (B) in Fabry–Perot resonator ($t = 240$, $\tilde{k} = 9.73 \approx \mathrm{Re}\bar{k}$, and $g = \{0, 9.85\}$).

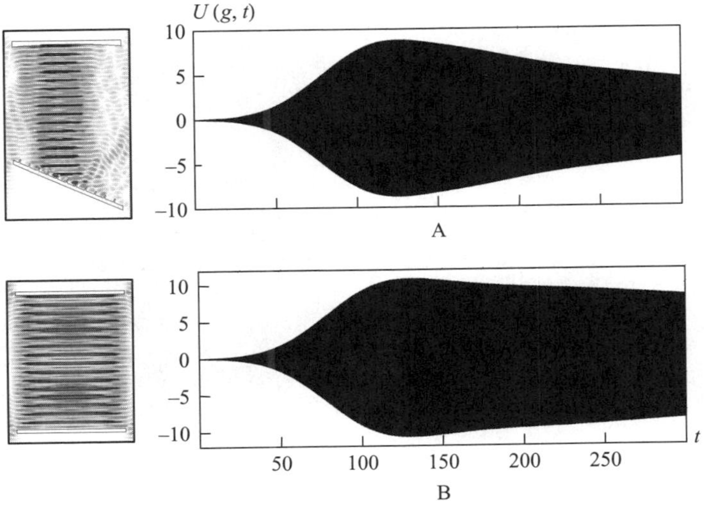

FIGURE 7.23. High-Q oscillations of $H_{0,1,33}$-type (A) in dispersive resonator $t = 240$, $\tilde{k} = 9.96 \approx \mathrm{Re}\bar{k}$, $R = 50$, and $g = \{0, 9.85\}$) and (B) in Fabry–Perot resonator ($t = 240$, $\tilde{k} = 10.03 \approx \mathrm{Re}\bar{k}$, and $g = \{0, 9.85\}$).

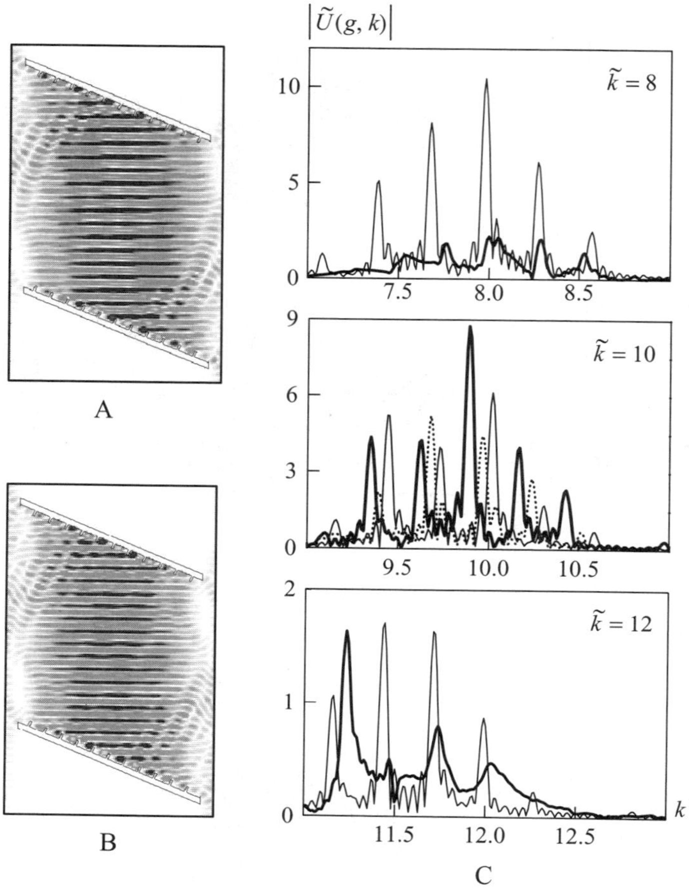

FIGURE 7.24. High-Q oscillations of (A) $H_{0,1,32}$-($\tilde{k} = 9.63 \approx \mathrm{Re}\bar{k}, t = 240$) and (B) $H_{0,1,33}$-type ($\tilde{k} = 9.89 \approx \mathrm{Re}\bar{k}, t = 240$) in resonator with two selective mirrors. (C) Spectral characteristics of dispersive open resonator and Fabry–Perot resonator (fine lines) in the frequency range $7 \leq k \leq 13$ ($g = \{0, 0.1\}$).

in choosing the optimal way from stating this kind of problems to their complete and well-grounded solution.

7.5. Transient Processes in the Near Zone of Pulsed Waves Radiators

Analysis and synthesis of the wideband signal radiators is a very extensive topic including a vast number of results and problems. Our contribution consists of several examples concerning the study of near pulse fields of some model structures by using the methods considered in previous sections of this book.

7.5.1. The Luneburg Lens

Unique wide-angle properties are inherent to antennas and radar reflectors designed on the basis of spherical dielectric lenses having central symmetry. Thus, for instance, while receiving television signals from satellites situated in different points of a geostationary orbit, one lens antenna can be as efficient as ten parabolic antennas of the same diameter. The wave dimension of the lenses varies from 2 (in television antennas) to 50 (in antennas of millimeter and centimeter wave ranges). The production technique (heat-resistant lenses are made of artificial dielectrics and composite bulk materials) is constantly being perfected and simplified. The theoretical problems arise sometimes when the conventional design of the Luneburg lens ($\varepsilon(r) = 2 - (r/R)^2$; R is the lens radius; r is the distance to the center) is changed to the actually manufactured one (with a piecewise constant distribution of the dielectric permittivity ε, presence of reflectors, and bracing components, etc.). The traditional methods of mathematical diffraction theory in such a case allow us to consider in detail some special questions (see, e.g., [230,231]), and the direct universal methods of the time domain are most fit to efficiently and adequately solve the constantly emerging diverse applied problems. Below we consider some results of an analysis of the variations in the characteristics of the radiated wide-band signal, which are conditioned by the replacement of the classic Luneburg lens by a layered structure having the same dimension (the approximation of a monotonous continuous function $\varepsilon(g)$ by averages).

The dielectric lens

$$\varepsilon(g) = \chi(R^2 - y^2 - z^2)[2 - (y^2 + z^2)/R^2]; \quad R = 3, \tag{7.65}$$

whose half-surface is occupied by a metal reflector

$$\begin{aligned} \sigma(g) = 2.19 \cdot 10^8 \chi(y^2 + z^2 - R^2)\{\chi[(R+0.5)^2 - y^2 - z^2]\chi(R - |z|)\chi(-y) \\ - \chi(y + 3.32)\chi(0.16 - |z|)\chi(-y - 2)\}, \end{aligned} \tag{7.66}$$

is excited by the source

$$F(g, t) = 10\chi(y + 3.32)\chi(0.16 - |z|)\chi(-y - 3)\cos(\tilde{k}t), \quad \tilde{k} = 6$$

of quasi-harmonic waves. This situation is modeled in Figure 7.25A. Both fragments here are replicas of the original continuous-tone image, with the contrast and brightness appropriately adjusted. In one of the variants, the filter is tuned to the strength of the field $U(g, t)$ in the principal lobe of the diagram; in the second one the limit is put down to the level of field amplitudes in the main side lobes. The capability of the system (7.65), (7.66) to generate a sharply directed radiation field with quite a high quality of the wave front in the principal lobe is not to be doubted.

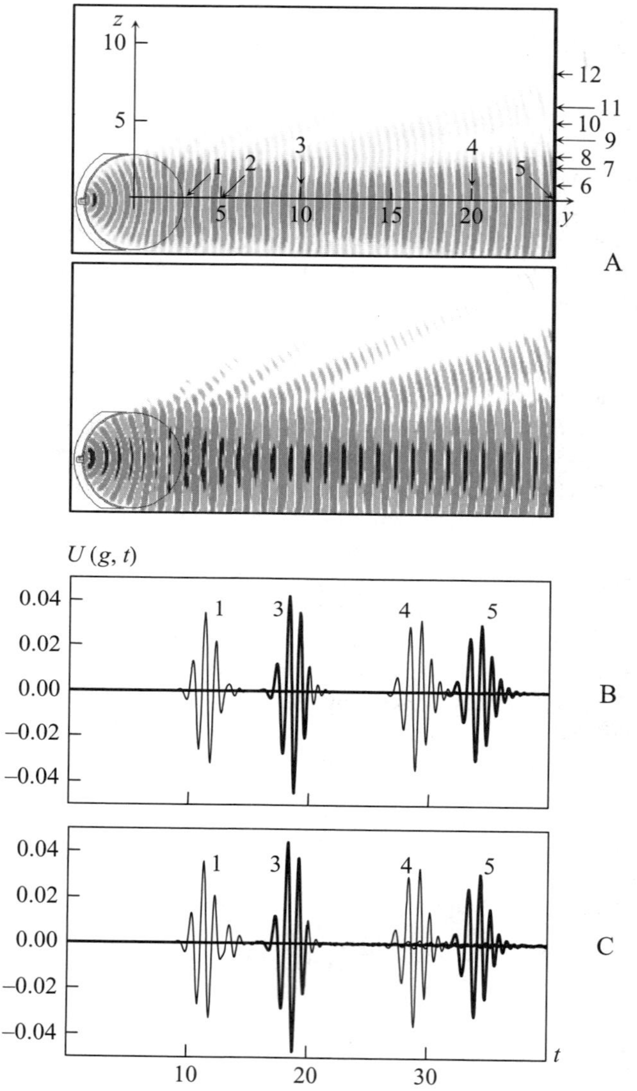

FIGURE 7.25. General geometry of the problem and near field pattern of the antenna in the case of (A) quasi monochromatic ($t = 34$) and (B, C) pulse excitation. Systems with Luneburg lenses (7.65) (A, B) and (7.68) (C).

The excitation of the system by the current pulse

$$F(g,t) = 10\chi(y+3.32)\chi(0.16-|y|)\chi(-y-3)\exp[-(t-\tilde{T})^2/4\tilde{\alpha}^2] \\ \times \cos[\tilde{k}(t-\tilde{T})]; \quad \tilde{\alpha}=0.5, \quad \tilde{T}=3, \quad \tilde{k}=6 \qquad (7.67)$$

(see Figs. 7.25B and 7.26: the functions $U(g,t)$ at the points 1,2, …,12 of the closure of the domain $\mathbf{Q}_L$) causes no qualitative changes in these results. The

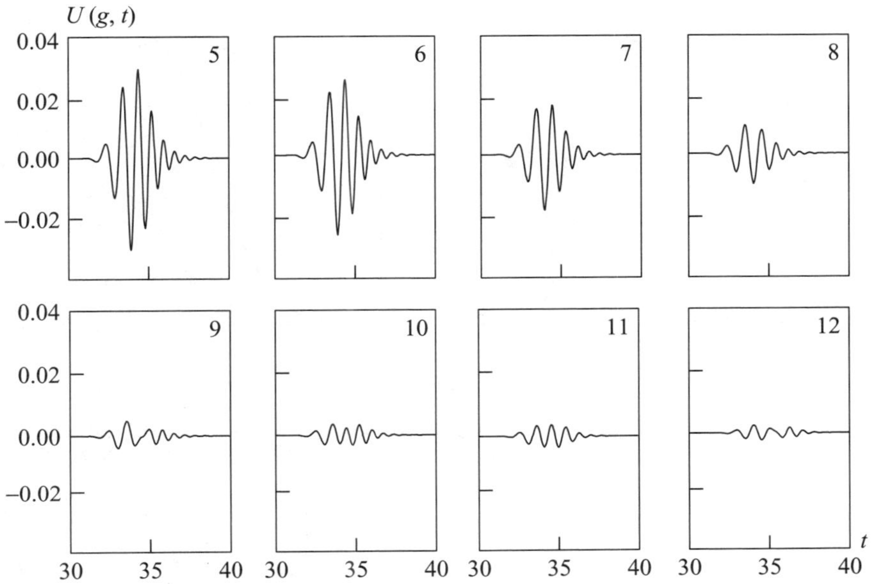

FIGURE 7.26. Field $U(g, t)$in certain point g of the axis $y = 25$ for pulse excitation of antenna.

structure of the diagram is completely formed in the interval $-R < y < 5R$, where the width of the free space band is $4R$. Beyond the interval, the values $U(g, t)$ at the point of axes $z = 0$ and $y = \text{const}$ determine all the characteristics of the principal lobe of the diagram. Thus, for example, the field spot of this lobe (outside the spot the normalized level of the absolute values of $U(g, t)$ is less than 0.15) at the distance of $7.3R$ from the antenna ($y = 25$) just slightly exceeds the lens size.

The system (7.65), (7.66) is a wideband system. It imparts to the radiation field $U(g, t)$ all the key spectral characteristics of the source $F(g, t)$ (see Fig. 7.27A). Only the maximum of the spectral amplitudes of the generated signal (points 4, 5, and so on) turns to be shifted into the domain of large frequencies: the long-wave components of the field $U(g, t)$ are radiated more intensively to the side directions from the interval $-R < y < 5R$. The distinctions in the smoothness of functions $\tilde{U}(g, k)$ for the points 1, 2 and 5, 11 is a computational effect: during the observation time $0 \leq t \leq 40$, for which the transform (7.60) is implemented, the tails of the pulses $U(g, t)$ have not enough time to pass the peripheral points of the domain $\mathbf{Q}_L$, and are artificially eliminated.

Replace the lens (7.65) by a layered structure

$$\begin{aligned}\varepsilon(g) = {} & 1.16\chi(R^2 - y^2 - z^2) + 0.28\chi[(R-0.5)^2 - y^2 - z^2] \\ & + 0.22\chi[(R-1)^2 - y^2 - z^2] + 0.17\chi[(R-1.5)^2 - y^2 - z^2] \\ & + 0.11\chi[(R-2)^2 - y^2 - z^2] + 0.06\chi[(R-2.5)^2 - y^2 - z^2]. \quad (7.68)\end{aligned}$$

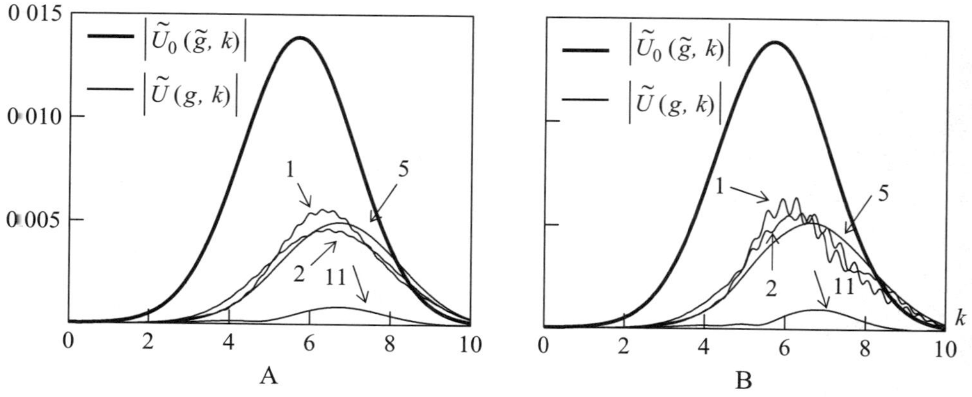

FIGURE 7.27. Frequency characteristics of pulse source (7.67) ($\tilde{g} = \{-2.5, 0\}$) and antenna with lenses (7.65) (A) and (7.68) (B): $\tilde{U}_0(\tilde{g}, k) \leftrightarrow U_0(\tilde{g}, t)$; $\tilde{U}(g, k) \leftrightarrow U(g, t)$; $U_0(g, t)$ is the field of the source in a free space.

The system (7.66), (7.68) roughly approximates the system (7.65), (7.66). However, the difference in their characteristics can hardly be noticed (see Figs. 7.25B,C and 7.27). Figure 7.26 shows the dependences $U(g, t)$ for an antenna with the classical Luneburg lens; still, if we superpose them with the corresponding dependences for an antenna with the lens (7.68), the graphs will just coincide. The system (7.66), (7.68) is not physically smooth. Nevertheless, the range properties of a smooth system remain. The main distinctions are the amplitude levels in the tails of the pulses $U(g, t)$, but they have practically no influence on the integral signal characteristics.

7.5.2. Radiation from Open Periodic Waveguides

Plane pattern-forming structures, such as that shown in Figure 7.28, have been the center of attention for a long time. However, in related research, theory is

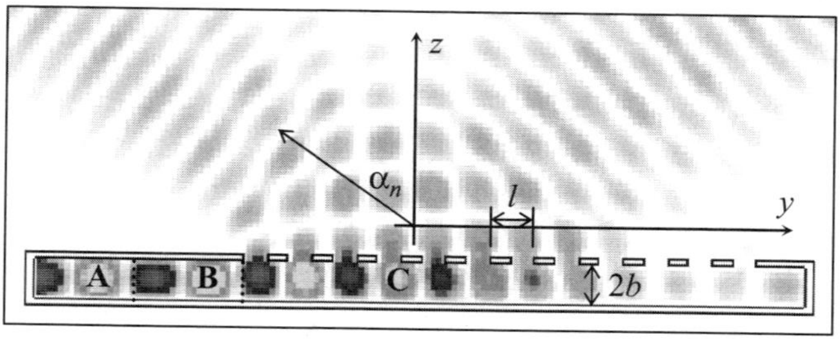

FIGURE 7.28. Analysis of grating type PFSs; **A** is the domain, containing source support; **B** and **C** are intervals of parallel-plane waveguide and open periodic waveguide.

still given minor, secondary roles. The difficulty is that the solutions obtained to the specific model problems cannot be properly summed up by the analytical description of the practically implemented situations. Here an experiment is required that will be sensitive to the quality of the involved theoretical data. If the respective package includes the data on the propagation constants $\gamma_m^{\mathbf{B}}$ and $\gamma_m^{\mathbf{C}}$; $m = 1, 2, \ldots$ of several main partial components ($\mathrm{Im}\,\gamma_m^{\mathbf{B}} = 0$, $\mathrm{Re}\,\gamma_m^{\mathbf{B}} > 0$ and $\mathrm{Im}\,\gamma_m^{\mathbf{C}} > 0$, $\mathrm{Re}\,\gamma_m^{\mathbf{C}} > 0$) of the harmonic fields $\tilde{U}\,(\mathbf{B}, k)$ and $\tilde{U}\,(\mathbf{C}, k)$ in closed (**B**) and open periodic (**C**) waveguides (the source of monochromatic waves is situated in the domain **A**), it is possible to make some preliminary evaluation of the PFS. Thus, for instance, in the case when only the first ($m = 1$) component of the field $\tilde{U}\,(\mathbf{C}, k)$ is propagating with a slight attenuation in the waveguide **C** (single-wave mode), the spatial orientation of the main lobe is determined by the angle $-\alpha_0$ such that $k \sin\alpha_0 = \mathrm{Re}\,\gamma_1^{\mathbf{C}}$. The number of additional lobes and their orientation are determined by the values of the relative frequency parameter $\kappa = l/\lambda = kl/2\pi$; l is the length of the grating period: all n satisfying $\kappa > |n + \kappa \sin\alpha_0|$ correspond to a separate lobe having the orientation $\alpha_n = -\arcsin\left(n\kappa^{-1} + \sin\alpha_0\right)$ (see [6,9] and Section 6.3.2; the angles are measured in the plane $y0z$, anticlockwise from the z-axis). The concentration of the radiated energy in each particular lobe depends on the parameters of the periodic structure (on the efficiency of diffraction radiation in the corresponding spectrum order; see an example with the synthesis of reflecting gratings for the devices of relativistic diffraction electronics in Section 6.3.2).

All these lobes are associated with the direct passing of the first wave propagating in the waveguide **C**. Lobes having the opposite orientation ($\alpha_n \to -\alpha_n$) will also be present in the radiation zone of the structure. The field amplitude level there can be different depending on how wellmatched the transfer from domain **B** into domain **C** is, as well as on the magnitude of the loss of the first wave passing from the left to the right edge of section **C**. The good quality of the wave fronts can be reached only by rather uniform decrease of the wave amplitude in this interval, that is, only by sufficiently small values of $|\mathrm{Im}\,\gamma_1^{\mathbf{C}}|$. For frequencies k, for which waveguide **C** is open for two, three, and more waves, the number of energy radiation channels increases two, three, or more times, respectively. It is obvious that it is impossible for such types of radiators to be wideband. However, their selective properties, the capability to form strictly ordered systems of spaced channels with mono-frequency signals are excellently fit for solving a very important applied problem, that is, the determination of short pulse parameters.

Computational experiments in the time domain are based on the same theoretical data as those used in the actual experiments. The major goal of the computational experiment is the final analysis and optimization of the model of the quasi-optical system including all its resonance elements, described by the real materials and geometric parameters (see Section 7.1). The comprehensive solution of only one such problem requires thorough preparations and is worth launching only if there is an interested party showing a real interest in it, as well as a clear and well-grounded specification. The results presented in Figures 7.29 and 7.30 are to be considered just as illustrations. All the key units of the system under analysis are

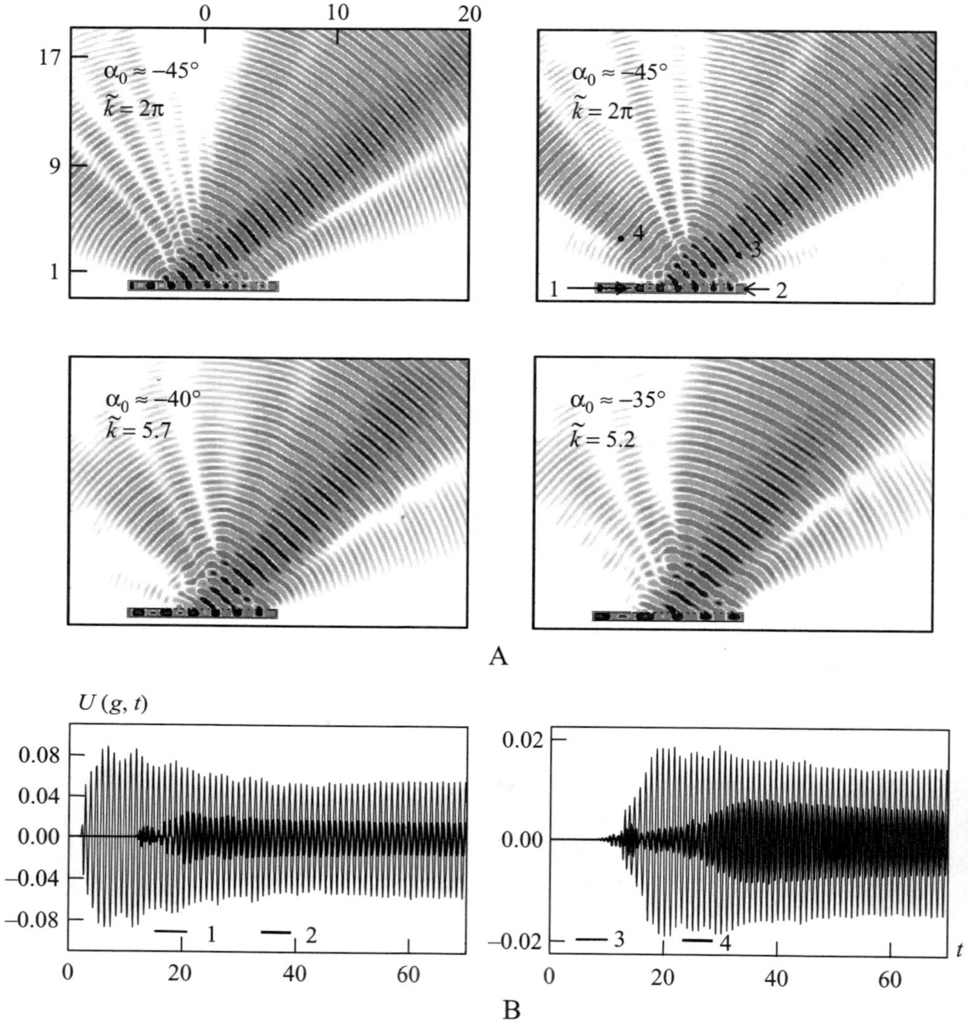

FIGURE 7.29. Radiation field of grating type PFSs, excited with a source $F(g,t) = 10\chi(y+5.54)\chi(-y-5.4)\chi(0.3-|z|)\cos(\tilde{k}t)$ of quasi-harmonic waves: $b = 0.31$; period of strip gratings is $l = 0.5$; slot width is $l\theta = 0.35$: (A) field pattern in near zone of PFS for different values of $\tilde{k}$, $t = 40$; (B) time dependencies of $U(g,t)$ in the points 1 to 4 of near zone of PFSs, $\tilde{k} = 2\pi$.

elementary ones (the metal plane-parallel waveguide and an open waveguide of the same size, whose translucent wall is made in the form of a thin metal ribbon grating) and operate in the simplest, single-wave mode.

The interiors of the unit, whose analysis is illustrated in Figure 7.29, is a rectangle of the dimension $(-5.54; 5.54) \times (-0.31; 0.31)$. In the interval $-3 \leq y \leq 5$

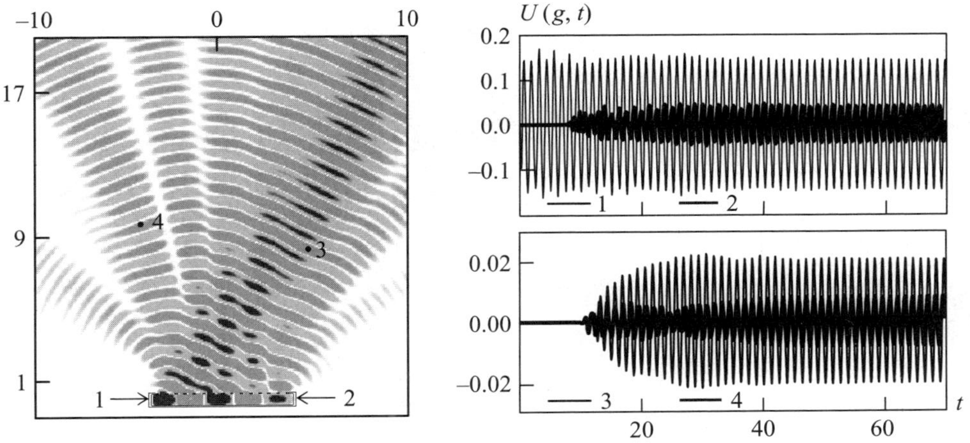

FIGURE 7.30. Excitation of the piece of open periodic waveguide with a source $F(g,t) = 10\chi(y+3.8)\chi(-y-3.6)\chi(0.3-|z|)\sin[\pi(z+0.3)/0.6]\cos(\tilde{k}t)$: Total length of grating is equal to 7.6, $l = 0.5$, $l\theta = 0.3$, $\tilde{k} = 5$, $\alpha_0 \approx -30°$.

in the plane $z = 0.31$, a grating is situated: the regular one (with all slots being equal; see the first part of Fig. 7.29A) and the matching one (at the first six periods the slots are widened up to the standard size $l\theta$; see other parts of Figs. 7.29A and 7.29B). The smaller values of $\tilde{k}$ correspond to the larger lengths of the waveguide waves in the domains **B** and **C** and the lower values of $|\alpha_0|$. The further lifting of the principal lobe is, therefore, limited by the critical frequency $k_1^{\mathbf{B}} \approx 5.07$ of waveguide **B** ($k_1^{\mathbf{B}} > k_1^{\mathbf{C}}$). In the system, whose analysis is presented in Figure 7.30, this limitation is removed: its interior area $(-3.8; 3.8) \times (-0.31; 0.31)$ along the whole length is closed from above by the periodic ribbon structure. The linear graphs in Figures 7.29B and 7.30 show the variation of the field strength $U(g,t)$ as a function of time in the characteristic points 1 to 4 of the analyzed domain $\mathbf{Q}_L$. The dependences in the pair of points 1 and 2 allow us to estimate the efficiency of the energy extraction from the wave, propagating in an open waveguide. Observing the process at the points 3 and 4, one obtains the data on the energy distribution between the main channels of radiation into free space.

In the next (and last) example we try to reproduce in a plane PFS one of the effects of a complete transformation of waves by periodic gratings (a complete nonmirror reflection). A metal comb grating having a slot width $\theta = 0.8$, filled with a material whose relative dielectric permittivity is $\varepsilon = 3.89$, in value ranges of $\kappa = l/\lambda$ and $\delta = h/l$, that are close to $\bar{k} = 0.9024$ and $\bar{\delta} = 0.245$, gives almost all the energy of the minus first harmonics of the spatial spectrum, with an angle of incidence of $\alpha_{-1} = 85.8°$, to the zero one: $\alpha_0 = 6.4°$; the telescoping transformation coefficient is $\cos\alpha_0/\cos\alpha_{-1} \approx 13.6$. Such a mode of resonance scattering of homogeneous plane waves by infinite periodic structures has been considered in Section 6.3.2

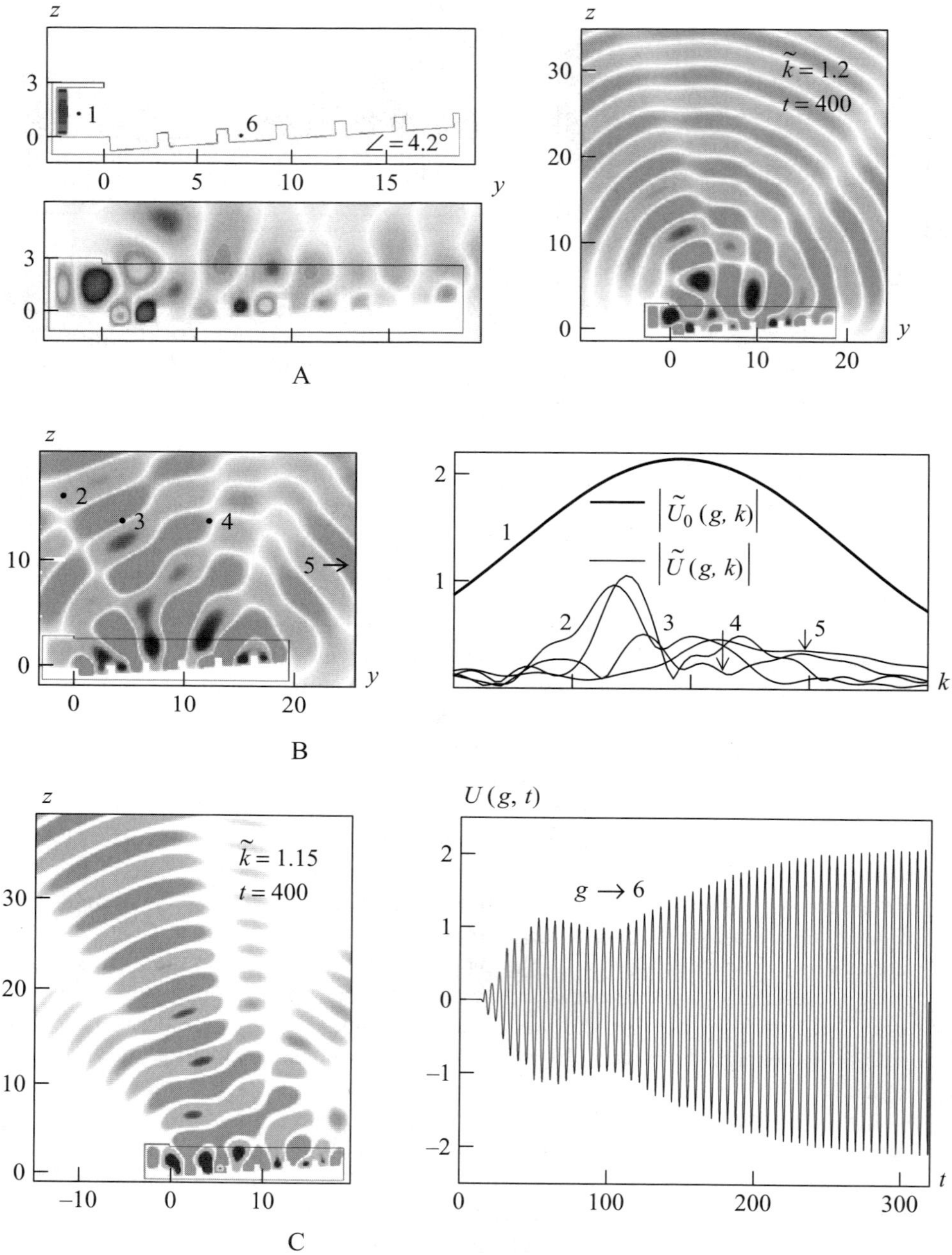

FIGURE 7.31. Total wave transformation by grating in antenna structure. Antenna is excited with sources of (A, C) quasi-harmonic and (B) pulsed signals.

(see Fig. 6.19). The choice of the antenna parameters (see Fig. 7.31A: the left bottom part), whose metal frame (the top left part) is partially covered by a teflon ($\varepsilon_1 = 2.2$) rod, is made in attempts to reproduce in the considered model all the key conditions necessary for the effect to take place. The grating slots are filled with a dielectric having $\varepsilon_2 = 8.56$ ($\varepsilon_2/\varepsilon_1 = \varepsilon = 3.89$); the grating period ($l = 3.185$) and the jut height ($h = 0.78$) are determined by the central frequency $\tilde{k} = 1.2$ of the source

$$F(g,t) = 10\chi(y+2.5)\chi(-y-2)\chi(z)\chi(2.7-z)\sin(\pi z/2.7)\cos\left(\tilde{k}t\right) \tag{7.69}$$

of quasi-harmonic waves and by the approximate value $2\pi/\tilde{k}\sqrt{\varepsilon_1} = 3.53$ of the length $\tilde{\lambda}(\varepsilon_1)$ of the wave radiated by the teflon rod (see Fig. 4.11).

The result of a computation presented in Figure 7.31A (the right part: the field strength distribution $U(g,t)$ in the region $\mathbf{Q}_L$), shows that the assumed, as the initial approximation, dependence $\tilde{\lambda}(\varepsilon_1) = 2\pi/\tilde{k}\sqrt{\varepsilon_1}$ should be specified as the effect of the concentration of radiated energy in the direction that was theoretically expected is not pronounced clearly enough. In order to solve this auxiliary problem, the excitation of the antenna by a wideband signal source

$$F(g,t) = 10\chi(y+2.5)\chi(-y-2)\chi(z)\chi(2.7-z)\sin(\pi z/2.7)e^{-(t-\tilde{T})^2/4\tilde{\alpha}^2} \times \cos(\tilde{k}(t-\tilde{T})); \quad \tilde{k} = 1.2, \quad \tilde{T} = 25, \quad \tilde{\alpha} = 5 \tag{7.70}$$

has been modeled (see Fig. 7.31B) and, judging by the shift of spectral amplitude centers of pulses $U(g,t)$ with respect to the central frequency in the field $U_0(g,t)$ (i.e., in the field generated by the source (7.70) in free space), a new value of frequency $\tilde{k} = 1.15$ was found for the source (7.69). This value corresponds mainly to the reference wavelength $\tilde{\lambda}(\varepsilon_1) = 3.53$ that has been used to determine the geometric parameters of the grating structure. The distribution of spectral amplitudes of the pulse $U(g,t)$ in point 3, the location of which corresponds to the expected direction of the main lobe of the radiation field, confirms these statements. All the dependences $\tilde{U}(g,k) \leftrightarrow U(g,t)$ ($\tilde{U}_0(g,k) \leftrightarrow U_0(g,t)$) and $U(g,t)$ presented in Figure 7.31 are obtained after the observation results for the characteristic points $g \in \mathbf{Q}_L$, indexed 1 through 6. The results of the experiment with the refined value of $\tilde{k}$ are shown in Figure 7.31C: the orientation of the principal lobe of the radiation diagram having quite a high level of radiated energy corresponds to the expected one, and the value $\tilde{k} = 1.15$ is indeed close to the real part of eigenfrequency $\bar{k}$, which is an ideal case for the observed effect to take place. This conclusion is based on the growth of the field amplitude in the grating slot (in the point 6).

References

1. Samarsky, A.A.: Mathematical modeling and numerical experiment. *Bull. Acad. Sci. of USSR*, No. 5, 38–49 (1979) (in Russian).
2. Shestopalov, V.P.: *The Method of the Riemann-Hilbert Problem in the Theory of Electromagnetic Wave Diffraction and Propagation*. Kharkov State Univ. Press, Kharkov (1971) (in Russian).
3. Shestopalov, V.P., Litvinenko, L.N., Masalov, S.A., Sologub, V.G.: *Wave Diffraction by Gratings*. Kharkov State Univ. Press, Kharkov (1973) (in Russian).
4. Shestopalov, V.P.: *Adder Equations in the Modern Diffraction Theory*. Naukova Dumka, Kiev (1983) (in Russian).
5. Shestopalov, V.P., Kirilenko, A.A., Masalov, S.A.: *Matrix Convolution-Type Equations in the Diffraction Theory*. Naukova Dumka, Kiev (1984) (in Russian).
6. Shestopalov, V.P., Kirilenko, A.A., Masalov, S.A., Sirenko, Y.K.: *Resonance Wave Scattering. Vol. 1. Diffraction Gratings*. Naukova Dumka, Kiev (1986) (in Russian).
7. Shestopalov, V.P., Kirilenko, A.A., Rud', L.A.: *Resonance Wave Scattering. Vol. 2. Waveguide Discontinuities*. Naukova Dumka, Kiev (1986) (in Russian).
8. Shestopalov, V.P.: *The Morse Critical Points of Dispersion Equations*. Naukova Dumka, Kiev (1992) (in Russian).
9. Shestopalov, V.P., Sirenko, Y.K.: *Dynamic Grating Theory*. Naukova Dumka, Kiev (1989) (in Russian).
10. Shestopalov, V.P., Tuchkin, Y.A., Poyedinchuk, A.Y., Sirenko, Y.K.: *New Solution Methods for Direct and Inverse Problems of the Diffraction Theory. Analytical Regularization of the Boundary-Value Problems in Electromagnetic Theory*. Osnova, Kharkov (1997) (in Russian).
11. Petit, R. (ed): *Electromagnetic Theory of Gratings*. Springer-Verlag, New York (1980).
12. Mittra, R., Lee, S.: *Analytical Techniques in the Theory of Guided Waves*. Macmillan, New York (1971).
13. Voytovich, N.N., Kazenelenbaum, B.Z., Sivov, A.N.: *Generalized Method of Eigen Oscillations in the Diffraction Theory*. Nauka, Moscow (1977) (in Russian).
14. Litvinenko, L.N., Prosvirnin, S.L.: *Spectral Scattering Operators in the Problems of the Wave Diffraction by Plane Screens*. Naukova Dumka, Kiev (1984) (in Russian).
15. Colton, D., Kress, R.: *Integral Equation Methods in Scattering Theory*. Wiley-Interscience, New York (1983).
16. Nazarchuk, Z.T.: *Numerical Analysis of Wave Diffraction by Cylindric Structures*. Naukova Dumka, Kiev (1989) (in Russian).

17. Georgakopoulos, S.V, Birtcher, C.R., Balanis, C.A., Renaut, R.A.: Higher-order finite-difference schemes for electromagnetic radiation, scattering, and penetration, Part I: Theory. *IEEE Ant. Propag. Mag.*, **44**, No. 1, 134–142 (2002).
18. Maikov, A.R, Sveshnikov, A.G., Yakunin, S.A.: Difference scheme for the Maxwell transient equations in waveguide systems. *J. Comput. Math. Math. Phys.*, **26**, No. 6, 851–863 (1986) (in Russian).
19. Maikov, A.R, Poezd, A.D., Sveshnikov, A.G., Yakunin, S.A.: Difference scheme of initial boundary-value problems for Maxwell equations in unlimited domain. *J. Comput. Math. Math. Phys.*, **29**, No. 2, 239–250 (1989) (in Russian).
20. Perov, A.O., Sirenko, Y.K., Yashina, N.P.: Explicit conditions for virtual boundaries in initial boundary value problems in the theory of wave scattering. *J. Electromag. Waves Appl.*, **13**, No. 10, 1343–1371 (1999).
21. Sirenko, Y.K.: *Simulation and Analysis of Transient Processes in Open Periodic, Waveguide, and Compact Resonators*. EDENA, Kharkov (2003) (in Russian).
22. Engquist, B., Majda, A.: Absorbing boundary conditions for the numerical simulation of waves. *Math. Comput.*, **31**, No. 139, 629–651 (1977).
23. Mur, G.: Absorbing boundary conditions for the finite difference approximation of the time-domain electromagnetic-field equations. *IEEE Trans.*, **EMC-23**, No. 4, 377–382 (1981).
24. Berenger, J.-P.: A perfectly matched layer for the absorption of electromagnetic waves. *J. Comput. Phys.*, **114**, No. 1, 185–200 (1994).
25. Berenger, J.-P.: Three-dimensional perfectly matched layer for absorption of electromagnetic waves. *J. Comput. Phys.*, **127**, No. 2, 363–379 (1996).
26. Sacks, Z.S., Kingsland, D.M., Lee, R., Lee, J.F.: A perfectly matched anisotropic absorber for use as an absorbing boundary condition. *IEEE Trans.*, **AP-43**, No. 12, 1460–1463 (1995).
27. Taflove, A., Hagness, S.C.: *Computational Electrodynamics: The Finite-Difference Time-Domain Method*. Artech House, Boston (2000).
28. Yee, K.S.: Numerical solution of initial boundary value problems involving Maxwell's equations in isotropic media. *IEEE Trans.*, **AP-14**, No. 4, 302–307 (1966).
29. Sirenko, Y.K., Velychko, L.G., Erden, F.: Time-domain and frequency-domain methods combined in the study of open resonance structures of complex geometry. *Progress Electromag. Res.*, **44**, 57–79 (2004).
30. Sirenko, Y.K., Yashina, N.P.: Nonstationary model problems for waveguide open resonator theory. *Electromagnetics*, **19**, No. 5, 419–442 (1999).
31. Vinogradova, M.B., Rudenko, O.V., Suhorukov, A.P.: *The Theory of Waves*. Nauka, Moscow (1979) (in Russian).
32. Borisov, V.V.: *Electromagnetic Fields of Transient Currents*. St. Petersburg State Univ. Press, St. Petersburg (1996) (in Russian).
33. Levin, L.: *Theory of Waveguides: Techniques for Solution of Waveguide Problems*. Newnes-Butterworth, London (1975).
34. Vladimirov, V.S.: *Equations of Mathematical Physics*. Dekker, New York (1971).
35. Ladyzhenskaya, O.A.: *The Boundary Value Problems of Mathematical Physics*. Springer-Verlag, New York (1985).
36. Vladimirov, V.S. (ed): *Collected Example Equations of Mathematical Physics*. Nauka, Moscow (1974) (in Russian).
37. Krein, S.G. (ed): *Functional Analysis*. Nauka, Moscow (1972) (in Russian).

38. Romanov, V.G.: On the structure of the fundamental solution of the Cauchy problem for the Maxwell equations. *Rep. Acad. Sci. USSR*, **281**, No. 5, 1052–1055 (1985) (in Russian).
39. Bateman, H.: *The Mathematical Analysis of Electrical and Optical Wave-Motion.* Dover, New York (1955).
40. Petrashen', G.I., Molotkov, L.A., Krauklis, P.V.: *Waves in Layered Homogeneous Isotropic Elastic Media*. Nauka, Leningrad (1982) (in Russian).
41. Tretyakov, V.V.: New analytic solutions of wave equation and the diffraction problem. *Appl. Math. Mech.*, **39**, NÓ. 1, 80–85 (1975) (in Russian).
42. Vekua, I.N.: *New Methods for Solving Elliptic Equations*. North-Holland, Amsterdam (1967).
43. Smirnov, V.I.: Solution of the limiting problem for wave equation for circle and sphere. *Rep. Acad. Sci. USSR*, **14**, No. 1, 13–16 (1937) (in Russian).
44. Miller, W. Jr.: *Symmetry and Separation of Variables*. Addison-Wesley, London (1977).
45. Borisov, V.V.: *Transient Fields in Waveguides*. Leningrad State Univ. Press, Leningrad (1991) (in Russian).
46. Tretyakov, V.V.: Mode basis method. *Radio-Eng. Electron.*, **31**, No. 6, 1071–1082 (1986) (in Russian).
47. Waynberg, B.R.: *Asymptotic Methods in the Equations of Mathematical Physics.* Moscow State Univ. Press, Moscow (1982) (in Russian).
48. Gold, B, Rader, C.M.: *Digital Processing of Signals*. McGraw-Hill, New York (1969).
49. Muravey, L.A.: Asymptotic character of solutions to the second outer problem for a two-dimensional wave equation. *Differential Equations*, **6**, No. 12, 2248–2262 (1970) (in Russian).
50. Muravey, L.A.: Analytical extension on the Green function parameter of the outer boundary-value problem for two-dimensional Helmholtz equation. III. *Coll. Papers Math.*, **105**, No. 1, 63–108 (1978) (in Russian).
51. Muravey, L.A.: On the asymptotic character at large time parameters of the Green function of the first outer boundary-value problem for waveguide equation with two space variables. *Rep. Acad. Sci. USSR*, **220**, No. 6, 1271–1273 (1975) (in Russian).
52. Morawetz, C.S.: The decay of solutions of the exterior initial-boundary value problem for the wave equation. *Comm. Pure Appl. Math.*, **14**, 561–568 (1961).
53. Andronov, V.D.: Some evaluations of the Green function of the Helmholtz equation. *J. Comput. Math. Math. Phys.*, **5**, No. 6, 1006–1023 (1965) (in Russian).
54. Ramm, A.G.: Exponential decrease of the solution of hyperbolic equation. *Differential Equations*, **6**, No. 11, 2099–2100 (1970) (in Russian).
55. Babich, V.M.: On the short-wave asymptotic of the solution to the problem about a point source in an inhomogeneous medium. *J. Comput. Math. Math. Phys.*, **5**, No. 5, 949–951 (1965) (in Russian).
56. Tikhonov, A.N., Arsenine, V.Y.: *Solutions of Ill-Posed Problems*. Winston, Washington, DC (1977).
57. Tikhonov, A.N., Goncharsky, A.V., Stepanov, V.V., Yagola, A.G.: *Numerical Methods for the Solution of Ill-Posed Problems*. Kluwer Academic, London (1995).
58. Brianzi, P., Frontini, M.: On the regularized inversion of the Laplace transform. *Inverse Problems*, **7**, No. 3, 355–368 (1991).
59. Hosono, T.: Numerical inversion of Laplace transform and some applications to wave optics. *Radio Sci.*, **16**, No. 3, 1015–1024 (1981).

60. Kwok, Yu., Barthez, D.: An algorithm for the numerical inversion of Laplace transform. *Inverse Problems*, **5**, No. 6, 1089–1095 (1989).
61. Baum, C.E.: Emerging technology for transient and broad-band analysis and synthesis of antennas and scatters. *Proceedings IEEE*, **64**, No. 11, 1598–1617 (1976).
62. Felsen, L.B. (ed): *Transient Electromagnetic Fields*. Springer-Verlag, New York (1976).
63. Marin, L.: Natural-mode representation of transient scattered fields. *IEEE Trans.*, **AP-21**, No. 6, 809–818 (1973).
64. Dolph, C.L., Cho, S.K.: On the relationship between the singularity expansion method and the mathematical theory of scattering. *IEEE Trans.*, **AP-28**, No. 6, 888–897 (1980).
65. Ramm, A.G.: Theoretical and practical aspects of singularity and eigenmode expansion methods. *IEEE Trans.*, **AP-28**, No. 6, 897–901 (1980).
66. Riley, D.J.: A simple and accurate resonance expansion for the electromagnetic field scattered by a lossy dielectric sphere. *IEEE Trans.*, **AP-34**, No. 5, 737–741 (1986).
67. von Hurwitz, A.: *Allgemeine Funktionentheorie und Elliptische Funktionen*. von Courant, R.: *Geometrische Funktionentheorie*. Springer-Verlag, Berlin (1964) (in German).
68. Hokhberg, I.Z., Krein, M.G.: *Introduction into the Theory of Linear Not Self-Adjoint Operators*. Nauka, Moscow (1965) (in Russian).
69. Keldysh, M.V.: On the completeness of eigenfunctions of some classes of not self-adjoint linear operators. *Advances Math. Sci.*, **26**, No. 4, 15–41 (1971) (in Russian).
70. Hokhberg, I.Z., Seagul, Y.I.: Operator generalization of the theorem about logarithmic residue and the Rouche theorem. *Coll. Papers Math.*, **84**, No. 4, 607–629 (1971) (in Russian).
71. Poyedinchuk, A.Y., Tuchkin, Y.A., Shestopalov, V.P.: On the regularization of the spectral problems of wave scattering by open screens. *Rep. Acad. Sci. USSR*, **295**, No. 6, 1358–1362 (1987) (in Russian).
72. Koshparenok, V.N., Melezhik, P.N., Poyedinchuk, A.Y., Shestopalov, V.P.: Spectral theory of two-dimensional resonators with dielectric inserts. *J. Comput. Math. Math. Phys.*, **25**, No. 4, 621–637 (1985) (in Russian).
73. Koshparenok, V.N., Melezhik, P.N., Poyedinchuk, A.Y., Shestopalov, V.P.: Wave interaction in open resonators. *Rep. Acad. Sci. USSR*, **279**, No. 5, 1114–1117 (1984) (in Russian).
74. Shlager, K.L., Schneider, J.B.: A selective survey of finite-difference time-domain literature. *IEEE Ant. Propag. Mag.*, **37**, No. 4, 39–57 (1995).
75. Ladyzhenskaya, O.A.: *Mixed Problem for Hyperbolic Equation*. Gostechizdat, Moscow (1953) (in Russian).
76. Sheen, D.M., Ali, S.M., Abouzahra, M.D., Kong, J.A.: Application of the three-dimensional FDTD method to the analysis of planar microstrip circuits. *IEEE Trans.*, **MTT-38**, No. 7, 849–857 (1990).
77. Moore, J., Ling, H.: Characterization of a 90° microstrip bend with arbitrary miter via the FDTD method. *IEEE Trans.*, **MTT-38**, No. 4, 405–410 (1990).
78. Fang, J., Ren, J.: A locally conformed FDTD algorithm of modeling arbitrary shape planar metal strips. *IEEE Trans.*, **MTT-41**, No. 5, 830–837 (1993).
79. Furse, C.M., Mathur, S.P., Gandhi, O.P.: Improvements to the FDTD method for calculating the radar cross section of a perfectly conducting target. *IEEE Trans.*, **MTT-38**, No. 7, 919–927 (1990).

80. Katz, D.S., Piket-May, M.J., Taflove, A., Umashankar, K.R.: FDTD analysis of electromagnetic wave radiation from systems containing horn antennas. *IEEE Trans.*, **AP-39**, No. 8, 1203–1212 (1991).
81. Dib, N.I., Katehi, L.P.B.: Analysis of the transition from rectangular waveguide to shielded dielectric image guide using the FDTD method. *IEEE Microwave Guided Wave Lett.*, **3**, No. 9, 327–329 (1993).
82. Krupezevic, D.V., Brankovic, V.J., Arndt, F.: The wave-equation FDTD method for the efficient eigenvalue analysis and S-matrix computation of waveguide structures. *IEEE Trans.*, **MTT-41**, No. 12, 2109–2115 (1993).
83. Jurgens, T.G., Taflove, A., Umashankar, K.: FDTD modelling of curved surfaces. *IEEE Trans.*, **AP-40**, No. 4, 357–366 (1992).
84. Yee, K.S., Shlager, K., Chang, A.H.: An algorithm to implement a surface impedance boundary condition for FDTD. *IEEE Trans.*, **AP-40**, No. 7, 833–837 (1992).
85. Luebbers, R.J., Hunsberger, F.: FDTD for N-th order dispersive media. *IEEE Trans.*, **AP-40**, No. 11, 1297–1301 (1992).
86. Hunsberger, F., Luebbers, R., Kunz, K.: FDTD analysis of gyrotropic media–I: Magnetized plasma. *IEEE Trans.*, **AP-40**, No. 12, 1489–1495 (1992).
87. Aoyagi, P.H., Lee, J.-F., Mittra, R.: A hybrid Yee algorithm/scalar-wave equation approach. *IEEE Trans.*, **MTT-41**, No. 9, 1593–1600 (1993).
88. Mittra, R., Harms, P.H.: A new FDTD algorithm for efficient field computation in resonator narrow-band structures. *IEEE Microwave Guided Wave Lett.*, **3**, No. 9, 316–318 (1993).
89. Mei, K.K., Fang, J.: Superabsorbtion—a method to improve absorbing boundary conditions. *IEEE Trans.*, **AP-40**, No. 9, 1001–1010 (1992).
90. Railton, C.J., Daniel, E.M., Paul, D.L., McGeehan, J.P.: Optimized absorbing boundary conditions for the analysis of planar circuits using the FDTD method. *IEEE Trans.*, **MTT-41**, No. 2, 290–297 (1993).
91. Betz, V., Mittra, R.: A boundary condition to absorb both propagating and evanescent waves in a FDTD simulation. *IEEE Microwave Guided Wave Lett.*, **3**, No. 6, 182–184 (1993).
92. Fang, J.: Absorbing boundary conditions applied to model wave propagation in microwave integrated circuits. *IEEE Trans.*, **MTT-42**, No. 8, 1506–1513 (1994).
93. Umashankar, K.R., Taflove, A.: A novel method to analyze electromagnetic scattering of complex objects. *IEEE Trans.*, **EMC-24**, No. 4, 397–405 (1982).
94. Moore, T.G., Blaschak, J.G., Taflove, A., Kriegsmann, G.A.: Theory and application of radiation boundary operators. *IEEE Trans.*, **AP-36**, No. 12, 1797–1812 (1988).
95. Tirkas, P.A., Balanis, C.A., Renaut, R.A.: Higher order absorbing boundary conditions for FDTD-method. *IEEE Trans.*, **AP-40**, No. 10, 1215–1222 (1992).
96. Katz, D.S., Thiele, E.T., Taflove, A.: Validation and extension to three dimensions of Berenger PML absorbing boundary condition for FD-TD meshes. *IEEE Microwave Guided Wave Lett.*, **4**, No. 8, 268–270 (1994).
97. Chen, B., Fang, D.G., Zhou, B.H.: Modified Berenger PML absorbing boundary condition for FD-TD meshes. *IEEE Microwave Guided Wave Lett.*, **5**, No. 11, 399–401 (1995).
98. Reuter, C.E., Joseph, R.M., Thiele, E.T., Katz, D.S., Taflove, A.: Ultrawideband absorbing boundary condition for termination of waveguide structures in FD-TD simulations. *IEEE Microwave Guided Wave Lett.*, **4**, No. 10, 344–346 (1994).
99. De Moerloose, J., Stuchly, M.A.: Behavior of Berenger's ABC for evanescent waves. *IEEE Microwave Guided Wave Lett.*, **5**, No. 10, 344–346 (1995).

100. Olivier, J.C.: On the synthesis of exact free space absorbing boundary conditions for the FDTD method. *IEEE Trans.*, **AP-40**, No. 4, 456–460 (1992).
101. Tromp, E.N.M., Olivier, J.C.: Synthesis of absorbing boundary conditions for the FDTD method: Numerical results. *IEEE Trans.*, **AP-43**, No. 2, 213–215 (1995).
102. De Moerloose, J., De Zutter, D.: Surface integral representation radiation boundary condition for FDTD method. *IEEE Trans.*, **AP-41**, No. 7, 890–896 (1993).
103. Mittra, R. (ed): *Computer Techniques for Electromagnetics*. Pergamon, New York (1973).
104. Miller, E.K.: Time-domain modeling in electromagnetics. *J. Electromag. Waves Appl.*, **8**, No. 9&10, 1125–1172 (1994).
105. Walker, S.P.: Scattering analysis via time-domain integral equations: method to reduce the scailing of cost with frequency. *IEEE Ant. Propag. Mag.*, **39**, No. 5, 13–20 (1997).
106. Rao, S.M. (ed): *Time Domain Electromagnetics*. Academic, San Diego (1999).
107. He, S., Strom, S., Weston, V.: *Time Domain Wave-Splittings and Inverse Problems*. Oxford Univ. Press, Oxford (1998).
108. Krueger, R.J., Ochs, R.L.: A Green's function approach to the determination of internal fields. *Wave Motion*, **11**, No. 6, 525–543 (1989).
109. He, S.: A 'compact Green function' approach to the time domain direct and inverse problems for a stratified dissipatice slab. *J. Math. Phys.*, **34**, No. 10, 4628–4645 (1993).
110. Jonsson, L.: Directional decomposition in anisotropic heterogeneous media for acoustic and electromagnetic fields. Stockholm, Sweden: Thesis. Division of Electromagnetic Theory, Royal Institute of Technology (2001).
111. Tretyakov, O.A.: Essentials of nonstationary and nonlinear electromagnetic field theory. In: Hashimoto, M., Idemen, M., Tretyakov, O.A. (ed.) *Analytical and Numerical Methods in Electromagnetic Wave Theory*. Science House, Tokyo (1993).
112. Tretyakov, O.A.: Evolutionary wave equations. *Radio-Eng. Electron.*, **34**, No. 5, 917–926 (1989) (in Russian).
113. Marchenko, V.A.: *The Sturm-Liouville Operators and Their Applications*. Naukova Dumka, Kiev (1977) (in Russian).
114. Ling, H., Moore, J., Bouche, D., Saavedra, V.: Time-frequency analysis of backscattered data from a coated strip with a gap. *IEEE Trans.*, **AP-41**, No. 8, 1147–1150 (1993).
115. Carin, L., Felsen, L.B.: Wave-oriented data processing for frequency- and time-domain scattering by nonuniform truncated arrays. *IEEE Ant. Propag. Mag.*, **36**, No. 3, 29–43 (1994).
116. Poularikas, A.D.: *The Transforms and Applications Handbook*. CRC Press & IEEE Press, Boca Raton, FL (1996).
117. Sirenko, Y.K., Sukharevskiy, I.V., Sukharevskiy, O.I., Yashina, N.P.: *Fundamental and Applied Problems in the Scattering Theory of Electromagnetic Waves*. Krok, Kharkov (2000) (in Russian).
118. Sirenko, Y.K., Shestopalov, V.P., Yashina, N.P.: New methods in the dynamic linear theory of open waveguide resonators. *Comput. Math. Math. Phys.*, **37**, No. 7, 845–853 (1997).
119. Sirenko, Y.K.: Exact 'absorbing' conditions in outer initial boundary-value problems of electrodynamics of nonsinusoidal waves. Part 3. Compact inhomogeneities in free space. *Telecomm. Radio Eng.*, **59**, No. 1&2, 1–31 (2003).
120. Abramowitz, M., Stegun, I.A. (ed): *Handbook of Mathematical Functions*. Dover, New York (1972).

121. Bateman, H., Erdelyi, A.: *Tables of Integral Transforms.* Vol. 1. McGraw-Hill, New York (1954).
122. Korn, G.A., Korn, T.M.: *Mathematical Handbook for Scientists and Engineers.* McGraw-Hill, New York (1961).
123. Gradshteyn, I.S., Ryzhik, I.M.: *Table of Integrals, Series, and Products.* Academic, New York (1994).
124. Fryazinov, I.V.: On difference schemes for the Poisson equation in polar, cylindrical, and spherical coordinates. *J. Comput. Math. Math. Phys.*, **11**, No. 5, 1219–1228 (1971) (in Russian).
125. Bamberger, A., Joly, P., Roberts, J.E.: Second order absorbing boundary conditions for the wave equation: A solution for the corner problem. *SIAM J. Numer. Anal.*, **27**, No. 2, 323–352 (1990).
126. Collino, F.: Conditions absorbantes d'ordre eleve pour des modeles de propagation d'onde dans des domaines rectangulaires. Rocquencourt, France: Report I.N.R.I.A. No. 1790 (1993) (in French).
127. Prudnikov, A.P., Brychkov, Y.A., Marichev, O.I.: *Integrals and Series.* Vol. 1, Vol. 2. Gordon and Breach, New York (1986).
128. Balanis, C.A.: Antenna Theory: *Analysis and Design.* Wiley & Sons, New York (1982).
129. Bateman, H., Erdelyi, A.: *Higher Transcendental Functions.* Vol. 1, Vol. 2. McGraw-Hill, New York (1953).
130. Lavrent'yev, M.A., Shabat, B.V.: *Methods for the Theory of Functions of Complex Variables.* Nauka, Moscow (1973) (in Russian).
131. Babushkina, L.V., Kerimov, M.K., Nikitin, A.I.: Algorithms for calculating the Bessel functions with semi-integer indices and complex arguments. *J. Comput. Math. Math. Phys.*, **28**, No. 10, 1449–1460 (1988) (in Russian).
132. Mikhailov, V.P.: *Partial Differential Equations.* Nauka, Moscow (1976) (in Russian).
133. Van, V., Chaudhuri, S.K.: A hybrid implicit-explicit FDTD scheme for nonlinear optical waveguide modeling. *IEEE Trans.*, **MTT-47**, No. 5, 540–544 (1999).
134. Kristensson, G.: Transient electromagnetic wave propagation in waveguides. *J. Electromag. Waves Appl.*, **9**, No. 5&6, 645–671 (1995).
135. Stutzman, W.L., Thiele, G.A.: *Antenna Theory and Design.* Wiley & Sons, New York (1998).
136. Maloney, J.G., Smith, G., Scott, W.R.: Accurate computation of the radiation from simple antennas using the FDTD method. *IEEE Trans.*, **AP-38**, No. 7, 1059–1068 (1990).
137. Shestopalov, V.P., Shcherbak, V.V.: Matrix operators in the diffraction problems. *Higher School News. Radio-Phys.*, **11**, No. 2, 285–305 (1968) (in Russian).
138. Weinstein, L.A.: Pulse propagation. *Adv. Phys. Sci.*, **118**, No. 2, 339–367 (1976) (in Russian).
139. Kirilenko, A.A., Tkachenko, V.I.: System for electromagnetic simulation of SHF-SWF devices. *Higher School News. Radio-Electron.*, **39**, No. 9, 17–28 (1996) (in Russian).
140. Sirenko, Y.K.: On the validation of the method of semi-inversion of matrix operators in the problems of wave diffraction. *J. Comp. Math. Math. Phys.*, **23**, No. 6, 1381–1391 (1983) (in Russian).
141. Sirenko, Y.K.: Some mathematical questions in the problems of wave diffraction on waveguide gratings. Kharkov, Ukraine: Academy of Sciences of Ukraine, Preprint IRE No. 103 (1978) (in Russian).
142. Verlan', A.F., Sizikov, V.S.: *Integral Equations: Methods, Algorithms, Software. Handbook.* Naukova Dumka, Kiev (1986) (in Russian).

143. Sirenko, Y.K., Shestopalov, V.P., Yashina, N.P.: Free oscillations in coaxial-waveguide resonator. *Soviet J. Commun. Technol. Electron.*, **32**, No. 7, 60–67 (1987).
144. Pochanina, I.Y., Shestopalov, V.P., Yashina, N.P.: Interaction and degeneration of eigenoscillations in open waveguide resonators. *Rep. Acad. Sci. USSR*, **320**, No. 1, 90–95 (1991) (in Russian).
145. Hessel, A., Oliner, A.A.: A new theory of Wood's anomalies on optical gratings. *Appl. Optics*, **4**, No. 10, 1275–1297 (1965).
146. Sologub, V.G., Shestopalov, V.P.: Resonance effects by a diffraction of a plane H-polarized wave on a grating of metal bars. *J. Tech. Phys.*, **38**, No. 9, 1505–1520 (1968) (in Russian).
147. Dolph, C.L.: Recent developments in some non-self-adjoint problems of mathematical physics. *Bull. Amer. Math. Soc.*, **67**, No. 1, 1–69 (1961).
148. Reed, M., Simon, B.: *Methods of Modern Mathematical Physics. III: Scattering theory*. Academic, New York (1979).
149. Reed, M., Simon, B.: *Methods of Modern Mathematical Physics. IV: Analysis of operators*. Academic, New York (1978).
150. Sanchez-Palencia, E.: *Non-Homogeneous Media and Vibration Theory*. Springer-Verlag, New York (1980).
151. Shestopalov, Y.V.: On the validation of spectral method of computing eigen waves of micro-strip lines. *Differential Equations*, **16**, No. 8, 1504–1512 (1980) (in Russian).
152. Shestopalov, Y.V.: Properties of spectrum of one class of non-self-conjugate boundary value problems for the Helmholtz equation set. *Rep. Acad. Sci. USSR*, **252**, No. 5, 1108–1111 (1980) (in Russian).
153. Pochanina, I.E., Shestopalov, V.P., Yashina, N.P.: Rigorous models of physical effects in the regions of concentration of the spectrum of open waveguide resonators. *Radio-Eng. Electron.*, **37**, No. 10, 1787–1795 (1992) (in Russian).
154. Poyedinchuk, A.E., Shestopalov, V.P., Yashina, N.P.: On the spectral theory of coaxial waveguide resonator. *J. Comput. Math. Math. Phys.*, **26**, No. 4, 552–562 (1986) (in Russian).
155. Yashina, N.P.: Accurate analysis of coaxial slot bridge. *IEEE Microwave Opt. Tech. Lett.*, **20**, No. 5, 345–349 (1999).
156. Pochanina, I.E., Yashina, N.P.: Electromagnetic properties of open waveguide resonator. *Electromagnetics*, **13**, No. 3, 289–300 (1993).
157. Rud', L.A.: On the nature of resonance effects in T-junctions of rectangular waveguides. *J. Tech. Phys.*, **55**, No. 6, 1213–1215 (1985) (in Russian).
158. Sirenko, Y.K., Shestopalov, V.P.: Uniqueness of solutions of spectral problems for one-dimensional periodic lattices. *Soviet Phys. Doklady*, **30**, No. 11, 928–930 (1985).
159. Sirenko, Y.K.: Analytical extension of diffraction problems and threshold effects in electromagnetics. *Rep. Acad. Sci. Ukrain. SSR*, Ser. A, No. 8, 65–68 (1986) (in Russian).
160. Sirenko, Y.K., Shestopalov, V.P.: Free oscillations of electromagnetic field in one-dimensional periodic gratings. *J. Comput. Math. Math. Phys.*, **27**, No. 2, 262–271 (1987) (in Russian).
161. Sirenko, Y.K., Shestopalov, V.P.: On the extraction of physical solutions to the wave diffraction problems on one-dimensional periodic gratings. *Rep. Acad. Sci. USSR*, **297**, No. 6, 1346–1350 (1987) (in Russian).
162. Sveshnikov, A.G.: On the radiation principle. *Rep. Acad. Sci. USSR*, **73**, No. 5, 917–920 (1950) (in Russian).

163. Sveshnikov, A.G.: A principle of limiting absorption for a waveguide. *Rep. Acad. Sci. USSR*, **80**, No. 3, 341–344 (1951) (in Russian).
164. Ladyzhenskaya, O.A.: On the principle of limit amplitude. *Advances Math. Sci.*, **12**, No. 3, 161–164 (1957) (in Russian).
165. Mikhailov, V.P.: On the principle of limit amplitude. *Rep. Acad. Sci. USSR*, **159**, No. 5, 750–752 (1964) (in Russian).
166. Eidus, D.M.: The principle of limit amplitude. *Advances Math. Sci.*, **24**, No. 3, 91–156 (1969) (in Russian).
167. Akimov, A.B., Iskanderov, V.A.: The principle of limiting absorption, limit amplitude and the partial conditions for a boundary-value problem in n-metric layer for the Helmholtz equation. *Differential Equations*, **13**, No. 8, 1503–1505 (1977) (in Russian).
168. Sirenko, Y.K., Shestopalov, V.P.: The principles of radiation, limiting absorption and limit amplitude in the wave diffraction problems on one-dimensional periodic gratings. *J. Comput. Math. Math. Phys.*, **27**, No. 10, 1555–1562 (1987) (in Russian).
169. Kirilenko, A.A., Senkevitch, S.L., Tysik, B.G., Sirenko, Y.K.: On the recovering of scattering matrices of waveguide and periodic structures on the spectrum of complex eigen frequencies. *Radio-Eng. Electron.*, **34**, No. 3, 468–473 (1989) (in Russian).
170. Sirenko, Y.K.: A grating in the field of a compact monochromatic source. *Electromagnetics*, **13**, No. 3, 255–272 (1993).
171. Sirenko, Y.K.: New results in the theory of open periodic directing structures. *Higher School News. Radio-Phys.*, **32**, No. 3, 331–338 (1989) (in Russian).
172. Elachi, C.: Waves in active and passive periodic structures. *Proc. IEEE*, **64**, No. 12, 1666–1698 (1976).
173. Sirenko, Y.K., Velychko, L.G.: Model synthesis of the grating type absorbing pattern forming structures. *Electromag. Waves Electron. Syst.*, **7**, No. 2, 45–59 (2002) (in Russian).
174. Galishnikova, T.N., Ilyinsky, A.S.: *Numerical Methods in the Diffraction Problems*. Publ. of Moscow State University, Moscow (1987) (in Russian).
175. Ilyinsky, A.S., Lebedeva, O.A.: Numerical method of calculating plane semi-infinite structures. *Comput. Meth. Program.*, No. 32, 155–164 (1982) (in Russian).
176. Steinschleiger, V.B.: *Effect of Wave Interaction in Electromagnetic Resonators*. Oboronizdat, Moscow (1955) (in Russian).
177. Svezhentsev, A.E.: Effect of inter-mode coupling of surface waves in partially screened dielectric rod. *Rep. Acad. Sci. Ukrain. SSR*, Ser. A, No. 7, 58–62 (1986) (in Russian).
178. Krein, M.G., Lyubarsky, G.Y.: On the theory of the pass bands of periodic waveguides. *Appl. Math. Mech.*, **25**, No. 1, 24–37 (1961) (in Russian).
179. Kirilenko, A.A., Senkevitch, S.A.: New mode of Q-factor oscillations in open waveguide resonators. *Lett. J. Tech. Phys.*, **12**, No. 14, 876–879 (1986) (in Russian).
180. Poston, T., Steward, I.: *Catastrophe Theory and Its Applications*. Pitman, London (1978).
181. Arnold, V.I., Varchenko, A.N., Gussein-Zade, S.M.: *Typical Features of Differentiable Images*. Nauka, Moscow (1982) (in Russian).
182. Melezhik, P.N., Poyedinchuk, A.E., Tuchkin, Y.A., Shestopalov, V.P.: On the analytical nature of the inter-mode coupling of eigenoscillations. *Rep. Acad. Sci. USSR*. **300**, No. 6, 1356–1359 (1988) (in Russian).
183. Sirenko, Y.K., Shestopalov, V.P., Yatsik, V.V.: The Morse critical points of dispersion equations of diffraction gratings. *Ukrain. Phys. J.*, **36**, No. 8, 1156–1162 (1991) (in Russian).

184. Masalov, S.A., Sirenko, Y.K., Shestopalov, V.P.: Conditions of the Malyuzhinets effect in the frequently-periodical gratings. *Lett. J. Tech. Phys.*, **6**, No. 16, 998–1001 (1980) (in Russian).
185. Kusaykin, A.P., Sirenko, Y.K.: Essential regularities of variation of selective properties of periodic waveguide gratings. *Higher School News. Radio-Phys.*, **26**, No. 2, 240–245 (1983) (in Russian).
186. Masalov, S.A., Sirenko, Y.K.: Resonance effects by electromagnetic wave diffraction by periodic waveguide gratings. *Ukrain. Phys. J.,* **23**, No. 9, 1439–1446 (1978) (in Russian).
187. Kirilenko, A.A., Kusaykin, A.P., Sirenko, Y.K.: Non-mirror wave reflection by the waveguide gratings. General regularities. *Higher School News. Radio-Phys.*, **28**, No. 11, 1450–1461 (1985) (in Russian).
188. Kirilenko, A.A., Kusaykin, A.P., Sirenko, Y.K.: Non-mirror wave reflection by the waveguide gratings. Specific scattering modes. *Higher School News. Radio-Phys.*, **29**, No. 10, 1182–1191 (1986) (in Russian).
189. Sirenko, Y.K.: Resonance scattering of plane inhomogeneous waves by reflective gratings. *Rep. Acad. Sci. Ukrain. SSR*, Ser. A, No. 9, 56–59 (1986) (in Russian).
190. Kusaykin, A.P., Sirenko, Y.K.: Effect of quasi-complete non-mirror reflection with large telescoping coefficient. *J. Tech. Phys.*, **55**, No. 6, 1241–1243 (1985) (in Russian).
191. Sirenko, Y.K., Yashina, N.P., Schuenemann, K.F.: Synthesis of mode converters in waveguides and gratings based on spectral theory. *J. Electromag. Waves Appl.*, **16**, No. 5, 611–628 (2002).
192. Shestopalov, V.P.: *Diffraction Electronics*. Publ. of Kharkov National University, Kharkov (1976) (in Russian).
193. Kirilenko, A.A., Kusaykin, A.P., Sirenko, Y.K.: Effect of resonance enhance of near field of waveguide gratings. *Lett. J. Tech. Phys.*, **10**, No. 7, 405–408 (1984) (in Russian).
194. Masalov, S.A.: On the possibility of using an echelette in diffraction radiation sources. *Ukrain. Phys. J.*, **25**, No. 4, 570–574 (1980) (in Russian).
195. Deryugin, L.N., Friedman, G.H.: Resonance curves of a double resonance at a reflective grating. *Rep. Acad. Sci. USSR*, **111**, No. 6, 1209–1211 (1956) (in Russian).
196. Bliokh, P.B.: Radiation pulse compression in a dispersive medium with random discontinuities. *Higher School News. Radio-Phys.*, **7**, No. 3, 460–470 (1964) (in Russian).
197. King, R.W.P.: The propagation of a Gaussian pulse in sea water and its application to remote sensing. *IEEE Trans.*, **GRS-31**, No. 3, 595–605 (1993).
198. Weinstein, L.A.: *Open Resonators and Open Waveguides*. Golem, Boulder, CO (1969).
199. Colton, D., Kress, R.: *Inverse Acoustic and Electromagnetic Scattering Theory.* Springer-Verlag, New York (1992).
200. Kirsch, A.: *An Introduction to the Mathematical Theory of Inverse Problems*. Springer-Verlag, New York (1996).
201. Poyedinchuk, A.Y., Sirenko, Y.K., Shestopalov, V.P.: The uniqueness of the solution of inverse problems of the electrodynamic theory of gratings. *Soviet Phys. Doklady*, **37**, No. 1, 39–40 (1992).
202. Sirenko, Y.K., Velychko, L.G.: Inverse two-dimensional boundary-value problems of the wave diffraction theory. *Foreign Radioelectron. Successes Modern Radioelectron.*, No. 2, 2–19 (1996) (in Russian).
203. Velychko, L.G. , Poyedinchuk, A.Y., Sirenko, Y.K., Shestopalov, V.P.: One inverse diffraction problem for periodic dielectric layer. *Rep. Nat. Acad. Sci. Ukraine*, No. 2, 21–26 (1996) (in Russian).

204. Velychko, L.G.: Quasi-linearization as a method of developing algorithms of numerical solution to the inverse boundary-value diffraction problems. *Rep. Nat. Acad. Sci. Ukraine*, No. 3, 84–90, (1997) (in Russian).
205. Sirenko, Y.K., Velychko, L.G.: Diffraction grating profile reconstruction: simple approaches to solving applied problems. *Electromagnetics*, **19**, No. 2, 211–221 (1999).
206. Sirenko, Y.K., Velychko, L.G., Karacuha, E.: Synthesis of perfectly conducting gratings with an arbitrary profile of slits. *Inverse Problems*, **15**, No. 2, 541–550 (1999).
207. Anokhov, S.P., Marusiy, T.Y., Soskin, M.S.: *Tunable Lasers*. Radio and Commun., Moscow (1982) (in Russian).
208. Avtonomov, V.P., Beltyugov, V.N., Ochkin, V.N., Sobolev, N.N., Udalov, Y.B.: Study of the selective properties of an optical resonator with a reflective grating. Moscow: Academy of Sciences of USSR, Preprint PhI; No. 80–29 (1980) (in Russian).
209. Belous, O.I., Fisun, A.I.: E-polarized oscillations in an open resonator with an echelette mirror. *Lett. J. Tech. Phys.*, **22**, No. 2, 81–86 (1996) (in Russian).
210. Belous, O.I., Kirilenko, A.A., Fisun, A.I.: Quasi-single-frequency spectra of an open resonator with a comb grating. *Higher School News. Radio-Electron.*, **41**, No. 4, 8–13 (1998) (in Russian).
211. Belous, O.I., Fisun, A.I., Tkachenko, V.I., Kirilenko, A.A.: Excitation of oscillations in open resonators with echelette and corner-echelette mirrors. *Radio-Eng. Electron.*, **45**, No. 5, 632–639 (2000) (in Russian).
212. Roger, A.: Newton-Kantorovitch algorithm applied to an electromagnetic inverse problem. *IEEE Trans.*, **AP-29**, No. 2, 232–238 (1981).
213. Murch, R.D., Tan, D.G.H., Wall, D.J.: Newton-Kantorovitch method applied to two-dimensional inverse scattering for an exterior Helmholtz problem. *Inverse Problems*, **4**, No. 4, 1117–1128 (1988).
214. Tobocman, W.: Inverse acoustic-wave scattering in two-dimensions from impenetrable targets. *Inverse Problems*, **5**, No. 6, 1131–1144 (1989).
215. Mönch, L.A.: Newton method for solving inverse scattering problem for a sound-hard obstacle. *Inverse Problems*, **12**, No. 3, 309–323 (1996).
216. Ramm, A.G.: *Multidimensional Inverse Scattering Problems*. Longman, New York (1992).
217. Yeryomin, Y.A., Zakharov, Ye.V.: *About Certain Direct in Inverse Problems in Diffraction Theory*. Appendix in Russian edition of the book [15]. Mir, Moscow (1987) (in Russian).
218. Yeryomin, Y.A., Sveshnikov, A.G.: The synthesis and recognition problems in diffraction theory. *J. Comput. Math. Math. Phys.*, **32**, No. 10, 1594–1607 (1992) (in Russian).
219. Hettlich, F., Kirsch, A.: Schiffer's theorem in inverse scattering theory for periodic structures. *Inverse Problems*, **13**, No. 2, 351–361 (1997).
220. Lax, P.D., Phillips, R.S.: *Scattering Theory*. Academic, New York (1967).
221. Zhdanov, M.S.: *Analogs of Cauchy Type Integrals in the Theory of Geophysical Fields*. Nauka, Moscow (1984) (in Russian).
222. Arsenin, V.Ya.: *Methods of Mathematical Physics and Special Functions*. Nauka, Moscow (1974) (in Russian).
223. Wombell, R.J., De Santo, J.A.: The reconstruction of shallow rough-surface profiles from scattered field data. *Inverse Problems*, **7**, No. 1, L7–L12 (1991).
224. Krutin', Y.I., Tuchkin, Y.A., Shestopalov, V.P.: Diffraction of E-polarized waves by smooth periodic wavy surface. *Radio-Eng. Electron.*, **37**, No. 2, 202–210 (1992) (in Russian).

225. Hutson, V.C.L., Pym, J.S.: *Applications of Functional Analysis and Operator Theory*. Academic, New York (1980).
226. Weinstein, L.A.: *Electromagnetic Waves*. Radio and Commun., Moscow (1988) (in Russian).
227. Koshparenok, V.N., Melezhik, P.N., Poyedinchuk, A.Y., Shestopalov, V.P.: The method of the Riemann-Hilbert in the spectral theory of open two-dimensional resonators. 1. Mathematical model and spectrum characteristics. *Radio-Eng. Electron.*, **31**, No. 2, 271–278 (1986) (in Russian).
228. Koshparenok, V.N., Melezhik, P.N., Poyedinchuk, A.Y., Shestopalov, V.P.: The method of the Riemann-Hilbert in the spectral theory of open two-dimensional resonators. 2. Spectral characteristics. *Radio-Eng. Electron.*, **32**, No. 2, 238–247 (1987) (in Russian).
229. Melezhik, P.N.: Mode conversion in diffractionally coupled open resonators. *Telecomm. Radio Eng.*, **51**, No. 6&7, 54–60 (1997).
230. Katsenelenbaum, B.Z., Golubyatnikov, A.V.: The Luneburg lens. Geometrooptical calculation. *Lett. J. Tech. Phys.*, **24**, No. 15, 69–72 (1998) (in Russian).
231. Boriskin, A.V., Nosich, A.I.: Whispering-gallery and Luneburg-lens effects in a beam-fed circularly-layered dielectric cylinder. *IEEE Trans.*, **AP-50**, No. 9, 1245–1249 (2002).

Appendix
List of Symbols and Abbreviations

$\mathbf{R}^n$ and $\mathbf{G} \subset \mathbf{R}^n$	n-dimensional Euclidean space and domain $\mathbf{G}$ in it
$\mathbf{C}$	Plane of complex variable w or s
$\mathbf{C}^n(\mathbf{G})$	Class of functions that are continuous together with their derivatives up to the order n, including, in $\mathbf{G}$
$\mathbf{L}_n(\mathbf{G})$	Space of functions $f(g)$, $g \in \mathbf{G}$, for which function $\|f(g)\|^n$ is integrable in $\mathbf{G}$
g and p	Points of the space $\mathbf{R}^n$; x, y, z are Cartesian coordinates; ρ, ϕ, z are cylindrical coordinates; r, ϑ, ϕ are spherical coordinates
t and τ	Time variables; $(0;T)$, $T < \infty$ is time interval
$\mathbf{Q}$	Unbounded domain of analysis in initial boundary value problems; $\mathbf{Q}^T = \mathbf{Q} \times (0;T)$
$\mathbf{R} = \left\{g \in \mathbf{R}^2 : 0 < y < l\right\}$ and $\mathbf{R} = \left\{g \in \mathbf{R}^3 : 0 < x < l_x;\ 0 < y < l_y\right\}$	Parallel-plane and rectangular Floquet channels
$\mathbf{Q}_a$	Bounded subdomain of the domain $\mathbf{Q}$; ${}_a\mathbf{Q} = \mathbf{Q} \backslash \overline{\mathbf{Q}}_a$ is a complement of $\overline{\mathbf{Q}}_a$ up to $\mathbf{Q}$
$\overline{\mathbf{G}}$, $\mathbf{G} \cup \mathbf{Q}$, $\mathbf{G} \cap \mathbf{Q}$ and $\mathbf{G} \backslash \mathbf{Q}$	Closure, union, intersection, and difference of sets
$\mathbf{D}(\mathbf{G})$	Set of finite- and infinitely differentiable in $\mathbf{G}$ functions
$\tilde{\mathbf{D}}(\mathbf{G})$	Space of generalized functions (linearly continuous functionals) on the space of fundamental functions $\mathbf{D}(\mathbf{G})$

(f, γ)	Value of functional (generalized function f) on fundamental function $\gamma \in \mathbf{D}(\mathbf{G})$
$\tilde{\mathbf{D}}_r(\mathbf{G})$	Class of regular (locally integrable) generalized functions
$\mathbf{W}^l_m(\mathbf{G})$	Set of all the elements $f(g)$ from $\mathbf{L}_m(\mathbf{G})$, having generalized derivatives up to the order l, including, from $\mathbf{L}_m(\mathbf{G})$
$\mathbf{L}_{2,1}(\mathbf{G}^T)$	Space containing all elements $f(g, t) \in \mathbf{L}_1(\mathbf{G}^T)$ with finite norm $\\|f\\| = \int_0^T \left(\int_{\mathbf{G}} \|f\|^2 dg\right)^{1/2} dt$;
$\overset{\circ}{\mathbf{W}}{}^1_2(\mathbf{G})$	Subspace of space $\mathbf{W}^1_2(\mathbf{G})$, where $\mathbf{D}(\mathbf{G})$ is a dense set
$\mathbf{W}^1_{2,0}(\mathbf{G}^T)$	Subspace of space $\mathbf{W}^1_2(\mathbf{G}^T)$, where smooth functions, equal to zero in the vicinity of $\mathbf{P}^T = \mathbf{P} \times (0, T)$ ($\mathbf{P}$ is a boundary of domain $\mathbf{G}$) is a dense set
$\mathbf{S}(g, a)$	Open in $\mathbf{R}^n$ circle of radius a centered in point g
$\varnothing$	Empty set;
$l_2 = \{a = \{a_n\} : \sum_n \|a_n\|^2 < \infty\}$ and $\tilde{l}_2 = \{a : \sum_n \|a_n\|^2(\|n\| + 1) < \infty\}$	Spaces of infinite sequences $a = \{a_n\}$
$U(g, t)$ and $\tilde{U}(g, k)$	Unknown functions of initial boundary value problems and boundary value problems, defining components of vectors of field strength
$\mathbf{S}$ and $\mathbf{S}_\varepsilon$	Lines (surfaces) of break of properties of medium where the excitation propagates or boundaries of perfectly conducting and dielectric scatterers
$\overline{\text{int}\,\mathbf{S}}$	Closure of the domain, filled with perfectly conducting scatterer or domain, bounded with perfectly conducting surface $\mathbf{S}$
$k = 2\pi/\lambda$ and λ	Wave number and length of wave in a free space
$\tilde{k}$ and $\tilde{\kappa}$	Amplitude centers (central frequencies) of momentary and current sources, signals, and pulsed waves
$\tilde{\alpha}$, $\tilde{\beta}$, $\tilde{T}$	Parameters of signals

$\varphi(g)$, $\psi(g)$, and $F(g,t)$	Momentary ($\varphi(g)$ and $\psi(g)$) and current ($F(g,t)$) sources of signals and pulsed waves
ε_0, μ_0, $\eta_0 = (\mu_0/\varepsilon_0)^{1/2}$, and σ_0	Electromagnetic parameters of vacuum, impedance of free space, and conductivity
$\varepsilon(g)$ and $\tilde{\varepsilon}(g)$	Real and complex relative permittivities of media
κ	Dimensionless frequency parameter
$a_n, b_n(A_n, B_n), n = 0, \pm 1, \ldots$	Amplitudes of harmonics of the scattered by grating field in the case of E-polarization (H-polarization)
$W_n^a = \|a_n\|^2 \mathrm{Re}\Gamma_n/\Gamma_0$	Partial of scattered energy of nth harmonic of the field, having amplitude a_n
K and **F**	Infinitely sheeted Riemann surfaces of variation of complex (nonphysical) values of frequency parameter κ (k) and parameter Φ of Floquet channel
κ_n and Φ_n	Branching points of surfaces **K** and **F** (threshold points)
$\Omega_\kappa(\Omega_\kappa)$ and Ω_Φ	Spectra of open resonators and waveguides (eigenfrequencies and eigenpropagation constants)
$\overline{\kappa}_n(\overline{k}_n)$ and $\overline{\Phi}_n$	Elements of spectral sets Ω_κ (Ω_k) and Ω_Φ
$\tilde{U}_n(g, \overline{\kappa}_n)$, $\tilde{U}_n(g, \overline{k}_n)$, and $\tilde{U}_n(g, \overline{\Phi}_n)$	Eigenmodes (or natural modes) and eigenwaves
$(f * g)$; $(f \times g)$, $(\vec{a} \cdot \vec{b})$, $\|\vec{a} \times \vec{b}\|$; and f^*	Operation of convolution, direct, scalar, and vector products; and complex conjugation
$[a;b]$ and $(a;b)$, $\{a_n\}$	Close and open intervals, set of elements a_n
$L[\ldots](\ldots)$ and $F[\ldots](\ldots)(L^{-1}[\ldots]$ $(\ldots)$ and $F^{-1}[\ldots](\ldots))$ and others	Direct (inverse) Laplace and Fourier transforms and other integral transforms
$E_{(\ldots)}$ and $H_{(\ldots)}$	Components of vectors of electrical ($\vec{E}$) and magnetic ($\vec{H}$) fields
TE- (or H-) and TM-waves (or E-waves)	Transverse electrical (or magnetic) and transverse magnetic (or electrical) fields and waves
$\delta(\ldots)$ and $\delta^{(m)}(\ldots)$	δ-Dirac function and its derivative of mth order
δ_m^n	Kronecker symbol

$\chi(\cdots)$	Heaviside step function
$\chi[f_1(g)]\chi[f_2(g)]\ldots\chi[f_m(g)]$	Generalized step function, equal to unit on the intersection $\mathbf{G}$ of sets $\mathbf{G}_j = \{g \in \mathbf{R}^n : f_j(g) \geq 0\}$, $j = 1, 2, \ldots, m$ and equal to zero on $\mathbf{R}^n \backslash \mathbf{G}$
$G(\cdots)$	Fundamental solution (Green function) of differential operator
$J_n(\cdots)$, $N_n(\cdots)$, and $H_n^{(1)}(\ldots)$	Bessel, Neumann, and Hankel cylindrical functions
$P_v(\ldots)$ and $Q_v(\ldots)$ ($P_n^m(\ldots)$ and $Q_n^m(\ldots)$)	Legendre functions (adjoint Legendre functions) of the first and second kind
$\mathrm{Re}\,(a)$ and $\mathrm{Im}\,(a)$	Real and imaginary parts of the complex value of a
$\mathrm{Res}\ f(w_n)$	Residue of $f(w)$ in a point $w = w_n$
ABC	Absorbing boundary condition
FDTD-method	Finite-difference time-domain method
PFS	Pattern-forming structure
PDE	Partial differential equation
OR	Open resonator
EF	Eigenfrequency; element of the set Ω_κ (Ω_k)
FD	Frequency domain
TD	Time domain

Index

Springer Series in
OPTICAL SCIENCES

Volume 1

1 **Solid-State Laser Engineering**
By W. Koechner, 5th revised and updated ed. 1999, 472 figs., 55 tabs., XII, 746 pages

Published titles since Volume 85

85 **Sensing with Terahertz Radiation**
By D. Mittleman (Ed.), 2003, 207 figs., 14 tabs., XVI, 337 pages

86 **Progress in Nano-Electro-Optics I**
Basics and Theory of Near-Field Optics
By M. Ohtsu (Ed.), 2003, 118 figs., XIV, 161 pages

87 **Optical Imaging and Microscopy**
Techniques and Advanced Systems
By P. Török, F.-J. Kao (Eds.), 2003, 260 figs., XVII, 395 pages

88 **Optical Interference Coatings**
By N. Kaiser, H.K. Pulker (Eds.), 2003, 203 figs., 50 tabs., XVI, 504 pages

89 **Progress in Nano-Electro-Optics II**
Novel Devices and Atom Manipulation
By M. Ohtsu (Ed.), 2003, 115 figs., XIII, 188 pages

90/1 **Raman Amplifiers for Telecommunications 1**
Physical Principles
By M.N. Islam (Ed.), 2004, 488 figs., XXVIII, 328 pages

90/2 **Raman Amplifiers for Telecommunications 2**
Sub-Systems and Systems
By M.N. Islam (Ed.), 2004, 278 figs., XXVIII, 420 pages

91 **Optical Super Resolution**
By Z. Zalevsky, D. Mendlovic, 2004, 164 figs., XVIII, 232 pages

92 **UV-Visible Reflection Spectroscopy of Liquids**
By J.A. Räty, K.-E. Peiponen, T. Asakura, 2004, 131 figs., XII, 219 pages

93 **Fundamentals of Semiconductor Lasers**
By T. Numai, 2004, 166 figs., XII, 264 pages

94 **Photonic Crystals**
Physics, Fabrication and Applications
By K. Inoue, K. Ohtaka (Eds.), 2004, 209 figs., XV, 320 pages

95 **Ultrafast Optics IV**
Selected Contributions to the 4th International Conference
on Ultrafast Optics, Vienna, Austria
By F. Krausz, G. Korn, P. Corkum, I.A. Walmsley (Eds.), 2004, 281 figs., XIV, 506 pages

96 **Progress in Nano-Electro Optics III**
Industrial Applications and Dynamics of the Nano-Optical System
By M. Ohtsu (Ed.), 2004, 186 figs., 8 tabs., XIV, 224 pages

97 **Microoptics**
From Technology to Applications
By J. Jahns, K.-H. Brenner, 2004, 303 figs., XI, 335 pages

98 **X-Ray Optics**
High-Energy-Resolution Applications
By Y. Shvyd'ko, 2004, 181 figs., XIV, 404 pages

99 **Mono-Cycle Photonics and Optical Scanning Tunneling Microscopy**
Route to Femtosecond Ångstrom Technology
By M. Yamashita, H. Shigekawa, R. Morita (Eds.) 2005, 241 figs., XX, 393 pages

100 **Quantum Interference and Coherence**
Theory and Experiments
By Z. Ficek and S. Swain, 2005, 178 figs., XV, 418 pages

101 **Polarization Optics in Telecommunications**
By J. Damask, 2005, 110 figs., XVI, 528 pages

102 **Lidar**
Range-Resolved Optical Remote Sensing of the Atmosphere
By C. Weitkamp (Ed.), 161 figs., XX, 416 pages

103 **Optical Fiber Fusion Splicing**
By A.D. Yablon, 2005, 137 figs., XIII, 306 pages

104 **Optoelectronics of Molecules and Polymers**
By A. Moliton, 2005, 229 figs., 592 pages

105 **Solid-State Random Lasers**
By M. Noginov, 2005, 131 figs., XII, 238 pages

106 **Coherent Sources of XUV Radiation**
Soft X-Ray Lasers and High-Order Harmonic Generation
By P. Jaeglé, 2005, 150 figs., approx. 264 pages

107 **Optical Frequency-Modulated Continuous-Wave (FMCW) Interferometry**
By J. Zheng, 2005, 137 figs., XVIII, 254 pages

108 **Laser Resonators and Beam Propagation**
Fundamentals, Advanced Concepts and Applications
By N. Hodgson and H. Weber, 2005, 497 figs., approx. 790 pages

109 **Progress in Nano-Electro Optics IV**
Characterization of Nano-Optical Materials and Optical Near-Field Interactions
By M. Ohtsu (Ed.), 2005, 123 figs., XIV, 206 pages

110 **Kramers–Kronig Relations in Optical Materials Research**
By V. Lucarini, J.J. Saarinen, K.-E. Peiponen, E.M. Vartiainen,
2005, 37 figs., X, 162 pages

111 **Semiconductor Lasers**
Stability, Instability and Chaos
By J. Ohtsubo, 2005, 169 figs., XII, 438 pages

112 **Photovoltaic Solar Energy Generation**
By A. Goetzberger and V.U. Hoffmann, 2005, 139 figs., XII, 234 pages

113 **Photorefractive Materials and Their Applications 1**
Basic Effects
By P. Günter and J.P. Huignard, 2005, 169 figs., approx. XII, 300 pages

114 **Photorefractive Materials and Their Applications 2**
Materials
By P. Günter and J.P. Huignard, 2005, 100 figs., approx. XII, 300 pages

115 **Photorefractive Materials and Their Applications 3**
Applications
By P. Günter and J.P. Huignard, 2005, 100 figs., approx. XII, 300 pages

116 **Spatial Filtering Velocimetry**
Fundamentals and Applications
By Y. Aizu and T. Asakura, 2006, 112 figs., approx. XII, 220 pages

117 **Progress in Nano-Electro-Optics V**
Nanophotonic Fabrications, Devices, Systems, and Their Theoretical Bases
By Motoichi Ohtsu, 2006, 122 figs., 3 tables, approx. 193 pages

118 **Mid-infrared Semiconductor Optoelectronics**
By A. Krier, 2006, 443 figs., approx. 776 pages

119 **Optical Interconnects**
The Silicon Approach
By L. Pavesi and G. Guillot, 2006, 265 figs., approx. 410 pages

120 **Relativistic Nonlinear Electrodynamics**
Interaction of Charged Particles with Strong and Super Strong Laser Fields
By H.K. Avetissian, 2006, 23 figs., XIII, approx. 333 pages

121 **Thermal Process Using Attosecond Laser Pulses**
When Time Matters
By M. Kozlowski and J. Marciak-Kozlowska, 2006, 46 figs., XII, approx. 232 pages

122 **Modeling and Analysis of Transient Processes in Open Resonant Structures**
New Methods and Techniques
By Y.K. Sirenko, S. Ström, N.P. Yashina, 2007, 110 figs., XIV, approx. 362 pages

Printed in the United States of America.